WITHDRAWN

# The Concise Illustrated Dictionary of Science and Technology

*Stan Gibilisco*

**TAB** TAB BOOKS
Blue Ridge Summit, PA

FIRST EDITION
FIRST PRINTING

© 1993 by **TAB Books**.
TAB Books is a division of McGraw-Hill, Inc.

Printed in the United States of America. All rights reserved. The publisher takes no responsibility for the use of any of the materials or methods described in this book, nor for the products thereof.

**Library of Congress Cataloging-in-Publication Data**

Gibilisco, Stan.
    The concise illustrated dictionary of science and technology / by Stan Gibilisco.
      p.    cm.
    Includes index.
    ISBN 0-8306-4152-1         ISBN 0-8306-4153-X (pbk.)
     1. Science—Dictionaries.  2. Technology—Dictionaries.
I. Title.
Q123.G54   1992
503—dc20                                         92-22545
                                                           CIP

Editorial team: Roland Phelps
                    Melanie D. Brewer
Design team: Jaclyn J. Boone
                Brian Allison
Production team: Katherine G. Brown, Director
                    Lisa M. Mellott, Typesetting
                    Tara Ernst, Proofreading
                    Linda King, Proofreading
                    Wanda S. Ditch, Layout
                    Toya B. Warner, Layout
                    Wendy L. Small, Layout
                    Brenda S. Wilhide, Layout
                    Donna M. Gladhill, Final look
Paperbound cover design: Graphics Plus, Hanover, Pa.            DICT

# Editorial Board

Jon Erickson, Geologist and Chairman of the Editorial Board

J. Christopher Benetti, Instructor, Woonsocket Vocational-Technical Facility

Joseph J. Carr, Electrical Engineer

Dr. Dean G. Duffy, Department of Mathematics, U.S. Naval Academy

Dr. M.Y. Han, Professor of Theoretical Science, Duke University

R. Jesse Phagan, Electronics Technology Instructor, Woonsocket Vocational-Technical Facility

Charles A. Vergers, Professor of Electronics and Telecommunications, Capitol College

*To Tony and Tim*
*from Uncle Stan*

# Contents

Preface   viii

| | | |
|---|---|---|
| **A** | A—azimuth-elevation mount | 1 |
| **B** | B—byte | 35 |
| **C** | C—cytosine | 55 |
| **D** | D—dystrophy | 103 |
| **E** | E—eyewall | 129 |
| **F** | F—fusion | 157 |
| **G** | G—gyroscope effect | 179 |
| **H** | H—Hz | 199 |
| **I** | I—IV | 223 |
| **J** | J—juvenile water | 243 |
| **K** | K—kWh | 247 |
| **L** | L—lysozyme | 253 |
| **M** | M—myxovirus | 271 |
| **N** | N—nymph | 307 |
| **O** | O—ozone pollution | 323 |
| **P** | P—Pythagorean triple | 337 |
| **Q** | Q—quintillion | 371 |

**R**　　R—Ryle, Martin　375

**S**　　S—Szilard, Leo　397

**T**　　T—tyrosine　443

**U**　　U—uvula　467

**V**　　V—Vulcan　473

**W**　　W—wrought iron　483

**XYZ**　X—zygote　493

APPENDICES

Animal classification　499

Atomic number and atomic weight　500

Beaufort scale　502

Boolean algebra　503

Constants　503

Conversion factors　504

Electromagnetic spectrum　504

Energy units　505

Geological time　505

Melting points　506

Metals　506

Morse code　507

Multiplication table    508

Periodic table of the elements    509

Plant classification    511

Prefix multipliers    512

Primate family tree    513

Radio spectrum    514

Recommended daily allowances    515

Richter scale    515

Rocks    516

Schematic symbols in electronics    516

Solar system data    517

Space probes    517

Standard international system of units    518

Trigonometric identities    518

Universe data    519

Vitamins and minerals    520

# Preface

This book is written mainly for students at the junior and senior high level. While there are several science dictionaries currently in existence, none heretofore have addressed young people specifically and directly. This book is intended to fill that void.

Science is important to our prosperity in a fast-changing world. A knowledge of our world, and of our universe, helps us to enjoy and appreciate our place in nature. This book contains often-used terms in science and technology. I hope this volume can help reinforce the science knowledge and enthusiasm of young people, who are our future.

The terms in this book have been chosen according to their usefulness for students age 12 to 18 years. They need not know the definitions of all these terms, but they might encounter them in their reading and schooling. Only science-related definitions are given.

Terms are always stated in **boldface.** They always precede a definition and/or cross references. There are many cross references; these are in *italic* and appear within the definitions. A cross reference in CAPITALS means that it is the title of one of the appendices in the back of this book. This enhances the completeness of the volume without making it excessively large. The student should use his or her imagination. Cross references, when given, are just suggested terms that might be looked at; there often are many more related terms in the book that space wouldn't allow listing. Terms are easy to look up, because they are in alphabetical order without confusing categorizations. I have done everything possible to use simple but concise language. Examples are included often, and in that sense, this book is encyclopedic. But a true science encyclopedia would easily take up a dozen volumes of this size.

Efforts have been made to be fairly informal, so definitions are easy to read without forcing the student to go to another dictionary to understand what's said here. This isn't meant to be a completely rigorous work. Such high-level science dictionaries already exist, intended for use by professors and engineers. Rather, I have tried to make this fun to read and use; it addresses a younger, less experienced audience.

I have included the names of some famous scientists. Science is more than formulas and machines; it is people. How easily we forget this, hiding behind test tubes or electronic dials, or flying in the sky or diving under the sea! The greatest scientists were good at nothing if not at being themselves and having independent minds.

There are many drawings and tables. Often the drawings are functional, to convey how something works or what it is used for, rather than to serve as an exact physical rendition. Some simplification is necessary if students are to grasp the material.

There is extensive back matter (in the form of appendices) in this book. It is my desire that this material be not only useful, but interesting. I have taken precautions to ensure that the definitions, while clear, are accurate.

Suggestions for future editions are welcome.

Stan Gibilisco
Editor-in-Chief

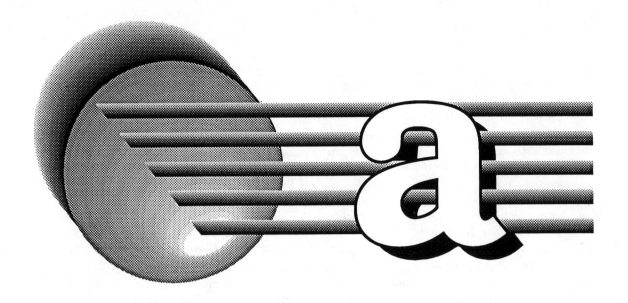

**A** Abbreviation for ampere, Angstrom unit, argon.

**a** Abbreviation for atto-. See PREFIX MULTIPLIERS appendix.

**abacus** A mechanical device used for calculating by ancient Orientals. The abacus consists of rods on a rigid frame with movable beads. Each rod represents a certain decimal place, and the beads represent units of one or five. Calculations are made by moving the beads. The abacus might have ten or more rods, allowing for calculations with large numbers. It is still used today in the Far East; experts on the abacus can make complex calculations with amazing speed. The drawing is a simplified rendition of an abacus.

**abdomen** 1. In a human being, the trunk of the body, recognizable as the belly containing the intestines and other organs. 2. The central part of the body of an animal or insect. See *insect*.

**aberration** 1. A defect in a device or component for a machine. 2. An event that is out of the ordinary; a fluke. 3. A defect in a lens. See *defect*.

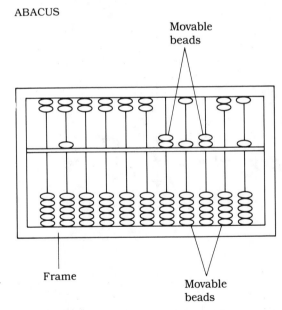

ABACUS

**abort** 1. A command to a computer to cancel an ongoing routine. 2. The canceling or interruption of a mission, such as a trip to the moon in a spacecraft.

**abortion** 1. The removal of a fetus from the womb of a pregnant woman, either by means of surgery or, in some countries, by the use of a pill that causes a miscarriage. See *fetus, miscarriage*. 2. A miscarriage of pregnancy. 3. The aborting of a computer command or a mission.

**abrasion** 1. The removal of solid matter by means of friction. Sanding wood is a good example. 2. A superficial skin wound; a "scrape."

**abscess** An infection on the skin or mucous membrane, like a large pimple. It is filled with pus, and it can become ulcerated. See *skin, mucous membrane, pus, ulcer*.

**abscissa** In a Cartesian coordinate system, the axis where the independent variable is shown. This is usually the horizontal axis. See the accompanying drawing. Compare to *ordinate*. See *Cartesian coordinates, Cartesian plane, dependent variable, function, independent variable, ordinate*.

ABSCISSA

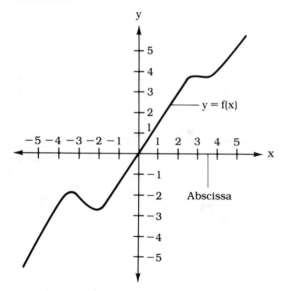

**abscission** In a plant, the process leading to the falling of a leaf in the autumn. Usually, the cells at the base weaken so that the leaf falls off.

**absolute** 1. A quantity or measurement that never varies; a constant. See *constant*. 2. A reference on which measurements are universally based, such as the National Bureau of Standards time and frequency radio stations. 3. Complete; total.

**absolute magnitude** 1. The brightness of a star relative to other stars, as if seen from a standard distance. See *apparent magnitude, magnitude*. The sun is a fairly dim star in absolute magnitude. There are stars that are thousands of times brighter than our sun.

**absolute temperature** The temperature as measured relative to absolute zero, usually in degrees Kelvin. See *absolute zero, Kelvin temperature scale, Rankine temperature scale*.

**absolute value** 1. For a real number, the distance of the number from zero on the number line. For example, the absolute value of +4 is 4; the absolute value of −6 is 6. 2. For a complex number, the distance of the point in the complex plane from the origin (0+0i) in the plane. See *complex number*. 3. A constant that never changes. See the CONSTANTS appendix.

**absolute zero** The coldest possible temperature; the absence of all heat, and the temperature at which all molecular motion ceases. This is about −273 degrees centigrade, or −459 degrees Fahrenheit. It is represented by zero Kelvin. See *Kelvin temperature scale, Rankine temperature scale*.

**absorbed dose** The total amount of radiation absorbed by a human or an animal in a given period of time. Unit: gray, abbreviated Gy.

**absorptance** The extent to which light energy is converted to heat as it passes through a substance. Measured in decibels per unit length. See *decibel, loss*. A glass fiber might have an absorptance of 10 decibels per kilometer. See also *fiberoptics*.

**absorption** 1. The conversion of energy or current to some unusable form. See *absorptance*. 2. The incident energy minus the reflected energy for an object, such as the earth. 3. The extent to which a material will retain a liquid or gas. 4. The acceptance of a nutrient or drug by the digestive system.

**absorption line** In a spectrum, a dark band caused by atoms preventing certain wavelengths from passing. The illustration shows a hypothetical example of absorption lines in a spectrum of light. Each element or compound has a unique "signature" of absorption lines, allowing astronomers to identify substances in outer space.

Absorption Line

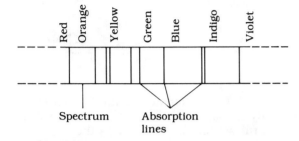

**abstract algebra** Algebra without involving specific numbers, only variables. See *algebra*.

**abyssal current** The movement of water deep in the ocean. See *ocean current*.

**abyssal plain** A flat, deep region on the ocean floor. It might cover many thousands of square miles.

**abyssal zone** The region in and around a deep part of the ocean floor.

**ac (alternating current)** An electrical current that reverses its polarity over and over. Ordinary household alternating current (ac) has a sine-wave shape and a frequency of 60 cycles per second, or 60 hertz (Hz). Alternating currents might have waveshapes other than sinusoidal, and the frequency might range from one cycle in millions of years to trillions or more cycles per second. The drawing shows a typical ac wave. See *direct current, frequency, sawtooth wave, sinusoid, square wave, wavelength*.

**Ac** Chemical symbol for actinium.

**acceleration** A change in the speed and/or direction of a moving object. Measured in units of velocity per unit time, such as meters per second per second or miles per hour per second. See *velocity*.

**accelerator** 1. The gas pedal in an automobile. 2. The throttle in any moving land vehicle. 3. An atom smasher. See *particle accelerator*.

**accelerometer** A device for measuring acceleration. It works by measuring the force created in the direction opposite the acceleration. The force is directly proportional to acceleration. See *acceleration*.

**acceptor** 1. An atom that tends to attract electrons into its outermost shell. 2. A semiconductor substance of this kind, added to silicon to produce certain electrical effects. See *doping, semiconductor*.

**access time** 1. The time it takes to get certain data out of a computer. 2. The time required to get onto a telephone line, so that you can dial a number.

**accretion** A buildup of particles that results in growth over a period of time. An example is the accumulation of molecules to form a crystal. Another example is the buildup of sand on a river bottom in a place where the current is sluggish.

**acellular** A type of living matter made up of a large mass with many nuclei instead of individual cells with their own nucleus. See *cell*.

**acetal** An organic substance, created when alcohol combines with aldehyde. See *alcohol, aldehyde*.

**acetaldehyde** A substance formed during the metabolism of alcohol. It is highly toxic and is usually metabolized further without causing ill effects.

**acetominophen** An aspirin substitute. Some people tolerate it better than they tolerate aspirin. See *acetylsalicylic acid, ibuprofen*.

**acetone** A simple ketone compound that is used as a solvent. It mixes with water easily, and it is flammable. You will recognize its odor in fingernail-polish remover. It is commonly used in making plastic. See *ketone*.

**acetylsalicylic acid** Common aspirin. Some people tolerate substitutes better than aspirin. See *acetominophen, ibuprofen*.

**Achilles tendon** The tendon in the heel, important for walking and running. The invincible Achilles of Greek legend was defeated by an arrow that hit him in this tendon; this is how the tendon got its name.

**achromatic lens** A lens specially designed so that it refracts light of all colors to the same extent. This lens prevents "color blurring" often observed with cheap lenses.

**acid** A compound containing hydrogen, and which is almost always corrosive. Acids react with bases to form salts and by-products such as gas or heat. Common acids include acetic (vinegar), hydrochloric (found in the stomach), nitric and sulfuric (found in acid rain). See *acid rain, base*.

**acid-base reaction** The process and results of combining an acid with a base. This reaction produces salts, gas, and/or heat, sometimes suddenly in large amounts. Combining a concentrated acid and base can cause an explosion. A familiar and fairly harmless demonstration is the combining of baking soda and household vinegar. See *acid, base*.

**acidity** The extent to which a substance is acid. This is usually given by specifying the pH, or concentration of hydrogen ions. Neutral is pH = 7; the most acid possible is pH = 0; and the most basic possible is pH = 14 (in theory). See *alkalinity, pH*.

**acidophilus** See *lactobacillus acidophilus*.

**acidosis** An abnormal and undesirable condition where the body becomes excessively acid. It might occur in such diseases as diabetes, or as a result of a bad diet.

**acid rain** A form of pollution caused mainly by sulfur compounds from the burning of fossil fuels. The pollutants are carried up into the atmosphere, where sulfuric acid is formed in clouds, and this makes the rain acid (see the illustration). Acid rain has caused fish to die in some lakes, especially in the northeastern United States. A few lakes have become as acid as vinegar. Acid rain can cause other problems, such as corrosion and plant deaths.

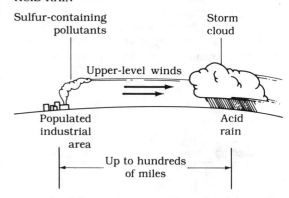

ACID RAIN

**acne** A fairly common condition affecting mostly teenage boys. Excessive or thickened oil from the sebaceous glands clogs the hair follicles. Blackheads and pimples result. See *follicle, sebaceous gland*.

**acoustics** The science of sound. Applications include design of concert halls, musical instruments, and even office buildings. See *sound, sound pressure, sound wave*.

**acoustic wave** See *sound wave*.

**acquired immune deficiency syndrome** See *AIDS*.

**acrylic** A substance commonly used to make hard plastic. This substance is manufactured artificially. Acrylic plastics are used in many ways by modern societies.

**ACTH (adrenocorticotropic hormone)** A substance that controls the activity of the adrenal glands (see *adrenal glands*). Usually, ACTH is secreted by the pituitary gland, but it also can be given as a drug. See also *endocrine glands*.

**actin** A form of protein fiber found in muscle. Actin and myosin are interwoven to form the muscle tissue itself. See *muscle, muscle fiber, myofibril, myosin*.

**actinic keratosis** A scaly skin lesion believed to precede basal cell and squamous cell types of skin cancer. It is like a sore that won't heal. See *basal-cell carcinoma, squamous-cell carcinoma*.

**actinide** A class of chemical elements following actinium, with atomic numbers 89 through 103. See ATOMIC NUMBER AND ATOMIC WEIGHT, PERIODIC TABLE OF THE ELEMENTS appendices.

**action-reaction** The well-known principle "For every action, there is an equal and opposite reaction." You might have noticed this when stepping out of a boat (see drawing), or when firing a rifle (recoil). See *Newton's Laws of Motion*.

ACTION-REACTION

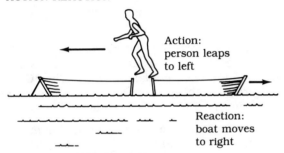

Action: person leaps to left

Reaction: boat moves to right

**activated carbon** A form of carbon used in water-filtration devices, and also in some filter cigarettes. It removes harmful impurities. Charcoal is a common example of activated carbon.

**activated sludge** Sewage to which bacteria have been added to speed up the process of decomposition. This allows waste water to be recycled. When chlorine is added to the treated water, it becomes drinkable.

**activation** 1. The addition of an ingredient, such as bacteria, to a substance to change its properties or to hurry up a desired process. 2. Making a device ready to use. An example is switching on a computer and loading the software.

**active absorption** Sometimes called reverse osmosis. In the intestine, certain nutrients pass from a region of low concentration to a region of higher concentration. This is opposite to the usual osmosis process. See *osmosis*.

**acupuncture** A form of medicine, used widely in Oriental countries such as China and Japan. Fine needles or lasers are injected into, or directed at, certain spots on the body causing predictable things to happen in the body.

**adaptation** 1. The adjustment of a species to changing environment. For example, man's ancestors adapted to the Ice Age thousands of years ago by inventing clothing, changing their diet, and moving to warmer places. 2. The adjustment of an individual to changes in the environment.

**adaptor** A connector used to hook up cords or cables when the connectors provided do not fit. The drawing shows a "cheater" plug. This device can be dangerous. Unless the grounding lug is connected to a good electrical ground, it should not be used, because it can produce a shock hazard.

ADAPTOR

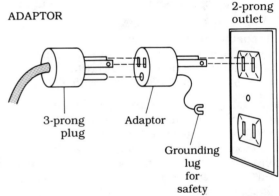

3-prong plug

Adaptor

2-prong outlet

Grounding lug for safety

**addiction** A physical and mental process where a person acquires a need for a drug or other substance to function normally. The result is almost always unhealthy. See *addictive substance*.

**addictive substance** A chemical that causes addiction (see *addiction*). Common addictive substances are alcohol, cocaine, heroin, marijuana, and nicotine. These drugs are responsible for great suffering and many deaths each year. For some people, coffee, sleeping pills, amphetamines (pep pills or diet pills), and other drugs can be addicting. Sometimes food, especially sugary food, is considered addictive.

**Addison's disease** An illness in which the adrenal glands do not function normally. It is not common or contagious. Symptoms include tiredness, weakness, darkened skin, and a loss of appetite and weight. See *adrenal glands*.

**addition** 1. The mathematical summing of two numbers. 2. A component or device attached to an existing device. An example is an expanded memory board for a computer.

**additive** 1. Producing a total effect greater than any of the parts. For example, the effects of two drugs might be called additive. 2. Pertaining to mathematical addition. See *addition*.

**address** A location of memory data in a computer. Might be indicated by a number, a letter, or a combination of numbers and letters.

**adenine** A protein important in DNA and RNA. See *deoxyribonucleic acid, ribonucleic acid*.

**adenosine** The substance resulting from combining adenine with D-ribose. It is intermediate in the formation of adenosine triphosphate. See *adenine, adenosine triphosphate*.

**adenosine triphosphate (ATP)** A substance that is important in living organisms. Chemical energy is converted to other forms, such as heat and movement. ATP is crucial in the process of converting food to energy. Without it, metabolism would not be possible. See *Krebs cycle, metabolism*.

**adhesion** 1. The extent to which a substance will stick to surrounding objects. 2. The tendency for a substance or particle to stick to surrounding objects or particles.

**adhesive** 1. The property of having adhesion. See *adhesion*. 2. A substance having a high degree of adhesion.

**adiabatic** Having constant heat. The temperature might change, but no heat energy leaves or comes in. When a gas is pressurized, its temperature rises, but if no heat energy is brought in, it is an adiabatic process.

**adipose** 1. Fatty. For example, "He has a lot of adipose tissue in his belly." 2. Pertaining to lipids. See *lipid*.

**adrenal glands** Two endocrine glands, situated just above the kidneys, responsible for producing the hormone adrenalin. See *adrenalin, endocrine glands*.

**adrenalin** 1. A hormone produced by the adrenal glands. It causes an immediate, dramatic rise in blood glucose and blood pressure, and makes a person or animal temporarily stronger than normal. Adrenalin is released in response to fright, certain kinds of physical stress, and after taking certain drugs. 2. The temporary increase in energy caused by the stimulation of the adrenal glands. See *adrenal glands, endocrine glands*.

**adrenocorticotropic hormone (ACTH)** A substance that controls the activity of the adrenal glands (see *adrenal glands*). Usually, ACTH is secreted by the pituitary gland, but it also can be given as a drug. See also *endocrine glands*.

**adsorption** The adhesion of a thin film of liquid or gas to the surface of a solid substance. The film might be as thin as a single atom or molecule.

**advection** Horizontal movement of a liquid or gas. The drawing shows a simplified picture of how wind transfers heat by advection from the

ADVECTION

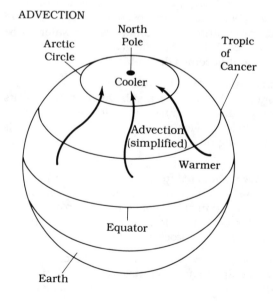

tropics to the arctic. This example shows the general (simplified) wind pattern in the northern temperate zone. See *convection*.

**aeration** The dissolving of air in water. This process is helped by agitating the water, or by "bubbling" air through the water. Fish need this process to survive. Their gills extract the air (and therefore the oxygen) dissolved in lake, stream, and ocean water.

**aeration zone** A level of the earth just below the surface where the soil contains air as well as water. Its depth might be just a few inches or many feet. When rain falls, the water content increases. This region of the soil is important in plant growth. It is sometimes called the A zone. See *soil profile*.

**aerial** 1. Antenna. See *antenna*. 2. Airborne. For example, we might use aerial photography to spy on enemies in a war. See *aerial photography*.

**aerial photography** The taking of pictures from an airborne object, such as a balloon, airplane, or satellite. This technique is useful for surveying, making maps, determining the extent of erosion, observing currents in bodies of water, looking at clouds and storms, and for many other purposes.

**aerobic** Involving oxygen. A process that takes place only when there is oxygen available.

**aerobic respiration** A process of metabolism where carbohydrates or fats are oxidized. Complete metabolism results in the production of energy, water, and carbon dioxide. Incomplete metabolism also yields byproducts such as ketones. See *anaerobic, metabolism, metabolite, oxidation*.

**aerodynamics** The science of the interaction of moving objects with the air. Important in the design of aircraft, and also in the streamlining of cars and other vehicles.

**aeronautical engineering** A field of engineering involving the design of aircraft and spacecraft, and their propulsion systems such as jets, propellers, and rockets.

**aeronautics** The study of the properties of flying objects, and of the effects of flying on people.

**aerosol** A device that allows a chemical to be sprayed from a pressurized container. Recently, aerosol products have been criticized because some of them are thought to damage the earth's protective ozone layer. See *ozone, ozone depletion, ozone hole, ozone layer*.

**aerospace** 1. Pertaining to the atmosphere and space just above the atmosphere of our planet, including modes of travel to and from outer space. 2. The atmosphere of our planet up to perhaps 300 miles, where it thins out to the near vacuum of outer space.

**AF** Abbreviation for audio frequency.

**aflatoxin** A poisonous substance that interferes with the action of deoxyribonucleic acid (DNA). This poison occurs in the form of a mold in some spoiled foods. See *deoxyribonucleic acid, mold*.

**afterbirth** Matter that is no longer of any use in the womb after a baby is born. This matter is ejected. It includes the placenta and umbilical cord. See *placenta, umbilical cord*.

**aftershock** Tremors following an earthquake. They might happen for hours, days, or weeks after a major quake.

**Ag** Chemical symbol for silver.

**agar** A substance used in the culturing of bacteria in the laboratory. It is jelly-like, rich in nutrients, and can serve as an easily absorbed food. Agar is obtained from seaweed and cow's blood.

**agate** A form of rock sometimes used for ornamental purposes. When polished, agates show unique patterns and colors. They can be artificially colored.

**agglutination** The tendency of cells to stick to each other under certain conditions. The most common example is the clumping of red blood cells in the presence of foreign substances. This reduces the ability of the red cells to transport oxygen to the tissues, because of reduced surface area, and also because the clumps cannot pass through tiny capillaries. See *red blood cell.*

**aggregate**   1. A combination of two or more different kinds of mineral into one mass of rock. 2. A combination, or amassing, of particles, objects, or cells into a single, larger object.

**aging**   1. A natural process, believed to be inevitable, in which the body processes of any living thing gradually deteriorate. It is a good example of how order decays into chaos. 2. A process used to enhance the quality of certain foods or beverages.

**Agricultural Revolution**   A major event in the history of humanity, sometimes called the greatest achievement and the worst mistake humans ever made. According to one theory, several thousand years ago humans began farming, and formed communities rather than living as nomads. It is possible to feed more people from crops than by hunting, fishing, and gathering. So the Agricultural Revolution caused a population explosion. Most of the world's problems today can be traced to the world's overpopulation. See *Malthusian model, population explosion.*

**agriculture**   The controlled raising of plants for food; crop farming. Common food plants include wheat, rye, soybeans, corn, barley, millet, potatoes, squash, and others. Crops might be used to feed people, livestock, or both.

**Ah**   Abbreviation for ampere-hour.

**AIDS**   Abbreviation for acquired immune deficiency syndrome. A disease caused by a virus known as Human Immunodeficiency Virus (HIV). There are many misconceptions about this disease. AIDS is spread via body fluids. It can be caught through any form of sex, from blood transfusions, and from sharing contaminated hypodermic needles. Everyday contact, insects, and sharing of clothing are not believed to transmit AIDS. Researchers are trying to find a cure or a vaccine. But as of this writing, the disease is always fatal. AIDS causes the victim to gradually lose his or her immunity to common infection. Death occurs from these illnesses, usually pneumonia. Normally, several years pass from the time of first exposure to the development of actual infections, but AIDS can be spread as soon as one has been exposed. A test can be done for the presence of HIV; a doctor should be consulted for this.

**aileron**   A movable device on an aircraft wing (see illustration). There is one on each wing. They control banking, and this controls how the aircraft turns.

AILERON

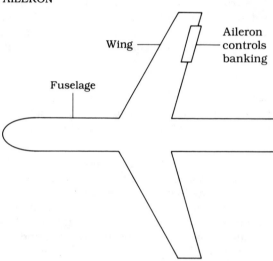

**air brakes**   Also called pneumatic brakes. Air pressure is used to obtain far greater braking power than would be possible using ordinary mechanical brakes. Compressed air provides the force. Air brakes are used in large vehicles and in trains.

**airflow**   The movement of air through, around, or past a device designed specifically to control where the air goes. An example is the motion of air around an airfoil. See *airfoil.*

**airfoil**   The shape of an airplane wing (see illustration). Because of the curved wing top, air flows faster over the top than along the bottom. This faster flow causes lower pressure on top of the wing. The pressure underneath is relatively higher, pushing the wing upwards.

**air mass**   A region of the atmosphere with certain properties. For example, the polar air mass

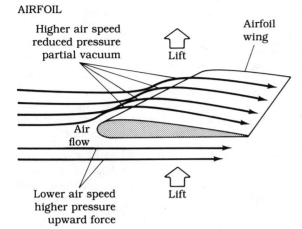

AIRFOIL

Higher air speed reduced pressure partial vacuum

Lift

Airfoil wing

Air flow

Lower air speed higher pressure upward force

Lift

is cold, and a continental air mass is drier than a maritime (ocean) air mass. The boundary between two air masses is called a front. See *weather front*.

**air pollution** The contamination of the atmosphere by manmade chemicals. Air pollutants are classified as particulate, photochemical, and gaseous. Examples respectively are smoke from coal burning, ozone from reaction of sunlight with automobile exhaust, and carbon monoxide, also from automobile exhaust. Sometimes severe air pollution makes breathing difficult, especially for sick or elderly people. See *carbon dioxide, carbon monoxide, ozone pollution, particulate, photochemical haze, smog, sulfur dioxide*.

**air pressure** 1. The force exerted per unit area, caused by air compressed to a greater density than normal atmospheric pressure. Measured in pounds per square inch (PSI), or in kilograms per square meter. Might also be given in atmospheres (atm) where 1 atm = 14.7 PSI. 2. Normal atmospheric pressure is about 14.7 PSI, or capable of supporting about 30 inches (760 millimeters) of mercury in a barometer. See *barometer*.

**air resistance** 1. The extent to which the air impedes the progress of a moving object. It might or might not be desirable to minimize this resistance. Also called drag.

**airscrew** See *propeller*.

**airspeed** The speed of an object, usually an aircraft, with respect to the air. Airspeed is often much different than ground speed because of the wind. See *ground speed*. Airspeed is easier to measure than ground speed in an aircraft. A device like an anemometer can determine low airspeeds. See also *anemometer*.

**airstream** The path followed by air as it moves around, past, or through an object or device. Can be seen by introducing smoke into the air, as long as turbulence is not too great. See *turbulence*.

**air-traffic control** The coordination of air flight schedules, routes, altitudes, and landing/takeoff patterns. Necessary to ensure safety of passengers and crew.

**airwaves** Radio-frequency energy; the range of the electromagnetic spectrum in which radio communications takes place. See *electromagnetic spectrum*.

**airway** The windpipe and throat, forming the passage for air coming into the lungs.

**Al** Chemical symbol for aluminum.

**alanine** An amino acid. In humans, it can be manufactured from other amino acids, so it is not necessary to get it in the diet. See *amino acid, essential amino acids*.

**albedo** The extent to which an object reflects light. Expressed as a number from 0 to 1, or as a percentage from 0 to 100. A black object has albedo of 0 or 0%. A white object has albedo of 1 or 100%. Most objects fall in between these extremes. See the illustration.

**albedo effect** A feedback mechanism where ice tends to sustain itself. Ice and snow have high albedo. They reflect sunlight and don't heat up easily. During the Ice Ages, this kept the earth cool. Conversely, if there was no ice, the earth's albedo would be lower, and it would absorb more heat and stay warmer. See *albedo*.

**albino** A rare hereditary condition in which an animal or human lacks normal pigmentation. An albino animal has white fur and usually pink eyes. An albino human cannot suntan and has extremely fair skin and hair.

ALBEDO

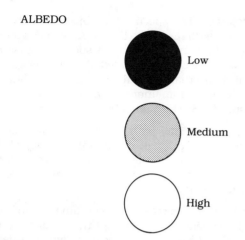

**albumin** A protein found in living organisms. Egg white is essentially pure albumin. It is also found in whey. Albumin is an essential component of blood plasma.

**alchemy** An ancient form of chemistry. Alchemy was regarded by some as magic, by some as science, and by some as sorcery and therefore evil. Alchemists were always looking for a way to turn lead into gold, and to transform various elements into other elements.

**alcohol** Any of various chemical solvents, similar in some ways to carbohydrate, and in some ways to fat. Alcohol molecules contain atoms of carbon, oxygen, and hydrogen in various arrangements. The ethanol in liquor, wine, and beer is a form of alcohol. See *ethanol, methanol*.

**alcoholism** Chronic addiction to alcohol as a drug. It is thought of as a disease by many doctors and psychologists. As with any form of addiction, it causes physical and mental degeneration.

**aldehyde** A compound resulting from the oxidation of alcohol. See *acetaldehyde, alcohol, formaldehyde*.

**Aldrin, Edwin** The second human to set foot on the moon during the Apollo 11 mission in 1969. See *Apollo* and *Armstrong, Neil*.

**aleph-null** The number of natural or counting numbers in the set N = {0,1,2,3,...}. This is an infinite number, also called a transfinite cardinal number (see *transfinite cardinal*). There are larger infinite numbers. See also *aleph-one* and *Cantor, Georg*.

**aleph-one** The number of real numbers. This is an infinite number, a transfinite cardinal (see *transfinite cardinal*) and it is, in a certain sense, "larger" than the infinity of the counting numbers. See *aleph-null* and *Cantor, Georg*.

**algae** A water plant consisting of one or more cells, having chlorophyll, and found in lakes, rivers, and oceans. It is usually green or blue-green but it might vary in color. Algae is responsible for the "blooming" of lakes during the warmer months.

**algebra** A branch of mathematics in which equations are solved for the values of unknowns. The unknowns are called variables (see *variable*). Algebra ranges in complexity from the simple single-variable linear equations to arrays of simultaneous equations having hundreds or thousands of variables. See also *abstract algebra, cubic equation, equation, linear algebra, nonlinear equation, quadratic equation, quadratic formula, quartic equation*.

**algebraic** Pertaining to algebra. For example, we might speak of an algebraic expression when referring to an equation to be solved.

**algebraic number** A real number (see *real number*) that can be defined by algebraic means according to polynomial equations. There are real numbers that cannot be defined this way; they are called transcendental. See also *polynomial, transcendental number*.

**algorithm** 1. In a computer, a process that is executed step by step to carry out a desired function. A program. 2. Generally, any process that can be broken down into steps and can be carried out by following concise instructions.

**alimentary tract** See *digestive system*.

**alkali** 1. Sodium carbonate and/or potassium carbonate, especially when they occur naturally. 2. Soil containing various salts that hinder plant growth. 3. See *alkali metal*.

**alkali metal** The elements lithium, sodium, potassium, rubidium, cesium, and francium. They are reactive; lithium burns in the air and sodium blows up when immersed in water. These elements get their name from the fact that they all can form hydroxides that are very alkaline or basic. See *hydroxide*.

**alkaline** Having the property of being a base. See *base*.

**alkaline-earth metal** The elements beryllium, magnesium, calcium, strontium, and barium. These metals do not react with oxygen as violently as do the alkali metals (see *alkali metal*). They all can form hydroxides, but these compounds are somewhat less alkaline, in general, than the hydroxides of the alkali metals.

**alkalinity** The extent to which a chemical is alkaline, or base. The most common measure of this is the pH. The higher the pH, the more alkaline the chemical. Neutral is pH = 7; the maximum alkalinity possible (in theory) is pH = 14. See also *acidity, pH*.

**alkaloid** Any of various compounds that can be derived from plants and used as drugs. Caffeine is the most common example. Small amounts of this substance cause increased alertness. Others are cocaine, like caffeine but stronger; atropine, used to ease intestinal cramping and to counteract certain poisons; morphine, used to relieve pain; quinine, to treat malaria. See *atropine, caffeine, cocaine, morphine, quinine*.

**alkalosis** An abnormal condition of metabolism in which the body pH is abnormally high. It might result from certain illnesses, from the abuse of certain medications or from a poor or unbalanced diet. See *acidosis*.

**alkane** Also known as the methane series. These are hydrocarbon compounds, many of which are abundant in nature and are widely used as fossil fuels. The first few molecules are shown in the table. Methane, also known as natural gas, is the simplest; alkane along with ethane, propane, and butane are gases at normal atmospheric pressure. Pentane, hexane, heptane, and octane are liquids; paraffin (one example formula is shown in the table) is solid. Refer to the terms listed in the table for further information.

ALKANE

| Compound name | Formula |
| --- | --- |
| Methane | $CH_4$ |
| Ethane | $C_2H_6$ |
| Propane | $C_3H_8$ |
| Butane | $C_4H_{10}$ |
| Pentane | $C_5H_{12}$ |
| Hexane | $C_6H_{14}$ |
| Heptane | $C_7H_{16}$ |
| Octane | $C_8H_{18}$ |
| . | . |
| . | . |
| . | . |
| Paraffin | $C_{30}H_{62}$ |
| (General formula) | $C_nH_{2n+2}$ |

**alkene** A hydrocarbon with at least one double carbon-to-carbon bond. There are always twice as many hydrogen atoms as carbon atoms. The example in the drawing is butene. The alkene series is named similarly to the alkanes: ethene, propene, butene, and so on.

ALKENE

$$H-\underset{\underset{H}{|}}{\overset{\overset{H}{|}}{C}}-\underset{\underset{H}{|}}{\overset{}{C}}=\underset{\underset{H}{|}}{\overset{}{C}}-\underset{\underset{H}{|}}{\overset{\overset{H}{|}}{C}}-H$$

H: Hydrogen atom
C: Carbon atom
Formula $C_4H_8$ (example)

**alkyne** A hydrocarbon with at least one triple carbon-to-carbon bond. The series begins with ethyne; then comes propyne, butyne, and so on, named similarly to the alkanes.

**allele** That part of a gene responsible for certain characteristics of an individual. Alleles can be either dominant or recessive. See *chromosome, dominant trait, gene, recessive trait*.

**allergen** A substance, such as a food, drug or contaminant, that causes an allergic reaction. One person's allergen may not affect someone else. Reactions to allergens can involve difficulty in breathing, increased heart rate, digestive problems, skin eruptions and, in severe cases, shock. See *allergy, shock*.

**allergy** The tendency to get sick from certain substances, such as drugs, foods, or contaminants. This is usually an inherited trait, but it can occur as a result of illness, drug abuse, or from other environmental causes. See *allergen*.

**allometric** A form of growth in which all the parts or organs of an individual get bigger in the same proportion.

**allopatric** Pertaining to species that could, and probably would, interbreed, but they do not because they live in different parts of the world and never come into contact with each other.

**allotrope** Different molecular forms for an element. Oxygen can group in twos (the usual way) or in threes (forming ozone). Carbon can exist either as graphite (the familiar black substance) or as diamond. Allotropes differ from ions or isotopes. See *ion, isotope*.

**alloy** A combination of two or more metals, mixed to obtain properties that no metal has by itself. Brass is a common example. Steel is another. Less common alloys are used in the manufacture of transistors and integrated circuits.

**alluvium** Gravel, sand, silt, and clay deposits on a riverbed or floodplain.

**Alnico** An alloy made from aluminum, nickel, iron, and cobalt that is used in permanent magnets. Alnico is easy to magnetize and demagnetize.

**alphabet** 1. A set of symbols that make it possible to communicate in written form. The English alphabet has 26 letters. Other alphabets might have more or less letters. 2. The set of symbols used by a computer. 3. The set of symbols used by an electronic communications system.

**Alpha Centauri** A star about 4.3 light years from the Solar System, and one of the closest to our planet besides the sun. See *Proxima Centauri*.

**alphanumeric** 1. Consisting of letters and numbers. 2. Arranged in sequence according to the scheme ABC...XYZ012...789.

**alpha particle** A helium nucleus consisting of two neutrons and two protons.

**alpha rays** High-speed alpha particles ejected in nuclear reactions. They are generally less of a hazard than other forms of radiation, although they can be dangerous. See *beta particle, beta rays, gamma rays, neutron radiation, X rays*.

**alpha tocopherol** A form of vitamin E. See *vitamins* and the VITAMINS AND MINERALS appendix.

**alpine glacier** A glacier in the mountains and mountain valleys. Such glaciers exist even near the equator, because it is cold at high altitudes all year long. See also *glacier*.

**alternating current** An electrical current that reverses its polarity over and over. Ordinary household alternating current (ac) has a sine-wave shape and a frequency of 60 cycles per second, or 60 hertz (Hz). Alternating currents might have waveshapes other than sinusoidal, and the frequency might range from one cycle in millions of years to trillions or more cycles per second. The drawing shows a typical ac wave. See *direct current, frequency, sawtooth wave, sinusoid, square wave, wavelength*.

ALTERNATING CURRENT

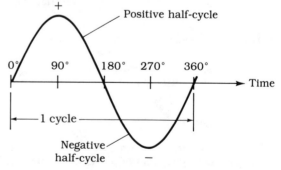

**alternator** 1. A generator that produces alternating current. See *alternating current, generator*. 2. A small generator in a motor vehicle used to charge the battery and to operate the electrical system when the engine is running.

**altimeter** A device in an aircraft that determines the altitude, either above the ground or with respect to sea level. It can operate like sonar (see *sonar*) or by using a barometer to measure air pressure (see *barometer*).

**altitude** 1. Height above the earth's surface. 2. Height above sea level. 3. The angle that an object in the sky subtends with respect to the horizon.

**altocumulus clouds** Clouds that form at medium altitude, and that look something like milk curds. See the illustration.

**altostratus clouds** Clouds that form at medium altitude and vary in consistency from a thin, milky veil to a gray overcast. See the illustration.

ALTOCUMULUS CLOUDS
ALTOSTRATUS CLOUDS

**altruism** A form of behavior in which an individual puts itself at risk, or even sacrifices its own life, for the sake of another of the same species.

**alum** A metallic salt, most often a compound of aluminum, potassium, and sulfur. It is known for its astringent properties. Alum is often used in the manufacture of dye. It is also used to remove solid waste from water.

**alumina** Aluminum hydroxide or oxide. Found in some antacid medicines. See *antacid*.

**aluminum** A light, rather brittle metal. An excellent conductor of electricity, noted for its strength and cheapness. Used to make wire, antennas, sheet metal, and for many other industrial purposes.

**alveola** One of the tiny air sacs in the lung responsible for the exchange of oxygen, carbon dioxide, and other substances between the air and the blood. A human lung has millions of them.

**Alzheimer's disease** A progressive, degenerative disease in which a person loses memory and mental ability earlier than usual. As a person gets old, some degeneration is normal, but with Alzheimer's, it takes place sooner and is much worse. Severe Alzheimer's disease requires that a person have professional care.

**AM** Abbreviation for amplitude modulation.

**A.M.** Abbreviation meaning after midnight and before noon, local time.

**Am** Chemical symbol for americium.

**amalgam** An alloy of mercury. Mercury is a liquid at room temperature. So are some amalgams. Liquid metals are useful in certain industrial applications. See *alloy, mercury*.

**amateur radio** A hobby in which participants communicate with each other by radio, and do electronic experiments involving radio. Requires a license issued by the government.

**amber** A clear or translucent rock, yellow to brown in color, used to make decorative items or jewelry. Fossilized tree sap often containing insects and even small tree frogs.

**ambient** 1. Surrounding, especially pertaining to the environment. Thus, a scientist might speak of the *ambient* temperature.

**ameba** See *amoeba*.

**American Morse code** See *Morse code*.

**American wire gauge** A standard for specifying sizes of wire. The gauge number increases as the diameter gets smaller.

**americium** Chemical symbol, Am. A heavy element, atomic number 95. Might have several isotopes; the most common has an atomic weight of 243. Americium does not occur in nature, but is made by scientists in the laboratory.

**amethyst** Quartz with impurities that make it appear violet-colored. It is used in jewelry because of its exotic appearance. See *quartz*.

**amicable numbers** Any pair of natural numbers a and b, such that a is the sum of the proper divisors of b, and b is the sum of the proper divisors of a. See *proper divisor*.

**amine** A chemical compound that is formed when plant and animal remains decompose. They consist of carbon, hydrogen, and nitrogen. Amines are chemically close to ammonia. See *ammonia*.

**amino acid** A substance that the body gets from protein in food. There are certain amino acids that must actually be in the food one eats. The others can be manufactured in the body from these. See *essential amino acids*. Amino acids can be isolated and used as drugs.

**ammonia** A compound of nitrogen and hydrogen, occurring naturally as a gas. It is found in abundance in the atmospheres of the large outer planets, Jupiter, Saturn, Uranus, and Neptune. Ammonia is also thought to have been a major gas in the atmosphere of the primordial Earth. Ammonia is important in biologic cycles, especially the nitrogen cycle. See *nitrogen cycle*.

**amnesia** A loss of memory following a serious illness or accident. The person might forget events over just a short span of time, or he/she might forget practically everything, even his/her own name. Amnesia can be temporary, or it can be permanent.

**amnion** In humans and many animals, a membrane housing the embryo or fetus before birth. Also called the amniotic sac, it is contained within the uterus. See *uterus*.

**amoeba** A protozoan recognizable by its bloblike appearance when seen under a microscope. It consumes its food by surrounding it (see drawing).

**ampere** The standard unit of electric current. It is represented by a flow of one coulomb of electrons past a point in one second. See *coulomb*.

**ampere-hour** A unit of electrical quantity, equivalent to a current of one ampere flowing for one hour, or 3600 coulombs of electrons. See *coulomb, kilowatt-hour*.

**Ampere's Law** A rule for determining the direction of magnetic flux around a current-carrying wire. If the electrons travel toward you, the flux is in a clockwise circle around the wire.

**ampere-turn** A unit of magnetic flux equivalent to the field produced by one ampere of current flowing through a one-turn loop of wire. See *ampere, magnetic flux*.

**amphetamine** A stimulant drug, often abused. It causes increased alertness. In high doses or with excessive use it can produce weight loss, sleeplessness, hallucinations, psychosis, heart palpitations, and digestive trouble.

**amphibian** The frogs, toads, salamanders, and newts. They have a backbone, are cold-blooded, and reproduce in water. They are thought to have been first to live on land. For evolutionary place, see ANIMAL CLASSIFICATION appendix.

**ampholyte** A chemical that can behave either as an acid or a base. When around a strong acid, it acts alkaline; when associated with a strong base, it acts acid.

AMOEBA

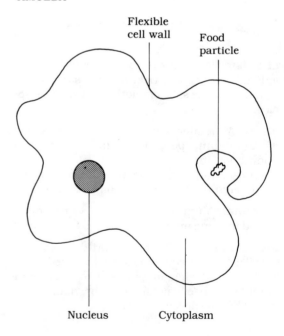

**amplification**  1. In electronics, the process of making a signal stronger, or a current or voltage larger. 2. The tendency for certain effects to be magnified under special conditions.

**amplifier**  An electronic circuit that makes a signal stronger. There are different types of amplifiers for different purposes. See *preamplifier*.

**amplitude**  The strength or intensity of a signal, current, or voltage. It can be specified in watts, amperes, volts, or large multiples or small fractions of these units. It can also be given in decibels relative to a certain reference level. See *decibel*.

**amplitude modulation**  In communications, a means of sending complex information. The carrier wave strength is made to change in sync with voice patterns or other data. The drawing shows an amplitude-modulated (AM) radio wave as it might look on an oscilloscope. See *carrier, frequency modulation, modulation, pulse modulation*.

AMPLITUDE MODULATION

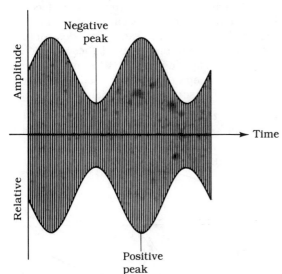

**amputation**  The surgical removal of a limb of the body or part of a limb. Sometimes necessary after a serious accident when a limb has been damaged beyond repair. Occasionally needed when infection or cancer in a limb threatens to spread to the rest of the body.

**amu**  Abbreviation for atomic mass unit.

**amylase**  Any of several enzymes that digest amylose and other complex sugars and starches. The long starch chains are broken down into simple sugars that are easily utilized: glucose and maltose. See *amylose, glucose, maltose, polysaccharide*.

**amylose**  A complex sugar found in starchy vegetables such as potatoes. See *glucose, maltose, polysaccharide*.

**anabolic steroid**  A hormone that is used by some athletes to increase strength and muscle mass. It can endanger the user's health, and might give a temporary advantage to those willing to harm their bodies so they can do better in sports.

**anabolism**  A building up form of metabolism, where proteins and other complex substances are made from simpler molecules. See *catabolism, metabolism*.

**anaerobic**  Not involving oxygen. Anaerobic respiration takes place in muscles under certain conditions. Some bacteria can thrive in the absence of oxygen.

**analgesic**  A mild pain reliever. Common aspirin and aspirin substitutes are analgesics. Stronger analgesics can be obtained with a doctor's prescription. Some analgesics are available in creams and ointments. See *acetaminophen, acetylsalicylic acid, ibuprofen*.

**analog**  1. Pertaining to continuous measurement, rather than measurement in discrete intervals. See *analog meter, analog-to-digital converter, digital*. 2. Something that is similar to a given object, substance, or process. For example, a waterspout is the maritime analog of a tornado.

**analog computer**  See *computer*.

**analog meter**  A device for measuring a variable quantity, such as electrical current or voltage, giving a continuous display. The illustration shows a typical analog meter.

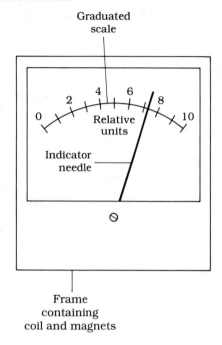

ANALOG METER — Graduated scale, Relative units, Indicator needle, Frame containing coil and magnets

**analog-to-digital converter** In data communications, a device that converts analog signals to digital signals. See *analog, digital, digital-to-analog converter.*

**analysis** 1. Scientific scrutiny of an object, phenomenon, or process. Often involves quantitative as well as qualitative study. 2. A branch of mathematics dealing with functions and their graphs, and the properties of point sets. See *analytic geometry, calculus, function.*

**analytic chemistry** Quantitative chemistry, involving the manipulation of chemical equations and expressions to predict reactions and other chemical processes.

**analytic geometry** First developed by the French mathematician Rene Descartes (see *Descartes, Rene*). The mathematical art of graphing functions and deriving functions from curves on the coordinate plane. See *analysis, calculus, Cartesian coordinates, Cartesian plane, function, polar coordinates.*

**anastigmatic lens** An optical lens that has been corrected, so that it has the same focal length for all colors of visible light, and so that images are not distorted.

**anatomy** 1. The physical body and biological processes of an animal or human. 2. A branch of medicine involved with the study of the structure and processes of the human body.

**andesite** A form of volcanic rock, intermediate between basalt and rhyolite. Takes its name from the Andes Mountains in South America. See *basalt, rhyolite.*

**Andromeda galaxy** A spiral galaxy about 2.2 million light years from the Milky Way. Located in the constellation Andromeda, it can be seen with a good pair of binoculars or a small telescope on a moonless night in rural areas. It is the most easily visible galaxy that is entirely separate from ours.

**anechoic chamber** A soundproof room padded with absorbent tile so that practically no echoes occur. Useful for conducting certain kinds of acoustical tests and experiments. Also called an acoustic chamber or quiet room.

**anemia** A disease affecting the blood, where there is a deficiency of healthy red cells (see *red blood cell*). Can occur because of insufficient red cell production, excessive destruction, blood loss, or combinations of these. Deficiencies of iron, copper, vitamin B-12, and folic acid can cause anemia. This can result from poor diet or poor absorption of nutrients.

**anemometer** A device for measuring wind speed. A set of three or four rotating cups is

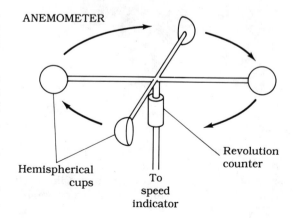

ANEMOMETER — Hemispherical cups, To speed indicator, Revolution counter

attached to a device that converts the revolution speed to miles per hour (see drawing). Most anemometers are accurate up to about 100 miles per hour. Special instruments are needed to measure higher speeds.

**anemone** A coelenterate, among the most primitive of all animals, living in the ocean. See *coelenterate*.

**anesthesia** A medical procedure where part or all of the body is numbed so that surgery will be painless. Local anesthesia is the numbing of part of the body, such as the arm, using drugs similar to novocaine. General anesthesia involves putting the patient "to sleep."

**anesthesiology** The medical practice of anesthesia.

**anesthetic** A drug that causes anesthesia. Usually refers to a local anesthetic. Some over-the-counter creams and ointments contain mild anesthetic chemicals. See *anesthesia*.

**aneurism** Also spelled aneurysm. An enlargement of a blood vessel in a certain place. In the extreme case, called a dissecting aneurism, an artery splits open, usually lengthwise. This can be fatal, especially if it occurs in the brain. The chances of having this happen increase as one gets very old.

**angina pectoris** Pain in the chest associated with heart disease. Occurs when the heart muscle receives insufficient oxygen. This pain can cause great distress. It might signify a heart attack. See *heart attack*.

**angiosperm** Trees, shrubs, grasses, and herbs. They evolved along with the pollinating insects during the middle Cretaceous period. They are at the top of the plant evolutionary scale. See PLANT CLASSIFICATION appendix.

**angle** 1. For two intersecting lines, the extent to which they run in different directions, given as a numerical measure. See *degree, radian*. 2. Slant. For example, we might say that light strikes a surface at an angle. This means it does not come straight down onto the surface. See *angle of incidence, angle of reflection*. 3. A numerical measure of the extent to which a line, such as the path of a light ray, changes direction. See also *angle of refraction*.

**angle measure** See *degree, minute, radian, second*.

**angle of incidence** The angle at which a beam of light or other ray of energy strikes a surface or barrier. Usually measured with respect to a normal (perpendicular) line at the point where the ray hits. See the drawing. See also *angle, angle of reflection*.

ANGLE OF INCIDENCE
ANGLE OF REFLECTION

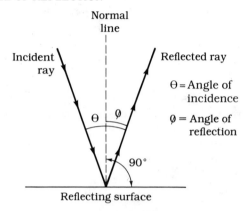

**angle of reflection** The angle at which a beam of light or other ray of energy, having struck a reflective surface, leaves it. For a shiny reflective surface, this angle is always equal to the angle of incidence (see drawing). Usually measured with respect to the normal line at the point where the ray is reflected. See also *angle, angle of incidence*.

**angle of refraction** 1. For a ray of light or other energy refracted at a boundary, the angle at which the ray leaves the boundary. Usually measured with respect to a normal (perpendicular) line at the boundary where the ray changes direction. See the drawing. 2. Less commonly, the extent to which a refracted ray changes direction. See *index of refraction, refraction*.

**Angstrom unit** An expression of measure equal to one ten-billionth (0.0000000001) meter or one ten-millionth (0.0000001) millimeter. Used to specify wavelengths for infrared,

ANGLE OF REFRACTION

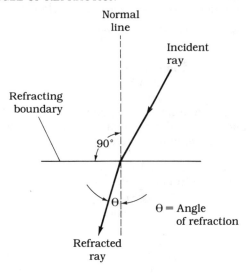

visible light, ultraviolet, X rays, and gamma rays. Sometimes also employed to express tiny sizes of objects such as bacteria, viruses and molecules.

**angular momentum** For a rotating object, the product of angular velocity and moment of inertia. See *angular velocity, moment of inertia*.

**angular motion** 1. Movement in a circle, around a common center point. 2. Rotation. 3. Revolution.

**angular velocity** An expression of how fast and in which direction an object rotates or revolves. For example, the second hand of a clock moves clockwise at 360 degrees per minute, or about one radian every 9.55 seconds, or 6.28 radians per minute, or six degrees per second. Given in degrees or radians per unit time.

**anhydrous** Devoid of water. Usually pertains to a chemical from which all the water has been removed.

**animal** 1. Any of a class of living things characterized by an ability to move around under its own power. 2. The class of living things able to move around under their own power.

**animal classification** The categorization of animals according to characteristics, or according to evolutionary process. See ANIMAL CLASSIFICATION appendix.

**animalia** See *animal, 2*.

**animal protein** Protein from animal sources. Meat, milk, eggs, and fish provide animal protein. Animal protein is often, but not always, better used by the body than vegetable protein. See *protein, vegetable protein*.

**anion** An ion with a surplus of electrons. See *cation, ion*.

**annelid** An earthworm, marine worm, or leech. A phylum of segmented invertebrates. Fairly high on the evolutionary scale of animals. They are cold-blooded. See ANIMAL CLASSIFICATION appendix.

**annihilation** 1. The process where matter and antimatter come into contact, and their masses are converted into energy. See *antimatter*. 2. Destruction; mass death. For example, a big nuclear war could cause the annihilation of humanity. 3. The extinction of a species. See *extinction*.

**annual** Occurring once a year, every year, usually at about the same time of year.

**annual flood** For a river, a flood that takes place (usually) once a year, every year. The Nile in Egypt is famous for this. Annual flooding is most commonly caused either by spring runoff from melting snow, or by the monsoon. See *monsoon*.

**annual plant** A plant that lives for only one season. See *perennial plant*.

**annular eclipse** See *eclipse*.

**annulus** 1. A ring or ring-shaped object. 2. A growth ring on a plant, such as in the cross section of a tree trunk (see drawing).

ANNULUS

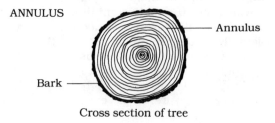

Cross section of tree

**anode** 1. A positively charged electrode that attracts electrons. 2. The positively charged terminal of a diode. See *cathode, diode, electrode.*

**anodizing** A process in which aluminum is given a protective coat. The metal is oxidized by electrolysis. See *electrolysis.*

**anomaly** 1. A strange or unusual event, such as a tornado in Alaska. 2. An angle that astronomers use to locate the position of a planet.

**anorexia** A lack of appetite for food. Can be caused by a malfunction of the appestat, a sickness such as "the flu," or a complex neurosis. See *anorexia nervosa, appestat.*

**anorexia nervosa** A complex physical/emotional disease in which a person always thinks he or she is too fat. This leads to chronic dieting, unnatural thinness, and in the extreme, malnutrition and death. Affects mainly young women in advanced countries. See *anorexia, appestat, bulimia.*

**antacid** A medicine used to neutralize stomach acid. Consists of weak or diluted alkaline compounds such as sodium hydroxide, calcium hydroxide, aluminum hydroxide, and magnesium hydroxide.

**antagonism** 1. A process in which two substances act in an opposing manner cancelling out each other's effects. 2. The tendency of certain organisms to interfere with each other's growth. 3. A condition in which body processes operate against each other.

**Antarctic** 1. The region around the earth's South Pole. 2. The region of the earth within the Antarctic Circle. See *Antarctic Circle.*

**Antarctic Circle** The latitude circle on the earth at 66.5 degrees south. South of this parallel, there are times in summer when the sun never sets, and times in winter when the sun never rises. On June 22 the entire region within this circle is in continuous darkness. On December 22 the whole region is in continuous light. See *Arctic Circle.*

**antenna** A device used for transmitting and receiving radio signals. It is a form of transducer (see *transducer*). A transmitting antenna converts high-frequency alternating current to an electromagnetic field; a receiving antenna does the reverse.

**anterior** Pertaining to the front side or part of an animal, the part facing forward as the animal moves along. See *posterior.*

**anthracite** A high grade of coal. It is hard, burns clean, and is fairly expensive compared to other forms of coal. See *bituminous, lignite.*

**anthrax** A disease affecting livestock. It does not occur often in people, except where sanitation is poor. If it does appear in a human, it causes a pustular rash on the skin. It can also affect the lungs. Antibiotics are used in treatment.

**anthropology** 1. Cultural anthropology: The study of the behavior, customs, religions, sciences, and beliefs of a society. 2. Physical anthropology: Study of a society by analyzing artifacts found by digging up remains.

**anti-** 1. Acting against. 2. Opposite to.

**anti-antibody** An agent that kills or interferes with the action of antibodies. See *antibody.*

**antiballistic missile** Abbreviation, ABM. A defensive weapon designed to intercept and disable offensive missiles before the offensive weapons reach their targets. Also called an antimissile missile. See *ballistic missile.*

**antibiotic** A drug that acts as an antibody. These drugs have been called "miracle drugs" because they make survival possible in illnesses that otherwise might be fatal, such as pneumonia. Common antibiotics include penicillin, tetracycline, and erythromycin. They can be taken by mouth, as a shot, or as a skin cream. They usually require a doctor's prescription.

**antibody** A protein that an animal cell manufactures to combat an unwanted organism or chemical. Most bacteria and viruses cause the production of antibodies. This enables the body to fight disease. See *antibiotic, antigen.*

**anticarcinogen** An agent that slows and possibly even prevents the growth of cancer. Certain foods and drugs have been called anticarcinogens. But there is a lot false information in this field.

**anticline** A fold in the earth's crust caused by movement of the crust. The rocks are pressed upwards into an inverted-V shape. See *syncline*.

**anticoagulant** 1. A substance that inhibits the coagulation of some other chemical. 2. A substance that interferes with the coagulation and clotting of blood. It can reduce the chance of a heart attack. Such medication should only be used with a doctor's recommendation, because there are risks.

**anticyclone** A high-pressure system, spinning clockwise in the Northern Hemisphere and counterclockwise in the Southern Hemisphere (see drawing). Generally associated with good weather. See also *cyclone*.

ANTICYCLONE

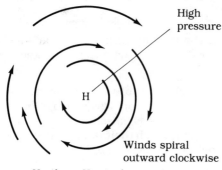

Winds spiral outward clockwise
Northern Hemisphere

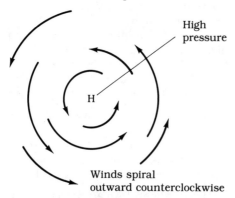

Winds spiral outward counterclockwise
Southern Hemisphere

**antidepressant** 1. Any medication that acts to relieve physical and/or mental depression. 2. A stimulant. A mild antidepressant effect is produced by strenuous aerobic exercise. See *stimulant*.

**antigen** An organism or chemical that causes the body to manufacture antibodies, specifically designed to kill or destroy the invading substance. See *antibody*.

**antihistamine** Any of a class of drugs that suppress the histamine-manufacturing processes in the body. Useful in relieving allergy symptoms. Might cause drowsiness and difficulty in concentrating. Many are available over the counter.

**antimatter** A substance or particle with opposite charge and the same mass that annihilates when it comes into contact with a particle of matter. Some theories about antimatter are strange. For example, it has been suggested that there are whole planets, stars, and galaxies made of antimatter. If we were to land on an antimatter planet we would instantly be annihilated, our mass would combine with an equal amount of the antimatter to produce a flash of energy more powerful than millions of atom bombs. See *annihilation, antineutron, antiparticle, antiproton, positron*.

**antimony** An element, with atomic number 51. The most common isotope has atomic weight 121. In pure form it is a metal with a slightly blue tinge. Used in the manufacture of some electronic semiconductors. Also a constituent of pewter.

**antineutron** A particle of antimatter having neutral charge, and having the same mass as a neutron. Has other properties that are opposite from those of a neutron. See *antimatter, neutron*.

**antioxidant** A substance that inhibits or prevents oxidation. An example is vitamin E (tocopherol), believed to slow down harmful oxidation processes in the body.

**antiparallel** Running alongside each other, but in opposite directions (see drawing). If direc-

ANTIPARALLEL

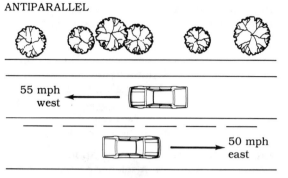

 Parallel paths
opposite directions

tion is not given, antiparallel has the same meaning as parallel. See *parallel*.

**antiparticle** 1. A particle of antimatter such that, upon contact with its counterpart of matter, the two annihilate to yield energy. See *annihilation*. 2. A particle of matter seen from the perspective of antimatter; the exact reverse of the preceding. 3. Energy-carrying and force-carrying particles are their own antiparticles. See also *antimatter, antineutron, antiproton, graviton, photon, positron*.

**antiproton** A particle of antimatter having negative charge and having the same mass as a proton. Upon contact with a proton, annihilation is total. See *antimatter, proton*.

**anus** The opening through which solid waste passes out of the bodies of most animals.

**anxiety** A psychological disturbance in which a person feels afraid or tense, often without any apparent reason. This might occur suddenly and be incapacitating. Then it is called an anxiety attack. It is accompanied by physical symptoms, such as changes in appetite and bowel habits, weight gain or loss, and even reduced immunity to infections such as colds and flu. This condition often requires psychiatric help. It sometimes occurs along with depression. See *depression*.

**aorta** The large artery leading out of the heart and to the arteries going to the body below the chest.

**aortic arch** The arch-like formation in the aorta of humans near the heart. It leads to other arteries going to the head, arms and upper torso.

**aperture** 1. The opening of an optical instrument, such as a camera or telescope. 2. The effective diameter or radius of an optical instrument. 3. The effective diameter of a radio-telescope antenna. 4. An expression of the effective size of an antenna.

**aperture synthesis** In radar or radio astronomy, a method of making an antenna effectively much larger than it really is. Two or more antennas are placed far apart, or one antenna is moved and observations are made at intervals. The signals are combined to form a composite radio image in great detail.

**apex** 1. The place where a cone or pyramid comes to a point. 2. The point of highest elevation.

**aphelion** 1. The maximum distance of the earth, or a planet or solar-orbiting satellite, from the sun (see drawing). 2. The condition of a planet or solar satellite being at its greatest distance from the sun.

APHELION
APOGEE

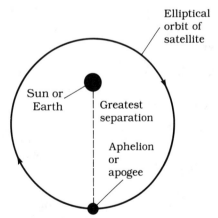

**aplastic anemia** A disease involving the bone marrow, resulting in a shortage of red blood cells. It occurs most often in teenagers and young adults. Can be caused by chemical agents and radioactivity. Treatment requires getting rid of the causal agent(s). Then blood transfusions are given until the bone marrow makes blood cells normally.

**apogee** 1. The maximum distance of the moon, or an earth-orbiting satellite, from the earth (see drawing). 2. The condition of the moon or an earth satellite being at its greatest distance from the earth.

**Apollo** 1. A constellation named after the Greek and Roman god of the sun and the arts. 2. The space program of the 1960s and 1970s in which astronauts traveled to and landed on the moon. The first landing was made by Neil Armstrong and Edwin Aldrin in the summer of 1969 during the mission of Apollo 11. See *Aldrin, Edwin* and *Armstrong, Neil*. See also *Collins, Michael*.

**apoplexy** See *stroke*.

**apparent magnitude** The brightness of a star as seen with the unaided eye. Might vary from less than zero to about 6. Originally, stars were catalogued by magnitude according to their brightness on a scale from 1 (brightest) to 6 (dimmest visible).

**appeasement** Behavior of certain animals in which one individual displays weakness and vulnerability. It is an attempt to avoid a fight, and sometimes goes along with courtship. It can also be a survival tactic.

**appendicitis** An acute and severe inflammation of the appendix (see *appendix*). Accompanied at first by nausea, then by fever and severe pain in the lower right part of the abdomen. Can be fatal if not treated.

**appendix** A part of the intestine in humans that is apparently without any function. Some believe it is an evolutionary accident. It can become inflamed and require surgical removal. See *appendicitis*.

**appestat** A mechanism in the brain that governs an animal's or human's desire for food. Normally, this device works in such a way that weight is kept constant and metabolic needs are met. If it malfunctions, anorexia or obesity result. See *anorexia, obesity*.

**appetite** The desire for food, regulated by metabolic needs and by the appestat. See *appestat*.

**application** 1. Use or purpose. For example, "The application of mathematics to astronomy has been going on for thousands of years," or "That is not its intended application." 2. The act of placing onto. For example, allergy cream might come with instructions for application.

**applications program** A computer program for solving a problem in science or engineering. For example, we might write a program for someone who has just bought a computer, teaching them how to use it, video-game style. Or, we might write a program for the navigation of a rocket to Mars.

**applied mathematics** The manipulation of equations and the solving of mathematical problems useful in the sciences. Special mathematical techniques might be developed and perfected for use in physics, chemistry, astronomy and engineering, as well as other fields. In this way, applied mathematics differs from pure mathematics, the study of mathematics as an end in itself. See *pure mathematics*.

**applied physics** Physics as it can be used in other endeavors such as engineering and medicine.

**applied science** The use of science for some concrete purpose. An example is the application of physics to obtain the optimum trajectory for a ballistic missile. Another example is the use of biological science to immunize the population against an epidemic.

**approximation** A value that is not exact, but is accurate enough for the task at hand. A number might be rounded off to a certain number of significant figures. See *significant figures*. For example, 3.14 is an approximation of pi (see *pi*). A closer approximation is 3.14159. In this case, an approximation is all we can ever get, because pi is a nonterminating, nonrepeating decimal. In the physical universe, all experimentally determined values are approximations. We can get more and more accurate, but never truly exact, when we measure physical quantities.

**aqua regia** A mixture of three parts hydrochloric acid to one part nitric acid. Used to dissolve metals such as gold and platinum, it is one of the strongest acids. See *hydrochloric acid, nitric acid*.

**aqueduct** A long pipeline or artificial river in which water is transported from a reservoir to a city. The ancient Romans built the first aqueducts. Some of these still work today. The trough or pipeline gradually runs downhill (see drawing). In this way large amounts of water can be supplied over long distances without mechanical pumps.

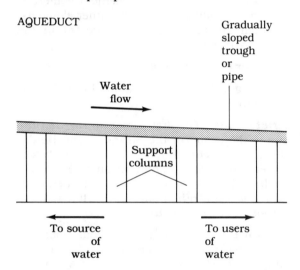

AQUEDUCT

**aqueous** 1. Containing water. 2. Dissolved in water.

**aqueous humor** The fluid between the cornea and the iris in the eye. See *eye*.

**aquifer** 1. Rock strata through which groundwater flows. See *water table*. 2. The water supply that serves a region.

**Ar** Chemical symbol for argon.

**Arabic numerals** Also called Hindu-Arabic numerals. The numbering scheme we use today. It has 10 digit symbols and works in powers of the number we know as 10.

**arachnid** The set of land-dwelling animals including mites, scorpions, spiders, and ticks. They are all carnivorous and are high on the animal evolutionary scale in one of the two main branches. See ANIMAL CLASSIFICATION appendix.

**arc** 1. A portion of a circle. It has a defined center and its span can be given in degrees according to the angle that its ends make with the center point. 2. In electricity, a spark that occurs when a voltage gets high enough. This spark might be continuous, and often gives off brilliant visible and ultraviolet light. It is hot and can be used for welding. See *arc lamp*.

**arch** 1. A structure shaped like an inverted U or V, used as a building support. First used by the Romans in aqueducts (see *aqueduct*). Later used in the Middle Ages to build cathedrals. It is still used today. 2. Any object bent into a U-shape.

**archaeocyanhid** A primitive species that some scientists believe evolved into the fishes, reptiles, birds, amphibians, and mammals. For position on the animal evolutionary scale see ANIMAL CLASSIFICATION appendix.

**Archean era** The time of initial life on the earth, from about 4.6 to 2.5 billion years ago. The earth was hotter and the crust thinner than it is now. There were more volcanoes and meteorite impacts. Our planet was in turmoil, and some believe that the earliest living molecules formed because of lightning, volcanoes, meteorites, or all of these. See GEOLOGIC TIME appendix.

**archeology** The study of fossil relics, especially remnants of past human societies.

**Archimedes** One of the greatest mathematicians in ancient times. He lived in the Third Century B.C. He also might be called an engineer. He was responsible for devising the first pulleys. He is also believed to have invented the water screw for pumping fluids upward. Some farmers still use this device today.

**Archimedes' principle** When an object floats in a liquid, the weight of the object is the same as the weight of the liquid that it displaces.

**arc lamp** A device that generates visible and ultraviolet light. Two carbon electrodes are brought close together and supplied with a high voltage. At the critical separation, an arc occurs (see *arc*). The drawing shows the structure of an arc lamp. The arc is dangerous to look at because of its brillance and ultraviolet content. Arc lamps were employed as street lights before other efficient lamps were developed. See also *fluorescent lamp, incandescent lamp, mercury-vapor lamp, sodium-vapor lamp.*

ARC LAMP

**arc length** The span of an arc, measured as if it were straightened out. See *arc.*

**Arctic** 1. The region around the earth's North Pole. 2. The region within the Arctic Circle. See *Arctic Circle.*

**Arctic Circle** The latitude circle on the earth at 66.5 degrees north. North of this circle there are times in summer when the sun never sets, and times in winter when the sun never rises. On December 22 the whole region within this circle is in continuous darkness. On June 22 the whole region is in continuous light. See also *Antarctic Circle.*

**area** 1. The extent of a specified two-dimensional region. For example the area of the United States, or of a football field, or of a postage stamp. Usually expressed in square units, such as square miles, square feet, or square millimeters. 2. Region or zone.

**arginine** An amino acid. It is not manufactured from other amino acids by the body in sufficient quantities for good health. Therefore, it must be obtained from dietary protein. See *amino acid, essential amino acid.*

**argon** A chemical element with atomic number 18. The most common isotope has atomic weight 40. It is a gas at room temperature and atmospheric pressure. This element is inert; it does not readily combine with other elements. A small amount of argon is present in the earth's atmosphere.

**argument** See *variable.*

**aril** A growth in a seed that might totally surround the seed.

**Aristarchus** The first person to say that the sun is the center of the Solar System. He did this in ancient times, 1700 years before Copernicus. Few took him seriously. Most people believed that the earth was the center, until Copernicus and others showed that it is actually the sun. See *Copernicus, Nicholas.*

**Aristotle** One of the most famous scientists and philosophers of all time. He lived in ancient Greece in the fourth century B.C. He was one of the first to say that the earth is round like a ball, instead of flat. He developed theories of matter and is responsible for the geocentric theory of the universe that was accepted for centuries. See *geocentric theory.*

**arithmetic** A branch of mathematics concerned with manipulation of known numbers. The simplest arithmetic consists of addition, subtraction, multiplication, and division. More complicated arithmetic involves exponentiation and various functions such as the logarithm, sine, and cosine. See also *function.*

**arithmetic logic unit** Abbreviation ALU. A main part in the central processing unit (CPU) of a computer. The ALU does arithmetic operations. It also does logical operations. See *arithmetic, central processing unit, computer, logic.*

**arithmetic mean** See *mean.*

**arithmetic progression** A series or sequence in which the terms increase by the same value every time. An example is the sequence 4,7,10,13,16, ..., where each term is larger than its predecessor by 3. See *geometric progression, sequence, series.*

**Arkwright, Sir Richard** An English engineer, 1732-1792. He is famous for his work in the cotton industry. He also had much to do with developing factories as we know them today.

**Armageddon** 1. The ultimate catastrophe; the destruction of humanity as we know it. 2. A science-fiction scenario, where every person is at war with every other person in the whole world, all at the same time. Total chaos.

**armature** 1. In an electric motor or generator, the part that rotates, usually a coil. See *generator, motor, motor/generator.* 2. In a relay, the lever that moves in response to the action of the electromagnet. See *relay.*

**Armstrong, Edwin H.** An American scientist and inventor, 1890-1954. He invented frequency modulation (FM) as a means of radio communication. He is also known for various other inventions related to radio. See *frequency modulation.*

**Armstrong, Neil** The first astronaut to walk on the moon. He flew the landing module for Apollo 11 with Edwin Aldrin in July, 1969. See *Apollo* and *Aldrin, Edwin.* When Armstrong first stepped on the lunar surface he radioed the message, "That's one small step for a man, one giant leap for mankind."

**arousal** 1. In an animal or human, the transition between sleeping and waking. 2. Stimulation. 2. Sexual excitement and the accompanying physiological changes that prepare an animal or human for intercourse.

**array** 1. An orderly arrangement of objects. 2. A set of objects or effects. 3. An antenna comprised of multiple elements. 4. A large store of data, arranged in an orderly way. See *matrix.*

**array processor** A computer that works with large arrays of data (see *array*). Used for systematic processing of information, where repeated calculations at high speed are necessary. See also *computer.*

**arrhythmia** A condition where the heartbeat is irregular, either for short periods or all the time.

**arroyo** Channels of intermittent streams, mostly in the desert Southwest of the United States. In this way a region of land gets rid of excess water when it rains. After a rainstorm, there might be dozens of such streams in just a few acres of land.

**arsenic** A chemical element, with atomic number 33. The most common isotope has atomic weight 75. Known as a poison, it has an enormous variety of scientific uses. For example, it can be combined with gallium to make a special kind of transistor.

**arteriole** A small artery. It is intermediate between the arteries and the capillaries. See *artery, capillary.*

**artery** A major blood vessel that transports blood from the heart to the rest of the body. Blood in most arteries is bright red because it is rich in oxygen. The exception is the pulmonary artery leading from the heart to the lungs. See *capillary, circulatory system, vein.*

**artesian spring** A place where naturally pressurized ground water comes to the surface. Common where there is a basin, or depression, in the terrain with subsurface water. This often forms lakes. The water might be cold or warm. Sometimes you can feel the cold and warm springs when skin diving or SCUBA diving in a lake. See *artesian well, spring.*

**artesian well** An artesian spring tapped for use by humans. This is done by drilling till the aquifer is struck (see drawing). A true artesian well does not need a pump, because the water is naturally under pressure. See *artesian spring.*

**26    arthritis • asbestos**

ARTESIAN WELL

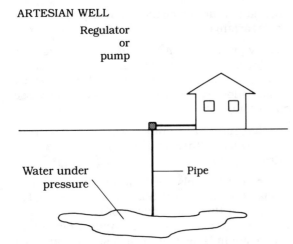

**arthritis** A progressive, degenerative disease of the joints. Some people don't get it at all, and others get it so bad that they're practically crippled. The causes appear to be mostly inherited, although lifestyle plays some role.

**arthropod** A phylum of animals, including arachnids, crustaceans, insects, centipedes, and millipedes. They all have exoskeletons. Some dwell on land, and some in the water. See *arachnid, centipede, crustacean, exoskeleton, millipede.* See also the ANIMAL CLASSIFICATION appendix.

**artificial gravity** A means of generating a gravitational field in space. According to Einstein's principle of equivalence, the force from acceleration is the same as gravity. The drawing shows a space station that rotates to obtain this acceleration. The force could be derived from straight-line acceleration during interstellar or intergalactic voyages. See *acceleration, equivalence principle.*

**artificial heart** A mechanical device that performs the functions of a human heart. Not useful for long periods, because the body tends to reject it as a foreign object. Can be used in emergency situations, such as during heart surgery. See *heart, heart-lung machine.*

**artificial insemination** A technique that allows conception without sexual intercourse. The sperm are placed in the female uterus. This is commonly done with animals. With humans it can be done under medical supervision. See *conception.*

**artificial intelligence** Sophisticated computer intelligence. It is now possible to build computers that can learn from their mistakes. Often used along with robots to perform menial labor that used to be done by people. This is a rapidly growing field of technology. It is thought by some that we will eventually build computers that are as smart as or smarter than we are. See *robot, robotics.*

**artificial kidney** A machine that performs all of the functions of a human kidney. Nowadays this is done by a dialysis machine. See *dialysis.*

**artificial limb** A mechanical device that performs some of the functions of an arm or leg. It might be stronger than a human arm or leg. But no artificial limb is as agile as a real one, and artificial limbs don't have any sensation. See *bionics.*

**As** Chemical symbol for **arsenic.**

**asbestos** A flame-retardant substance, commonly used in the construction of buildings

ARTIFICIAL GRAVITY

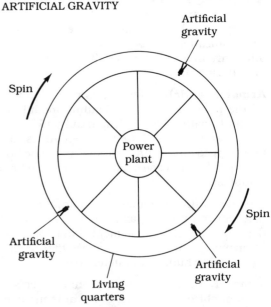

before the middle 1970s. Made from natural minerals spun into fibers. It has been used as a fireproofing agent and an insulator, but recently it has been linked to lung disease, including cancer.

**ASCII (American Standard Code for Information Interchange)** A communications code used by computers and teletype terminals. Speeds range from 110 to 19,200 words per minute. See *BAUDOT code*.

**ascorbic acid** See *vitamins* and the VITAMINS AND MINERALS appendix.

**asexual reproduction** A form of reproduction common in plants, bacteria, viruses, and some animals. There need be only one parent. An example is cell division, another is budding. Still another example is reproduction by means of seeds and spores. See *cell division, budding, seed, spore*.

**Asimov, Isaac** A well-known writer and scientist. A magazine is named after him. He has written numerous short stories and novels.

**asparagine** An amino acid. In humans it is manufactured from other amino acids, and need not be supplied in dietary protein. See also *amino acid, essential amino acids*.

**aspartame** An artificial sweetener having practically no calories. In recent years it largely has taken the place of saccharin. See *saccharin*.

**aspartic acid** An amino acid. In humans it is manufactured from other amino acids, and need not be supplied in dietary protein. See *amino acid, essential amino acids*.

**aspect ratio** 1. The ratio of width to height for an object or image. 2. In electronics, the ratio of width to height for a video frame. In the United States, for television signals, it is standardized at 4 to 3. See *television*.

**asphyxia** Suffocation. Lack of oxygen causes loss of consciousness and possibly death.

**aspiration** 1. The removal of fluid from an abscess or from some body cavity, usually by drawing it out through a needle. 2. Accidental inhalation of foreign matter into the lungs. Can cause pneumonia.

**aspirin** See *acetylsalicylic acid*.

**assembler** The program that translates between a computer's machine language and the user's language, like BASIC, COBOL, or FORTRAN. See *BASIC, COBOL, computer, FORTRAN, machine language*.

**assembly language** A computer language, used by programmers to write the assembler program. Each user language, such as BASIC or FORTRAN, has a somewhat different assembly language. See *assembler*.

**assembly line** A factory technique that allows far greater production than is possible if units are built separately. Developed in the nineteenth century. Each worker does one specific task over and over. The whole unit moves through the line and comes out complete at the end. The main problem is the monotony of each worker's job. This is being resolved by having robots do much of the tedious work on the line.

**assimilation** The absorption and utilization of nutrients or drugs by a living organism.

**association** 1. A tendency for two or more events to happen in conjunction with each other. For example, obesity is associated with heart disease. 2. A situation where different species live close to each other for some reason besides pure accident.

**associative law** A law in mathematics that applies to some operations but not to others. Addition is associative; this means that for any three numbers x, y, and z, $(x+y)+z = x+(y+z)$. Multiplication is associative; however, subtraction is not. See *commutative law, distributive law*.

**astable circuit** A circuit that switches between two states at a constant rate. An oscillator is a common form of astable circuit; so are some multivibrators. See *multivibrator, oscillator*.

**astatine** A chemical element with atomic number 85. The most common isotope has atomic weight 210. It is radioactive, being the result of the decay of uranium and thorium. Astatine itself decays quite rapidly.

**asteroid** An object too big to be a meteoroid and too small to be a planet. Also called a planetoid. The largest, Ceres, is over 400 miles long. Some asteroids are moons, caught by the gravity of a planet. The term means star-like. Asteroids, like stars, are seen through telescopes as points of light. Most asteroids are between Mars and Jupiter. A few have eccentric orbits. Several come close to the earth. The earth has been struck occasionally by asteroids. See *asteroid belt, meteor, meteoroid.*

**asteroid belt** The circular region between Mars and Jupiter where most of the asteroids in the Solar System orbit. The drawing is a scale rendition of planetary orbits (except for Neptune and Pluto) and the asteroid belt. See *asteroid.*

ASTEROID BELT

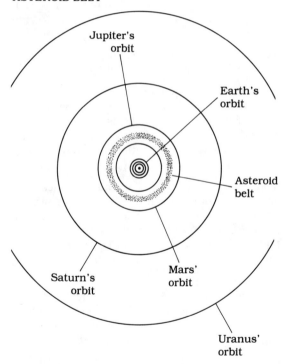

**asthenosphere** A part of the mantle of the earth, from about 50 miles to 200 miles down. It is thought that this layer, somewhat more fluid than the rock above and below, might be in convective motion. See also *mantle.*

**asthma** An allergic reaction in which the sufferer has difficulty breathing. This is caused by constriction, or spasm, of the bronchi. See *allergen, allergy, bronchi.*

**astigmatism** 1. In a convex lens or concave mirror, a condition where the focal length is not the same in all planes. This is normally considered a defect. See *concave mirror, convex lens.* 2. A defect in the lens of the eye resulting in distorted and/or blurred vision. See *eye.*

**astro-** Pertaining to space science or outer space.

**astrobleme** A blemish on the earth that results from the impact of a large meteorite, asteroid, or comet. A good example is Meteor Crater in Arizona. Some geographic features, such as Hudson Bay in Canada, might be astroblemes. See *crater, meteorite.*

**astrolabe** An instrument used by ancient astronomers to determine latitude and time of day by sighting the positions of the stars. Some of these devices were embellished with fancy engravings by craftsmen who made them.

**astrology** A system of forecasting events, based on the belief that the movements of the stars and planets govern what happens to people. Most scientists do not believe in this, except for phenomena caused by the moon and by planetary alignment. The full moon can have some psychological effect, and tidal forces might affect human cells in ways we do not yet understand. See *syzygy, tide.*

**astrometry** The science involved with measuring distances in space and with determination of the positions of stars, planets, and other heavenly objects. See *celestial sphere.*

**astronaut** A person who travels and works in space.

**astronautics** 1. The science of navigation in space, and of dealing with problems that face astronauts. 2. The calculations, usually done by a computer, necessary for navigation in space. These calculations must be precise. An error of a few parts in a million could send a spacecraft off course by thousands of miles.

**astronomical unit** Abbreviated AU. The average distance of the earth from the sun, or about 93 million miles. Used as a yardstick for measuring distances in space. See *light year, parsec.*

**astronomy** The science involved with outer space. In ancient times this was simply a matter of observing heavenly objects. Nowadays astronomy has numerous subfields. See *astrophysics, cosmogony, cosmology, infrared astronomy, radio astronomy, ultraviolet astronomy, X-ray astronomy.*

**astrophysics** A branch of physics or astronomy concerned with the behavior of celestial bodies and the structure and evolution of stars, galaxies, and the universe in general. There are many specialties within this field. For examples see *black hole, blue shift, cosmogony, cosmology, gravitational collapse, gravitational constant, gravitational red shift, gravity waves, neutron star, radio galaxy, red shift, relativity, unified field theory.*

**asymptote** A straight line on the coordinate plane, towards which certain curves get closer and closer as they are extended. The drawing shows asymptotes for the tangent function. See *Cartesian plane, function, limit.*

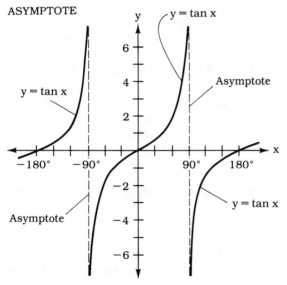

ASYMPTOTE

**At** 1. Abbreviation for ampere-turn. 2. Chemical symbol for astatine.

**atherosclerosis** A degenerative disease of the arteries. Deposits form on the inner walls of the vessels. This interferes with blood flow and increases the chances of clot formation. Complications include heart attacks, strokes, kidney problems, and various other illnesses. This disease is thought to be partly natural, partly inherited, and partly the result of bad living habits such as smoking, air pollution, heavy drinking, inactivity, stress, and overeating. There are probably other causes that we don't know about yet. See *heart attack, heart disease, hypertension, plaque, stroke.*

**Atlantis** 1. A legendary continent believed by some to exist thousands or millions of years ago in the North Atlantic Ocean. There was supposedly a catastrophe and the continent sank to the bottom of the sea. Proponents of this theory say that the effects of the Bermuda Triangle are caused by the remnants of the ancient society of Atlantis, whose powerful machines might still produce magnetic fields and other phenomena. Most scientists doubt this theory. See *Bermuda Triangle.* 2. The name of one of the space shuttles. See *Space Shuttle.*

**atm** Abbreviation for atmosphere. See *atmospheric pressure.*

**atmosphere** 1. The envelope of gases held close to the earth by gravitation (see drawing).

ATMOSPHERE

2. The envelope of gases held close to a planet by its gravitation. 3. The pressure at sea level on the earth. Abbreviation, atm. See *atmospheric pressure*.

**atmospheric modeling**  The use of a computer to simulate large-scale patterns in the atmosphere. See *meteorology*.

**atmospheric pressure**  1. The force per unit area of the atmosphere at sea level. This is about 14.7 pounds per square inch (PSI) or 760 millimeters of mercury (mmHg). Also called one atmosphere (atm). 2. The barometric pressure. See *barometric pressure*.

**atmospheric pump**  A device designed in the eighteenth century that was part of a steam engine used to drain water from mines. James Watt (see *Watt, James*) later refined the device and is well known for his work with steam engines. See *steam engine, steam pump*.

**atoll**  In the ocean, a lagoon surrounded by reefs, usually circular in shape. The drawing shows a cross section of an atoll at low tide. At high tide the whole reef may be submerged.

ATOLL

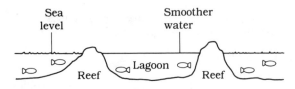

Reef encloses lagoon
at low tide

**atom**  At one time believed to be the smallest possible material particle. Composed of electrons, neutrons, and protons in various combinations. See *Bohr atom, electron, element, molecule, neutron, proton, quark, Rutherford atom*. See also the ATOMIC NUMBER AND ATOMIC WEIGHT, PERIODIC TABLE OF THE ELEMENTS appendices.

**atomic bomb**  A powerful weapon based on nuclear reactions to produce a burst of destructive energy. See *hydrogen bomb, neutron bomb*.

**atomic charge**  The electrical charge carried by an atom. This is usually neutral. The protons and electrons have equal and opposite charge. When there is a shortage of electrons, an atom is positively charged; when there is an excess of electrons, the atom is negatively charged. This charge is in multiples of the charge of one electron (or proton). See *ion*.

**atomic clock**  A device that keeps time by counting the oscillations of stable atoms. The cesium atom is commonly used as a standard for this purpose. Such clocks are accurate to within billionths of a second.

**atomic energy**  See *nuclear energy*.

**atomic fallout**  See *radioactive fallout*.

**atomic fission**  See *fission*.

**atomic fusion**  See *fusion, hydrogen fusion*.

**atomic mass**  See *atomic weight*.

**atomic mass unit**  Abbreviation, amu. Approximately, the mass of a proton. Specifically, $1/12$ of the mass of one atom of carbon-12, the most common isotope of carbon. See *isotope*. Atomic weights are given in amu. See also *atomic weight*.

**atomic nucleus**  The heavy central part of an atom. It carries a positive charge because it always contains protons (see *proton*). The hydrogen nucleus consists of just one proton and nothing else. The nuclei of all the other elements have neutrons as well as protons. See also *neutron*.

**atomic number**  The number of protons in the nucleus of an atom. In nature, there are elements for all the possible atomic numbers up through 92. Even larger atomic numbers are possible, but these are artificial atoms, made by humans. See the ATOMIC NUMBER AND ATOMIC WEIGHT, PERIODIC TABLE OF THE ELEMENTS appendices.

**atomic physics**  See *nuclear physics*.

**atomic reactor**  See *nuclear reactor*.

**atomic weight**  Also called atomic mass. The total mass of an atom including the protons, neutrons, and electrons. The electrons contribute only a tiny part of this mass. For most practical purposes, the atomic weight can be thought of as the number of protons plus the number of neutrons. Because the number of neutrons in an atom might vary (see *isotope*), there can be several different atomic weights for most elements. See the ATOMIC NUMBER AND ATOMIC WEIGHT, PERIODIC TABLE OF THE ELEMENTS appendices.

**atomizer**   A pump-spray device, such as that used in inhalers for asthma patients. It works without the propellants usually employed in aerosols. See *aerosol*.

**atom smasher**   See *cyclotron, particle accelerator*.

**atopic**   Pertaining to an allergy. See *allergen, allergy*.

**ATP**   Abbreviation for adenosine triphosphate.

**atrium**   See *auricle*.

**atrophy**   Wasting of tissue. For example, we might speak of muscle atrophy, perhaps caused by lack of exercise or by aging. It might be localized or generalized. It can take place in almost any part of the body.

**atropine**   A drug used to relieve muscle spasms. Also sometimes added to narcotics to prevent the patient from abusing the drug. Atropine in moderate doses has unpleasant side effects, such as dry mouth and sensitivity to light. In large doses it is a poison.

**attenuation**   A decrease in the strength of a signal, light beam, sound wave, or other effect, especially if it is introduced on purpose. See *absorption, gain*.

**attenuator**   A device that is designed to provide a certain amount of attenuation. Often used in electronics test labs. See *attenuation*.

**atto-**   See PREFIX MULTIPLIERS appendix.

**AU**   Abbreviation for astronomical unit.

**Au**   Chemical symbol for gold.

**audibility**   A measure of the loudness of sound. Usually given in decibels with respect to the threshold of hearing, or weakest detectable sound, which is zero decibels (0 dB). Sounds louder than about 120 dB are painful to the ears. See *decibel, sound*.

**audio**   1. Pertaining to sound. 2. A synonym for sound. 3. Pertaining to frequencies of about 20 to 20,000 hertz (cycles per second). 4. Pertaining to the ears and hearing. See *audio frequency, sound*.

**audio frequency**   An alternating current with a frequency between about 20 hertz (cycles per second) and 20,000 hertz. This is approximately the range of human hearing.

**audiology**   The science of human hearing. Especially, the development of techniques and devices to help people with hearing problems.

**audiometer**   A meter used for measuring sound intensity or audibility. Usually calibrated in decibels. Consists of a precision microphone, an amplifier, and a calibrated meter. See *audibility, decibel, sound*.

**audiovisual**   1. Consisting of pictures (video) and/or sound (audio) information, usually at the same time. 2. Pertaining to information that consists of video and/or audio, usually at the same time.

**auditory nerve**   The nerve leading from the inner ear to the brain's hearing center. There is one nerve for each ear. See *ear*.

**aura**   1. A peculiar feeling, sometimes taking the form of a mild hallucination, that occurs in an epileptic person before a seizure. See *epilepsy, seizure*. 2. A visible glow.

**aural**   Pertaining to the ears and hearing. See *ear*.

**auricle**   Also called atrium. Either of two upper chambers in the heart. The left auricle receives blood from the lungs, oxygen-enriched. The right auricle receives blood from the body, oxygen-poor because the body has used the oxygen. See *heart*.

**aurora**   The interaction of charged solar particles with the earth's upper atmosphere. The particles are emitted by solar flares. These particles take several hours to get to the earth where the geomagnetic field focuses and accelerates them; this causes ionization that can be seen on moonless nights. Aurora upsets the ionosphere and interferes with radio communications at some frequencies. Extreme auroras might cause disruption of power transmission. The ionization is concentrated above the North Magnetic Pole (Aurora Borealis) and the South Magnetic Pole (Aurora Australis). The drawing illustrates the phenomenon. See also *geomagnetic field, solar flare*.

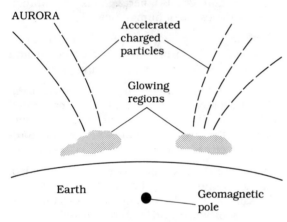

**Aurora Australis**   See *aurora*.

**Aurora Borealis**   See *aurora*.

**auroral propagation**   The reflection of radio waves at certain frequencies by the aurora (see *aurora*). Radio amateurs sometimes use this effect to communicate over long distances.

**Australopithecus**   A primate somewhat like modern humans, but with a smaller brain. Walked upright rather than on all-fours. Lived during the Pliocene epoch. See *Pliocene epoch, primate* and the PRIMATE FAMILY TREE appendix.

**auto-**   Synonym, self-. Pertaining to something that regulates itself, is self-contained, or is self-propelled.

**autoimmunity**   A condition where the body's immune system acts against the body itself. See *immune system*.

**automatic**   Not requiring a human operator; self-controlled. A human operator might be needed to set such a device up, but after that, it controls itself and completes a given task.

**automatic pilot**   A device in an airplane that allows the pilot to temporarily leave the controls. The plane continues to fly straight and level, unless there are atmospheric disturbances.

**automation**   The use of machines to replace people. This frees people to do more interesting things, leaving mundane chores to computers and robots. See *robot*.

**automaton**   See *robot*.

**automotive engineering**   A field of engineering, involved with the design of motor vehicles including cars, trucks, and motorcycles.

**autonomic**   Having to do with body processes that are controlled without conscious effort, such as heartbeat, respiration, and gland function.

**autonomic nervous system**   A part of the nervous system of an animal or human that operates without conscious effort. See *autonomic*.

**autonomy**   1. The ability to stand alone, without depending on something else. A home computer has autonomy, because it need not be tapped into a larger mainframe (although this can be done with some home computers to expand their memory or to gain access to data). 2. The property of being automatic. See *automatic*.

**autopilot**   See *automatic pilot*.

**autotrophism**   In nutrition, a process where an organism makes its own complex nutrients out of simple, inorganic compounds. Plants and some microorganisms have this ability.

**autumnal equinox**   See *equinox*.

**auxiliary**   1. Supplementary; not normally used, but available in case of special need. A spacecraft might have an auxiliary computer, for example. 2. Backup; used in case the main device fails.

**auxin** A substance essential to the growth of plants. They can occur naturally or be man-made. In a sense, auxins are the plant counterparts of amino acids in animals.

**avalanche** A snowslide. Might be triggered by an earth tremor, by an explosion, or even by a skier passing by. The drawing illustrates the start of the process. Once a slab of snow breaks loose, more snow is carried with it. Large avalanches produce violent wind gusts and can cause considerable destruction. People can be buried for hours or days before they are rescued; sometimes they are never found.

AVALANCHE

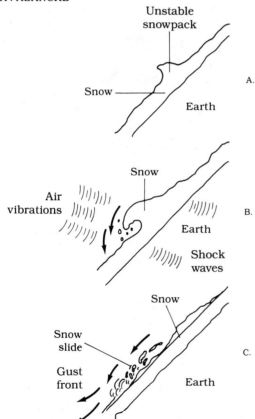

**avalanche breakdown** Also called avalanche or avalanche effect. In a semiconductor diode, a condition where large reverse bias causes conduction. See *bias, diode, P-N junction.*

**average** 1. The arithmetic mean of two or more numbers. 2. The integral of a function over some interval, divided by the width of the interval. See *integral*. 3. The geometric mean of two or more numbers. See also *mean*. 4. Typical.

**Aves** The class of birds. See *bird*.

**avian** Pertaining to birds. See *bird*.

**aviation** 1. The pursuit of flying in aircraft, including blimps, balloons, and gliders. 2. The science involved with the operation and maintenance of aircraft.

**Avogadro constant** See *mole*.

**Avogadro's Law** A rule concerning gases. All ideal gases have the same number of molecules at a given temperature, pressure, and volume. A liter of oxygen at 1 atm (atmospheric pressure) and 20 degrees Celsius (room temperature) would have the same number of molecules as a liter of hydrogen, helium, argon, or any other gas under these same conditions. See *ideal gas*.

**axial** Along an axis; lengthwise. We might speak of axial leads in an electronic component. This means that the leads lie along a common line. See *radial*.

**axil** In a plant, the measure of the angle between the stem and a branch or leaf. This angle tends to be fairly constant for a given species of plant.

**axillary** 1. Leading into or out of an arm. 2. Pertaining to the arm or the nerves or blood vessels of the arm.

**axillary artery** The large artery supplying blood to the arm. A human has one for either arm.

**axillary vein** The large vein that guides blood out of the arm. A human has one for each arm.

**axiom** 1. In mathematics, something that is assumed true without proof; a postulate. A mathematical system always has axioms. Mathematicians try to keep the number of axioms to a minimum. 2. A scientific principle.

**axiomatic** 1. Proceeding by formal logic from a set of assumed principles, called axioms (see *axiom*), and from precise, unambiguous definitions. 2. Always true.

**axis** A line for determining coordinates. Might be straight or curved, and the divisions might all be of equal size or they might be of different size. There might be just one axis, or two; or there might be hundreds, thousands, or millions of axes. See *abscissa, coordinate system, ordinate*.

**axon** The part of a nerve cell that carries impulses. It is like a thin fiber. The impulses are quite similar to electric currents in a wire, but they travel more slowly. See *nerve, nerve fiber, nerve impulse, synapse*.

**az-el** Abbreviation for azimuth-elevation.

**azimuth** One component for locating an object in the sky. Measured in degrees clockwise from true north. See *azimuth-elevation, azimuth-elevation mount, elevation*.

**azimuth-elevation** The location of an object in the sky according to its compass bearing (azimuth) and the angle relative to the horizon (elevation). See *azimuth, elevation*.

**azimuth-elevation mount** A pair of bearings for use with a telescope, photographic camera, or video camera. One bearing allows rotation around the horizon, and the other allows up-and-down adjustment from the horizon to the zenith. This is shown in the drawing. In this way the whole sky can be covered. Most sophisticated telescopes use the equatorial mount, because this makes it possible to follow the movement of stars resulting from the earth's rotation, by moving only one of the bearings. See *equatorial mount*.

AZIMUTH-ELEVATION MOUNT

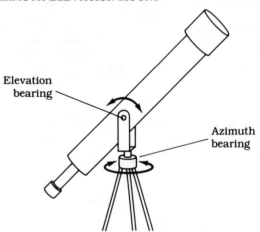

**B** 1. Chemical symbol for boron. 2. Abbreviation for Bel.

**Ba** Chemical symbol for barium.

**babbit metal** A durable alloy made from tin, antimony, and lead. Used in a variety of industrial applications, such as bearings and printing.

**Babo's Law** When a substance is dissolved in a liquid, the vapor pressure of the liquid goes down. The more of the substance is dissolved, the more the vapor pressure is reduced. The reduction occurs in direct proportion to the amount of substance dissolved. See *vapor pressure*.

**bacillus** Rod-shaped bacteria. See *bacteria*.

**background noise** 1. In radio communications, the level of noise that would be heard if there were no manmade sources of noise. 2. In radio astronomy, the noise that is heard coming from all directions at once, as a result of the Big Bang. See *Big Bang*.

**background radiation** Radioactivity in the environment coming from rocks in the earth and from outer space. There has always been some of this. See also *radioactivity*.

**backscatter** In radio communication, the waves that are bounced back from the ionosphere towards the transmitting station. These might fall within the skip zone, making communications possible that otherwise wouldn't be possible. See *skip, skip zone*.

**Bacon, Francis** The seventeenth century originator of the work concept for scientists: "Observe, measure, explain, verify."

**bacteria** A microorganism consisting of one cell. Bacteria vary in shape, as shown in the

BACTERIA

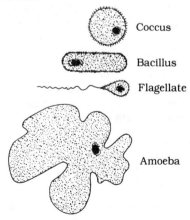

drawing. They reproduce by dividing (see *cell division*). Some bacteria are essential for health. Others cause infections.

**bacteriocide** A chemical used to kill bacteria on surfaces. In the body, an antibacterial agent is called an antibiotic (see *antibiotic*).

**bacteriology** The science involved with the study of bacteria, their behavior and control.

**bacteriophage** A virus that infects and kills bacteria, just as some bacteria infect animals or people. Used in the laboratory to identify bacteria because these viruses attack only certain bacteria.

**badlands** Heavily eroded regions with short, steep slopes and little or no plant growth.

**Baekeland, Leo** The inventor of the plastic we now call Bakelite. He first made it in 1907 by heating formaldehyde and phenol. See *Bakelite*.

**Bakelite** A synthetic plastic known for its durability and hardness. Used in a great variety of industrial and consumer products and applications.

**baking soda** See *sodium bicarbonate*.

**balance** 1. A condition in which different elements of a system are in ideal proportion. 2. A precision scale. 3. A sense governed by the middle ear that keeps a person upright and aware of where he or she is. 4. Exact canceling out of two currents or signals having the same frequency.

**balanced diet** An eating pattern that includes adequate amounts of all essential nutrients, without excesses in any. See the RECOMMENDED DAILY ALLOWANCES appendix.

**ballast** 1. Water pumped into and out of special tanks, to control the buoyancy of a submarine. 2. Weight in the bottom of a ship that helps keep the ship upright in heavy seas.

**ballistic missile** A rocket with an explosive warhead, often nuclear, capable of hitting a target from a long distance. See *intercontinental ballistic missile*.

**ballistic pendulum** A device that allows measurement of the speeds of projectiles, such as bullets. When the projectile hits the pendulum, the pendulum weight is displaced (see drawing). The speed can be calculated knowing the mass of the projectile.

BALLISTIC PENDULUM

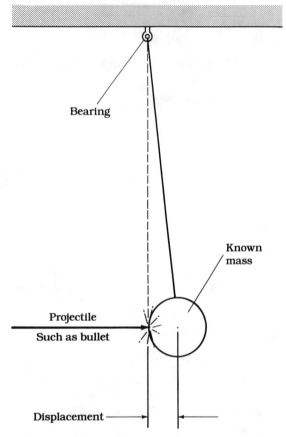

**ballistics** The physics of objects that fly from one place to another, usually without propulsion except the boost they get at the start of their flight. An example is the flight of a bullet.

**band** 1. A range of frequencies or wavelengths in radio communications. See *electromagnetic spectrum, radio spectrum*. 2. A range of energy levels, for example, in the orbits of electrons in an atom. See *band gap, electron*.

**band gap** In an atom, a range of energy levels that an electron cannot have. When the electron changes levels, it "skips" over these gaps. See *atom, electron.*

**bandwidth** The amount of spectrum space taken up by a signal in radio communications. Measured from the highest to the lowest frequency, or from the shortest to the longest wavelength. Generally the faster the rate of information sent the greater the bandwidth needed.

**bar** A unit of atmospheric pressure equal to 29.53 inches or 750.062 millimeters of mercury. This is a little less than normal sea-level pressure, which is about 30 inches of mercury. See *atmospheric pressure, barometric pressure.*

**barbiturate** A class of drugs, prescribed sometimes as sleeping aids. These are addictive substances and should be used only with a doctor's recommendation. See *addictive substance.*

**barium** Chemical symbol, Ba. An element with atomic number 56. The most common isotope has atomic weight 138. In medicine, a radioactive isotope of this element is used for diagnosis of certain gastrointestinal problems. In electronics, barium is used to remove the last traces of gas from vacuum tubes and other devices requiring a vacuum.

**bark** 1. The protective outer layer on the trunk and branches of a tree or shrub. 2. A sailing ship with three or four masts, common before gas and steam engines became widely used.

**barn** A tiny unit of area used in atomic physics. Equal to the area of a square just one one-hundred-billionth of a millimeter on a side. The term is derived from the saying, "You couldn't hit the side of a barn."

**Barnard's Star** A nearby star, about 1/6 as big as the sun. Astronomers noticed that the path of this star wobbles, indicating that there is at least one planet orbiting it. This has been taken to mean that planets are common in the universe.

**Barnett Effect** When an iron, nickel, or steel rod is rotated around its long axis, it becomes permanently magnetized. The effect is weak. See also *magnetic polarization, permanent magnet.*

**barograph** An aneroid barometer, attached to a pen recorder (see drawing). This allows a record to be kept of barometric pressure over a period of hours, days, or even weeks. See *barometer.*

BAROGRAPH

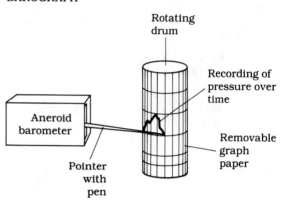

**barometer** A device used for measuring air pressure, usually in inches or millimeters of mercury. The drawing shows a mercury type barometer. The pressure of the atmosphere supports the column of mercury up to about 30 inches, sometimes more and sometimes less depending on weather conditions. Another type of instrument, called an aneroid barometer,

BAROMETER

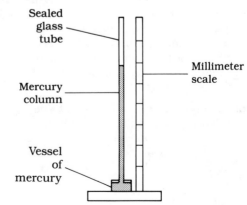

uses a hollow cavity on which air exerts pressure. Variations in this pressure cause a needle to move along a calibrated scale. A pen recorder might be used to make a graph of barometric pressure versus time. See also *barograph*.

**barometric pressure** The air pressure as adjusted to sea level, given in inches of mercury or in millibars. See *atmospheric pressure, millibar*. Measured with a barometer (see *barometer*). Usually between 29 and 31 inches of mercury. In the eyes of hurricanes it might drop to as low as about 26.3 inches. Under dry conditions it might rise to a little over 31 inches.

**barrel cactus** A desert plant that can retain large amounts of water for a long time. Known to desert travelers as a lifesaver.

**barrier reef** A massive coral growth that occurs in the tropics off some coasts. The largest such reef is probably the Great Barrier Reef off of northeastern Australia. It is about 1200 miles long, 90 miles wide, and 400 feet high. Many types of marine life dwell near barrier reefs.

**Barrow, Isaac** Newton's teacher at Cambridge University, England. Newton took his place as Lucasian Professor of Mathematics. See also *Newton, Isaac*.

**barycenter** See *center of mass*.

**basal cell** A cell in the lower epidermis of the skin.

**basal-cell carcinoma** A form of skin cancer that appears as a lesion something like a sore that will not heal, or like a pimple that won't go away. More common in sunny regions. It is feared that this type of cancer will become more common as the ozone layer is depleted. See *malignant melanoma, ozone depletion, squamous-cell carcinoma*.

**basal metabolism** The rate of energy burning that occurs with minimal physical activity, just sleeping and lying around. In men it is 1500-2000 kilocalories a day; in women about 20 to 30 percent less. See *kilocalorie, metabolism*.

**basalt** Dark volcanic rock that is fluid when it is molten. Found on the ocean floor and in extensive deposits in the earth's crust. See *igneous rock*.

**base** 1. A substance with a pH greater than 7 that reacts with acids to form salts, various gases, and sometimes heat. See *acid, pH*. 2. The middle controlling part of a bipolar transistor. See *transistor*. 3. In a number system, the number of different possible digit symbols. We commonly use base-10. Computers use base-2 and base-16. See *modulo*. 4. The supporting understructure of a device or geological formation.

**base metal** 1. Any common metal such as copper, lead, iron, or zinc. 2. The principal metal in an alloy; for example, the copper in bronze.

**base unit** A standard unit or a unit that is used as a reference for other units for the same effect. See *Standard International System of Units*, and the STANDARD INTERNATIONAL SYSTEM OF UNITS appendix.

**BASIC** A computer language useful for elementary calculations in algebra, trigonometry, and other fields. Usually it is the first computer language learned by students. See *COBOL, FORTRAN*.

**basilar membrane** A membrane in the inner ear, forming part of the cochlea and responsible for transmission of sound vibrations to the auditory nerve. See *cochlea*.

**bass** Low-frequency sound. This term is used by musicians more than by scientists. See *sound, treble*.

**batholith** A mass of rock beneath the surface on which mountains rest. Usually massive, having a surface area of 40 square miles or more.

**bathysphere** A vessel for deep-sea diving. Spherical in shape because this shape can withstand the greatest amount of pressure. Lowered from a ship on a cable, through which air is supplied.

**battery** 1. A group of two or more cells connected in series to provide greater voltage. 2. A single cell. This is technically not a correct use of the term, but it has become common. Zinc-carbon and other cells are sold often as "batteries." See *cell*.

**BAUDOT code** A binary code used by teleprinters. Nowadays it has been largely replaced by ASCII. See *binary code, ASCII*.

**bauxite** The raw ore from which aluminum is refined. It is a light, crumbly rock in its natural state. See *aluminum*.

**Bayeaux Tapestry** A woven-cloth document and artistic work showing the Norman French invasion of England in 1066 A.D. Halley's Comet made an appearance at this time, and the comet is shown on the tapestry. The Anglo-Saxons and the invading French both believed that the comet foretold of great events to come. They were right in a sense, the invasion had a permanent effect on the English language and customs. See *Halley's Comet*.

**Be** Chemical symbol for beryllium.

**beam** A narrow shaft of energy, sometimes called a ray. 2. A radio antenna with directional properties. See *Yagi antenna*. 3. The width of a ship at its widest point.

**bearing** 1. Azimuth, especially when referring to an object on the horizon as seen from a ship or an aircraft. See *azimuth*. 2. A device that allows rotation in one or two planes.

**beat** 1. The effect of combining two musical notes or radio signals. New notes or signals are produced at the sum and difference frequencies. 2. Tempo, rhythm, or frequency.

**beat frequency** The sum and difference frequencies produced when two notes or signals beat (see *beat*). In music, these are an important component in the sound we hear. In radio, beat frequencies are used in superheterodyne receivers. They also can cause problems, called intermodulation. See *intermodulation, superheterodyne radio receiver*.

**Beaufort Scale** A scheme for estimating wind speed from calm to hurricane force, based on visible effects. See the BEAUFORT SCALE appendix.

**Beckmann thermometer** A type of mercury thermometer with two bulbs designed so that it is sensitive to tiny changes in the temperature.

**becquerel** Abbreviation, Bq. A unit of radiation. More often in practice, the rad, rem, or roentgen are used. See *rad, rem, roentgen*.

**bedrock** A solid layer of rock, generally a few feet to about 50 feet below the surface. Bedrock is limestone, granite, or basalt, depending on the location.

**behavior** The response of an organism to various stimuli.

**behavioral disorder** A problem or illness in which the responses are inappropriate to the stimuli. In humans, usually thought of as mental or emotional illness. See *depression, mania, manic-depressive psychosis, neurosis, schizophrenia*.

**behavioral science** See *psychology*.

**bel** A unit of sound or signal intensity, equivalent to 10 decibels. See *decibel*.

**Bell, Alexander Graham** Usually credited with the invention of the telephone. He realized that voice impulses and simple on/off electric currents could be carried via wire, and he conducted experiments and devised the equipment needed to prove it. The Bell System gets its name from him.

**Bell, Jocelyn** A graduate student who discovered the first pulsar in 1967 with her professor, Anthony Hewish. See *pulsar*.

**Bell Laboratories** The research facilities of the Bell System, which we commonly associate with the telephone network. One of these labs, in Holmdel, New Jersey, was where two scientists, Arno Penzias and Robert Wilson, first observed the background radiation from what they believed to be the Big Bang. See *Big Bang*. The Bell Labs have been involved with many other important scientific discoveries.

**bends** A serious problem that can occur when a SCUBA diver has been down deep and comes up too fast. Nitrogen bubbles out of the blood and interferes with blood flow. Slow decompression is needed to prevent this. See *SCUBA*.

**Benedict's Test**  A method of determining whether or not there are sugars in a solution. It is a sensitive test and can be used to test the urine to see if someone is diabetic.

**beneficiation**  A technique for refining ore, first perfected by Reserve Mining in Minnesota. Taconite, or crude iron ore, is pulverized, and the iron bits removed using magnets. This makes it possible to use taconite to obtain iron, whereas before, only higher grades of ore could be used. Variations of this process can also be used with other ores besides iron ore. See *taconite*.

**benzene**  A compound consisting of six carbon atoms and six hydrogen atoms. It is hexagonal in shape. See *benzene ring*.

**benzene ring**  The hexagonal pattern formed by the carbon atoms in a benzene molecule. See the drawing. Benzene rings have actually been photographed.

BENZENE RING

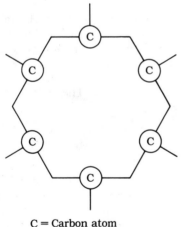

C = Carbon atom
| = Bond for attachment of hydrogen atom

**benzoate**  A form of food preservative produced from naturally occurring acids in certain plants. Found in beer, some cheeses, and some soft drinks. It is thought that some people have allergic reactions to these compounds. But they might also be allergic to the plants from which the compounds are made.

**benzyne**  A compound consisting of six carbon atoms and four hydrogen atoms. Found only in certain organic processes, it does not last long because it is extremely reactive. Similar in shape to benzene. See *benzene ring*.

**beriberi**  A deficiency disease that occurs when there is not enough thiamine (vitamin B-1) in the diet. Rare in civilized countries. Symptoms include irritability, constipation, and heart trouble. Usually occurs along with other deficiency problems. See the RECOMMENDED DAILY ALLOWANCES and VITAMINS AND MINERALS appendices.

**berkelium**  Chemical symbol, Bk. An element, made in the laboratory, with atomic weight 97. Gets its name from where it was first made, the University of California at Berkeley.

**Bermuda high**  A large weather system that develops and persists over the North Atlantic ocean during the summer and fall. Affects weather all along the east coast of the United States (see drawing).

BERMUDA HIGH

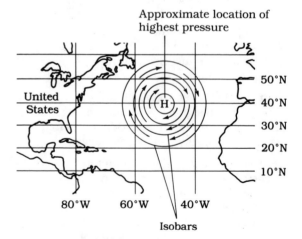

Isobars

Arrows show direction of prevailing wind

**Bermuda Triangle**  A region in the North Atlantic Ocean and the Caribbean Sea where strange events have happened. It is thought that maybe a meteorite landed in the area thousands or millions of years ago, and its residual magnetism still exists there, upsetting the geomagnetic field (see *geomagnetic field*). There are also other, much stranger theories about the nature of the area.

**berry** A fruit with many seeds, and usually having a soft outer covering. See *fruit*.

**beryl** Beryllium in its crude natural form. Sometimes used in jewelry, especially the green type, known commercially as emerald. See *beryllium*.

**beryllium** An element with atomic number 4. The most common isotope has atomic weight of 9. Beryllium is a light metal that has a variety of commercial uses. In its raw form it is sometimes gem-like (see *beryl*). Oxides of beryllium are used in electronics because they insulate against electricity but conduct heat very well. Beryllium and all its compounds are toxic, especially when inhaled or when they get on the skin.

**Bessemer Process** A way of getting steel from iron and carbon. The iron is heated and oxygen is blown into the furnace to maximize the heat. Steel gets its strength from the added carbon, but it must be put in without any other impurities that would weaken the metal.

**beta decay** A form of atomic decay in which the atomic number changes but the atomic mass remains almost exactly the same. A high-speed electron might be emitted, and a neutron changes into a proton. Or a high-speed positron might be ejected, and a proton changes into a neutron. The high-speed electron or positron is a beta particle; hence the name of this reaction. See also *electron, neutron, positron, proton*.

**beta particle** A high-speed electron or positron. See *beta rays, electron, positron*.

**beta rays** A type of radioactivity caused by a stream of high-speed electrons. Less frequently, it consists of high-speed positrons (see *positron*). This type of radiation is not very penetrating but it can still be dangerous. See also *alpha rays, electron, gamma rays, positron*.

**betatron** A device for accelerating electrons to high speed so that they become beta particles. Used in various scientific experiments. A functional diagram is shown. See also *beta particle, beta rays, electron, particle accelerator*.

BETATRON

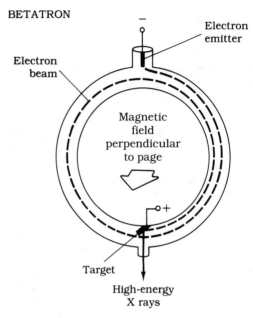

**BHT** Abbreviation for butylated hydroxytoluene.

**Bi** Chemical symbol for bismuth.

**bi-** Denoting pairs, doubles, or multiples of two.

**bias** 1. Prejudice in interpreting or formulating data so that the results come out a certain preplanned way, even though this might not be true to life. This happens in science more often than we would like to think. 2. A voltage or current in a semiconductor device applied for a certain purpose. Usually applied to the emitter-base junction of a transistor. See *transistor*.

**bicarbonate of soda** See *sodium bicarbonate*.

**biceps** The muscles that assist in the bending of the arm (biceps brachii) or the bending of the leg (biceps femoris).

**biennial** 1. Occurring every two years. 2. A plant that goes through its growing cycle in two years.

**Big Bang** 1. The theory that the universe originated in a dense, incredibly hot fireball about 10 billion to 20 billion years ago. 2. The original explosion from which the universe was made, according to the Big Bang hypothesis. See *Big Bounce, Big Crunch.*

**Big Bounce** A rebounding of a collapsed universe causing another Big Bang and the evolution of another universe. See *Big Bang, Big Crunch, Oscillating Universe Theory.*

**Big Crunch** The collapse of the universe sometime in the future. This will happen only if there is enough matter to create the needed gravitational pull. Scientists are not sure yet whether or not there is enough matter in the universe. See *Big Bang, Big Bounce.*

**bilateral symmetry** Mirror-image symmetry. Seen from the outside, a human being is almost perfectly bilaterally symmetric (see drawing). So are most insects and many other creatures.

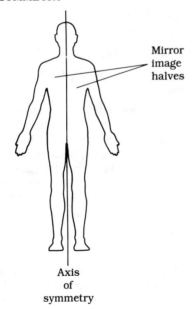

BILATERAL SYMMETRY

Mirror image halves

Axis of symmetry

**bile** A substance produced by the liver, and that aids in digestion, especially the assimilation of fats. Also helps the body to get rid of excess cholesterol. Salts of this substance (bile salts) are also important in digestion.

**bilge pump** A device that gets rid of excess water accumulating in the bottom of a boat or ship. Some water comes into a boat from waves spilling over the gunwales; some also comes in when it rains.

**billion** 1. In America, the number 1000 million or 1,000,000,000. 2. In England, the number one million million, or 1,000,000,000,000.

**bimetal strip** Two pieces of different metals glued or pressed together. They have different rates of expansion when they are heated. So when the strip gets hot, it bends. Used in thermometers and in the switching mechanisms of thermostats. See *thermostat.*

**binary** 1. Having two states, such as on and off, high and low, or 1 and 0. 2. A number in base-2, with only digits 0 and 1 (see *modulo*). 3. Branching off by splitting in two at each branch point. 4. Occurring in pairs (see also *binary star*). 5. Pertaining to certain acids (see also *binary acid*).

**binary acid** An acid in which the hydrogen ions are attached to some element(s) besides oxygen. Ordinary stomach acid (hydrochloric) is one such acid. See *acid.*

**binary code** A code in which there are just two states: on and off. Examples include Morse code, BAUDOT, and ASCII. See *ASCII, BAUDOT code, Morse code.*

**binary number** A number in base-2. Such a number has only the digits 0 and 1 in it. An example is 1001001. Used by computers because they can always be represented by sets of switches that are either on or off. See *modulo.*

**binary star** A double star. The stars might or might not be similar to each other, but they always orbit around each other held close by gravity.

**binoculars** A pair of low-power telescopes, designed to provide a wide angle of view (usually) and binocular vision (always). The illustration is a cut-away view of a typical pair of binoculars. See *binocular vision.*

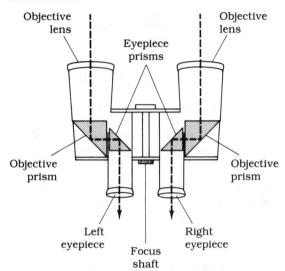

BINOCULARS

**binocular vision** Literally, seeing with two eyes. One eye sees a view slightly different from the other eye, except for extremely distant objects. The parallax effect makes it possible to gauge depth. See *parallax*.

**binomial** A mathematical expression having two elements, or terms, that are either added to each other or subtracted one from the other.

**Binomial Theorem** A mathematical formula for determining the expansion of a binomial raised to an integer power. The expression is fairly complicated, and a mathematics text should be consulted for a full explanation of it. See *binomial*.

**bioastronautics** The science concerned with how space travel affects the human body.

**biochemistry** The science that studies the chemistry of living cells, plants, and animals. This can become extremely complicated and is an important science involved with the control of diseases.

**bioenergetics** The science concerned with metabolism and other energy pathways in cells, plants, and animals. Might be considered a subspecialty within biochemistry. See *biochemistry*.

**bioengineering** A field of engineering concerned with technology and its effects on, and uses in, the human body. Bioengineers might design artificial organs and artificial limbs. See *bionics, human engineering, medical engineering*.

**biogenesis** An axiom in biology stating that a living thing can only be produced from other living things resembling it. Thus we cannot, for example, have spontaneous generation as was once thought (see *spontaneous generation*). Nor can a human have a horse for a mother.

**biological assay** The use of a living plant or animal to find out how much of a certain drug or hormone is present in a substance. The substance is given (by injection, for example) and the effects are observed. This has created controversy when done on test animals, and especially when done on human beings (such as in prisons).

**biological clock** A timekeeping mechanism within animals often geared to the tides because of the changes in gravitation that are caused by the sun and moon. Governs such activities as feeding and sleeping.

**biological pest control** A method of increasing crop production by getting rid of pests using nonharmful, nontoxic devices. A good example is the use of one type of bug that eats harmful bugs in stores of grain. The "good bugs" don't eat the grain, but only eat the "bad bugs." So no chemicals are needed.

**biology** The study of living things, from cells to human beings, and the processes that occur in living things. This is a major science with many subspecialties. Sometimes also used interchangeably with the term "life science."

**bioluminescence** A biological process that generates visible light. The glowing of a firefly is an example.

**biomass** 1. The total mass of all the organisms in a certain place, such as in a lake. This might range up to many thousands of kilograms. 2. The total mass of all the organisms of a certain kind, in the entire world. We might say that the biomass of tropical trees is decreasing because of deforestation.

**biome** A group of species unique to a certain part of the world. We might speak of the temperate forests in the Southeast of the United States, for example. Or we might refer to the various life forms native to Antarctica.

**bionics** The science of developing artificial organs and limbs. Although we might make an arm, for example, that is stronger than a real human arm, it is doubtful that artificial body parts will ever be as good as the real thing. The sense of touch is always compromised in artificial limbs; the body tends to reject some artificial organs. This is an expanding field of science.

**biophysics** The science concerned with the physical nature of living things, especially at the atomic level.

**biopsy** A sample of tissue taken from a living organism, especially a human being, for analysis. This procedure is especially valuable in the early detection of cancers inside the human body.

**bioregion** A certain part of the earth set apart by the types of life that are naturally found there. We might speak of a mountain lake as a small bioregion and the Pacific Ocean as a large bioregion.

**biorhythm** 1. See *biological clock*. 2. Cycles of activity in living things, such as the menstrual cycle in women. 3. A science, not widely accepted, based on regular cycles of intellect, emotion, and physical function that begin at birth. Some scientists believe that such cycles exist, but they doubt that they are so regular or uniform that they can be used to predict events in a person's life years ahead of time.

**biosphere** 1. The life on a planet, such as the earth, integrated with the land, air and water. 2. The whole earth as an organism. See *Gaia Hypothesis*.

**biota** All living things on the surface of the earth, or of a planet where there is life.

**biotechnology** The use of biological processes to make industrial goods. A good example is penicillin, which in its natural form is a mold. Another example is yogurt, made by adding certain bacteria to milk and then letting the bacteria multiply at the right temperature.

**biotin** A substance manufactured by intestinal bacteria, and also available from certain foods. Considered essential, and available in many vitamin preparations. Important in the synthesis of certain enzymes. See RECOMMENDED DAILY ALLOWANCES and VITAMINS AND MINERALS appendices.

**biotite** A black mineral similar to mica (see *mica*), occurring in crystals that form thin, flat plates.

**biotron** A gigantic greenhouse with various rooms having different earthlike environments. One room might simulate the desert; another, Antarctica; another, a jungle. All of the plant and animal life is added, so that it is as realistic as possible. See also *greenhouse*.

**biped** An animal that propels itself on two legs, such as a gorilla or a human.

**biplane** An airplane of early design with two front wings. The two wings were thought to provide more lift than would be possible with one wing; however, this also slowed the aircraft down. After World War I, monoplanes (with just one front wing) became more common until, by the time of World War II in the 1940s, biplanes were obsolete. See the drawing.

BIPLANE

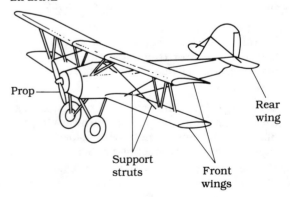

**biprism** A special prism that creates a double image. Used as a beam splitter in optical experiments.

**bird** A class of animals characterized by their beak, wings, and feathers. Not all birds can fly: the ostrich and the penguin are examples. Birds are warm-blooded. Small birds must eat large amounts of food in relation to their weight. Known for their collective instincts, such as migration. See *migration.*

**birth control** See *contraception.*

**bisection** 1. The act of cutting into two equal pieces. 2. A geometrical construction where a line segment or arc is divided into two equal parts.

**bismuth** Chemical symbol, Bi. An element with atomic number 83. The most common isotope has atomic weight 209. Used in making certain alloys. Bismuth subsalicylate is used to relieve stomach upset. Other compounds of this element have a great variety of industrial and scientific applications.

**bistable circuit** See *flip-flop.*

**bit** A single element in a binary code. Can be either a 0 (low, or off) or 1 (high, or on). See *binary code.*

**bituminous** 1. A form of coal, of lower grade than anthracite but better than lignite. See *anthracite, lignite.* 2. A form of asphalt made using bituminous coal. Extensively used for repaving roadways.

**biuret test** A chemical test used to determine whether or not a solution contains protein. Useful in medical labs.

**Bk** Chemical symbol for berkelium.

**blackbody** A theoretically ideal radiator and absorber of energy at all wavelengths. So called because it appears visually black in most circumstances.

**blackbody radiation** Energy emitted from a blackbody. It displays a curve of intensity versus frequency (see diagram) that is dependent on the temperature of the object. The higher the temperature, the higher the frequency at which intensity is greatest. Scientists can determine the temperature of a distant object in space on the basis of this curve, if the object behaves as a blackbody. See *blackbody.*

BLACKBODY RADIATION

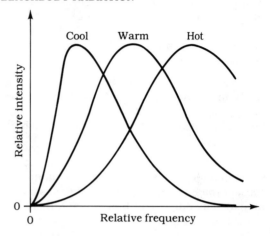

**black box** In electronics, a circuit or device that has unknown contents, but specific and predictable characteristics.

**black dwarf** A dead star that has exhausted its fuel, cooled off, and contracted. It is incredibly dense and is about the size of a planet, but with the mass of a star. Larger stars contract to form neutron stars, pulsars, and perhaps even black holes. See *black hole, collapsar, neutron star, pulsar.*

**black hole** The final product of gravitational collapse (see *gravitational collapse*). The gravity is so strong that even light can't escape. Time comes to a stop, and whatever goes in never comes out again, at least not for trillions of years. These objects have been predicted to exist by astrophysicists, and it is thought that X-ray stars might contain them. One theory says our universe is a huge black hole. See also *black dwarf, collapsar, neutron star, pulsar.*

**Black, Joseph** A professor that taught James Watt at the University of Glasgow. Watt later developed the steam engine. See *steam engine* and *Watt, James.*

**black light** See *ultraviolet.*

**blackout** 1. A major power failure covering a large population or geographic area. 2. A loss of memory covering a certain period of time (see *amnesia*). 3. In flight, a momentary loss of sight caused by high g force (see *g force*).

**black smoker** An undersea geyser or hydrothermal vent from which water is ejected. The water is rich with sulfide minerals that make it black, so it looks like black smoke. See *white smoker*.

**bladder** 1. The part of the urinary tract that holds urine until it is discharged. 2. In some fish, a sac or set of sacs that can be filled with air to allow floatability. In this way the fish can control its buoyancy.

**blast furnace** An oven in which iron ore is smelted and steel is made. Gets its name from the fact that pressurized air or oxygen is used to increase the rate of combustion, and thereby to make the oven hotter.

**blastula** In an animal, the earliest embryonic stage, just after the fertilized egg has first divided.

**blimp** See *dirigible*.

**blind spot** 1. A region on the retina of the eye, about 1/16 inch across, where the optic nerve is attached. This region is blind. 2. A sightless portion of the visual field about 5 degrees wide. This results from insensitivity in the retina where the optic nerve joins it. The illustration lets you see this effect.

BLIND SPOT

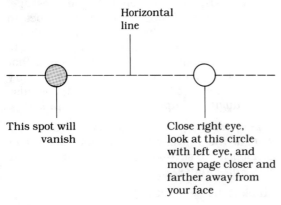

**block diagram** An illustration of a device or process in simplified form, with each major unit in a box, triangle, or circle. The drawing shows an amplitude-modulation (AM) radio receiver. See also *flowchart*.

BLOCK DIAGRAM

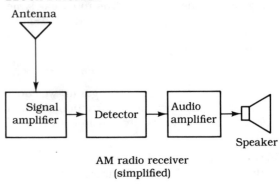

AM radio receiver (simplified)

**block glide** A type of landslide in which a section of earth falls because it is soaked with water. It is similar to a mudslide. See *landslide*.

**blocking** 1. A weather phenomenon in which a pattern persists because one system keeps others from moving along (see *blocking high*). 2. In electronics, a method of preventing one type of signal or current from flowing, while allowing others to pass. A capacitor might be used to block direct current (dc) but pass alternating current (ac).

**blocking high** A high-pressure system that enlarges, becomes stationary, and causes storms to be diverted around it. The Bermuda high is an example (see *Bermuda high*).

**blood** The substance that transports oxygen and nutrients to the cells of the body, and also that removes wastes and carbon dioxide from the cells. A full-grown man has about 6 quarts of blood; a woman has about 3 1/2 quarts.

**blood plasma** The liquid part of the blood. It carries nutrients to the cells. See also *platelet, red blood cell, white blood cell*.

**blood pressure** The force with which the blood presses against the walls of the arteries. Systolic pressure is the highest pressure when the heart muscle contracts. Diastolic pressure is the lowest pressure in between heartbeats. Given as a fraction with systolic pressure on top, such as 110/70 (stated as "110 over 70"), an example of ideal blood pressure.

**blood type** A means of determining whether or not two people's blood are compatible. The simplest method is by protein types. Type O contains no clumping proteins; type A one kind; type B another kind; type AB both kinds. There are more complex factors in blood also. See *Rh factor*.

**blood urea nitrogen** Abbreviation, BUN. A measure of the amount of urea in the blood (see *urea, uremia*). This level is high in certain diseases and abnormal metabolic states.

**blue shift** A change in the wavelength of electromagnetic radiation, especially visible light, coming from distant celestial objects. The spectral absorption lines are shifted towards the shorter wavelengths, or blue end (see drawing). This happens when the emitting object is moving towards us. See also *absorption line, red shift*.

BLUE SHIFT

Reference spectrum

Red  Orange  Yellow  Green  Blue  Indigo  Violet

Observed spectrum

Absorption lines shifted toward blue end (shorter wavelength)

**BMEWS** Abbreviation for Ballistic Missile Early Warning System.

**Bode Law** See Titius-Bode Law.

**body** 1. The physical organism of an animal or human. 2. A large set. We might speak of a body of data, for example. 3. The main part of an object.

**body cavity** The inside of a human body or an animal's body, where the organs are located.

**body temperature** The temperature, usually given in degrees Fahrenheit or centigrade, of the core of a human or animal body, usually measured at the rectum. In humans it is about 99 degrees Fahrenheit or 37 degrees centigrade, but it might vary depending on factors such as illness, amount of exercise, exposure to the weather, amount of food eaten, and other things.

**Bohr atom** A model of the atom in which negatively charged electrons orbit in defined levels or shells around the positively charged nucleus (see drawing). See *Rutherford atom*.

**Bohr magneton** A unit of magnetic moment for particles. See *magnetic moment*.

**Bohr, Niels** The physicist who developed a model of the atom in 1913 that we still find useful today. See *Bohr atom*.

**bogie** 1. A false echo on radar or sonar. 2. An unidentified flying object on radar. See *radar, sonar*.

**boiling point** The temperature at which a given substance changes from liquid to gaseous state. For water it is 212 degrees Fahrenheit or 100 degrees centigrade at sea level. As the air

BOHR ATOM

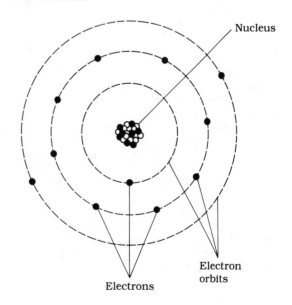

Nucleus

Electrons

Electron orbits

pressure goes down, so does the boiling point for many substances.

**bolide** A meteorite or comet that strikes the earth. See *comet, meteorite*.

**bolometer** A heat-sensitive resistor that can be used for any of various purposes in electronics, such as temperature measurement, circuit protection, and power measurement.

**Boltzmann constant** A constant in physics having to do with thermionic emission and thermal noise. Roughly equal to $k = 0.0000863$ volts per degree Kelvin.

**bond** 1. A place where two elements are securely joined. 2. An attractive force that holds atomic particles or atoms together. All compounds have bonds that keep the different atoms from separating under normal conditions. See *compound*.

**bond energy** The energy needed to break the bond that holds atoms together in a compound.

**Bondi, Hermann** One of the astronomers who first proposed the Steady State theory of the universe. See *Steady State theory*.

**bone** A rigid material that has many functions in a vertebrate animal, the most obvious being physical support. Bone is made mostly of calcium phosphate. In the center of most bones is the marrow (see *bone marrow*). An adult human has 206 bones. Bone makes up about 18 percent of the body weight.

**bone marrow** The soft interior portion of some bones, especially the long bones of the arms and legs. Responsible for manufacturing blood cells and antibodies. See *antibody, bone, red blood cell*.

**Boolean algebra** A system of mathematical logic in which statements are treated as variables. Certain rules are used, but they are different than the rules of ordinary algebra. Statements are proven in a manner similar to the solving of equations in algebra. Invented by George Boole, a self-taught British mathematician, around 1847. See *logic, mathematical logic* and the BOOLEAN ALGEBRA appendix.

**borax** A complex chemical containing boron, oxygen, hydrogen, and sodium. It has many uses in industry and medicine. It is an antiseptic and astringent, and when combined with water, boric acid forms (see *boric acid*). Perhaps the best known of its uses is in doing the laundry, because it aids the effectiveness of detergents.

**bore hole** A hole drilled into the earth's crust for any purpose, but especially for scientific research. Samples from various depths are brought to the surface and analyzed.

**boric acid** An acid containing oxygen and boron used in medicine as an antiseptic and astringent for treating skin irritation. It also has many industrial uses, including the manufacture of detergents. It might also be added to food to increase the shelf life.

**boride** Any compound in which boron is combined with a metal. See *boron*. These compounds form hard metals useful in industry.

**bornite** A raw ore of copper. In natural form it is mixed with iron sulfide.

**boron** Chemical symbol, B. A chemical element with atomic number 5. The most common isotope has atomic weight 11. Might occur either as a powder or a black, hard metal. The metal is a poor electrical conductor. Boron is used in nuclear reactors and in the manufacture of electronic semiconductor devices, and in various industrial applications.

**Bose, Satyendrenath** A physicist from India who worked with Albert Einstein to formulate a theory of quantum statistics in the 1920s. See *Einstein, Albert* and *quantum statistics*.

**boson** Any subatomic particle that behaves according to the theory of Bose and Einstein. See *quantum statistics*.

**botany** A branch of biology concerned with the study of plant life.

**botulism** A type of poisoning that occurs when contaminated food is eaten. Caused by a certain bacteria in canned or bottled food that has been improperly processed. Rare in civilized countries. Symptoms include sensory disturbances, such as problems seeing right. It might become difficult to swallow. There is usually no fever. This is a serious type of poisoning and requires immediate medical care. Death occurs in up to two-thirds of untreated cases.

**bound** 1. Held near by forces such as electricity, magnetism, or gravitation. See *bound electron*. 2. Limit. We might speak of an upper bound for a function, for example.

**boundary** 1. A line or surface that denotes a limiting value or a maximum extent. We might speak of the boundary beyond which one cannot go, or beyond which one must stay. 2. A specific line or surface where substances change composition or behavior. See *boundary layer*.

**boundary layer** 1. In a fluid flowing along a solid, the set of fluid molecules right next to the solid that do not move with the rest of the fluid. It is where most of the heat transfer occurs between the solid and the fluid. The nature and thickness of this layer determine friction and other properties of the fluid flow. 2. In flight, the layer of air right next to the wing or fuselage. Turbulence in this layer increases drag.

**bound electron** An electron that stays within the influence of a certain atomic nucleus, because of the electrical attraction between the electron and the nucleus. See also *free electron*.

**Bourbaki** A society of mathematicians that arose in France just after World War I. Emphasis was on formal proofs and defined structure. The use of pictures was discouraged. Member mathematicians regarded their pursuit as detached from the rest of science, and as an end in itself. See *pure mathematics*.

**bowel** The colon and rectum. See *colon, rectum*.

**Boyle's Law** At a given, unchanging temperature, the pressure of an ideal gas is inversely proportional to the volume. That is, if you double the volume, the pressure goes down to half; if you compress it to $1/10$ its former volume, the pressure gets 10 times as great.

**Bq** Abbreviation for becquerel.

**Br** Chemical symbol for bromine.

**brachial** Pertaining to the region where the arm joins the torso.

**brachial artery** A major artery running into the arm from the torso. A human being has two of these arteries.

**brachial plexus** The network of nerves branching off of the spinal cord and running into the arm. Humans have two of these networks, one for each arm. See the illustration.

BRACHIAL PLEXUS

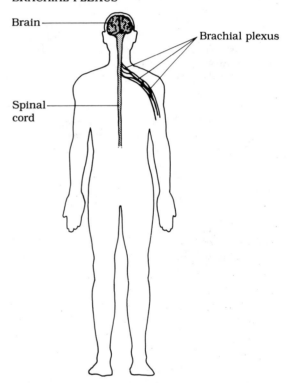

**brachiopod** An early sea animal with two saucerlike shells something like a clam, that fitted together and opened and closed. They were anchored to the ocean bottom by means of rootlike appendages. They prospered during the Paleozoic era. See the ANIMAL CLASSIFICATION and GEOLOGIC TIME appendices.

**Bradbury, Ray** A well-known fiction writer. His work often reflects the strangeness, the seriousness, and the fascinating nature of humankind's predicament. Some of his work can be called science fiction (such as an early book *The Martian Chronicles*) but his philosophical outlook is always apparent.

**Brahe, Tycho** A Danish astronomer who lived during the sixteenth century. He devised more precise instruments than had been available before his time, and used them to plot the positions of stars in the heavens. A large crater on the moon is named for him.

**brain** A complex nerve center in most animals that controls the activities of the body, the instincts, and in higher animals, the consciousness. See the illustration. It is thought that humans actually make use of only about 10–20 percent of the learning power and storage capability of their brains.

BRAIN

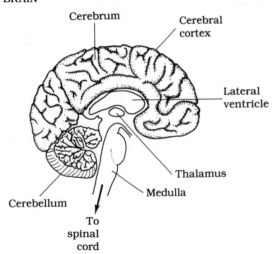

**brain cell** A single nerve cell in the brain. See *brain*.

**brain death** A biological condition in which brain function ceases. There have been cases of people being brain dead for a short time, and then being revived. Brain death is not the same as coma. See *coma*.

**brain stem** The part of the brain that connects with the spinal cord. Responsible for reflexes and unconscious life functions such as breathing. See *brain*.

**brain waves** Electrical impulses generated by the brain and graphically displayed on an electroencephalogram (see *electroencephalogram*). The illustration shows brain waves in various states of consciousness. Medical doctors can detect various problems by looking at these waves.

BRAIN WAVES

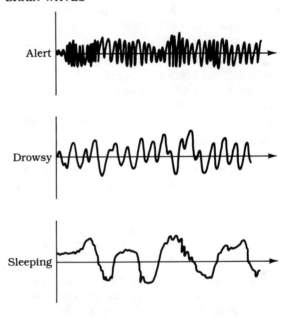

**bran** The outer covering of the kernel of a grain such as wheat, rye, oats, or rice. The bran is mostly undigestible carbohydrate. Some cereals use bran to help with bowel function. There has been some concern in recent years that we throw out the nutritious bran and germ in making flour. See *endosperm, germ*.

**brass** An alloy of zinc and copper. Typically has a yellowish color.

**breadboard** A phenolic, plastic or wooden board, usually perforated in a grid pattern or equipped with a matrix of terminals for constructing electronic circuits in the laboratory. Used for experiments and for designing prototypes.

**break** 1. A gap or a pause. 2. A discontinuity. 3. An interruption in a circuit. 4. A pause in a wire or radio conversation so that a third party can enter. 5. The transition between the "on" and "off" conditions in a digital signal. See also *make*.

**breakdown**  1. Decomposition; rendering of a complex substance into its parts. 2. Failure, as with a computer or machine. 3. The effect that results from a force becoming great enough to overcome resistance. See *avalanche, avalanche breakdown*.

**breaker**  A swell on an ocean, sea, or large lake that arrives at shore. The lower part of the wave loses speed because of friction with the bottom, and the top part falls over, forming a curl.

**breath**  The air exhaled from the lungs. In a human it contains about 16 percent oxygen, 78 percent nitrogen and 5 percent carbon dioxide. The other 1 percent is trace gases and waste products. Inhaled air contains about 21 percent oxygen; the main difference between air and breath is the amount of carbon dioxide.

**breathalyzer**  A device used to determine how much, if any, alcohol is in a person's bloodstream. The test subject exhales into the device, and if there is alcohol in the blood, it comes out in the breath. See *breath*.

**breed**  1. To arrange for animals to be born under controlled conditions so that certain characteristics can be obtained. 2. Animals resulting from the foregoing process. 3. Less frequently, the foregoing definitions as applied to plants. 4. Species. 5. Type.

**breeding**  1. The controlled birth of generations of animals for a certain purpose. 2. Less frequently, the foregoing as applied to plants.

**Bremsstrahlung radiation**  Energy generated when a high-speed electron decelerates. When the electron loses speed, it loses energy; this energy is usually given off as X rays. See *X rays, X-ray tube*.

**Brewster's Law**  A principle governing polarization of reflected light. When a light ray is reflected from a transparent surface, like a calm lake or a pane of glass, the most polarization occurs when the angle of incidence is 45 degrees. See *angle of incidence*.

**bridge**  1. A structure spanning a river or canyon designed to carry a given amount of weight without failure. 2. An electronic measurement circuit (see *bridge circuit*). 3. A bridge rectifier (see *bridge rectifier*). 4. In a certain type of compound, an atom that serves to join two other atoms.

**bridge circuit**  A device that is used for precise measurement of electrical quantities such as capacitance, inductance, and resistance. See also *capacitance, inductance, resistance*.

**bridge rectifier**  A group of four rectifier diodes arranged to work on both halves of the alternating-current cycle. Does not need to have a center-tap on the secondary winding of the transformer. See *diode, rectification, rectifier*.

**Brindley, James**  An English engineer who lived during the eighteenth century. He designed a system of canals for his country. They were the main transportation network for industry until railroads replaced them.

**brine**  1. A solution of salt water (water and sodium chloride) with a somewhat greater concentration of salt than seawater. 2. Seawater. 3. Any saltwater solution.

**brine shrimp**  A microscopic organism that lives in salt water. Popular for home and school biology experiments and projects.

**British Standard wire gauge**  A means of specifying wire sizes, similar to (but not identical with) American wire gauge. See *American wire gauge*.

**British thermal unit**  Abbreviation, BTU or Btu. 1. The amount of energy needed to raise 1 pound of water by 1 degree Fahrenheit. 2. An energy of 1055 joules. See *joule*.

**broadcast band**  1. Any range of radio frequencies designated for broadcast. 2. The amplitude-modulation (AM) radio band; this is 535–1605 kilohertz (kHz) in the United States. 3. The frequency-modulation (FM) radio band; in the United States this is 88–108 megahertz (MHz). 4. Any of the television broadcast bands. 5. A shortwave broadcast band.

**bromide** A compound containing bromine. See *bromine*.

**brominated vegetable oil** A food additive, made by adding bromine to refined plant oils. Acts as an emulsifier (see *emulsion*). It is stored in body fat. Some environmentalists and naturalists are wary of its possible harmful effects if consumed over long periods of time.

**bromine** Chemical symbol, Br. An element with atomic number 35. The most common isotope has atomic weight of 79. It is a reddish liquid at room temperature. It is highly reactive and was used during the 1960s and 1970s to purify water in swimming pools. It has a foul smell when it vaporizes, so now chlorine is preferred for this. Bromine is one of several elements known as halogens (see *halogen*).

**bronchi** The large main openings branching off from the windpipe and leading to the alveoli (air sacs) in the lungs. See *lung*.

**bronchial asthma** See *asthma*.

**bronchitis** Inflammation of the bronchi (see *bronchi*). This causes difficulty in breathing, along with coughing and discharge. Severe cases can progress into pneumonia (see *pneumonia*).

**bronze** An alloy of copper and tin. Tends to corrode easily, turning an ugly bluish-green. But the oxide layer protects against further damage. Aluminum, magnesium, or other metals can be used instead of tin to make copper alloys that are sometimes called bronze. Used in various industrial applications as well as in monuments and statues.

**Brownian motion** First observed by the Scotsman Robert Brown in the early part of the nineteenth century. The apparently random movement of tiny cells and particles. Albert Einstein suggested, in an early paper (1905), that this motion is caused by molecules that bump into the particle as they move, jostling the particle around. His hypothesis was proven correct a few years later by Fletcher and Millikan in the United States.

**brownout** A reduction in the voltage supplied by electric utilities. Might occur when there is excessive demand, such as during a heat wave when everyone runs their air conditioners. Usually the effects are not very noticeable, although television pictures might appear to shrink, and electric cooking devices do not get as hot as they do normally.

**Brunel, Isambard** A nineteenth-century English engineer, who designed the first steamship to cross the Atlantic ocean on a regular basis. In 1866, one of his ships laid the first telegraph cable across the Atlantic.

**bryozoan** A form of undersea animal life that attach themselves to the ocean bottom. They resemble tiny coelenterates (see *coelenterate*) and live in large groups. For their place in the scale of animal life, see the ANIMAL CLASSIFICATION appendix.

**BTU, Btu** Abbreviation for British thermal unit.

**bubble chamber** A device for identifying subatomic particles. Hydrogen bubbles form in the wakes of the fast-moving particles (see the drawing); magnets deflect them if they are charged. Knowing the speed and charge, the

BUBBLE CHAMBER

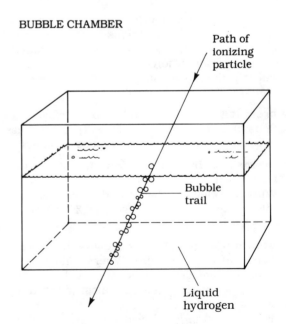

mass of a particle can be determined, and in this way it can be identified. See *cloud chamber*.

**bubble memory** Also called a magnetic bubble memory. Used to store data in computers. A magnetic film covers a semiconductor chip; the film is magnetized in tiny increments or "bubbles." Either the north or the south magnetic pole might face upwards, making a polarity of plus or minus, or "yes" and "no," or on and off. Millions of bits of data can be stored in a small space in this way, and it can be easily erased and changed.

**bubonic plaque** A severe infectious disease rarely observed today. It killed so many people during epidemics in Europe, before immunizations were available, that the distribution of population in the world has been affected to this day. Victims developed bruises all over their bodies and died from hemorrhage (internal bleeding) or exhaustion.

**bud** 1. A shoot on a plant, in which the leaves have not yet opened out. Usually found on the stem or on a branch. 2. A growth that breaks away in a form of reproduction called budding. See *budding*.

**budding** A means of reproduction in certain primitive life forms, especially yeasts, coelenterates, and single-celled plants. A growth appears and eventually breaks off as a new unit. It is an asexual form of reproduction. See *coelenterate, sponge, yeast*.

**buffer** 1. A liquid added to dilute a chemical or to reduce its tendency to react. Used to prepare acids or bases so that they have a certain pH (see *pH*). 2. In electronics, a device that stores and holds information as it comes in, letting it out at a defined, precise speed. In this way it "smooths out" the data rate.

**bulb** 1. In a plant, a large organ that allows survival between growing seasons. 2. In electricity, a lamp. See *fluorescent lamp, incandescent lamp, neon lamp*. 3. An object that is bulb-shaped.

**bulimia** A physical and psychological disease in which a person deliberately induces vomiting after eating large amounts of food. This occurs often in conjunction with anorexia nervosa (see *anorexia nervosa*). If untreated it can cause damage to the tooth enamel because of the corrosive effects of stomach acids in the mouth. Also the person might accidentally inhale small amounts of vomit and get pneumonia as a result.

**Bunsen burner** A device used in the laboratory for heating. Gas, such as methane (natural gas) or hydrogen, is fed through a tube. Holes in the sides of the tube allow air to enter. The air flow can be adjusted to control the size and heating ability of the flame.

**buoyancy** 1. Floatability. An object has positive buoyancy if it floats; it has negative buoyancy if it sinks. 2. The upward force on a floating object.

**bursitis** Inflammation of a joint. Often occurs in athletes. A swimmer might get bursitis in the shoulders, for example.

**Bussard ramjet** A hypothetical spaceship for interstellar and intergalactic travel. Conceived by R. W. Bussard and others, it scoops up the atoms of hydrogen in space, and uses these atoms to fuel a fusion rocket. See the drawing.

BUSSARD RAMJET

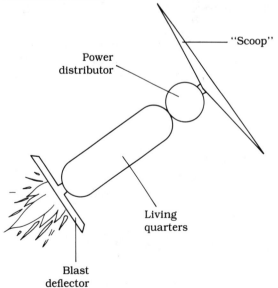

The faster it goes, the more hydrogen it gets and the better it works, like a ramjet. See also *ramjet*.

**butane** A hydrocarbon and a member of the alkane series with four carbon atoms and 10 hydrogen atoms, arranged as in the drawing. Highly flammable, it is a gas at room temperature and atmospheric pressure. It is liquid when pressurized, and is used in cigarette lighters and in various industrial applications.

BUTANE

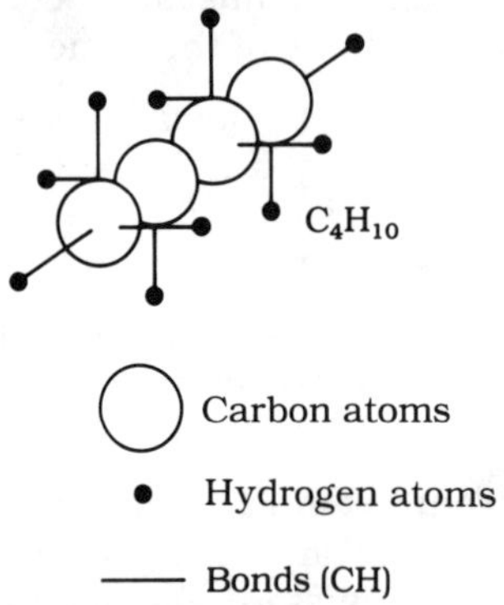

$C_4H_{10}$

◯ Carbon atoms

• Hydrogen atoms

—— Bonds (CH)

**buttress** An architectural support. When placed atop a pointed-arch support it is called a flying buttress. Used in cathedrals and other structures needing large space inside without columns getting in the way.

**butylated hydroxytoluene** Abbreviation, BHT. A food preservative used as an antioxidant. Thought by some to have possible harmful effects over long periods, it accumulates in body fat. It has been banned in some countries.

**Buys Ballot's Law** A weather rule. In the Northern Hemisphere, if you stand with your back to the wind, low pressure is to your left, and higher pressure is to your right. In the Southern Hemisphere this is reversed.

**BVO** Abbreviation for brominated vegetable oil.

**bypass** 1. To divert around an obstruction. 2. In electronics, pertaining to a component (such as a capacitor) that lets one form of signal or current through, while blocking another. 3. In medicine, pertaining to surgery in which a blood vessel is grafted into the body to let blood go around an occlusion. See *coronary occlusion*.

**by-product** A substance generated during a process, that has no use in the process itself but might be used for something else.

**Byrd, Harry** A senator who got famous in the 1960s by his loud objections to the use of vast sums of money for obscure scientific purposes that seem ridiculous. A hypothetical example might be to use $10 million to study the colonization instincts of red ants in south Floridian suburbs. This, when some people in the same areas don't even have housing.

**byte** A single binary digit (see *bit*) or a group of bits that form a unit. Usually eight bits long, corresponding to one character in computer memory.

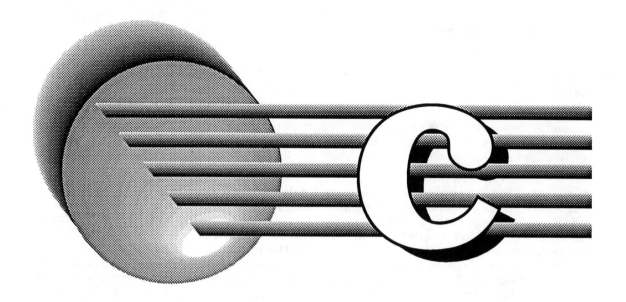

**C**  1. Chemical symbol for carbon. 2. Abbreviation for coulomb. 3. Abbreviation for capacitance or capacitor. 4. Abbreviation for centigrade or Celsius.

**c**  Abbreviation for centi-; constant speed of light. See PREFIX MULTIPLIERS appendix.

**Ca**  Chemical symbol for calcium.

**cable**  1. A set of two or more wires used for the purpose of transmitting electricity or communications signals. There are different kinds of cable for different applications (see drawings). 2. A strong wire used for structural purposes, such as in a suspension bridge. 3. A written message sent by wire.

**cable communications**  The transmission of data, either in one direction or in two directions, by means of wire or optical fibers. See *cable, fiberoptics*.

**cable television**  1. The transmission of commercial television signals via wire or optical fiber. It is possible to put dozens of channels onto a single wire; millions of signals can be put onto optical fibers. Optical-fiber television has not yet come into widespread use in residential areas. See *cable, fiberoptics*. 2. A network for

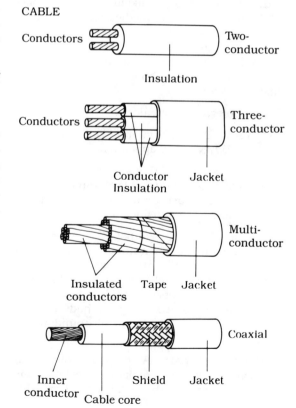

CABLE

the transmission of television signals via wire or optical fiber.

**cactus** A desert plant, sometimes as tall as two stories (about 20 feet), capable of surviving long periods without rain. It is noted for its protective covering of spines, which makes it unattractive as food for animals.

**cadmium** Chemical symbol, Cd. An element with atomic number 48. The most common isotope has atomic weight 114. Used in the manufacture of rechargeable electrochemical cells and in certain solders and platings. In its pure form it is a soft metal. Generally considered as a toxin in the body, it is found in cigarette smoke and other pollutants.

**caecum** The very beginning of the large intestine; the first part of the ascending colon. The small intestine empties into it. Also spelled cecum. See *colon, small intestine*.

**caffeine** A stimulant alkaloid found in coffee and added to some medications to reduce drowsiness. In low to moderate doses it causes an increase in alertness and seems to quicken the reflexes. In large doses it causes heart palpitations, muscle twitching, intestinal cramps, and sleeplessness.

**cal** Abbreviation for calorie.

**calciferol** A form of vitamin D. See the VITAMINS AND MINERALS appendix.

**calcite** Calcium carbonate in crystal form (see *calcium carbonate*). It is abundant in rocks in the earth's crust. Some varieties are transparent; others are translucent or opaque white.

**calcium** Chemical symbol, Ca. A chemical element with atomic number 20. The most common isotope has atomic weight 40. In its pure form it is a lightweight, grayish, rather brittle metal. It reacts with acids to form various salts. Calcium is needed by the body to make bones and teeth. It is found in abundance in rocks.

**calcium carbonate** A compound of calcium and carbon. Occurs in rocks such as limestone. Some "stones" in the body form from this. Found in "hard" water. Soluble in various acids.

**calculation** 1. A process in which an equation or arithmetic problem is solved, giving a specific number value or values. 2. The various techniques for solving equations and arithmetic problems.

**calculator** A small electronic device (often hand held or pocket sized) that performs calculations rapidly and accurately. Some calculators have scientific functions such as trigonometric and logarithmic. Others just do simple addition, subtraction, multiplication, and division problems. Most calculators are accurate to at least eight digits, and many have 10-digit accuracy. The drawing shows a solar-cell-powered calculator. Technology has made calculators cheap enough so that almost anyone can afford one.

CALCULATOR

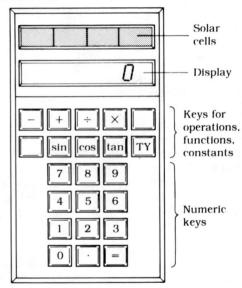

**calculus** 1. A branch of mathematics dealing with instantaneous rate of change, averaging of functions, and many other calculations. 2. Differential calculus: The mathematics of rate of change, first developed by Isaac Newton and others. See *derivative, differential, differential equation, differentiation*. 3. Integral calculus:

The mathematics of finding the area under a curve, developed by Riemann and others. See *integral, integration.* 4. A hard, calcium-containing deposit that can form in organs such as the kidneys, or on the teeth, and in many other places throughout the body.

**caldera**  A large depression at the top of some volcanoes. Often a mile or more in diameter. Left after the volcano erupts explosively and then collapses. Sometimes called a crater (this is a misnomer). See *volcano.*

**calibration**  1. Adjustment of a measuring device for maximum possible accuracy. This can be done using a standard such as the time signals from the National Bureau of Standards. 2. The extent to which a measuring device is adjusted correctly.

**californium**  Chemical symbol, Cf. An element not occurring naturally, but can be synthesized. Atomic number is 98. The most common isotope has atomic weight 251. It is radioactive, but the isotope Cf-251 is fairly stable, with a half-life of 700 years.

**callus**  1. A hard, horny thickening of the skin, seen on often-used, exposed body surfaces. 2. In plants, a protective growth over an injury.

**calorie**  1. The amount of energy needed to raise 1 gram of water by 1 degree centigrade at a pressure of one atmosphere. 2. The energy equivalent of 4.19 joules (see *joule*). 3. In nutrition, a kilocalorie (see *kilocalorie*), also called a diet calorie.

**calorimeter**  A device for measuring heat or heat capacity with high accuracy.

**calving**  The formation of icebergs as large pieces of glaciers break off into the sea. See *glacier, iceberg.*

**Cambrian period**  The earliest part of the Paleozoic era. Began about 575 million years ago, and lasted about 75 million years. The earliest known fossils, found in central Wales, come from this period. See the GEOLOGIC TIME appendix.

**camera**  1. A device that uses a lens and a photosensitive film to record a still image of a scene

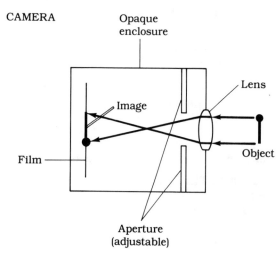

Functional diagram

(see drawing). 2. See also *iconoscope, image orthicon, vidicon.*

**camphor**  A ketone that can be manufactured and that also occurs naturally. Used in various medicinal preparations and also in mothballs. See *ketone.*

**canal**  1. An artificial waterway used for irrigation, flood control, transport of drinking water, or for shipping. 2. A passage in the body, for example the ear canal or the digestive canal. 3. Any of various straight-line markings occasionally seen on Mars as viewed through a telescope. Nowadays we know that these are rows of small craters.

**cancer**  Uncontrolled, destructive growth of body cells. Thought to be caused by pollutants, radiation, prolonged irritation, and also as a natural result of the aging process. See *carcinogen, carcinoma.*

**candela**  Abbreviation, cd. The standard unit of light-source intensity. Equivalent to one lumen per steradian (see *lumen, steradian*). An ordinary candle flame has an intensity of about 1 cd. A 40-watt incandescent bulb produces about 3000 cd.

**candida**  A yeast that can cause infection, especially in the mouth and throat, and sometimes all along the digestive tract. See *yeast.*

**candlepower** An archaic term for candela. See *candela*.

**canine** 1. Pertaining to dogs. 2. A dog. 3. A sharp tooth used for tearing food.

**canker sore** A virus-caused sore commonly occurring in the mouth. Some people get them so often and in such numbers that they cause great discomfort. They usually go away in a few days without medical attention; however, severe cases might require a doctor's treatment.

**cannabis** The hemp plant commonly associated with the drug marijuana. Smoking its leaves produces intoxication. Long-term use might cause psychological problems and dependence.

**Cannon, Annie** An astronomer of the late nineteenth century who worked at Harvard. She is known for compiling a catalog of the spectra of almost 500,000 stars. She called it the *Henry Draper Catalogue*.

**cantilever** A device used in bridges and other structures where support is needed (see drawing). Often used for architectural convenience or decoration.

CANTILEVER

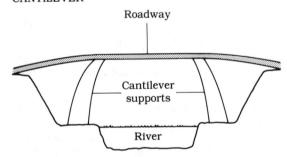

**Cantor, George** A mathematician known for his work with infinite sets in the late nineteenth century. He was the first to note that there are different "sizes" or "cardinalities" of infinity. See *aleph-null, aleph-one, transfinite cardinal*.

**canyon** A large ravine or gorge carved out of the landscape by a river by erosion over millions of years.

**capacitance** An electrical effect where energy is stored in the form of an electric field. The standard unit is the farad; more often it is given in microfarads or picofarads (units of one millionth and one trillionth of a farad). See *farad*.

**capacitive reactance** The opposition that a capacitance offers to alternating current. Given in units of ohms (see *ohm*). In general, the higher the capacitance, the lower the capacitive reactance. If C is the capacitance in farads, and f is the frequency in hertz, then the capacitive reactance X is given by the formula:

$$X = \frac{1}{(6.28 f C)}$$

Engineers put a minus sign before this, and consider capacitive reactance to have a negative value. See *capacitance, impedance, inductance, inductive reactance*.

**capacitor** An electronic device designed especially to have a certain amount of capacitance. Might be fixed or variable. See *capacitance, capacitive reactance*.

**capillarity** The property of a liquid that makes capillary action possible. See *capillary action*.

**capillary** 1. In the body, a tiny blood vessel that receives blood from the arteries and deliv-

CAPILLARY

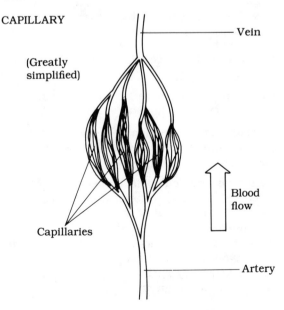

ers it to the tissues. Then it takes the blood back from the tissues, and to the veins. The illustration shows a simplified view of capillary structure. See *artery, vein*. 2. A fine tube through which liquid passes.

**capillary action** In a fine tube through which liquid passes, the tendency of the liquid to slowly travel along the tube because of interaction between the surface of the liquid and the inside surface of the tube. The liquid can go upwards against gravity.

**capillary fringe** In the soil, a region below the aeration zone (see *aeration zone*), where water from deeper down is lifted up by capillary action (see *capillary action*). The depth of this fringe depends on various factors, such as the rainfall and the nature of the soil.

**capsid** In a virus, the envelope or skin, made of protein. See *virus*.

**capsule** 1. A protective coating around some bacteria. 2. A protective enclosure around some body organs in humans and animals. 3. A form of fruit that breaks open when it ripens, scattering seeds. 4. An enclosure with life-support systems inside, such as a space vessel or an undersea station. 5. In general, any coat that serves to protect or isolate its contents.

**carapace** 1. The top shell of a turtle. 2. The top part of an exoskeleton.

**carat** 1. A mass of 0.2 gram (200 milligrams). 2. An indicator of the purity of precious metals, especially gold. The maximum possible is 24 carats (abbreviated 24K). This would be 100 percent pure.

**carbide** A compound of carbon and a metal. Might be metallic in character. Some carbides are excellent electrical conductors. Some are noted for their exceptional hardness and are used to make the cutting edges of drill bits and saw teeth.

**carbohydrate** A compound containing carbon and hydrogen, and often oxygen as well. Sugars and starches are examples of food carbohydrates used by the body as an efficient energy source. Foods high in carbohydrates include grains, beans, fruits, and some vegetables. See *hydrocarbon*.

**carbon** Chemical symbol, C. An element with atomic number 6. The most common isotope has atomic weight 12. Most often found in nature as graphite; also might occur as diamond. Carbon is the base for all life on this planet. The isotope C-14 is radioactive and can be used to determine the ages of fossils and other objects. See *radioactive dating*.

**carbonaceous chondrite** A type of meteor, made of carbon with small amounts of minerals. They are unique in the Solar System. When they are found on the earth, we know that they came from space.

**carbonate** A carbon-metal salt. An example is calcium carbonate, the compound that makes up some kidney stones.

**carbon black** 1. As black as black can be; albedo zero (see *albedo*). 2. Powdered carbon, coal-black in color.

**carbon cycle** A cycle that is important to all life on the earth. In the drawing, arrows show

CARBON CYCLE

carbon entering and leaving the earth by natural processes. Some carbon is dissolved in the sea and lakes and rivers from carbon dioxide in the air; some comes down in rain and snow. Carbon enters the atmosphere from the earth mainly in volcanoes; some is exhaled by animals and humans. Some comes from forest fires. And too much, perhaps, is created by the burning of fossil fuels like gasoline, coal, and oil. This has led to fears of global warming. See *carbon dioxide, global warming, greenhouse effect.*

**carbon dating** See *radioactive dating.*

**carbon dioxide** A compound of one atom of carbon and two of oxygen. Important in the carbon cycle (see *carbon cycle*). There is a small, but important, amount in the air. In recent years this amount has been increasing because of the burning of fossil fuels and because of other human activities such as deforestation. See *global warming, greenhouse effect.*

**carbonic acid** An acid formed when carbon dioxide dissolves in water, as happens in the carbon cycle (see *carbon cycle*). It is found also in soft drinks that have been "carbonated."

**Carboniferous period** A geologic period noted for its lush plant growth. At this time, about 350 million to 290 million years ago, the equator ran through North America and Europe. Much of the earth's deposits of coal were formed from the decay of these plants. The Carboniferous period is often broken chronologically into the Pennsylvanian and Mississippian periods. See the GEOLOGIC TIME appendix.

**carbon monoxide** A compound of one atom of carbon and one atom of oxygen. A byproduct of the incomplete burning of gasoline. It is dangerous because it is odorless and because it takes the place of oxygen in the blood. This deprives the body and brain of oxygen and can cause unconsciousness and death.

**carbon tetrachloride** A compound also known as chlorinated methane or tetrachloromethane. It is used as a cleaner and solvent, and is highly volatile. Its vapors are toxic.

**carbon-14 dating** See *radioactive dating.*

**carbuncle** A large abscess, often with pus, that is painful and often becomes infected. Occurs typically on the skin, also called a boil.

**carburetor** A device that combines droplets and vapor of gasoline or other liquid fuel with air for efficient burning in an engine. See the drawing. The ratio of air to gasoline is adjustable, so that as little gasoline as possible is wasted, and the most energy is available for the engine.

CARBURETOR

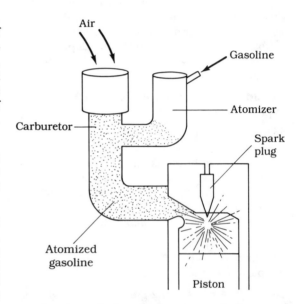

**carcinogen** A chemical or agent that causes, accelerates the progress of, or contributes to the development of cancer in humans or animals. In recent decades, there has been concern about manmade carcinogens in food and in the environment.

**carcinogenic** Acting as a carcinogen. See *carcinogen.*

**carcinoma** A cancerous tumor or growth. See *cancer.*

**cardiac** Pertaining to the heart. See *heart.*

**cardiac arrest** See *heart attack, heart failure.*

**cardiac arrhythmia** See *heart arrhythmia.*

**cardinality** The size of a set, as given by its cardinal number. The cardinality of the set of people at a football game, for example, might be 56,867. The cardinality of the set of stars in our galaxy is about 100,000,000,000. Must be a nonnegative integer or a transfinite number. See *cardinal number, integer, set, transfinite cardinal*.

**cardinal number** A number that indicates the size of a set. Can be finite, such as 34 or 45,999. Or can be infinite, such as aleph-null or aleph-one. See *aleph-null, aleph-one, cardinality, set, transfinite cardinal*.

**cardioid** 1. Heart-shaped. 2. A characteristic pattern that is heart-shaped or valentine-shaped.

**cardiopulmonary resuscitation** Abbreviation, CPR. A technique of resuscitation for unconscious people. Can be done by one person, but is more efficient when done by two people. The victim is given heart massage and mouth-to-mouth respiration at the same time. This has saved many lives. It is taught by the American Red Cross, and is required training for lifeguards, police officers, firefighters, and paramedics.

**cardiovascular** Pertaining to the heart and circulatory system. See *heart, circulatory system*.

**caries** Cavities in the teeth. Caused by bacteria and certain acids that are present when foods, especially carbohydrates, stick to the teeth for long periods. Brushing with fluoridated toothpaste greatly reduces the number of caries you will get.

**carnivore** An animal that consumes the flesh of other animals. A strict carnivore lives off only animal flesh and little or nothing else. Wolves are an example.

**Carnot cycle** A four-way cycle in which an efficient heat engine works. This is illustrated by the curve. A gas is made to expand and contract, and also to get warmer and cooler. But these do not necessarily happen along with each other. The cycle has four separate stages, as shown in the drawing.

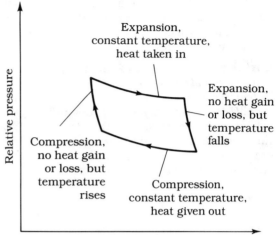

CARNOT CYCLE

**carotene** A yellowish substance found in certain vegetables, converted by the body into vitamin A. See the VITAMINS AND MINERALS appendix. Refers to the form more technically known as beta-carotene. In recent years this substance has been of interest because it might act to discourage cancer. See *cancer*.

**carotid artery** A major artery that supplies blood to the brain.

**carpus** The set of bones in the wrist. In a human being, eight wristbones make up each wrist.

**carrier** 1. A person or animal who harbors an infectious disease without showing symptoms. Although the animal or person does not seem sick, he or she can infect others and they can get sick. 2. A person or animal with a genetic defect that is not apparent until it shows up in descendants. 3. In communications, the signal that is modulated with the information to be sent. It is generally a radio wave but might be visible light or infrared. 4. A particle that has an electric charge and that moves in a medium. See *electron, hole*.

**Carroll, Lewis** See *Dodgson, C.L.*

**Cartesian coordinates** A scheme of locating points on a plane or in space, devised by the French mathematician Rene Descartes. There are two or three (or in some cases more than three) axes, all intersecting at a common point called the origin. Units are marked off positively and negatively along each axis. See *Cartesian plane* and *Descartes, Rene.*

**Cartesian plane** A two-dimensional Cartesian coordinate system (see *Cartesian coordinates*). This is the common coordinate plane taught in elementary-school or junior-high-school math courses. See the drawing. Two number lines are arranged at right angles and are usually marked off in units that are the same size, both positive and negative.

CARTESIAN PLANE

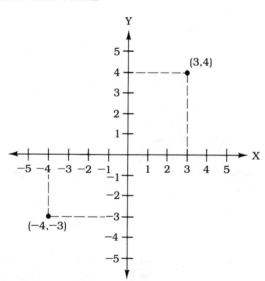

**cartilage** A plastic-like, durable body substance in joints that get much use. Found between the vertebrae of the backbone, in the wrists and ankles, and also in the elbows, shoulders, knees, and hips. In infants, some cartilage later becomes bone.

**cartography** The science and art of making maps.

**cascade** 1. To connect one after the other. We might, for example, put three amplifiers one after the other to get far greater loudness than we would get with just one amplifier. 2. To perform a task in steps that are similar to, or identical with, each other. 3. To do any task in steps, rather than all at once.

**cascade process** Any procedure done in discrete steps.

**cascode** A method of connecting transistors to obtain high amplification and low noise. Two transistors are connected directly together, one after the other. The emitter or source of the first transistor is grounded, and the base or gate of the second transistor is grounded. See *transistor.*

**casein** A protein in milk and milk products, such as cheese and cottage cheese. Important in nourishment for children. A few people have enzyme deficiencies that cause this protein to be undigestible.

**Cassegrainian telescope** See *telescope.*

**Cassini gap** A zone on either side of the central, and brightest, ring of Saturn. The inner and outer rings are separated from the middle ring by this empty zone.

**caste** A social position in an insect colony. Bees for example, are classified as drones, workers, and a queen. Ants also have a caste system. Some human societies, Ancient Rome for example, have caste systems.

**cast iron** Iron with a small amount of carbon added. It is heavy and is dull gray in color. Used in a variety of industrial applications. It is stronger than pure iron but not as strong as steel. See also *iron.*

**catabolism** A form of metabolism in which substances are broken down or assimilated into other substances, providing energy. For example, fat stores can be broken down; in starvation, even body muscle is used for energy. In the intestine, bacteria metabolize various substances such as gas; this is a form of catabolism. See *anabolism, metabolism, respiration.*

**catalyst** 1. In a chemical process, an agent that speeds things up. We might add a catalyst to make glue harden more quickly, for example. 2. Anything that tends to make a process go more quickly or more efficiently.

**catenary** A curve, shaped something like a parabola, that you get if you have a perfectly flexible, nonstretchable, uniform rope or chain and suspend it from two fixed points. See *parabola*.

**cathartic** A strong laxative, used to rid the bowel of all of its contents. Can be used in severe constipation or before surgery. Examples are milk of magnesia and Epsom salt. In large doses these cause cramping and diarrhea, and the large intestine is emptied within a few hours. These medicines can be dangerous if used excessively.

**catheter** A tube inserted into the body for a medical purpose, such as to pump the stomach, drain the bladder, or visually examine the lungs. The tube can be hollow, like a tiny hose. It can also have optical fibers to light up the interior of the body and to pass images back to a viewing screen. See also *endoscope*.

**cathode** The electron-emitting electrode in a vacuum tube or semiconductor device.

**cathode-ray oscilloscope** See *oscilloscope*.

**cathode rays** High-speed electrons given off by an electrode that is negatively charged with a high voltage. The electrons are accelerated by anodes (see *anode*). They travel in straight lines unless they are deflected by electric or magnetic fields. Cathode rays cause phosphors to glow in oscilloscopes and picture tubes. See *cathode-ray tube*, *kinescope*, *oscilloscope*.

**cathode-ray tube** A vacuum tube in which electrons are accelerated into beams. The beams might be deflected by means of charged plates or magnets. Then the electrons strike a screen where they cause a phosphor to glow. The drawing is a simplified cross section of a cathode-ray tube. See *cathode rays*, *kinescope*, *oscilloscope*.

**cation** An ion with positive charge. This happens when there is a deficiency of one or more electrons in an atom. See *anion*, *ion*.

**CATSCAN** Abbreviation for *computerized axial tomography*.

**causation** A phenomenon where one event is the result of one or more preceding events, and where this relationship can be scientifically proven. See *cause-effect*.

**cause-effect** A principle that tells us that one or more events are responsible for events that follow. For example, we might say that fossil-fuel burning causes carbon dioxide to enter the air; this causes a greater greenhouse effect; this causes global warming; this causes the ocean level to rise; this causes flooding in coastal cities. So we might say that the flooding is caused by the burning of fossil fuels, according to the rules of cause-effect. But there might be other, unknown causes too. This principle is often used imprecisely to explain things, making events seem simpler than they really are.

**caustic** Highly reactive or corrosive. Tending to react easily with other substances. For example, many acids are caustic because they react with metals.

**cave** An underground cavity or system of chambers, carved by an underground stream or river. Also called a cavern.

**Cavendish, Henry** A scientist who first showed in the eighteenth century that water can be synthesized from hydrogen and oxygen.

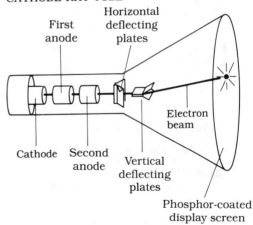

CATHODE-RAY TUBE

**cavitation** Improper operation of a hydrofoil (see *hydrofoil*), where the water flow breaks away from the hydrofoil, and a vapor cavity develops. Occurs at about 65 miles per hour. A special hydrofoil has been devised to reduce the problems of cavitation, allowing higher speeds.

**cavity** 1. A gap, hole or enclosure, either natural or artificial. 2. In a tooth, a region of decay. Also called caries or dental caries. See *caries*.

**cavity resonator** In electronics, a device that operates as a tank circuit (see *tank circuit*). Instead of a coil and capacitor, an adjustable, hollow metal enclosure is used. These devices are used mainly at ultrahigh frequencies and above. See also *microwave, ultrahigh frequency*.

**CB** Abbreviation for citizen band.

**cc** Abbreviation for cubic centimeter (archaic).

**CCD** Abbreviation for charge-coupled device.

**Cd** Chemical symbol for cadmium.

**cd** Abbreviation for candela.

**CDT** Abbreviation for Central Daylight Time.

**Ce** Chemical symbol for cerium.

**celestial coordinates** The latitude and longitude lines of the earth, projected up into the sky. You might have seen these lines in a planetarium. See *celestial equator, celestial pole, declination, right ascension*.

**celestial equator** The earth's equator, as projected indefinitely into space. Appears as an imaginary line in the heavens. Celestial latitude, more commonly called declination, is measured relative to this line. See *declination*.

**celestial latitude** See *declination*.

**celestial longitude** See *right ascension*.

**celestial mechanics** See *Newton's Laws of Gravitation, Newton's Laws of Motion, relativity*.

**celestial navigation** A means of ship navigation using the sun and stars. Useless when it is cloudy or foggy, but modern electronic devices have solved this problem. See *navigation, navigational satellite, radionavigation*.

**celestial pole** Either the North Pole or the South Pole, projected onto the celestial sphere. Polaris, the North Star, is almost exactly at the North Celestial Pole. The celestial poles are 90 degrees from the celestial equator. See *celestial coordinates, celestial equator*.

**celestial sphere** The grid of celestial latitude and longitude as they would appear on a huge sphere around the earth. This imaginary sphere rotates once every 23 hours and 56 minutes. See also *celestial coordinates, celestial equator, celestial pole, declination, right ascension*.

**cell** 1. A single unit of living matter, first named by the English scientist Robert Hooke in the seventeenth century, who noticed their resemblance to little rooms or jail cubicles. Most cells are so small that a microscope is needed to see them. Some types of cells, and their names, are shown in the drawing. 2. A device that produces electrical current from some other source, such as chemicals (electrochemical cell) or sunlight (solar cell). See *battery, solar cell*. 3. A single unit in a thunderstorm, represented by positive and negative electrical charges in the atmosphere. Individual cells last 15-30 minutes but often

CELL

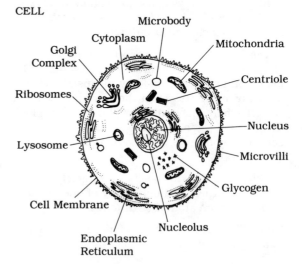

regenerate so that a storm has a life of several hours. Cells also merge together to create large storms. See *squall line, thundershower, thundersorm*. 4. A geographic sector in a mobile telephone network. See *cellular mobile radio telephone*.

**cell division** A process whereby a cell reproduces itself. The illustration shows the steps in this process. There are two types of cell division. See *meiosis, mitosis*.

CELL DIVISION

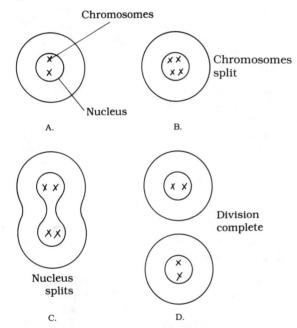

**cell membrane** The skin or coating of a cell. In plants this is usually thick and rigid; in animals it is generally more flexible. See *cell, cell wall*.

**cellophane** Cellulose formed into a transparent, thin, airtight material like plastic. Used for wrapping. See *cellulose*.

**cellular mobile radio telephone** Also called cellular radio or cellular telephone. A means of coordinating radio telephones so that it is possible to make or receive calls from anywhere at any time. The system uses many repeaters (see *repeater*). When a user drives into the range of a certain repeater (the area is called a cell), that repeater is automatically used. Computers control the system to make sure that a connection is almost always possible and is always maintained once it is made. All car phones nowadays are of this kind.

**cellular respiration** A process where cells give off carbon dioxide and water, and produce energy, from oxygen. It is the opposite of photosynthesis. See *photosynthesis*.

**celluloid** Cellulose nitrate and camphor (see *cellulose, camphor*) processed into a plastic-like sheet, often used to make film and transparencies.

**cellulose** The carbohydrate that comprises the walls of plant cells (see *cell wall*). It is undigestible in humans. Used for a variety of industrial purposes. See *cellophane, celluloid*. It is a major constituent of dietary fiber found in bran. Sometimes powdered cellulose is added to food to increase its bulk-producing and filling ability without increasing the number of usable calories.

**cell wall** In plants, the rigid coating of cells that keeps them separate from each other and gives the plant a certain amount of stiffness. Also serves to keep water and nutrients inside the cell. See *cell*.

**Celsius** See *centigrade temperature scale*.

**Cenozoic era** The time of recent life. Some believe this era began about 65 million years ago with the evolution of such species as alligators and horses. The Rocky Mountains were formed at the beginning of this time. See the appendix GEOLOGIC TIME.

**center of gravity** 1. The point about which an object will balance. Usually, objects are more stable when their centers of gravity are low, as compared with when they are high. But the opposite is true in certain cases, such as when a circus performer walks a tightrope. 2. See *center of mass*.

**center of mass** A point representing the average location of the mass of an object. If the object has uniform density, the center of mass is at the geographic or physical center of the object (see drawing). If the object has variable density, the center of mass must be determined by rather complex mathematical processes or by physical experimentation.

CENTER OF MASS

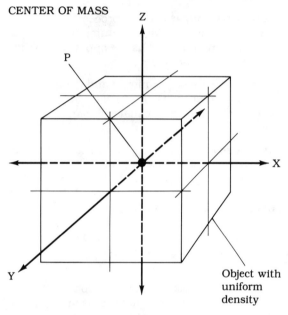

Object with uniform density

**centi-** See the PREFIX MULTIPLIERS appendix.

**centigrade temperature scale** A metric means of measuring temperature. The freezing point of pure water at one atmosphere is assigned the value zero; the boiling point is +100. Increments in centigrade are exactly 9/5 (1.8 times) the size of Fahrenheit degrees. Absolute zero is at −273 degrees centigrade. Generally abbreviated as C or degrees C. See *Fahrenheit temperature scale*.

**centimeter** A unit of length equal to 1/100 (0.01) meter. See *meter*.

**centipede** Generally, a poisonous millipede. See *millipede*. More specifically, centipedes can move faster, and they eat animal flesh, while the slower millipedes eat plants.

**Central Daylight Time** Central Standard Time plus one hour, or Coordinated Universal Time minus five hours. Generally used from early April until the last weekend in October in the United States. See *Central Standard Time, Coordinated Universal Time*.

**central nervous system** The brain and spinal cord in humans and other vertebrates. In lower species it consists of the brain and the nerves immediately branching from the brain. Abbreviation, CNS.

**central processing unit** Abbreviation, CPU. The part of a computer that supervises the running of the programs. Consists of the arithmetic logic unit (ALU), a control and timing unit, and registers. Acts as the computer's intelligence center. See *arithmetic logic unit, computer, register*.

**Central Standard Time** Coordinated Universal Time minus six hours. Used along and somewhat to either side of the longitude line 90 degrees west. This runs through the central United States. Used from the end of October until the beginning of April in most applicable places. See also *Central Daylight Time, Coordinated Universal Time*.

**centrifugal force** An outward force that occurs on or in an object or enclosure when it revolves around some central point. See *centrifuge, centripetal acceleration*.

**centrifuge** A device in which an enclosure or test chamber is made to revolve at high speed in a circle, producing centrifugal force (see *centrifugal force*). In the laboratory, a centrifuge can be used to separate liquids into their constituents. A large centrifuge, such as the one shown in the drawing, is used to subject astronauts to the forces they will encounter in space travel.

**centripetal acceleration** The acceleration of an object as it revolves around a central point. This is always a vector pointing straight inward toward the center when the object revolves at constant speed in a circle. This acceleration gives rise to the outward force called centrifugal force. See *centrifugal force*.

**centroid** The center of mass in an object if the density is uniform. If the density is not uniform,

CENTRIFUGE

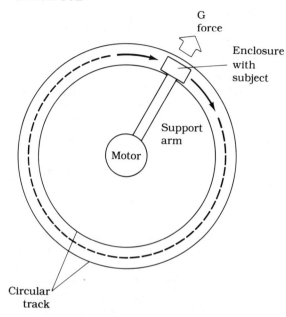

the point where the center of mass would be if the density were uniform. See *center of mass.*

**cephalization**  A characteristic of animals where the eyes, ears, mouth, and brain are together in the head or near the front of the body.

**cephalopod**  An animal life-form that some believe developed fairly early in the evolutionary process yet it was quite advanced, with sharp vision and other senses. They dwell in water and are flesh eaters. Examples include octopuses and squids. For their place on the evolutionary scale, see the ANIMAL CLASSIFICATION appendix.

**cephalothorax**  The head and thorax of certain arthropods, when these two parts are connected together as a unit. See *arthropod, thorax.*

**Cepheid variable star**  A unique type of star whose period, or length of cycle, is related to its brightness. This has made it possible for astronomers to measure large distances in space. This is how the distance to the Andromeda Galaxy was determined, for example. See *variable star.*

**ceramic**  An artificial compound made of aluminum oxide, magnesium oxide, titanium oxide, or similar substances. Ceramics are strong but brittle. They are employed as insulators and dielectrics in electrical and electronic devices. Ceramics are used in microphones and other transducers and are employed as coil forms. Emerging technology has ceramics being used as cutting edges for high-speed tools. See *dielectric.*

**cerebellum**  A major part of the brain. In humans it is located below the cerebrum (see *brain, cerebrum*). It is responsible for the actions of the muscles.

**cerebral cortex**  The outer part of the brain, responsible for movement and locomotion, senses and intelligence. Sometimes called gray matter. See also *brain, cerebrum.*

**cerebral palsy**  A brain defect that results in physical disability or deformity, but generally does not affect intelligence. People with this disability might have superior intelligence, perhaps because the physical problems give them more time for thought.

**cerebrum**  The large part of the brain with two halves or hemispheres in mammals. This is where the functions of learning, memory, feeling, and senses are located. It is the physical center of the so-called "conscious mind." Sometimes called white matter. See also *brain.*

**Cerenkov counter**  An instrument for measuring the number of high-speed charged particles in a given location. A transparent substance is brought near the source of charged particles, and Cerenkov radiation occurs (see *Cerenkov radiation*) as visible light. The intensity of this light is measured and this gives an indication of the number of charged particles passing by.

**Cerenkov radiation**  When electrons pass through a substance at a speed greater than the speed of light in the substance, visible light is produced. This is called Cerenkov radiation, after its discoverer. See the drawing. Other high-speed charged particles also produce this glow. See *Cerenkov counter.*

CERENKOV RADIATION

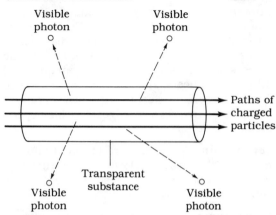

**Ceres** The first asteroid discovered. It is about 400 miles long. Since its discovery, thousands more have been catalogued. Most have orbits that lie between Mars and Jupiter. See *asteroid*.

**cerium** Chemical symbol, Ce. An element with atomic number 58. The most common isotope has atomic weight 140. There are numerous isotopes, some radioactive. In its pure form it is a shiny metal. It is used in various industrial applications.

**cermet** An alloy of ceramic (see *ceramic*) and a metal, usually nickel. Used as a resistive element in electronic circuits.

**cervix** 1. The outer end of the uterus in a female mammal. See *uterus*. 2. The neck or the neck portion of the spinal column in vertebrates.

**Cesarian section** An operation where a baby is removed from the womb surgically. This is done when a doctor thinks it would be harmful to the baby and/or the mother to attempt a normal delivery.

**cesium** Chemical symbol, Cs. An element with atomic number 55; the most common isotope has atomic weight 133. Used as a time standard in atomic clocks. Also used in phototubes and to eliminate residual gas in vacuum tubes.

**cesium time standard** An atomic clock that uses cesium oscillations as the basis for timekeeping. Accurate to a tiny fraction of a second per year.

**Cf** Chemical symbol for californium.

**CFC** Abbreviation for chlorofluorocarbon.

**CGS system** A system of units based on the centimeter, gram and second, hence the abbreviation CGS or cgs. See also the STANDARD INTERNATIONAL SYSTEM OF UNITS appendix.

**chain** 1. Atoms interconnected to form a molecule. A good example is the connection of carbon atoms in compounds such as propane, octane, or paraffin (see *alkane*). 2. A series of events related by cause. See *cause-effect, chain reaction*.

**chain reaction** An atomic process where one reaction leads to several others, over and over. The atomic bomb works on this principle. The drawing shows the phenomenon as neutrons break atomic nuclei apart, generating more neutrons, and so on. See *fission*.

CHAIN REACTION

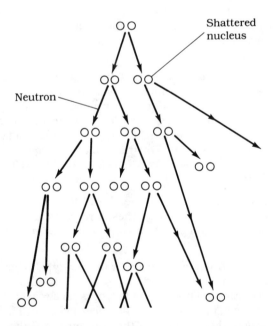

**chalk** 1. Calcium sulfate. A white powder used for writing on blackboards or for other temporary marking purposes. 2. A mineral, mainly calcium carbonate, found as a white rock that crumbles into powder.

**Challenger Space Shuttle** One of the United States' fleet of space shuttles. It exploded shortly after liftoff on January 28, 1986. The problem was in the seals in the solid-fuel booster engines. See *Space Shuttle*.

**chameleon** A lizard noted for its ability to change color depending on its surroundings. This color change can take place in seconds. It serves as camouflage.

**chancre** An open sore that develops, usually on the genitals, during the early stages of syphilis. See *syphilis*.

**channel** 1. In communications, a designated frequency for broadcast or two-way operation. Consists of a narrow band, just wide enough to allow the information to be conveyed. 2. In a field-effect transistor, the path that the charge carriers follow, running from the source to the drain. The flow is regulated by the voltage on the gate. See *field-effect transistor*. 3. The navigable part of a river, often dredged out to make it deep enough for boats to pass.

**chaos** 1. Complete absence of order. 2. A disorganized state. 3. See *chaos theory*.

**chaos-order** See *order-chaos*.

**chaos theory** A young science concerned with the behavior of systems as seen by observing general patterns. Chaotic phenomena often show identical shapes no matter what the scale. Despite the term, there is a certain kind of order apparent: the disorder is the same at every scale. See *fractal, Mandelbrot set*.

**character** 1. A single element in a digital communications signal. 2. Type or variety, or particular properties. For example, we might speak of the character of ionizing radiation. 3. An inherited trait.

**characteristic** A property of an object, device, or living thing that is predictable and repeatable.

**characteristic curve** A graph, usually a continuous line or curve, that shows the behavior of a certain device. In the drawing, the current through a semiconductor diode is shown as a function of the voltage across it. Graphs of this kind are commonly given for electronic components. See *diode, field-effect transistor, transistor*.

CHARACTERISTIC CURVE

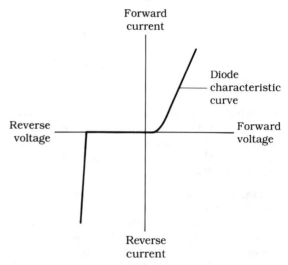

**characteristic triangle** See *differential triangle*.

**charge** In electricity, a surplus or deficiency of electrons. A surplus results in a negative charge; a deficiency in a positive charge. Can only exist in integral multiples of one electron. See *coulomb, electricity, electron, hole, potential difference*.

**charge carrier** See *carrier*.

**charge-coupled device** A digital circuit that can store and process signals. Used to enhance video detail. Important in astronomy and in computer vision. See *computer vision*.

**Charles' Law** A principle that applies to ideal gases. If the mass and pressure do not change, then the gas expands at a constant rate as it is heated. The volume increases by about 0.00366 of its volume at 0 degrees centigrade (C) for each degree C that the temperature rises.

**charm** A characteristic that some elementary particles are thought to have. In particular, certain quarks have this property. See *quark*.

**chart** 1. A graph, especially on the Cartesian plane. See *Cartesian plane, graph.* 2. A table. 3. A map used for navigation. 4. See *histogram.* 5. See *pie graph.*

**chassis** 1. In electronics the grounded, or common, usually metal, structure to which all components are attached. 2. The main rigid supporting structure in a motor vehicle.

**chelation** 1. A process where minerals are bound up with other substances to make them more easily absorbed by the body. 2. A means of breaking down plaques in atherosclerosis. See *atherosclerosis, plaque.*

**chemical** 1. In general, any element or compound. 2. Pertaining to elements, compounds, mixtures, and solutions. See *compound, element, mixture, solution.* 3. An artificially manufactured substance.

**chemical bond** The force that keeps atoms together in a compound. Generally, the atoms share electrons. Atoms with extra electrons are especially likely to bond with atoms having a deficiency of electrons in their outer shells. See *electron.*

**chemical combination** Two or more different elements bonded together as a compound. See *chemical bond, compound.*

**chemical dependency** See *addiction, addictive substance.*

**chemical engineering** That branch of engineering concerned with the applications of chemicals and chemical reactions. See *chemistry.*

**chemical equation** A means of expressing a chemical reaction in a form similar to a mathematical equation. We might say, for example, $C+O=CO$, meaning that one atom of carbon combines with one atom of oxygen to form a molecule of carbon monoxide.

**chemical reaction** 1. The combination of two or more elements to form a compound or compounds. This might be accompanied by a release of energy. 2. When two or more compounds are brought together, a conversion into new compounds and possibly also free elements. 3. Dissociation of compounds into the elements that make them up.

**chemistry** The branch of science dealing with the behavior of elements, compounds, mixtures, and solutions and with the nature of the reactions that occur among elements. See *inorganic chemistry, organic chemistry, physical chemistry.*

**chemosynthesis** The production of carbohydrates, proteins, and fats from chemicals such as carbon dioxide and hydrogen sulfide. This process occurs in deep-sea bacteria. It is important in the food chain for deep-sea life and, perhaps, for all ocean life and therefore for all life on the earth.

**chemotherapy** 1. A method of treating cancer by means of special chemicals introduced into the bloodstream. The chemicals slow down, and might even reverse, cancer growth. See *cancer.* 2. The use of chemicals to treat diseases. More often called drug therapy.

**chinook** A wind that blows down from mountains into valleys, especially in the springtime. These winds are warm and dry, and can reach hurricane force in gusts. They are known for the speed with which they melt snow. See the drawing.

CHINOOK

**chip** A piece of semiconductor material, such as silicon, on which circuits are etched. See *integrated circuit.*

**chlamydomonas** An algae with a flagellum that allows it to move toward light, where photosynthesis occurs. Thus it has characteristics of both an animal (self-propulsion) and a plant (photosynthesis).

**chloride** A compound of an element with chlorine. Examples are sodium chloride (table salt), potassium chloride (used as a salt substitute), copper chloride, cobalt chloride, and many others. See *chlorine*.

**chlorination** The addition of chlorine to water to kill bacteria and viruses. This is a common practice in drinking water reservoirs and swimming pools. See *chlorine*.

**chlorine** Chemical symbol, Cl. An element with atomic number 17. The most common isotope has atomic weight 35. It is highly reactive. It combines with many of the same elements that oxygen reacts with. It is a gas at normal temperatures and pressures. It is used to kill bacteria and viruses in swimming pools and in drinking water. See *chloride, chlorination*.

**chlorobenzene** A flammable liquid used as a cleaner. As its name implies, it is a compound of chlorine and benzene. See *benzene, chlorine*.

**chloroethane** A flammable gas, used in making the lead compounds in some gasolines. This practice is becoming less common because the burning of such gasolines releases lead into the atmosphere, and lead pollution has been found dangerous.

**chlorofluorocarbon** Abbreviation, CFC. Chemicals used in some aerosol sprays and for various industrial purposes. Believed to be causing depletion of the ozone layer (see *ozone depletion, ozone hole, ozone layer*). Efforts are underway to restrict the use of CFC.

**chlorophyll** A substance in plants that converts light into energy (see *photosynthesis*). It is usually green or blue-green, and gives plants their color. It is used in some breath deodorizers.

**chloroplast** An organelle in a plant cell containing chlorophyll. See drawing. Responsible for photosynthesis. See *chlorophyll, organelle, photosynthesis*.

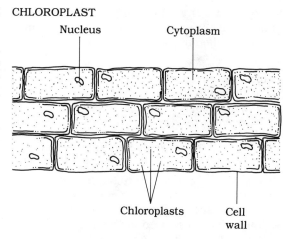

CHLOROPLAST

**choke** A device used to block the flow of alternating currents above a certain frequency. Usually it is a coil that has low loss for direct current, but greater and greater loss for alternating currents of increasing frequency. See *inductance, inductive reactance*.

**cholera** An infection of the small intestine caused by a bacillus type bacteria. Symptoms include diarrhea, fever, and dehydration. Severe cases can be fatal. More common in the less developed nations and in tropical climates.

**cholesterol** A waxy, complex substance found in animal tissues. It is manufactured in the body even if it is not eaten in the diet. Important in nerve function and hormone production. Also found in arterial plaques. See *plaque*.

**cholesterol level** The amount of cholesterol in the blood given in milligrams per 100 milliliters (mg%). The normal level is a debatable issue. A good level is 170 mg%. The cholesterol level is best evaluated by a doctor.

**choline** See *vitamins*.

**chroma** A combination of color hue and saturation. All shades of gray, pure white, and pure black do not have chroma. It is a subjective quality, depending on wavelength and bandwidth. See *color, hue, saturation*.

**chromaticity** See *chroma*.

**chromatography** A means of telling what is in a mixture or solution. This can be done in a variety of different ways. The lab procedures are rather complex.

**chrominance** A measure of color used in television. The difference between a given color and a standard color with the same brightness. See *chroma, color, hue, primary colors, saturation.*

**chromite** The raw ore of chromium. See *chromium.*

**chromium** Chemical symbol, Cr. An element with atomic number 24. The most common isotope has atomic weight 52. In its pure form it is a shiny metal. It is used for plating because it resists oxidation. Trace amounts are needed in the human diet; no recommended daily allowance has been established.

**chromosome** In a cell's nucleus, threadlike parts that contain the information necessary for cell duplication. See *cell, deoxyribonucleic acid, gene, genetic code.*

**chromosphere** The atmosphere of the sun, several thousand miles thick. It is above the photosphere but below the corona. See *photosphere, solar corona.*

**chronograph** A graph in which time is the independent variable. We might have temperature versus time, or current versus time, for example. The time scale might vary from millionths of a second to millions of years or more.

**chronometer** 1. Any device that measures and gives an indication of the time, such as a clock, a wristwatch, or an atomic time standard. Nowadays quartz oscillators are used in most battery-powered clocks and watches; electric clocks work from the 60-hertz utility line current. 2. The clock on a ship, used for navigation.

**chronon** An elementary unit of time, invented by physicists. If there is such a thing as an elementary particle, then a chronon is the time (t) that it takes for a photon to travel across the particle (distance d) at the speed of light, c, as shown in the drawing. Also called a jiffy.

**chyle** Lymphatic fluid and emulsified fats found in the small intestine. It is this substance that is absorbed and allows fat to be utilized. See *lymph, lymphatic duct.*

**chyme** Food that has been chewed, churned, and mixed with stomach acid, and that is ready to pass into the small intestine for absorption.

**Ci** Abbreviation for curie.

**cilium** 1. One of the many tiny, hairlike objects that move mucus in the bronchi and lungs (see *bronchi, lung*). 2. Any of the many microscopic, fingerlike projections in the intestine that increase available surface area for food absorption. 3. Hairlike objects on the outsides of some cells that help the cells move around. See *cell.*

**cinnabar** The raw ore from which mercury is obtained. See *mercury.*

**cipher** 1. The numeral zero. 2. A code, used to ensure that only the intended receiving party gets the message. Especially, in reference to a digital code.

**cipherization** The process of converting a message into a cipher. Especially applies to digital messages. See *cipher.*

**circadian rhythm** A daily cycle in activity of living things. Most plants and animals show some sort of daily pattern. Humans are a good example; we sleep and eat at just about the same times every day (24 hours).

**circle** The set of all points in a plane equidistant from some specified point. The specified point is the center, and the distance is the

CHRONON

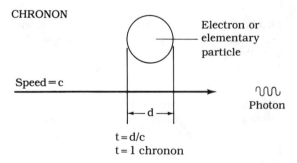

Speed = c
t = d/c
t = 1 chronon
Electron or elementary particle
Photon

radius of the circle. See *circumference, diameter, pi, radius*.

**circuit**  1. A complete path for the flow of electric current. 2. A set of electronic components interconnected to perform a certain function. We might have an oscillator circuit or an amplifier circuit, for example.

**circuit board**  A flat piece of phenolic, plastic, or glass-epoxy material on which electronic components are mounted. Nowadays the wiring is usually printed on the board. See *printed circuit*.

**circuit breaker**  A device that interrupts the flow of current in a circuit (opens the circuit) if the current exceeds a certain value. Protects components from damage by high current. Also prevents fire. Can be reset when it opens, so replacement is not needed. This makes it more convenient than a fuse, but it is also more expensive. See also *fuse*.

**circular measure**  The measure of an angle in radians. One radian is about 57.3 degrees of arc. See *radian*.

**circular mil**  An expression for the cross-sectional area of a wire or optical fiber. The area of a circle with a diameter of 0.001 inch. This is about 0.785 millionths of a square inch, or 0.000000785 square inch.

**circular polarization**  A means of transmitting an electromagnetic field to reduce fading. The electric flux lines go around and around, so a receiving antenna will pick them up fairly well no matter how it is oriented. See also *electromagnetic field, polarization*.

**circulation**  Any flow of fluid that eventually goes around in a complete cycle. We might speak of the circulation of water in the oceans or of blood in the blood vessels. Sometimes also refers to electric current in a closed circuit. See *circuit*.

**circulatory system**  The heart, blood vessels, and the blood that flows through them. Also can refer to the lymphatic system. In the drawing, the heart and major arteries are shown. The veins follow basically the same pattern as the arteries. Arteries run deeper and veins are closer to the skin. See *artery, blood, capillary, heart, lymphatic system, vein*.

CIRCULATORY SYSTEM

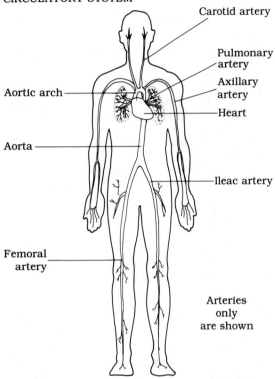

**circumference**  1. The distance around an object, such as a square or circle; the perimeter. 2. The distance around a circle, equal to about 3.14159 times the diameter. See *diameter, pi*.

**circum-Pacific belt**  See *ring of fire*.

**cirque**  A natural, roughly circular depression or excavation in a mountain. The term probably derives from the Latin "circus," or amphitheater. Usually the result of glacial action.

**cirrhosis**  A degenerative disease of the liver. Occurs when the liver has been injured, either all at once or over a long time. The normally smooth tissue becomes gristly and scarred. The patient develops various symptoms such as jaundice and a tendency to bleed internally. See *liver*.

**cirro-** High up. See *cirrocumulus clouds, cirrostratus clouds, cirrus clouds*.

**cirrocumulus clouds** Clouds at high altitude that resemble fine curds of milk. See drawing.

**cirrostratus clouds** Clouds at high altitude that make the sky appear washed-out or milky. The sun or moon is visible and is often surrounded by a halo. See drawing.

**cirrus clouds** Wispy clouds at high altitude, often called "Mares' tails" because of the way they look. Made up of ice crystals. See drawing.

CIRROCUMULUS CLOUDS
CIRROSTRATUS CLOUDS
CIRRUS CLOUDS

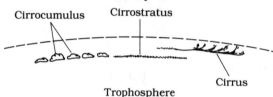

**Citizen Band** A channelized band for two-way communications between citizens of the United States. Ranges from 26.960 megahertz (MHz) to 27.410 MHz, in 40 channels, each 10 kilohertz wide. Amplitude modulation (AM) voice transmission is used.

**citrate** A compound formed with citric acid, sodium citrate, for example. Used in food processing.

**citric acid** A moderately weak acid present in citrus fruits. Important in the Krebs cycle. See *citrus fruit, Krebs cycle*.

**citric-acid cycle** See *Krebs cycle*.

**citrus fruit** Any of various acidic fruits, including oranges, grapefruit, lemons, and limes. Grown in subtropical climates. They are good sources of vitamin C. See *fruit*.

**civil engineering** A branch of engineering concerned with design, manufacture, and construction of roads, bridges, water distribution systems, and other essentials of civilization.

**Cl** Chemical symbol for chlorine.

**cladding** A coating or covering. In an optical fiber, it serves to keep the light beam inside the fiber (see *fiberoptics*). In wire, such as copper-clad steel, the coating has low electrical resistance and the inner core lends strength.

**clade** A class of living things all having the same individual ancestor or pair of ancestors, according to the theory of cladistics. We might say, for example, that humans are a clade, all descended from a single couple. See *cladistics*.

**cladistics** A method of classifying living things, based on the idea that new species are created in a single generation, rather than gradually over many generations. Popular with Creationists. See *Creationism, Evolutionism*.

**Clark cell** An electrochemical cell used as a standard, producing 1.4322 volts at a temperature of 15 degrees centigrade. Has zinc and mercury electrodes, and an electrolyte of mercurous sulfate. See *battery*.

**Clarke, Arthur C.** A noted writer of science fiction. He is the author of such bestselling novels as *Rendezvous with Rama*.

**class** 1. In mathematics a set, especially in the larger sense, such as a set of sets. See *set, set theory*. 2. A group with certain common characteristics. A class of animals, for example, or a class of star types.

**classical mechanics** See *Newtonian mechanics*.

**classical physics** See *Newtonian physics*.

**classification** 1. The process of dividing into classes. 2. See *class*.

**clavicle** A bone connecting the upper rib and the shoulder. Sometimes called the shoulder bone or shoulder blade.

**clay** Fine, moist, tightly packed particles, found in soils in some locations. Usually made up of iron silicate, aluminum silicate, or magnesium silicate. Ranges in color from bluish-gray to red. Clay has been used to make utensils for centuries. Can also be used to make insulators, and to help soil hold water where plants are cultivated. When heated it becomes hard and durable.

**cleavage**  1. In a crystal the plane, or pattern of planes, along which fracture occurs. Determined by the molecular structure of the crystal. See the illustration. See also *crystal, crystal lattice*. 2. Repeated cell division of a fertilized egg; the initial stages of embryo development. See *embryo*.

CLEAVAGE

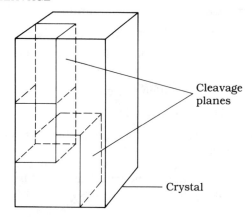

**clepsydra**  A device for keeping time, developed in the third century B.C. by a Greek named Ctesibius. The device used water that dripped through a hole into a system of vessels, where the rising water level indicated the passage of time. It was devised so evaporation wouldn't make any difference.

**climate**  1. The nature of the weather in an area over a very long time. Temperatures, rainfall, and amount of sunshine are all factors to consider. 2. The general nature of weather on the entire planet earth. 3. The nature of weather on some planet besides the earth.

**climate control**  1. Interior heating and air conditioning. 2. Attempts to affect the weather, for example by cloud seeding or by blowing away a fog.

**climate modeling**  The use of a computer to simulate large-scale weather patterns. The scientist can program the computer with data that results in an ongoing simulation, for hours, days, or even months. High-pressure areas, low-pressure areas, storms, and other weather phenomena develop and mature. Sometimes the computerized climate stabilizes, and sometimes it fluctuates chaotically like the real weather. Used in an effort to better understand and predict actual weather and climate patterns.

**climatology**  The study of climate and how it might be affected by the activities of humans. See *global warming, nuclear winter, ozone depletion*.

**clitoris**  A small, sensitive projection of tissue in the female, the counterpart of the male penis. Believed mainly responsible for female orgasm.

**clock**  1. A device that gives a continuous reading of time, usually down to the second but sometimes to 0.1 or 0.01 second. Modern clocks use quartz crystals and are accurate to better than 1 second per month. Electric clocks depend on the 60-hertz utility line frequency for their accuracy. Some mechanical (windup) clocks still exist. They are far less precise. 2. The time standard in a digital circuit, such as a computer. 3. An electronic pulse generator.

**clone**  1. A life-form produced by cloning. See *cloning*. 2. To reproduce by cloning.

**cloning**  Reproduction by artificial means, using a single cell to obtain the genetic data. The most common example is the use of plant cuttings to get new plants. Recently there have been attempts to reproduce more complex life forms in this way, using single cells in place of the eggs. Bizarre visions have arisen, such as the use of this technique to breed armies of strong soldiers that are all exactly alike. We're still a long way from doing things like this, however. Many scientists doubt that cloning of humans is possible.

**closed circuit**  1. A circuit in which current flows. Classical current (according to formal physics) flows from plus to minus. Electrons actually move the other way, from minus to plus. See *circuit*. 2. A television system in which signals are sent only over wires for a selected audience.

**close packing** With spherical objects, the method of packing them into a given volume so as to get the most spheres into the available volume. The result is an orderly geometric pattern. A good example is the way fruits are stacked in a grocery store.

**clot** A reddish-brown solid mass of dried or coagulated blood, consisting of fibrin and dead blood cells. Serves as a natural bandage. In an artery it might obstruct blood flow, causing tissue death. See *clotting, embolism, fibrin*.

**clotting** The process in which blood coagulates for the purpose of closing wounds. It is a natural defense mechanism to control bleeding. It also might occur when it is not wanted, such as in major arteries or capillaries. This can result in a heart attack or stroke. See *clot, embolism, fibrin, heart attack, stroke*.

**cloud** 1. A large group of water droplets or ice crystals, having condensed onto particles in the air. 2. Any of various formations in the atmosphere resulting from condensation. Refer to the definitions listed under *cloud types*.

**cloud chamber** A device for finding the charge and mass of an atomic particle. See the drawing. The chamber contains a vapor that causes the particle to leave a trail. A magnetic field makes charged particles follow curved trails. Negatively charged particles curve one way; positive ones curve the other way. Neutrons go straight. Heavier particles curve less than lighter ones. The physicist can tell, knowing the particle speed, what type of particle is passing through the chamber.

**cloud cover** 1. A layer of clouds that makes the sky partially or totally invisible. 2. The percentage of the sky that is obscured by clouds.

**cloud formation** 1. The condensation of water droplets or ice crystals on particles in the air. 2. Any of various shapes or types of clouds. Refer to the definitions listed under cloud types.

**cloud types** See *altocumulus clouds, altostratus clouds, cirrocumulus clouds, cirrostra-* *tus clouds, cirrus clouds, cumulonimbus clouds, cumulus clouds, fog, fractocumulus clouds, lenticular clouds, mammatocumulus clouds, nimbostratus clouds, noctilucent clouds, stratocumulus clouds, stratus clouds, vapor trail*.

**clubmoss** A type of plant that proliferated during the Mississippian and Pennsylvanian periods about 350 million to 290 million years ago. Still exists today. Grows close to the ground and is recognizable by its cones containing spores. See the GEOLOGIC TIME and PLANT CLASSIFICATION appendices.

**Cm** Chemical symbol for curium.

**cm** Abbreviation for centimeter.

**CMOS** Abbreviation for complementary metal-oxide semiconductor.

**CNS** Abbreviation for central nervous system.

**Co** Chemical symbol for cobalt.

**coagulation** Clumping together unevenly, such as curdling in milk or the initial stages of blood clotting. Might occur because of dehydration or because of cold temperatures, or in some chemical reactions.

CLOUD CHAMBER

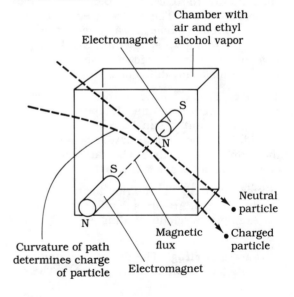

**coal** Carbon, in deposits beneath the earth's surface, that is the result of the decay of plants over many thousands or millions of years. Used as a fossil fuel. See *anthracite, bituminous, lignite, peat*.

**coalescence** Growing together or merging because of the growth of individual parts. Especially pertains to individual cells or components that work together as a unit.

**coal tar** A product of coal, usually bituminous, used in making steel and also in the manufacture of certain drugs. Resembles the tar used in highway construction or roofing. See *bituminous*.

**coastline** The irregular line marking the boundary between land and water, especially in the ocean. Coastlines are always changing because of erosion and because of continental drift. A coastline is fractal in that it appears irregular at every scale. The illustration shows this. See *continental drift, erosion, fractal*.

COASTLINE

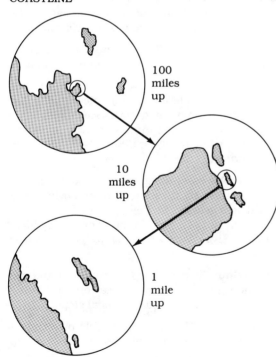

**coaxial** 1. Having the same axis. 2. Concentric in cross section. See *coaxial cable*.

**coaxial cable** A cable with a tubular shield surrounding a central conductor. The shield is generally grounded and the center conductor carries the signal. This generally forms an unbalanced transmission line. See *cable*.

**cobalt** Chemical symbol, Co. An element with atomic number 27. The most common isotope has atomic weight 59. In its pure form it is a gray metal. Used in the manufacture of stainless steel. The isotope Co-60 is radioactive and is used in radiology. See also *radiology*.

**cobalt chloride** A compound consisting of cobalt and chlorine. Known for its ability to absorb water vapor. When dry it is bluish and when saturated it is pink. Sometimes used as a primitive means of weather forecasting: blue means fair, pink means rain. (Of course, looking out a window works just as well.) Also used to absorb moisture in bottles of pills.

**COBOL** A computer language used in business. Acronym for Common Business Oriented Language.

**coca** A plant with leaves that contain a stimulant. When these leaves are chewed, a person can go for a long time without food seeming to have great stamina and strength. Grows in tropical places. Cocaine is refined from it. It is illegal to grow these plants in the United States. See *cocaine*.

**cocaine** A white powder, refined from coca plants. Causes intense stimulation. Highly addictive, it causes paranoia and depression with continued use. Dangerous because a single dose can sometimes cause death. It is illegal in the United States.

**coccus** A bacterium, spherical in shape (see *bacteria*). Some varieties cause infections such as strep throat (streptococcus) and pneumonia (pneumococcus).

**coccyx** The tip of the spinal cord in vertebrates that have no tails.

**cochlea** 1. In the inner ear, a snail-shaped cavity that converts the physical vibrations of sound into impulses transmitted by the auditory nerve to the brain. See the drawing. 2. A snail.

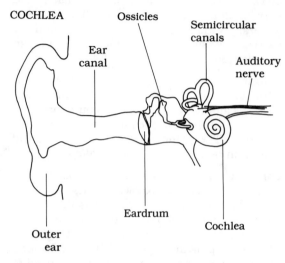
COCHLEA

**Cockcroft, Sir John** A physicist who worked along with Ernest Walton during the 1930s to build the first particle accelerator to split atoms. See *particle accelerator*.

**cocoon** 1. A pouch or housing for eggs of insects and various other primitive animals. 2. The silk housing in which a moth or butterfly matures. It enters as a worm and leaves as a flying insect.

**code** 1. A cipher, or a cryptic means of sending a message. See *cipher*. 2. A digital means of sending a message. See *ASCII, BAUDOT code, decoder, encoder, Morse code*. 3. Genetic information, contained in cells.

**codeine** A narcotic, derived from opium (see *opium*). Used to relieve pain. Available by prescription in some aspirin or aspirin-substitute tablets. Also sometimes used to suppress coughs.

**coding** The process of making up a code. Some codes use single words (groups of characters) such as "QRX" to represent entire sentences. Others use digital symbols to represent characters. See *code, decoder, encoder*.

**coefficient** 1. A multiplier in an equation, as $y = mx + b$ (slope-intercept form for a straight line). The slope, m, appears as a coefficient. 2. Degree or extent, quantitatively expressed for a physical phenomenon. We might have, for example, the coefficient of coupling between two coils, or the temperature coefficient of a capacitor.

**coelecanth** A fish that was thought to have been extinct for 70 million years, but was recently found alive off the coast of South Africa. Called a "living fossil." Shows that sometimes we mistakenly assume certain species are extinct.

**coelenterate** An invertebrate ocean-dwelling animal, including coral, jellyfish, hydra, and sea anemone. These dominated ocean life at the beginning of the Phanerozoic eon. See the evolutionary position of these animals; see the GEOLOGIC TIME appendix to locate the beginning of the Phanerozoic eon relative to other times.

**coelom** The major body cavity of most invertebrate animals.

**coenzyme** A chemical important in biochemical processes. Some of these substances are made from vitamins. See *enzyme, vitamins*.

**coercive force** The demagnetizing force needed to completely remove all of the magnetism from a substance.

**cofactor** 1. In biology, a chemical that is necessary for an enzyme to work. See *enzyme*. 2. A characteristic of a number matrix in matrix algebra or linear algebra. See also *linear algebra, matrix*.

**cog** 1. An essential part in a machine, as a gear, a ratchet, or some similar device. 2. A tooth on a gear or ratchet. 3. Slang for a component, one out of many similar to or exactly like it in a massive structure or machine. See *component*.

**cognition** In psychology, awareness and recognition by an animal or a human being. Thoughts that proceed by logical steps.

**coherent** 1. Clear; easily visible or recognizable. 2. Having wavefronts that are all lined up. See *coherent light, coherent radiation*.

**coherent light** Visible light of one color in which all the wavefronts are lined up. This produces maximum light intensity, and makes it easy to collimate. Lasers produce coherent light. See *coherent radiation, collimation, incoherent light, laser.*

**coherent radiation** Any radiation in the form of waves where the wavelengths are all identical and the wavefronts are exactly lined up (see drawing). This applies especially to electromagnetic radiation. Coherent radiation has theoretically zero bandwidth and travels with the greatest possible efficiency (least possible path loss). See also *incoherent radiation, laser, maser.*

COHERENT RADIATION

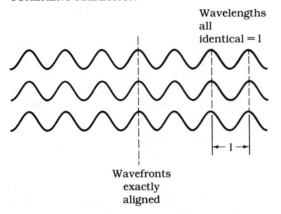

Wavefronts exactly aligned

**coil** 1. A helically-wound length of wire providing inductance. In general, the more turns in the helix, the larger the inductance. The inductance can be increased by putting powdered iron inside the helix. See *inductance, inductive reactance.* 2. A helically-wound length of tubing through which fluid flows. Used for heating and cooling because radiation or absorption is enhanced.

**coitus** Sexual intercourse, especially between human beings.

**coke** 1. A moderately refined form of coal, used for fuel in some applications, especially in industrial processing. See *anthracite, bituminous, lignite.* 2. Slang for cocaine. See *cocaine.*

**cold front** See *weather front.*

**colic** Cramps in the intestine, especially the colon (see *colon*). Often occurs along with diarrhea or constipation. Can be caused by nervousness, but might indicate a serious condition.

**coliform count** A common indicator of water pollution. The number of coliform bacteria per cubic centimeter (milliliter) of water. Pollutants cause this count to rise; the bacteria actually metabolize dangerous substances into harmless ones. See *water pollution.*

**colitis** Inflammation of the colon. Might be accompanied by diarrhea and rectal bleeding, colic and general exhaustion. The ulcerative type needs a doctor's care. See *colic, colon.*

**collagen** A protein that forms the major material making up a mammal's body. About 33 percent of the protein in the human body is bound up in this substance.

**collapsar** An object, usually a dead star, undergoing gravitational collapse. Might eventually become a black hole. See *black hole, gravitational collapse, neutron star.*

**collective dose** A measure of absorbed radiation in a large population. The average dose times the number of exposed people. Or the sum total of all the individual doses. Used for estimating probabilities that an individual will get cancer (or other illness) from radiation exposure. See *dosimetry.*

**collector** 1. In a bipolar transistor, the electrode towards which the charge carriers flow. Usually (but not always) the output is taken from this electrode. See *base, emitter, transistor.* 2. A basin or vessel in which samples accumulate for analysis later.

**collimation** Making rays of light parallel. This can be done with a convex lens or with a concave mirror (see drawing). It can also be accomplished with a laser. See *collimator, laser.*

**collimator** 1. A device, usually a lens or set of lenses, or a mirror that makes light rays parallel. 2. A reflector, such as a dish, that makes high-frequency radio waves nearly parallel for transmission over long distances.

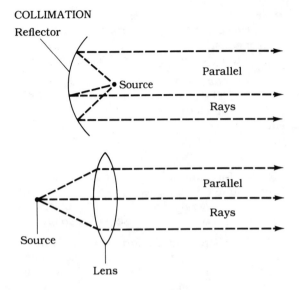

COLLIMATION

**Collins, Michael** The astronaut who manned the command module of the Apollo 11 mission. He orbited the moon while Neil Armstrong and Edwin Aldrin landed and carried out their experiments on the moon's surface. See *Aldrin, Edwin* and *Armstrong, Neil*.

**collision** The impact and resulting interchange of energy and momentum between two objects. 1. Elastic collision: The sort that occurs between steel bearings or hard balls like billiard balls. The total kinetic energy does not change. 2. Inelastic collision: The kind that takes place when one or both objects are soft, so that one of them absorbs some of the impact. As when a wad of wet gum hits a steel bearing. The total kinetic energy changes; some might be converted into heat or radiation.

**colloid** 1. A suspension in which the particles are extremely tiny and won't settle out. See *suspension*. 2. A mucus-like body fluid.

**colon** The large intestine. It is sometimes divided into three parts: the caecum and the ascending colon (see *caecum*), the transverse colon, and the descending colon or sigmoid (see *sigmoid*). The waste enters the caecum from the small intestine; from the sigmoid it passes into the rectum. See *rectum*.

**colony** 1. A set of animals of the same species, living in a community and working together for mutual survival. Beehives and anthills are good examples. 2. Bacteria, viruses, or yeasts grown from a few "starter" organisms, as in a lab culture.

**color** 1. The wavelength of visible light. 2. A subjective, perceived quality of visible light, that is a function of wavelength or combinations of wavelengths, and also of bandwidth or saturation. See *hue, saturation*.

**color blindness** 1. Total inability to perceive color. It is thought that rabbits see only in black and white and shades of gray. 2. A defect in vision that causes confusion between colors (such as red and green), but not total inability to perceive color. See *color, hue, saturation*.

**colorimeter** A device that measures color. Usually this is done by comparison. Red, blue, and green lights (the primary colors) can be adjusted in intensity and combined till the resulting color looks just like the one under test. Then the levels of red, blue, and green are read from calibrated dials. See *primary colors*.

**color television** The transmission and reception of television signals in color. This is basically the same as black-and-white television, but with the added color carrier and modulation needed to reproduce color from red, blue, and green (the primary colors). See *primary colors, television*.

**color vision** The ability to perceive different wavelengths of light. In true color vision, the spectrum cast by a prism seems to have constantly changing hue all along its length, with no two parts of the spectrum looking the same. Some animals, such as insects, perceive color differently than humans; insects can see ultraviolet light but are blind to the red end of the spectrum that we see. See *color, hue, saturation, visible spectrum*.

**colostomy** A surgical procedure in which part or all of the colon is bypassed. See *colon*.

**coma** A state of unconsciousness associated with some illnesses or serious injury. Can follow a seizure (see *seizure*). Can last for hours, days, weeks, or months. Sometimes patients recover without apparent reason.

**combustible** Capable of being burned or oxidized. Flammable. See *combustion*.

**combustion** Rapid oxidation. More commonly called burning. Can occur as an explosion. When controlled, it allows gasoline engines or other fossil-fuel devices to work. Our society is dependent on controlled combustion devices.

**comet** An interplanetary object, usually with an elongated orbit that periodically brings it near the sun. When it comes near the sun, the solar wind and heat cause a "tail" to develop. Halley's Comet was observed up close by a space probe during its 1985-86 visit. Astronomers said it looked like a big potato. The drawing shows a comet structure. Most comet heads are just a few miles across, but the tails might stretch for millions of miles. See *Halley's Comet*.

COMET

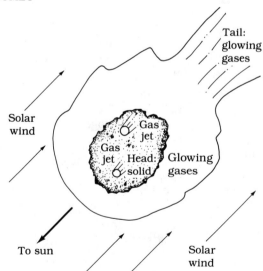

**command** 1. An input to a computer, telling it what to do. Might be from the operator, from another computer, or from some storage medium. 2. Control, such as when we speak of the command module in a space mission.

**command module** In the Apollo missions, the part of the spacecraft that remained in orbit around the moon while the landing craft descended and carried out its assignments. Future moon missions and trips to the planets will probably be done in a similar way. See *Apollo*.

**communications** Transmission of signals, in one direction, two directions, or via several or many paths. Can be electrical impulses sent by wire, or can be radio or fiberoptic signals.

**communications satellite** See *satellite communications*.

**community** The totality of animals and plants in a certain area or region. Usually named with reference to a certain species, such as the insect community. It is an informal categorization and can be combined with others or split up into smaller sets for various purposes.

**commutative law** In mathematics, a property that an operation might or might not have. Suppose we have an operation, $*$. Then $*$ is commutative if, and only if, for every two variables x and y, it is true that $x*y=y*x$. Addition and multiplication are commutative. Subtraction and division are not.

**commutator** A device that converts direct current (dc) to alternating current (ac) or vice-versa. Used in a dc motor; also can be used in a generator to produce dc from the generated ac. See *generator, motor, motor/generator*.

**compact disk** A small disk containing voice and music information, or digital codes. It is played, or read, by a device that uses a laser. The laser scans the fine grooves on the disk to recover the information. See *laser*.

**compandor** A communications device that compresses a signal at the transmitting end (see *compression, speech compression*), and then expands it back again to normal at the receiving end. This improves the intelligibility of some signals. See also *expansion, intelligibility*.

**comparative anatomy** A branch of medicine dealing with similarities and differences in human anatomy. See *anatomy*.

**comparator** In electronics, a circuit that compares two signal levels, voltages, frequencies, or other parameters. The output depends on which signal is louder or higher in frequency. The output can be analog or digital.

**compartmentalization** Division into units or cell-like enclosures. Plant cells are compartments separated by cell walls. Some life forms grow by adding on more and more compartments; an example is the chambered nautilus.

**compass** A device that indicates azimuth with respect to magnetic north. The needle is magnetized and lines up with the geomagnetic field. See the drawing. True north is usually not the same as magnetic north, so a correction factor must be included when using a compass for navigation. See *declination, geomagnetic field*.

COMPASS

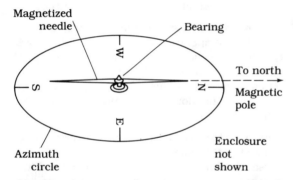

**compass plant** A type of plant that aligns its leaves to take the most advantage of the daily sunlight. You can determine true north from the directions the leaves are facing.

**competition** 1. A form of natural selection among animals with different traits or among plants in a given area. It is "survival of the fittest." This occurs especially when food is scarce or when populations are high. 2. In mating, a process that determines which male will win the favors of the female. (It sometimes works the other way around, but not very often.)

**compiler** In a computer, the program that converts the user language, such as BASIC or FORTRAN, into machine language for use by the digital circuits of the computer. See *BASIC, COBOL, FORTRAN, machine language*.

**complement** 1. The opposite of. 2. Something that completes a process or phenomenon. For an example see *complementary colors*.

**complementary colors** Two colors that produce white when they are shone together. See *complementary pigments, primary colors, primary pigments*.

**complementary pigments** Two pigments that produce black when mixed. See *complementary colors, primary colors, primary pigments*.

**complementary metal-oxide semiconductor** Abbreviation, CMOS. A technology used in integrated circuits. CMOS devices use very little current. But they are easily damaged by static electricity. See *integrated circuit, metal-oxide semiconductor, semiconductor*.

**complete protein** A protein that is 100-percent utilized by the body to build tissue. No protein is truly 100-percent complete when obtained from just one source, but foods can be mixed to obtain nearly complete proteins. We can mix beans and rice, or cheese and bread, for this purpose. Eggs have protein that is more than 90 percent complete all by itself.

**complex** 1. A complete, functional machine designed for a variety of tasks. 2. Interconnected machines working together. 3. Complicated. 4. A type of number. See *complex number*.

**complex number** The sum of an imaginary number and a real number, in the form $x+yi$, where $x$ and $y$ are real numbers and $i$ is the square root of $-1$. In this case, $x$ is called the real part of the complex number, and $yi$ is called the imaginary part. See *imaginary number, real number*.

**component** 1. A part that, along with others, makes up a whole. See, for example, *component vector*. 2. A unit with a defined function in a machine or electronic circuit. Examples are gears, resistors, levers, and meters.

**component vector** Any of two or more vectors that make up a vector. The drawing shows an example. If we are trying to fly 80 miles per hour (mph) north, but there is a wind towards the east at 60 mph, we will actually end up moving at 100 mph north by northeast. See *vector, vector diagram, vector sum.*

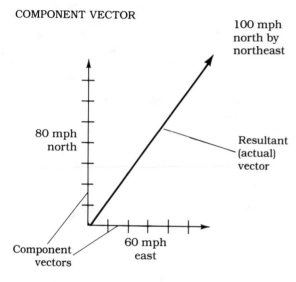

COMPONENT VECTOR

**composite** 1. A sum total. In the definition and drawing for component vector, the 100 miles per hour (mph) north by northeast is the composite, or sum, vector. 2. A combination or mixture.

**composite volcano** A volcano that erupts both cinder and lava. Also called a stratovolcano. The most famous, and largest, volcanoes in the world are of this type, such as Vesuvius, Mount Fuji, and Mount Saint Helens. See *volcano.*

**compost** The biological reduction of organic waste to humus. See *humus.*

**compound** Two or more elements held together by chemical bonds to make a molecule that might behave differently than any of the elements themselves. The molecules in water are an example. They have two hydrogen atoms combined with one oxygen atom. See *chemical bond, element.*

**compound eye** A visual organ consisting of many little eyes. The result is the ability to see in a wide angle. But this type of vision is not very detailed. The little individual eyes aren't as sophisticated as eyes like ours. Insects have compound eyes. See *eye.*

**compressibility** The ease with which a substance can be forced down into a smaller volume. Most gases are compressible; some solids are fairly compressible. But some substances are not. Water in liquid form is one example.

**compression** 1. Pressing into a smaller volume. Usually this causes increased temperature and pressure, but not always. 2. See *speech compression.* 3. Changing a signal so that the weak parts are amplified while the strong parts are not. See *expansion.*

**compressor** 1. A device that presses a substance into a smaller volume. An air conditioner has a compressor, for example. When the compressed gas expands again, it cools. 2. A device for speech compression. See *speech compression.* 3. A circuit that compresses a signal. See *compression.*

**Compton effect** When a photon strikes an electron, the photon loses some of its energy to the electron. The amount of change depends on the angle at which the photon hits (see drawing). This is seen as a wavelength increase (reddening). The more nearly head-on the collision, the greater the change. See *electron, photon.*

**computed tomography** See *computer-aided medical imaging.*

**computer** A high-speed calculator that does many logical functions. The range of applications is immense. See the computer-related definitions that follow, and also see *arithmetic logic unit, artificial intelligence, assembly language, BASIC, central processing unit, COBOL, compiler, firmware, FORTRAN, machine language, register, robot, robotics, software.*

**computer-aided medical imaging** A way to see cross sections of a human body, or of body parts, without cutting the body open. A set of X rays or magnetic images are taken from angles all around the body. Then these images are processed by a computer to get a crosswise-slice picture. This allows doctors to make diagnoses that would otherwise require surgery. See *nuclear magnetic resonance imaging, X-ray photography.*

**computer engineering** A branch of engineering dealing with the design, manufacture, and maintenance of computers. See *computer.*

**computer generations** A historical way to speak about computers. There are five generations usually mentioned:

> first generation 1946-1956
> second generation 1957-1963
> third generation 1964-1981
> fourth generation 1982-1989
> fifth generation 1990 and beyond.

Each generation has had computers that were faster, more compact, and more powerful than the ones before. See *computer.*

**computer graphics** The making of sketches, drawings, plots, maps, projections, and diagrams on a computer display. This is a help to scientists and engineers, because it enables them to see complex objects from various angles.

**computer image processing** See *computer vision.*

**computerized axial tomography** See *computer-aided medical imaging.*

**computer monitor** A cathode-ray tube or liquid-crystal display that shows the graphics and information from a computer. Some monitors have color and others are only gray or green. They have excellent detail (resolution) and resemble television sets.

**computer program** A procedure that tells a computer what to do. A computer has certain programs built into it, called firmware, and some that can be changed by the operator, called software. See *computer programming, firmware, software.*

**computer programming** The writing, debugging, and improvement of programs for computers at all levels. See, for example, *assembly language, BASIC, COBOL, FORTRAN, machine language.*

**computer science** A field of study concerned with the operation, application, and programming of computers. See *computer.*

**computer vision** The use of a computer to process visual images. A camera tube such as a vidicon (see *vidicon*) is used. A charge-coupled device can be used to improve detail. The signal is then fed to the computer, which can be programmed to recognize objects or to enhance certain details. Used especially in robotics.

**concave** Curved inward. See *concave lens, concave mirror.*

**concave lens** Also called a negative lens or a diverging lens. A lens with concave faces on one or both sides. Spreads the light rays as shown in the drawing.

CONCAVE LENS
CONCAVE MIRROR

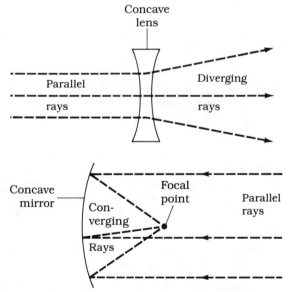

**concave mirror** A mirror with a concave surface. See the drawing. Used to focus parallel incoming rays; can also be used to collimate rays from a point source (see *collimation, collimator*). Used in reflecting telescopes as the objective, or main light-gathering mirror. See also *telescope*.

**concentrate** 1. A solution or chemical from which much or most of the inert ingredients, like water, have been removed. 2. To remove most of the inert ingredients from a chemical.

**concentration** A measure of the amount of a given chemical dissolved in a solution. Might be given as the number of grams per liter, for example, of the substance or as the number of milligrams per 100 milliliters (milligrams percent or mg%).

**conception** 1. Fertilization of an egg by a sperm cell. 2. An idea, or the formulation of an idea.

**condensation** Vapor that has liquified or solidified on a surface that is below the boiling point of the vapor. An example is the formation of water droplets on a cold tumbler.

**condensation nuclei** Airborne particles, usually less than 1 micron (0.001 millimeter) across, that act as collection points for water vapor in the atmosphere. Sources include natural dust and human-made particulate pollution.

**condenser** 1. A device that causes condensation (see *condensation*). 2. In optics, a lens or mirror that collimates or focuses light (see *collimation, collimator*). 3. A capacitor, especially the variable type (archaic use of the term). See *capacitor*.

**conditioning** 1. Processing for a certain use or purpose. 2. A learning process, where animals or humans are taught to respond in a certain way to stimuli. (When you see a hot stove, you don't touch the burners!) 3. Athletic training. Improves heart, lung, and muscle efficiency.

**conductance** 1. The extent (qualitatively) to which a material transfers heat. Steel has excellent thermal conductance. Wood is a poor thermal conductor. See *thermal conductivity*. 2. Symbol, G. A measure of the extent to which an object or component carries an electric current. The reciprocal of resistance. If the resistance is R, the conductance is:

$$G = \frac{1}{R}$$

Given in siemens. See *ohm, resistance, siem*ens.

**conduction** 1. The movement of charge carriers through a substance. See *electrical conductivity*. 2. The transference of heat through a material. See *thermal conductivity*.

**conductivity** See *electrical conductivity, thermal conductivity*.

**conductor** A material that has high conductance. Most metals are considered conductors, both of heat and of electric current. See *conductance*.

**conduit** A pipe through which electrical wiring is run. Can be metal or plastic. Common in houses and buildings. Reduces the chance of fire if a short circuit occurs, and also makes the wiring easy to repair or replace.

**condyle** In a joint, the rounded end of one bone that fits into the socket of the bone next to it.

**cone** 1. A three-dimensional geometric figure, obtained by rotating a line around an axis that intersects the line. The axis is at an angle to the line of more than 0 degrees but less than 90 degrees. 2. One half of this figure. 3. Any portion of this figure.

**cone of depression** A dip in the water table around a well. When water is taken from the well, the water table is lowered near the well, causing the cone-shaped dip. The faster water is pumped from the well, the sharper and deeper the cone gets.

**cone of protection** The region under a lightning rod where a strike is unlikely. See the drawing. This zone is anywhere in a cone whose apex is at the top of the rod, and within a circle whose radius is equal to the height of the rod.

**conference calling** In a telephone system, a connection arranged among three or more parties.

CONE OF PROTECTION

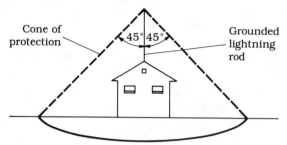

**congenital** Inherited. We speak of congenital traits, or defects, or behavior characteristics.

**congenital defect** A birth defect. It might not become apparent until some time after birth, but it can be traced to the genetic background of the person. Thus, doctors know that it isn't a sickness that was caught by means of germs.

**congestive heart failure** An illness where the heart does not pump enough blood. This causes fatigue, a feeling of coldness, and coughing as fluid accumulates in the lungs. This condition also might cause problems with the liver and kidneys. It is a serious condition and requires medical diagnosis and treatment.

**conglomerate** A type of rock with large hard fragments embedded. Might form as existing stones are caught up in volcanic lava that later cools and solidifies.

**congruence** The property of being exactly the same size and shape. For example, two triangles are congruent if the corresponding angles have the same measures, and if the corresponding sides have the same lengths.

**congruent** Having congruence; geometrically identical. See *congruence*.

**conic section** A circle, ellipse, parabola, or hyperbola. So named because a plane, cutting through a cone, always produces one of these shapes. You can make them for yourself, as shown in the drawing, by shining a flashlight at various angles onto a smooth pavement. See *circle, ellipse, hyperbola, parabola*.

**conifer** The pines, spruces, and fir trees. They have cones that contain spores for reproduction.

**conjugate** 1. To join together. 2. A unit resulting from the joining of two or more smaller units. 3. Chemicals in a compound, different only because one nucleus has one more proton than the other. 4. In complex numbers, the result of changing the sign of the imaginary part. The conjugate of $x+yi$ is $x-yi$. The square root of the product of a complex number and its conjugate equals the absolute value (magnitude) of the complex number. See *complex number*.

**conjugation** 1. Fertilization in lower animal forms; a joining of cells with exchange of matter between their nuclei. 2. A transfer of DNA (deoxyribonucleic acid) between bacteria.

**conjunction** 1. Lining up of planets with each other as seen from the earth. 2. Lining up of a planet with the sun. See *inferior conjunction, superior conjunction*. 3. The logical "and" operation.

**conjunctiva** The membrane joining the eyelid and the eyeball. Allows the eyelid to move over the eyeball, and also helps to lubricate the eyeball.

**connate water** Water deep in the earth's crust that has been there for millions of years. It is hot and under pressure; it contains dissolved minerals from ancient seas and from prolonged contact with rocks.

CONIC SECTION

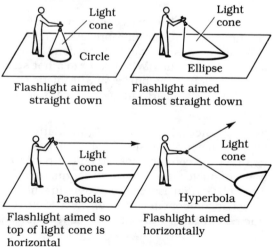

**connective tissue** Animal tissue giving support, bulk, and protection from trauma or cold. Holds the organs in place.

**connector** A device for electrically joining two cables, or for joining a cable to a piece of electrical or electronic equipment. There are many different types of connectors for various uses. Radio-frequency connectors must be made much more carefully than ordinary household plugs and outlets. See *cable*.

**consciousness** The quality of being aware of one's own existence, and to remember the past in detail. There is much debate over whether animals have consciousness, and at what age it begins in humans.

**conservation** 1. The property of neither gaining nor losing. For instance, the conservation of angular momentum or kinetic energy. 2. Taking care of the earth, its environment, and its resources so we can have a livable planet.

**conservation of energy** 1. A law of physics: When energy changes form, its capacity for doing work stays the same. 2. Use of energy resources in a nonwasteful manner.

**conservation of momentum** A law of physics: Momentum in a given direction always stays the same in a system, unless an external force intervenes. If the system is rotating, the angular momentum stays the same unless external torque intervenes. See *angular momentum, momentum*.

**Conshelf sphere** An underwater dwelling in which the pressure of the air inside is the same as the water pressure outside. Helium is used as the inert gas, rather than nitrogen. Makes it possible to use simpler construction techniques because there is no pressure difference between inside and outside.

**consonance** A form of resonance effect when one object vibrates because another nearby object is vibrating. This occurs when objects have identical or harmonically related resonant frequencies. See also *harmonic, resonance*.

**constant** 1. A quantity in physics that does not change. See the CONSTANTS appendix. 2. In mathematics, a quantity that is fixed as opposed to a variable. See *variable*.

**constantan** A copper-nickel alloy used in the manufacture of thermocouples and wirewound resistors. See *resistor, thermocouple*.

**constellation** A group of stars that form a pattern as seen from the earth. Ancient people assigned gods and animals to these patterns, to better remember them. Astronomers still use these names today. See *zodiac*.

**constipation** Sluggish passage of feces through the colon. Can be caused by dehydration, inactivity, or illness. See *colon*.

**consumption** 1. Use and processing for a specific purpose, such as production of energy. We might speak of fuel consumption in a car, for example. 2. Archaic term for tuberculosis. See *tuberculosis*.

**consumptive use** 1. Use of water so that it evaporates into the air. It does not return directly to its source, so the source supply drops. 2. Use of a natural resource (such as fossil fuel) so that the source supply diminishes.

**contact** 1. A terminal for an electrical connection. 2. A completed two-way communications circuit. 3. For physical objects, a condition of being so close that they are touching.

**containerization** A method of shipping in which items are packed in boxes of a precise size that all fit so as not to waste space. The contents are protected from jostling around and also from the weather.

**containment** 1. Confining of a hazardous substance within a defined area to protect people. Especially pertains to radioactive waste. 2. In a nuclear reactor, the use of a force field to keep the plasma away from the walls of the chamber.

**contaminant** 1. An unwanted chemical in a mixture or solution; an impurity. Might interfere with the desired behavior of the mixture or solution. 2. A foreign substance or poison in air, food, or water.

**contamination** The intrusion of a contaminant into a mixture, solution, or other substance. See *contaminant*.

**continent** A large landmass. On the earth there are six recognized major continents: North America, South America, Africa, Australia, Eurasia, and Antarctica. Eurasia is sometimes divided into Europe (west) and Asia (east). The continents float on the upper mantle and slowly drift around. See *continental drift, mantle*.

**Continental code** International Morse code. See *Morse code*.

**continental crust** The earth's crust in a continent. Generally thicker than the crust beneath the oceans, it is 25 to 30 miles deep in most places. The thickest crust on the planet (45 miles) is found in the region of the Himalaya Mountains in northern India, southwestern China, and Nepal.

**continental drift** 1. The theory that the continents were all joined sometime long ago, that they have drifted to their present positions by sliding over the earth's mantle, and that they will keep on drifting in the future. The drawing shows how the continents have moved to their present positions. 2. The moving of the continents according to the foregoing theory. See *Gondwanaland, Laurasia*.

CONTINENTAL DRIFT

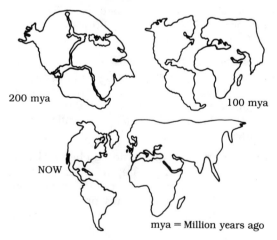

200 mya

100 mya

NOW

mya = Million years ago

From Erickson, THE LIVING EARTH (TAB Books, Inc., 1989).

**continental fault** A fracture line or boundary where different pieces of the earth's crust push or rub against each other. Mountains usually form in such places. In the United States, some of these places are southern Oregon, Nevada, western Utah, southern Arizona and New Mexico, and much of California.

**continental glacier** 1. A glacier that covers a large part of a continent, as in the ice ages. 2. A glacier on a continent. See also *glacier*.

**continental plate** A large section of basalt-like rock that floats on the earth's mantle, somewhat like a huge iceberg floats on the ocean, and comprises one of the land masses we call continents or large islands. See *continent*.

**continental shelf** A region of the ocean floor immediately surrounding a continent. The ocean is shallow here. Might extend just a few miles, or hundreds of miles, offshore, before the water deepens.

**continental slope** The ocean floor sloping from a continental shelf down into deep water farther out. Many different types of marine life are to be found here. The types of life vary as the water becomes deeper. See also *continental shelf*.

**continuity** 1. The extent to which a quantity is continuous. See *continuous*. 2. For a mathematical function, the property of being continuous. See *continuous function*. 3. In an electrical circuit, completeness, so that current flows.

**continuity tester** A device for testing a circuit to see if it is closed (current is flowing). An ohmmeter is probably the most common instrument used for this. See *ohmmeter*.

**continuous** 1. Having no abrupt changes or gaps; unbroken within a specified range. See *continuous function*. 2. Having no pauses; 100-percent duty cycle. See *duty cycle*.

**continuous function** A type of mathematical function. 1. A function $y = f(x)$ is "continuous at the point $(x,y) = (a,b)$" if and only if y approaches b as x approaches a, from either side. 2. A function is "continuous in an interval" if and only if it is continuous at every point in that interval. 3.

A function is "continuous" if and only if it is continuous at every point in its domain. (These are simplifications.) The illustration shows the qualitative meaning of this. See also *discontinuous function, function.*

CONTINUOUS FUNCTION

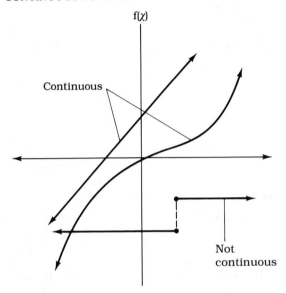

**continuous wave** Abbreviation, CW or cw. 1. An uninterrupted, unmodulated electromagnetic wave, usually at radio frequency. 2. Morse code sent by keying an otherwise unmodulated wave on and off.

**continuum** An unbroken line, plane, or space of three or more dimensions. Might be finite in extent, but has no boundaries or cuts. Space-time is an example of a four-dimensional continuum. A circle is a one-dimensional continuum; the surface of a sphere is a two-dimensional continuum.

**contour** 1. The elevation features of a landscape. 2. A line on a map joining points that have equal elevation. See *topography.*

**contour map** See *topographical map.*

**contour feathers** On a bird, feathers that are arranged in such a way that they help stabilize the animal's flight. According to the theory of natural selection, the bird's body has evolved by "trial and error" for optimum efficiency in the air, down to the last feather.

**contraception** The most common method of birth control in use today. Various drugs or mechanical devices are used to block the way of sperm in the uterus, or to render the female infertile, so that conception is unlikely. See *conception.*

**contract** Get smaller in volume. This can occur for various reasons, such as cooling, leakage of gas or fluid, or compression.

**contraction** To get smaller either because of an external force (compression) or on its own (as when a balloon is cooled down). See *expansion.*

**contradiction** 1. In mathematics or logic, the statement "X and not X," meaning truth and falsity at the same time. 2. A situation where the conclusions of different experiments conflict with each other.

**contraindication** In medicine, a term indicating that something should not be done or a medicine should not be used. For example, use of laxatives or cathartics is contraindicated when a patient has abdominal pain. (The appendix could rupture.) In other cases it might be recommended treatment.

**control** 1. In an experiment, a sample that is left alone so that effects of the variables can be identified and not mistaken for normal happenings. 2. Mechanical or electrical operation of a device by a human, a robot, or a computer. 3. An instrument, such as the pedals in an airplane.

**control system** 1. A set of instruments that allows an operator to make a device work properly and efficiently. Might consist of switches, knobs, dials, display screens, and in recent decades, one or more computers and possibly also robots.

**convection** 1. Rising and falling of air. Warm air rises and cool air falls. One example is shown in the drawing. This is one way in which heat is transferred. See *conduction, radiation.* 2. A circular flow of a fluid or gas caused by temperature differences. The mantle in the earth is thought to have convection currents as heated, molten rock rises and cooling, molten rock sinks. See also *mantle*.

CONVECTION

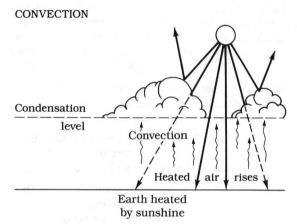

**convection current** The movement of a fluid or gas caused by convection. See *convection*.

**convection zone** 1. Any region where convection occurs. 2. The outer layer of hot gases in the sun. 3. A region of convection in the fluid core (outer core) of the earth or in the earth's mantle.

**convention** A standard method or notation for physical phenomena or mathematical equations. For example, a physicist would say that current flows from plus to minus by convention. A polynomial is expressed with exponents in descending size, by convention. See *conventional current*.

**conventional current** The physicist's expression for current flowing from plus to minus. The actual electrons move the other way, from minus to plus. Measured in amperes. See *ampere, current, electron*.

**conventional warfare** Armed conflict carried out using non-nuclear weapons, and not using any other types of weapons banned by international agreement.

**convergence** 1. Air movement into an area, such as takes place in a low-pressure weather system. 2. Focusing of light rays or charged-particle beams to a point, as in a telescope or a cathode-ray tube. 3. In a mathematical series, the property of being convergent (see *convergent series*). 4. The condition of a curve approaching an asymptote (see *asymptote*).

**convergent evolution** The fact that organs in different species can look very similar. The vascular system in a tree is shaped much like the circulatory system in a person, for example. Both systems do serve similar purposes.

**convergent series** An infinite series with a finite sum. An example is the series $1/2 + 1/4 + 1/8 + 1/16 + \ldots = 1$. As we add more terms, the sum converges or gets closer and closer to the final value. See also *divergent series*.

**conversion** 1. The expression of a quantity in different units. See *conversion factor* and the CONVERSION FACTORS appendix. 2. Changing one form of energy to another. See *energy conversion*. 3. Changing the frequency of a radio signal. See *superheterodyne radio receiver*.

**conversion factor** 1. A constant by which a quantity is multiplied, in order to express that same quantity in different units. See the CONVERSION FACTORS appendix. 2. A mathematical operation or formula for conversion between one type of unit and another. An example is conversion of electrical conductance C to resistance R. In this case, we take the reciprocal of one to get the other. That is,

$$R = \frac{1}{C}$$

and

$$C = \frac{1}{R}$$

**converter** 1. A device that changes one form of energy into another. An example is an electrochemical cell, converting chemical energy into electrical current. 2. A device that changes the frequency of a radio signal. See *superheterodyne radio receiver*. 3. A device for changing the format or speed of a signal. See *analog-to-digital converter, digital-to-analog converter*.

**convex** Curved outward. See *convex lens, convex mirror*.

**convex lens** Also called a positive lens or a converging lens. A lens with convex faces on one or both sides. Brings light rays together, or collimates rays from a point source. See the drawing. Used in refracting telescopes as the objective. See also *collimation, collimator, telescope*.

CONVEX LENS
CONVEX MIRROR

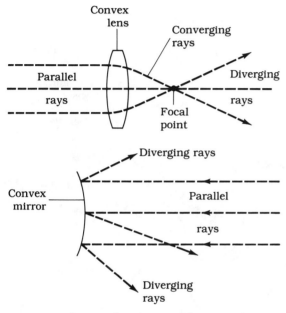

**convex mirror** A mirror with a convex surface. See the drawing. Used to spread light rays, just the way a concave lens does (see *concave lens*). Also used for wide-angle viewing in places such as convenience stores. Used in Cassegrainian reflector telescopes to direct the image to the eyepiece. See also *telescope*.

**coolant** 1. A substance that evaporates, or compresses and expands, easily and rapidly, and in so doing, lowers temperature. Freon serves this purpose in an air conditioner. 2. Any fluid that moves heat away from a machine to keep it cool. Water and air are often used for this.

**cooling** Any means of keeping a machine from overheating, by transferring excess heat away. This can be done by conduction, convection, radiation, or a combination of these. See *conduction, convection, radiation*.

**coordinate** 1. One of the numbers or variables in a system defining point locations. For example, 2 is the x coordinate of (x,y) = (2,6) in the Cartesian coordinate plane. 2. To cause to work together in an efficient way.

**coordinate geometry** See *analytic geometry*.

**Coordinated Universal Time** Abbreviation, UTC. Astronomical time based on the sun's position at 0 degrees longitude. The day starts at 0000 and ends at 2400 hours. The first two digits give the hour from 00 to 24, and the second two digits the minute from 00 to 59. UTC is based on the average solar rotational period of the earth. The earth is slightly behind UTC on June 1 and slightly ahead of UTC near October 1. The time in London, England is the same as UTC. In other parts of the world the time differs, usually by a whole number of hours, up to plus or minus 12 hours. In the United States, time is behind UTC by several hours. In Europe and Asia, time is ahead of UTC.

**coordinate system** Any means of uniquely defining points on a line, a plane, in three-space, or in space of more than three dimensions. Typically, in n dimensions (where n is a whole number), we need n coordinate values to uniquely define a point. See *Cartesian coordinates, curvilinear coordinates, log-log graph, polar coordinates, semilog graph, spherical coordinates*.

**coordination** 1. The smooth operation of a system, so that all parts function efficiently together. 2. The functioning of muscles in an efficient and smooth manner. 3. Alignment of a system so that it will work smoothly and efficiently.

**coordinatograph** A device used by car and truck manufacturers to help in building models of motor vehicles. The models are then tested for flaws.

**Copernicus, Nicholas** The Polish astronomer recognized for his theory that the sun, rather than the earth, is at the center of the Solar System. In his time, in the sixteenth century, this idea was resisted by the Church, then a powerful force. Aristarchus had recognized that the sun was the center of the Solar System long before. See *Aristarchus*.

**copper** Chemical symbol, Cu. A chemical element with atomic number 29. The most common isotope has atomic weight 63. In its pure form it is a shiny, reddish-gold metal. It oxidizes fairly easily. It is a good conductor of electricity and of heat and is employed in electrical wiring. It is fairly expensive, and for this reason, it has been replaced in some applications by aluminum.

**copper chloride** A compound of copper and chlorine. It is a white or brown solid. Forms on copper wires and utensils exposed to salt spray (sodium chloride) or chlorinated water.

**copper oxide** A compound of copper and oxygen. Generally it is red or dark brown in color. Forms as a coating on copper wires when they are exposed to the air for long periods. Humidity speeds up the corrosion process. Used to make some rectifier diodes.

**copper sulfate** A compound of copper and sulfur. It is bluish in color. Used in the manufacture of fungus killer. Has various industrial applications.

**coral** An ocean-dwelling animal that flourishes in tropical waters. Large coral reefs exist off the coasts of Australia, Madagascar, and various other places including the Florida Keys. In the Keys, human activity is killing the coral, which is causing concern among scientists and environmental activists. The reefs remain solid long after the coral dies. Coral tends to grow on its own remains, so that reefs are built up.

**coral reef** See *coral*.

**core** 1. Center or central region. 2. The interior of a coil, especially if it is powdered iron or ferrite material. 3. The inside part of a nuclear reactor.

**Coriolis effect** A phenomenon observed in large-scale wind patterns. Caused by the earth's rotation. In the Northern Hemisphere, inrushing air twists counterclockwise, and outflowing air twists clockwise (see drawing). There are small-scale exceptions. Tornadoes, for example, often result from wind shear (see *wind shear*), not from Coriolis force. In the Southern Hemisphere the wind twists in the opposite direction. Coriolis effect causes the generally east-west flow of the prevailing winds. Otherwise they would blow north-south.

CORIOLIS EFFECT

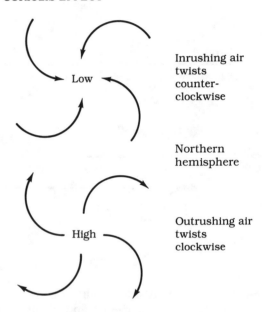

**cork** A light, durable substance formed by certain plants. Some forms of cork are used in consumer items, such as for bulletin boards and decorative ceiling tile.

**corm** In some plants, a means of survival between growing seasons. Similar to a bulb (see *bulb*). The unit remains underground over the winter.

**cornea** The transparent covering that protects the lens and iris of the eye. For illustration see *eye*.

**corner reflector** Also called a tricorner reflector. Three flat, reflecting surfaces brought together like the corner of a cube. Such a device reflects light or microwaves right back in the direction from which they came (see drawing).

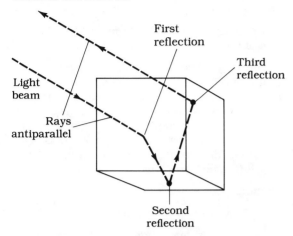
CORNER REFLECTOR

**corolla** The part of a flower that forms the visible, often brightly colored petals, which serves to attract flying insects for pollination.

**corollary** A minor mathematical theorem that is proved in addition to some other, more important theorem, when the main theorem is proved. It is like a bonus. See *theorem*.

**corona** 1. A region of glowing gases surrounding the sun out to a distance of several million miles. These gases are extremely hot, up to a million degrees Fahrenheit. The corona is seen during total solar eclipses. 2. A discharge from static electricity leaking off into humid air. Sometimes can be seen at night, and has been called Saint Elmo's fire.

**coronary** 1. Pertaining to the heart and the arteries supplying blood to the heart. See *heart*. 2. Slang for a heart attack. See *heart attack*.

**coronary artery** An artery that supplies the heart with blood. See *heart*.

**coronary occlusion** An obstruction in a coronary artery resulting in a heart attack. Usually caused by a blood clot. See *coronary thrombosis*, *heart attack*, *plaque*.

**coronary thrombosis** A blood clot that obstructs the flow of blood in a coronary artery. This might cause a heart attack. See *heart*, *heart attack*.

**corpuscle** 1. A particle of light or electromagnetic energy. See *particle model*. 2. In the body, a red or white blood cell. See *red blood cell*, *white blood cell*.

**corpuscular model** See *particle model*.

**correlation** 1. Positive correlation: A tendency for one quantity to increase when another increases. 2. Negative correlation: A tendency for one quantity to decrease when another increases. 3. Zero correlation: No relationship at all between an increase or decrease in one quantity and the value of the other. 4. The matching of similar rock strata over long distances.

**corrosion** Deterioration, especially of a metal such as copper or iron, because of reaction with the air, water vapor in the air, or pollutants or salt spray.

**corrugation** A means of bending a metal or other structural material back and forth to increase its ability to hold weight. A good example is the cardboard in heavy-duty boxes.

**cortex** 1. In an organ, the covering or outside layer. 2. The wrinkly outer layer of the brain. See *cerebral cortex*.

**corticosteroid** A hormone that is used in some skin creams as an anti-inflammatory agent. Useful in treating insect bites and allergic reactions. Related to adrenalin. Plays a role in the metabolism of foods. See *cortisol, cortisone*.

**cortisol** An adrenal hormone, similar to corticosteroid (see *corticosteroid*). Important in sugar (glucose) metabolism. An excess of this hormone produces body changes, such as a fat torso and thin arms and legs. Medically used as an anti-allergy agent. See also *cortisone*.

**cortisone** 1. Synthetically manufactured cortisol (see *cortisol*). 2. A hormone similar to corticosteroid (see *corticosteroid*). Used as an anti-inflammatory agent. Can be injected into sore joints, so that athletes can perform even when injured.

**cos** Abbreviation for cosine.

**cosecant** Abbreviation, csc. The reciprocal of the sine (see *sine*). On the unit circle, the cosecant is given by

$$\frac{1}{y}$$

See the drawing.

COSECANT
COSINE
COTANGENT

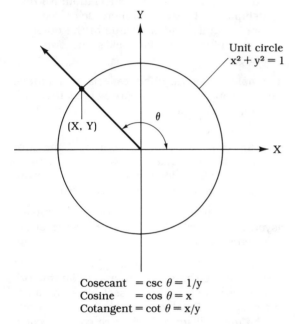

Cosecant = csc θ = 1/y
Cosine = cos θ = x
Cotangent = cot θ = x/y

**cosh** Abbreviation for hyperbolic cosine.

**cosine** Abbreviation, cos. On the unit circle, the cosine is given by the value x. See the drawing.

**cosmic noise** Electromagnetic waves generated by distant stars and galaxies. Some of this comes from the primordial fireball, the explosion from which the universe began. Radio astronomers analyze this noise. See *cosmology*, *radio astronomy*.

**cosmic rays** High-speed particles that come from the sun and from distant stars and galaxies. These (primary) particles bombard the upper atmosphere of the earth, producing other (secondary) particles and high-energy photons such as X rays and gamma rays. See *gamma rays*, *X rays*.

**cosmogony** The study of the origin and evolution of the universe. See *cosmology*.

**cosmology** The study of the structure and evolution of the universe. Cosmogony is sometimes considered a subfield of this. A popular theory nowadays is that the universe is expanding because of a primordial explosion that took place 10,000,000,000 to 20,000,000,000 years ago. See *Big Bang*, *Big Bounce*, *Big Crunch*, *oscillating universe theory*.

**cosmos** See *universe*.

**cot** Abbreviation for cotangent.

**cotangent** Abbreviation, cot. The reciprocal of the tangent (see *tangent*). On the unit circle, the cotangent is given by

$$\frac{x}{y}$$

See the drawing.

**coth** Abbreviation for hyperbolic cotangent.

**coulomb** A unit of electrical charge quantity. One coulomb of electrons flows past a point in one second, when there is a current of one ampere flowing. A coulomb of charge is about $6.24 \times 10^{18}$ electrons. See *ampere*.

**Coulomb's Law** A physical principle governing the force between electrically charged objects. The force is proportional to the product of the charges on the objects, and inversely proportional to the square of the distance between their centers. If the objects have like charges, they repel. If they have opposite charges, they attract.

**counter** 1. An electronic circuit that counts impulses per unit time. In this way it can measure frequency with great accuracy. 2. Acting against; anti-. We might speak of a counterforce, for example, acting against some force. 3. Balancing. For example, a counterweight balances a machine.

**counting numbers** See *natural numbers*.

**coupling** Interaction between coils or capacitors. Sometimes this is intended, as with transformers, and sometimes it is not wanted. The drawing shows how bringing a hand near a television antenna might degrade reception (electrostatic coupling between body and antenna), and also one of the ways coils are positioned to provide electromagnetic coupling.

COUPLING

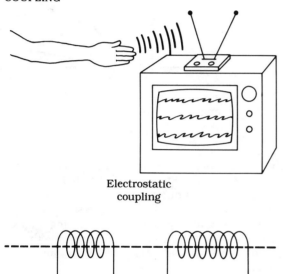

Electrostatic coupling

Electromagnetic coupling

**course** 1. The path taken by a ship or airplane, or by a spacecraft. Usually shown in two dimensions for a ship or airplane, and easy to draw on a map. For a spacecraft, the course is in three dimensions and computers are needed for precise calculations. 2. Life cycle or evolution. We might speak of the course of an illness, for example.

**course plotting** See *navigation*.

**Cousteau, Jacques** A well-known French oceanographer and environmentalist. Perhaps best known because of his television shows. He said in 1985 that our ocean ecology is being seriously disrupted by the activities of humans.

**covalent bond** A form of chemical bond in which atoms share electrons. The atoms might share just one electron, or two, or sometimes even more than two. The bonds tend to be stronger when more electrons are shared. This is the type of bond that generally holds atoms together in a compound. See *compound*.

**covalent crystal** A compound with a crystalline structure, in which the atoms are joined by covalent bonds.

**coxa** In an insect, the portion of the leg that is closest to the body. See *insect*.

**CPR** Abbreviation for cardiopulmonary resuscitation.

**CPU** Abbreviation for central processing unit.

**CQ** A code in ham radio and communications meaning "Calling anyone who will answer."

**Cr** Chemical symbol for chromium.

**Crab Nebula** A cloud of interstellar gas and dust, glowing because of a supernova that was seen on the earth about 1000 years ago. The cloud was thrown off by the violent explosion of a star at its center. It is expanding rapidly.

**crane** 1. A species of bird with long legs that inhabits shallow waters, especially in warmer climates. 2. A device used for construction, with a horizontal crosspiece and lifting pulleys and cables.

**cranium** The bones that make up the top of the skull and that protect the brain from trauma.

**crash** A rapid decline in the population of a species after it has increased far beyond the earth's ability to support it. Some scientists predict this will happen with humans, because of disease, war or mass starvation. See also *Malthusian model*.

**crater** 1. A depression left in the surface of the earth or of a planet caused by the impact of a meteorite. 2. A depression in the earth's crust that is often found at the top of an inactive volcano. 3. A gouged-out hollow in the ground caused by a large explosion.

**craton** The ancient interior of a continent, made up usually of Precambrian granitic rocks.

**creatine** A compound found in muscle tissue. It is used as an energy source for muscular movement.

**Creationism** The belief that humans are descended from Adam and Eve, who were put on the earth by God; and that the universe was created by the will of God. While much of modern science is based on Evolutionism (see *Evolutionism*), absolute proof of either theory is lacking. In the case of the universe, some astronomers have said that the Big Bang can be ascribed to God just as well as to any other cause, because the equations tell us nothing about what came before. See *Big Bang, cladistics, cosmology*.

**creep** 1. A slow change that is imperceptible over short periods of time, but very significant over long periods. 2. Slow movement of boulders, caused by alternate freezing and thawing.

**creosote** Coal tar, used as a preservative on utility poles and some structures.

**Cretaceous period** A part of the Mesozoic era. There were seas where we find land today, such as in the center of North America. This period began about 137 million years ago and ended with the beginning of the Cenozoic era, or the era of recent life, about 65 million years ago. See the GEOLOGIC TIME appendix.

**cretinism** A congenital defect in which a person fails to grow and thrive normally. Mental retardation is common. There is a deficiency of thyroid hormone.

**crevasse** A deep, narrow ravine in the earth or in a glacier. Might be caused by cracking as a result of lateral pressure.

**criminal psychology** A branch of psychology dealing with the behavior of criminals, especially repeat or habitual offenders.

**critical angle** 1. The largest angle measured with respect to the perpendicular line intersecting a boundary, at which light will be refracted from a certain substance into one with a lower index of refraction (see drawing). If the angle gets larger, the light is reflected at the boundary, back into the substance. See *index of refraction*. 2. For a radio wave sent up towards the ionosphere, the largest angle, measured between the wave and the ionosphere layer, at which the wave will be returned to the earth. At larger angles, it passes into space. See *ionosphere, skip, skip zone*.

CRITICAL ANGLE

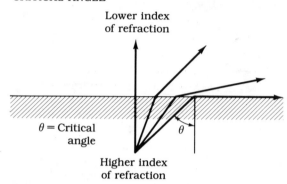

$\theta$ = Critical angle

**critical mass** The minimum mass of uranium-235 needed to produce the chain reaction that results in a nuclear explosion. The lump of uranium-235 would be about the size of a baseball. Two smaller pieces are pressed together to cause a blast that would have more force than a whole baseball stadium full of TNT ignited all at once.

**critical pressure** The pressure of a substance when it is in its critical state. See *critical state*.

**critical state** The condition of a substance in which its density is the same whether it is a gas or a liquid. See *critical pressure, critical temperature*.

**critical temperature** For a gas, the highest temperature at which pressurization can be used to make it liquid. Above this temperature, no amount of pressure can make the gas liquid. See *critical pressure, critical state*.

**Cro-magnon man** The first known creature considered to be a contemporary human being. Existed in Europe about 30,000 to 40,000 years ago. If a generation is 25 years, then this would be just 1200 to 1600 generations ago. On an evolutionary scale, this is not that long a time. They were cave dwellers.

**cross breeding** Controlled production of offspring from different breeds of animal. This produces a hybrid strain or breed.

**cross product** The vector product of two vectors. The product vector A×B is perpendicular to the plane containing the two vectors A and B to be multiplied. The length of A×B is equal to the product of the lengths of A and B, times the sine of the angle between A and B. See the drawing. See also *dot product, vector*.

CROSS PRODUCT

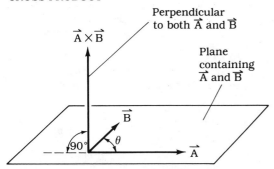

**crust** 1. A hard shell or outer coating. 2. The topmost layer of rocks on the earth, that floats on the mantle. Ranges from a few miles thick on the ocean floor to as much as 45 miles thick on the continents. See *continental crust, mantle*.

**crustacean** An ocean-dwelling creature with an exoskeleton, including the barnacles, shrimps, crabs, crayfish and lobsters. They shed their outer skeletons as they grow, so they might leave several fossils for each individual.

**cryo-** Pertaining to extremely low temperatures, of just a few degrees Kelvin (near absolute zero). See the definitions that follow.

**cryobiology** The study of the behavior of life forms at extremely low temperatures. In particular, the study of freezing body organs and perhaps even whole bodies, to be thawed out weeks, months or years later. It is doubtful whether it will ever be possible to freeze people and bring them back hundreds of years in the future, but it makes good science fiction.

**cryogenerator** In the Conshelf sphere (see *Conshelf sphere*), a device that recycles the oxygen-and-helium atmosphere.

**cryogenics** The study of the behavior of materials at very cold temperatures, especially near absolute zero (−459 degrees Fahrenheit or −273 degrees centigrade).

**cryology** The study of ice packs and glaciers on the earth. See *alpine glacier, continental glacier, glacier, ice pack*.

**cryometer** A device designed to measure temperatures near absolute zero.

**cryptic** In the form of a cipher; encoded so that only the intended recipient can tell what the message says. See *cipher*.

**crystal** 1. A structure that some elements and compounds have, with a regular geometric pattern. May be cubical, rhombic, hexagonal, octagonal, or of some other regular shape. Quartz is a good example of a substance that naturally forms crystals. Common table salt is a compound that forms crystals. So is water when it freezes. 2. A piece of an element or compound that forms crystals as in the foregoing definition.

**crystal lattice** The pattern formed by the arrangement of atoms in a crystal. See the drawing. See also *crystal*.

CRYSTAL LATTICE

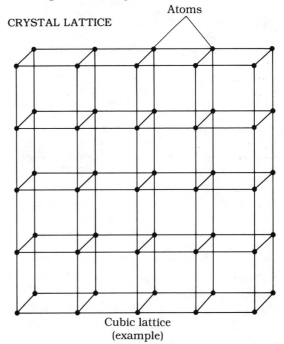

Cubic lattice (example)

**crystalline** Pertaining to a substance that forms crystals. See *crystal*.

**crystallization** The formation of crystals as an element or compound hardens or accumulates. Ice is a good example; it forms crystal patterns as liquid water freezes. See *crystal*.

**crystallography** A branch of physics and chemistry, concerned with the nature and behavior of crystals. See *crystal*.

**crystal oscillator** An electronic oscillator that uses a quartz crystal to determine its operating frequency. See *oscillator*.

**Cs** Chemical symbol for cesium.

**csc** Abbreviation for cosecant.

**csch** Abbreviation for hyperbolic cosecant.

**CST** Abbreviation for Central Standard Time.

**Ctesibius** The Greek engineer who, in the third century B.C., designed a water clock called the clepsydra. See *clepsydra*.

**Cu** Chemical symbol for copper.

**cube** 1. A geometric form with six identical square faces. All the edges intersect at right angles. 2. The third power; an exponent of 3.

**cubic centimeter** The volume equivalent to that of a cube measuring 1 centimeter on an edge. Also equivalent to 1 milliliter (ml). See *milliliter*.

**cubic equation** An equation in which one or more variables are raised to the third power (cubed); no variables are raised to any power higher than this.

**cultivar** Contraction of CULTIvated VARiety. A plant developed by special cultivation for a certain purpose. See *cultivation*.

**cultivation** The controlled raising of plants for food, for industrial products, for medicinal purposes or for other human needs.

**cultural anthropology** See *anthropology*.

**culture** 1. A controlled growth of bacteria or viruses or other microorganisms. Often used in medicine to make a diagnosis, such as strep throat (streptococcus grow in the culture). 2. To grow microorganisms in a controlled manner.

**culture medium** The substance, such as agar, in which a culture is grown. See *culture*.

**cumulative dose** The amount of radiation received by an individual over a period of time. This is a number that always increases; it never goes down. See *dosimetry*.

**cumulonimbus clouds** The large, "thunderhead" clouds typical of thundershowers. May rise to altitudes of more than 60,000 feet, or the top of the troposphere. Above this level they are sheared off by the jet stream. See the drawing.

**cumulus clouds** The puffy, low-level clouds associated with fair weather. They are also found near thundershowers. See the drawing.

CUMULONIMBUS CLOUDS
CUMULUS CLOUDS

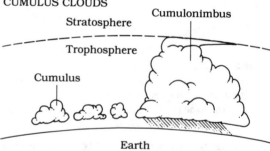

**cuneiform** An early form of writing, in and around the time of Sargon the Great in the twenty-third century B.C. Pressed into clay tablets with a stylus. The tablets were baked in the sun to harden and preserve them. Tablets have been found with laws, tax records, and stories on them in the region then known as Mesopotamia.

**curie** A unit of radioactivity, no longer widely used.

**Curie, Marie** Discovered three types of ionizing radiation in 1897, working with Pierre Curie. The first radioactive element to be isolated was radium (see *radium*). The radiation was called alpha, beta and gamma. These terms have stuck to this day (see *alpha rays, beta rays, gamma rays*). Won the Nobel prize for her work.

**Curie, Pierre**  Worked with Marie Curie in discovering radium and its behavior at the end of the nineteenth century.

**Curie point**  For a ferromagnetic substance, the highest temperature at which this property remains (see *ferromagnetic*). Above this temperature, it is no longer a ferromagnetic material. These temperatures are normally very high, and are encountered only under exceptional conditions.

**curium**  Chemical symbol, Cm. An element with atomic number 96. There are various different isotopes, the most common with atomic weight 247. This element is not found in nature, but must be artificially made in atom smashers. The most common isotope is very stable and takes millions of years to decay.

**curl**  An expression in vector calculus, given as a cross product. See *cross product*.

**current**  A flow of charge carriers. Measured in amperes. One ampere of current represents one coulomb of charge per second moving past a certain point. Physicists consider current to flow from plus to minus; this is called conventional or Franklin current. Electrons actually move from minus to plus. See *ampere, coulomb, electron, hole*.

**current density**  The electrical current flowing through a conductor or semiconductor, per unit cross-sectional area. Usually given in amperes per square meter, centimeter or millimeter. Might also be given in milliamperes per square millimeter, or per circular mil.

**current feed**  In an antenna, the connection of the feed line at a point on the radiating element where the current is maximum. See *voltage feed*.

**cursor**  The movable marker on a computer screen, especially in a word processor or communications terminal that indicates where the characters are.

**curvature**  1. The visible rounding of an object, such as the earth as seen from a spacecraft. 2. The extent to which an object is rounded. 3. Deflection from a straight line, as with electrons in a cathode-ray tube or atomic particles in a cloud chamber.

**curve**  1. The graph of an equation or function, when the graph is one-dimensional, even if it is linear (a straight line). 2. To travel in a path that is bent rather than straight.

**curvilinear coordinates**  A system of coordinates in which one or both axes are curved, rather than straight (see drawing). A one-to-one correspondence exists, nevertheless, between the curved system and a Cartesian system in the same number of dimensions. See *Cartesian coordinates*.

CURVILINEAR COORDINATES

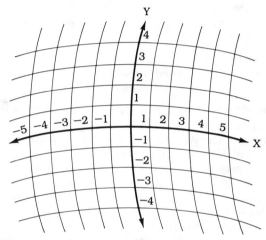

**cusp**  1. A transition point, zone or interval of time. 2. A point where two curves intersect, or where a curve abruptly changes direction (see *curve*). 3. A sharp place on a tooth. 4. A flap in a valve in the heart or in a vein.

**cutaneous**  Pertaining to the skin. See *skin*.

**cuticle**  A protective covering, such as the exoskeleton in an arthropod or crustacean (see *exoskeleton*), or the coating that protects plants from drying out and also prevents unwanted substances from getting in.

**cutin**  The chemical that forms the cuticle in plants. See *cuticle*.

**cutting**  1. A part of a plant that is removed and replanted, growing into a new plant. This is most often a small branch with some leaves attached. The leaves allow photosynthesis to occur in the new plant as it begins to grow. 2. The starting of new plants by the aforementioned means.

**cyanide** A compound with a carbon-nitrogen negative ion. Perhaps known best as a quick-acting poison. It works by preventing the blood from properly carrying oxygen.

**cyanobacteria** A very early life form, that flourished around 2,800,000,000 years ago in the primitive seas. Also called blue-green algae. These microorganisms used photosynthesis to generate their energy (see *photosynthesis*).

**cyanocobalamin** An essential nutrient, also called vitamin B-12. See *vitamins* and the VITAMINS AND MINERALS appendix.

**cybernetics** A branch of science and engineering concerned especially with artificial intelligence. See *artificial intelligence*.

**cyclamate** An artificial sweetener that was popular in the 1960s and 1970s, before researchers gave rats the equivalent of 700 bottles of diet soda daily for a lifetime, and some of the rats got cancer. Then this sweetener was banned. This raised a controversy over scientific methods. Some people criticize research that doesn't duplicate real life. Others say that they must give massive doses of a test substance, to make bad effects appear soon enough to make tests practical.

**cycle** 1. One complete rotation, or revolution, or event that is repeated in a sequence. 2. In alternating current (ac), the fluctuation of the current from zero, through maximum, back through zero, to minimum, and back to zero. See also *alternating current, hertz*.

**cycles per second** See *hertz*.

**cycloid** A curve that consists of arcs, connected together at cusps, and repeated along an axis, usually the abscissa. See the illustration. See also *abscissa, arc, cusp*.

**cyclone** A low-pressure weather system. This may be a typical "low" in temperate latitudes, a blizzard system, or a hurricane. In the Northern Hemisphere, the systems rotate counterclockwise; and in the Southern Hemisphere they rotate clockwise. Tornadoes are not true cyclones. See *hurricane, tornado*.

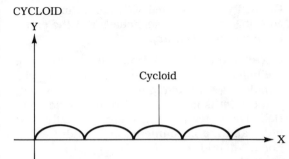

CYCLOID

**cyclostome** A water-dwelling animal, something like a fish, but without jaws. The sea lamprey, found in the Great Lakes, is an example. It attaches its suction-cup-like mouth to fish and lives from their blood like a leech. The species is nearly 500,000,000 years old.

**cyclotron** A particle accelerator (atom smasher) that works by using magnetic fields to bend the paths of the particles into a spiral. The particles gain energy with each revolution (see drawing). The main advantage of a cyclotron is

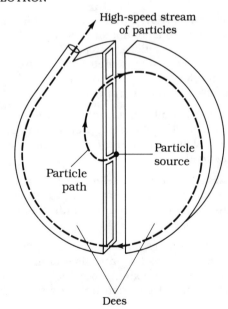

CYCLOTRON

that it allows high energy levels, without the large spaces needed by linear accelerators. See *linear accelerator, particle accelerator.*

**Cygnus A**  A strong source of radio waves in the constellation of Cygnus. The letter A means that this is the strongest radio source in the constellation. In fact, this object is one of the strongest radio sources in the whole sky. It is called a radio galaxy because it emits so much energy at radio wavelengths — far more than ordinary galaxies. See *radio galaxy.*

**Cygnus X-1**  An X-ray star in the constellation Cygnus. It is thought that this is a black hole in orbit around an ordinary star. The X rays are produced when matter from the star is torn away by the gravitation of the black hole. See *black hole, X-ray star.*

**cyst**  A noncancerous growth in the skin or in a body organ. The body closes it off and fluid accumulates in it. Skin cysts can sometimes last for months or years, and they may become inflamed or infected. See also *abscess, carbuncle.*

**cysteine**  An amino acid. The body can make this from other amino acids, so it is not necessary to get it in the diet. See *amino acid.*

**cystic fibrosis**  A fatal inherited disease, in which the lungs secrete a sticky mucus that clogs them and makes them susceptible to infection. The average victim of cystic fibrosis, called CF, lives to be only 26 years old. Recently, genetic researchers have been testing a therapy that seems to work by changing the defective gene.

**cystitis**  A urinary-tract problem in which the bladder becomes inflamed or infected. Symptoms include painful, frequent urination, and perhaps blood in the urine. Treated with antibiotics.

**cytogenetics**  A branch of genetics that is concerned with the way cell behavior affects reproduction of whole organisms.

**cytology**  The science of the behavior of living cells. A branch of biology, this field is also closely connected with medicine.

**cytolysis**  Cell destruction. This can be undesirable, as in various illnesses; it also can be done on purpose, as is the case with chemotherapy and radiation therapy in the treatment of cancer. See also *chemotherapy, radiation therapy.*

**cytoplasm**  A thick liquid inside a cell, outside the nucleus. Contains ribosomes that manufacture proteins. See *cell, ribosome.*

**cytosine**  A compound that joins with guanine in the deoxyribonucleic acid (DNA) molecule. See *adenine, deoxyribonucleic acid, guanine, thymine.*

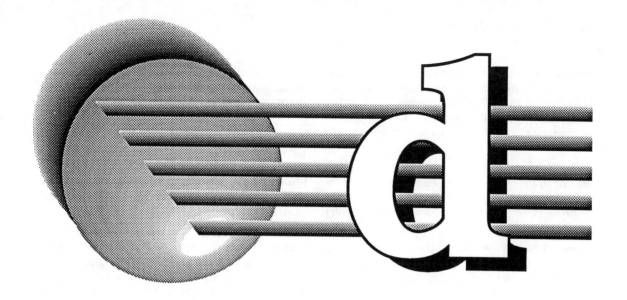

**D** 1. Abbreviation for density. 2. Abbreviation for deuterium. 3. Abbreviation for dissipation. 4. Abbreviation for drain in a field-effect transistor.

**d** Abbreviation for deci-. See appendix PREFIX MULTIPLIERS.

**da** Abbreviation for deca-, deka-. See appendix PREFIX MULTIPLIERS.

**Daedalus spacecraft** A hypothetical interstellar vessel proposed by the British Interplanetary Society. It would use fusion reactions in a continuous blast. It could reach speeds of about 18,000 miles per second, or 1/10 the speed of light. This would allow it to reach only the nearest stars. See the drawing.

**Da Gama, Vasco** The first explorer to find an oceanic route to India and Asia from Europe. The round trip, begun in 1497, took two years.

**dalton** See *atomic mass unit*.

**Dalton, John** A scientist remembered for his work in early atomic physics. He professed in the 1800s that atoms were the fundamental particles of all materials. We now know that he was right, in general.

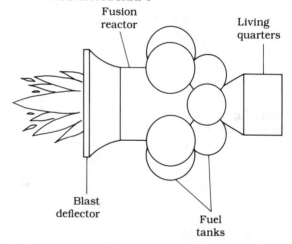

DAEDALUS SPACECRAFT

**Dalton's Law of partial pressures** The total pressure exerted by a mixture of gases is equal to the sum of the partial pressures of each of the gases. In the air, for example, the total pressure at sea level is about 14.7 pounds per square inch (PSI). Because air is about 78 percent nitrogen, the partial pressure from this gas is about 14.7 PSI × 0.78 = 11.5 PSI. Because air is about 21 percent oxygen, the partial pressure from this gas is about 14.7 PSI × 0.21 = 3.1 PSI. The other 1 percent of the air, made up of carbon dioxide, argon and other trace gases, gives about 14.7 PSI × 0.01 = 0.15 PSI partial pressure. Another way of stating this is to say that the proportions of the gases are the same as the proportions of their pressures.

**dam** A barrier in a river, used for flood control and/or to provide a difference in water level for the production of hydroelectric energy. See *hydroelectric energy, hydroelectric power plant.*

**damped wave** A wave that decreases in amplitude (level), approaching zero amplitude. The level follows the typical decay curve. See *decay curve.*

**Daniell cell** An electrochemical cell that uses zinc for the negative electrode, copper for the positive electrode, and electrolytes of zinc sulfate and copper sulfate. It is a wet cell. That means the electrolytes are liquids. It provides about 1.1 volts. See *battery.*

**Dart, Harry Grant** An American journalist of the early twentieth century, who was interested in the possibilities of air travel. He drew pictures of airliners of the future—many of them were surprisingly accurate. See *Rickenbacker, Edward.*

**Darwin, Charles** The originator of the concept of natural selection and evolution of species in the nineteenth century. Some people, especially those in the Church, rejected his ideas as lowering humans to the level of animals. His theory, detailed in his book *The Origin of Species*, is accepted by most scientists today; however, absolute proof in the case of humans is still lacking. See *Creationism, Evolutionism.*

**Darwinism** See *Evolutionism.*

**data** Information. Usually in digital form, such as ASCII or BAUDOT. See *ASCII, BAUDOT code, data acquisition, data conversion, data processing.*

**data acquisition** The monitoring, controlling, and recording of information from sensors or transducers. A data-acquisition system includes the transducers, data-conversion circuits, and displays or monitors. A computer terminal is an example of such a system. See *data, data conversion, data processing, monitor, transducer.*

**database** 1. Stored data, set up so that it can be used without changing its contents. An example is the microfilm card catalog at a library. 2. A system that uses data without changing its contents. See *data, data acquisition, memory.*

**data conversion** A process done by integrated circuits that changes data from one form or speed or code to another. We might convert, for example, from ASCII to Morse code or from analog to digital signals. See *analog-to-digital converter, ASCII, BAUDOT code, digital-to-analog converter, Morse code.*

**data processing** The work done on data once it has been acquired. Might be calculations, rearrangement, or both. See *data, data acquisition, database, data conversion.*

**dating** Determination of the age of a sample, especially a geologic sample or a fossil. This is most often done by using radioactivity measuring techniques. See *radioactive dating.*

**da Vinci, Leonardo** A Renaissance scientist and one of the earliest true engineers, mainly self-taught, who conceived and built various devices. Lived in the fifteenth and early sixteenth centuries. He studied bird movements and imagined that people could build flying machines.

**dayglo** A chemical often added to dye to produce a brilliant color.

**dayglow** An air glow seen during the day.

**dB** Abbreviation for decibel.

**dc** Abbreviation for direct current.

**DDT** Abbreviation for dichlorodiphenyltrichloroethane. A common and controversial pesticide.

**debris** 1. Miscellaneous matter, such as might be left over after flood waters recede. Includes timbers, human-made waste, dirt and sand, and sometimes ice. 2. The wreckage following a major disaster such as a hurricane or tornado. 3. Flying or floating material in a high wind or in active flood waters. This material can be dangerous because of the force of its impact.

**debris avalanche** Also called a mudslide. A type of avalanche (see *avalanche*), where loose dirt on a steep incline gets wet and breaks loose because of the extra weight of the water.

**debris flow** A mudflow, such as takes place after a heavy rainstorm in hilly or mountainous terrain. It is a slower form of debris avalanche. See *avalanche, debris avalanche*.

**de Broglie, Louis-Victor** A physicist who studied the wave-like properties of electrons. Showed how particles sometimes act like waves. See *de Broglie wavelength*.

**de Broglie wavelength** The wavelength of electromagnetic energy that is generated by a moving particle. Often associated with high-speed subatomic particles like electrons. If the wavelength is given by w in meters, and the particle mass is given by m in kilograms, and the velocity is given by v in meters per second, and h is Planck's constant (see *Planck's constant*), then:

$$w = \frac{h}{mv}$$

**debugging** 1. The process of getting a computer program to work exactly as desired, in all situations. 2. The process of getting an electronic circuit to work according to the design specifications.

**deca-** See the appendix PREFIX MULTIPLIERS.

**decay** 1. Decomposition, as the rotting of a dead plant or animal. See *decomposition*. 2. A decrease in amplitude, level or intensity with time. See *decay curve, decay time*. 3. For radioactive materials, a process in which the nucleus changes to a more stable form, giving off radiation. See *half life, time constant*.

**decay curve** A graphical expression of natural decay as a function of time. This curve has a characteristic shape (see drawing) based on logarithmic functions. It is also sometimes called a logarithmic decrement or exponential decrement. This is the natural curve of radioactive degeneration, damped oscillations, the cooling down of a hot object, and many other phenomena.

DECAY CURVE

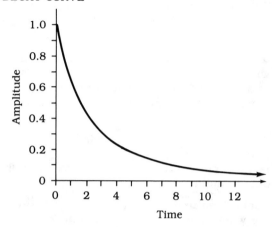

**decay time** The time required for a decay process to be completed or partly completed. Usually given as the time needed for the value to drop to some percentage of the starting value, such as 37 percent (electric time constant), or 50 percent (half life), or 1 percent (practically down to zero). See also *half life, time constant*.

**decca** A British radionavigation system that operates between 70 kilohertz (kHz) and 130 kHz. This is only about 1/10 the frequency of the standard amplitude-modulation (AM) broadcast band. See *LORAN, radionavigation*.

**deci-** See the appendix PREFIX MULTIPLIERS.

**decibel** 1. An expression of a difference in power level. If the power levels are given by P and Q, then the decibel gain or loss is given by dB = 10 log (P/Q), where "log" is the base-10 logarithm. 2. An expression of a difference in voltage or current level. If the voltage or current levels are given by X and Y, then the decibel gain or loss is given by dB = 20 log (X/Y). For power, current or voltage, loss is indicated by a negative result, and gain by a positive result. 3. A measure of sound intensity. The threshold of hearing is assigned the level zero or 0 dB. Sounds of 120 dB or more may injure the ears. See *sound*.

**deciduous tree** A tree with broad leaves, rather than needles. This type of tree sheds its leaves in the autumn at temperate latitudes. In warmer climates, some deciduous trees keep their leaves all year round, and others lose them during the dry season.

**decillion** The number one followed by 33 zeroes.

**decimal** 1. See *decimal point*. 2. A number written so that the fractional part is expressed as a period followed by digits representing $1/10$, $1/100$, $1/1000$ and so on. An example is 23.5565, which is $23 + 5565/10,000$.

**decimal number** A number in base-10, consisting of digits 0, 1, 2, 3, 4, 5, 6, 7, 8 and 9. This is the number system in common use throughout the world.

**decimal point** A period (.) that goes between the whole-number part and the fractional part of a numeral in decimal form. See *decimal*.

**declination** 1. Celestial latitude; that is, the number of degrees north or south of the celestial equator that an object is in the sky. Objects north of the celestial equator are given values from 0 to +90 degrees; objects south are given values 0 to −90 degrees. See *celestial coordinates, celestial sphere, right ascension*. 2. The difference between true north and magnetic north (compass north) in degrees. See drawing. In some places this is zero; in others it might be a very large angle, even 180 degrees in between the geographic and magnetic poles.

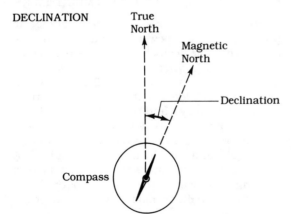

DECLINATION

**decomposition** A process in which organic matter is broken down into simpler compounds over a period of time. This happens when plants or animals die. After many thousands or millions of years, oil and coal are formed. Bacteria eat the dead organic matter and metabolize it; later it is compressed by the weight of the earth above it.

**decompression** A process that SCUBA divers must undergo when they have been under deep water for a long time. Otherwise, illness will occur (see *bends*). The process involves stops at various levels on the way up. The deeper and longer the dive, the more time the decompression takes. See also *SCUBA*.

**decongestant** A medication, usually in pill or liquid form, that dries out the nasal passages and provides relief from cold or allergy symptoms. Also can be in the form of a nasal spray.

**deduction** 1. A mathematical process of logic; a proof. 2. The removal of a certain value or quantity from a total. 3. Logical result; conclusion.

**deductive logic** Mathematical logic that is done in step-by-step form, so that a certain theorem is proven on the basis of axioms and definitions. See also *axiom, inductive logic, mathematical logic*.

**deep-sea diving** Diving to depths greater than usual in SCUBA. Various underwater vessels have been designed for this purpose. Some have normal atmospheric pressure inside, and some have greater than normal atmospheric pressure to reduce the pressure difference between inside and outside. When air is breathed at high pressure, an oxygen-helium mixture is used, because nitrogen at high pressure causes stupor. See also *SCUBA*.

**deep-sea trench** A place on the ocean floor where subduction has occurred (see *subduction*), so that the bottom is literally swallowed up into the earth. The crust is pushed down into the mantle. Most of these zones are in the Pacific Ocean at the edges of continents or along chains of volcanic islands.

**defect** 1. A malfunction, or a bad component, in a machine, device, or circuit. 2. A mistake in the design of a machine, device or circuit, that causes malfunction. 3. A flaw in a semiconductor chip. 4. An impurity or flaw in a crystal.

**deferent** The main part of the orbit of a planet according to the Ptolemaic model of the Solar System. According to this theory, the sun, moon, and planets all revolved around the earth. Some of the planets had little orbits in addition to the deferent or main, large, circular orbit. See *epicycle, Ptolemaic model.*

**defibrillation** The use of electrical stimulation to shock the heart into beating normally when it has gone into fibrillation. See *fibrillation.*

**deficiency** 1. A lack; an insufficiency; a state of not getting enough of something for normal functioning or normal health. 2. A state of having less than the usual number or amount; for instance when an atom loses an electron.

**deficiency disease** An illness that results from a lack of nutrients to the body. This lack might be because of bad diet, or because the digestive system is not absorbing food properly, or because of some problem with metabolism. It also could be a combination of these things. See *beriberi, pellagra, scurvy,* and the appendices RECOMMENDED DAILY ALLOWANCES and VITAMINS AND MINERALS.

**definite integral** See *integral.*

**definition** 1. In mathematics, a sentence or set of sentences that indicates the nature of something and its allowed uses or contexts. 2. Degree of clearness; resolution. We might speak of the definition of a photographic image for example.

**defoliation** The killing of vegetation, either by nature or by human activity. Sometimes this is accidental; sometimes it is deliberate. The gypsy moth is known for eating tree leaves. Certain chemicals are used in warfare to kill trees so the enemy cannot hide as easily.

**deforestation** The stripping of forest land for farming, lumber, and cattle-raising in various parts of the world (see drawing). This has given rise to fears of climate changes and species extinctions. See also *global warming.*

DEFORESTATION

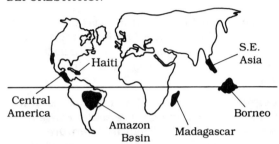

**de Forest, Lee** An engineer who worked towards the development of "wireless," now known as radio, starting in 1900. Later was involved with the development of vacuum tubes.

**degassing** 1. The process of removing any residual gases from a vacuum tube. The devices for this are a vacuum pump and a getter. 2. The process of removing poison gas from an area. 3. The loss of gases from a comet impacting the earth.

**degaussing** A process of demagnetizing a metal in which the molecules have been magnetically polarized. This destroys the magnetic properties of a permanent magnet.

**degenerate** 1. To decrease in amplitude or strength; to decay. See *decay, decay curve.* 2. To decompose or be destroyed.

**degenerate gas** A gas in which the electrons move freely among the nuclei, rather than being tied to one single nucleus. This is the state of matter in a white dwarf star.

**degeneration** 1. The process of degenerating. See *degenerate.* 2. Negative feedback. That is, the feeding back of a signal out of phase to control amplification. See *feedback.*

**degenerative disease** 1. Any disease that tends to get worse with time, but rarely better. 2. Certain diseases of modern living, including cancer, diabetes and atherosclerosis. These have always existed, but detection methods have improved. So we notice them more. Also, lifestyle is thought to play a role, such as overeating, smoking and lack of exercise. See *atherosclerosis, cancer, diabetes insipidus, diabetes mellitus.*

**degradation**  1. A continuous decrease in quality; as the slow deterioration in the gain of a vacuum tube as it ages. 2. The changing of a compound into simpler and simpler substances.

**degree**  1. Extent. 2. A unit of temperature. See *centigrade temperature scale, Fahrenheit temperature scale, Kelvin temperature scale, Rankine temperature scale.* 3. A unit of angular measure, equal to 1/360 of a full circle. See also *radian.*

**degree-days**  A measure of how hot a summer is, or how cold a winter is. Important in determining fuel costs and requirements. One degree-day means an average temperature one degree more than a certain value for one day. A shortfall is given a negative value. We might choose 78 degrees as the threshold past which we need air conditioning, and 62 degrees as the limit below which we need heating. Or the limits might be the same, such as 68 degrees. In the drawing, see how degree-days accumulate.

DEGREE-DAYS

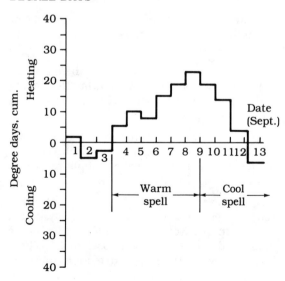

**degree of freedom**  1. The number of dimensions in a coordinate space. 2. The number of possible variables in a situation. For example, we might say a car can change speed or direction. This means there are two degrees of freedom for driving a car.

**dehydration**  1. The removal of most or all of the water from something, especially food. 2. An unhealthy and possibly dangerous condition in which the human body is depleted of water. Might occur after hard work in the hot sun. Along with the loss of water, mineral imbalances often develop.

**deka-**  See the appendix PREFIX MULTIPLIERS.

**delay**  1. The time difference between the input and output signals in a circuit. Might be tiny and insignificant. Sometimes important, especially in high-speed digital circuits or in radio-frequency circuits. 2. Condition of being late.

**delay circuit**  An electronic circuit that produces a delay (see *delay*), for a certain purpose. A length of transmission line will do this at radio frequency. Integrated circuits, or networks of coils and capacitors, also can be made to produce delay for various purposes.

**delay line**  A transmission line, cut to a certain length, so that it causes the output to be delayed by a predetermined length of time. The longer the line, the greater the delay. See *delay.*

**delay time**  See *delay.*

**delay timer**  A device, usually an electronic pulse counter, that is used for ensuring that a delay is always exactly the same. See *delay.*

**delirium**  A state of mental confusion or stupor. Occurs with high fever, in brain injury, and with overdose of some drugs. The person might not know what time it is, where he or she is, or even who he or she is. Hallucinations can take place.

**delirium tremens**  An effect of the withdrawal of alcohol from a chronic alcoholic. The person is shaky, cannot sleep, might have hallucinations, and is excitable to the point that he or she might need tranquilizers. Hospitalization is necessary.

**delta** 1. A land formation where a river meets the sea. Some rivers have deltas and some do not. The Mississippi Delta is large, but the Saint Lawrence does not have a delta. 2. A difference in a variable's value that can be made smaller and smaller mathematically to find the derivative of a function. See *derivative*. 3. The fourth letter of the Greek alphabet, used to denote various things.

**delta wing** In a high-speed aircraft, a triangle-shaped wing (see drawing). This type of wing has large surface area for plenty of lift, and also has low air resistance to allow high speed. Used in military aircraft, especially fighters.

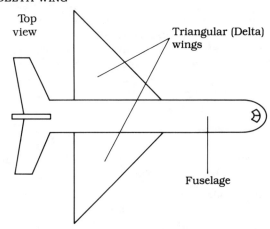

DELTA WING
Top view
Triangular (Delta) wings
Fuselage

**deltoid** A group of muscles on the tops of the shoulders. Helps lift the arms up.

**deluge theory** The idea that the oceans were created suddenly (on a geologic time scale) by a continuous, huge rainstorm or deluge.

**delusion** A mental condition in which a person believes things to be true when they are not. For example, the person might think that he or she is an agent of God. In paranoia, a person thinks certain others are out to get him or her.

**demagnetization** A process in which a permanent magnet is neutralized. See also *degaussing*. The term *demagnetization* is more often used in reference to tools like screwdrivers and needle-nosed pliers, while *degaussing* refers to powerful magnets.

**dementia** A mental condition of deterioration, usually because of old age. Often called senility.

**Democritus** A philosopher who, 2000 years ago, postulated that matter is made up of tiny particles. Perhaps one of the earliest atomic physicists.

**demodulation** See *detection*.

**de Moivre, Abraham** A mathematician of the 18th Century who contributed to the theories of probability and statistics. He wrote numerous papers and supported himself mostly by private tutoring.

**de Moivre's Problem** A question: What is the probability of rolling a certain number when you have n dice, each with m faces? For example, if you have four dice with six faces, what is the probability that, when you throw the dice, the numbers will add up to 13?

**de Morgan, Augustus** A logician who worked with Boolean algebra. See *de Morgan's Laws* and the appendix BOOLEAN ALGEBRA.

**de Morgan's Laws** Two related theorems in logic. If we let * mean "and" and + mean "or" and − mean "not," and if we let X and Y be statements (sentences), then $-(X+X) = -X*-Y$, and $-(X*Y) = -X + -Y$. Also can be expressed in terms of set union, intersection and complement. See the appendix BOOLEAN ALGEBRA for some common theorems in logic.

**denaturing** 1. The mixing of a bad-tasting chemical or a drug with bad side effects, to some substance, to prevent abuse. This is done especially with mood-altering chemicals like rubbing alcohol and opium. 2. Processing of food, such as cooking or dehydration.

**dendrite** 1. A thin fiber that forms part of a nerve cell. See also *axon, nerve, synapse*. 2. A crystal structure that forms branches, onto which more branches might be added as the crystal grows.

**density** 1. The mass per unit volume of a substance. Expressed, for example, in grams per milliliter. Pure water in liquid form has density 1.0 grams per milliliter. 2. The number of items per unit length, unit time, unit area, or unit volume. We might speak of the number of flux lines per square centimeter, for instance.

**density slicing** A process for enhancing contrast; a display. We might show the intensity of a thunderstorm in colors by means of computer-enhanced radar. You have probably seen this on television. Or we might use color to depict the surface temperature of the ocean in various parts of the world, in a spectrum, with blue coolest and red warmest, or vice-versa.

**dentin** The part of the tooth underneath the enamel, making up the bulk of the volume of a tooth. Softer than the enamel but harder than the pulp. See drawing.

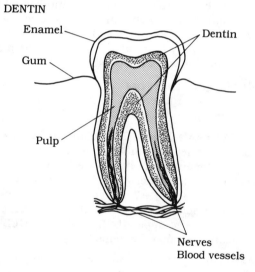

DENTIN

**deoxyribonucleic acid** Abbreviation, DNA. The substance important in reproduction of animal cells. In a human, the coiled-up DNA in a single cell would be three feet long if stretched out. It has two strands in a double helix connected by nitrogen compounds. Adenine joins with thymine, and cytosine with guanine. They lock like pieces in a jigsaw puzzle, holding the double helix structure in place. See also *double helix, ribonucleic acid*.

**dependent variable** The value of a function that depends on the value of the independent variable (see *function, independent variable*). For example, in a graph of temperature versus time of day, the temperature is the dependent variable.

**depletion** 1. A condition resulting from an ongoing deficiency. 2. Chemical destruction resulting in the lack of something essential. See, for example, *ozone depletion*. 3. An electrical condition in a semiconductor. See *depletion layer*.

**depletion layer** In a semiconductor P-N junction, a zone on either side of the actual plane where the P and N materials are joined. The width of this zone where there are no charge carriers, varies depending on the polarity and intensity of the voltage across the junction. See also *carrier, P-N junction*.

**deposit** 1. A layer of sediment that forms over a period of time. 2. A concentration of a certain mineral or fossil fuel. We might speak of an iron-ore deposit or a coal deposit. 3. The accumulation of a substance on another substance, under controlled conditions. 4. To cause a substance to accumulate on another material under controlled conditions.

**depressant** A chemical or drug that slows down the activity of the central nervous system. Alcohol is one such substance. Sleeping pills are another. See also *stimulant*.

**depression** 1. A spot in the terrain that is lower than the surrounding land. 2. A reduction. 3. A mental condition in which a person is in a bad mood much or all of the time. Sometimes can be caused by physical imbalances. Sometimes, it is just a normal reaction to disasters.

**depression cone** See *cone of depression*.

**depth charge** An explosive designed to go off when the pressure gets to a certain level. It is used in anti-submarine warfare. The device looks like a sealed barrel. It is rolled off the deck of a ship into the water, and when it gets to the specified depth, it explodes creating shock waves.

**depth of visual field** A term especially important to photographers. The difference between the farthest and closest points that appear in proper focus. This value gets larger as the lens diameter (aperture) is made smaller.

**depth sounding** The use of underwater sound waves to determine the distance to the bottom of a lake, river, or ocean. See the drawing. Some false echoes occur, in addition to the echo from the bottom. The depth is determined according to how long it takes for the echo to get back. See also *sonar*.

DEPTH SOUNDING

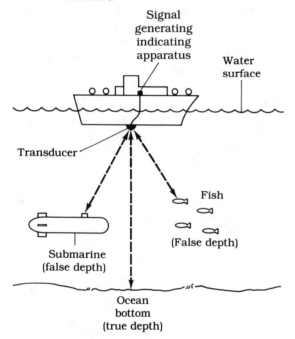

**derating** A means of giving the performance specifications for a device at increasing temperatures, currents, or duty cycles. This can be shown by a table or by a graph. For instance, a transistor might work at 100 watts output if the duty cycle is 50 percent, but only 50 watts when the duty cycle is 100 percent. See *duty cycle*.

**derivative** 1. The instantaneous rate of change of the value of a function. 2. The slope of the graph of a function at a specified point. 3. A function that denotes the slope of the graph of some other function, in general, for any given point. See *calculus, function*.

**dermatitis** A rash or inflammation of the skin. Examples include acne, poison ivy allergy, and sunburn.

**dermatology** A branch of medicine concerned with the diagnosis and treatment of skin disorders.

**dermis** The inner part of the skin, just beneath the epidermis. Consists of living cells. When these cells die they become the epidermis. See the drawing. See also *epidermis, skin*.

DERMIS

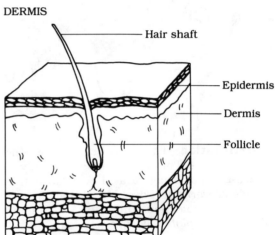

**desalination** The removal of salt from seawater. One method is distillation, in which all of the minerals are removed. Reverse osmosis also can be used; this has been done by U.S. forces in Saudi Arabia. One proposed idea is to use geothermal heat to provide the necessary energy. See *distillation, geothermal energy, geothermal power plant*.

**Descartes, Rene** One of the best-known mathematicians of all time. Also known as a philosopher, he lived and worked in France, and also traveled extensively in Europe during the seventeenth century. The Cartesian plane is named for him (see *analytic geometry, Cartesian coordinates, Cartesian plane*).

**desensitization** 1. A reduction in the sensitivity of a radio receiver, caused by overload from a nearby strong signal. 2. A decrease in the sensitivity of an organism to a certain stimulus or stimuli. This can be physical, psychological, or both. See *tolerance*.

**desert** A place where the annual rainfall is less than 10 inches (about 25 centimeters). The drawing shows the deserts of the world. The polar regions are also deserts, although they are not shown on the map.

DESERT

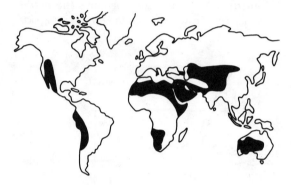

**desertification** The enlargement of the world's deserts. This is happening especially in Africa. It might be aggravated by deforestation and global warming. Some of it also might be a natural phenomenon unrelated to human activities. See *global warming*.

**dessiccant** A substance used to remove water or water vapor from a material. Cobalt chloride works well for this purpose. You might have seen this in little bags in bottles of pills. Various other compounds are also used.

**desiccation** The removal of all of the water from something, especially food, without altering any of its other chemical properties. We might, for example, buy tablets made from liver, dried out and ground into powder and pressed into pill form. These are sometimes used as a nutritional supplement.

**de Sitter, Willem** During the time of Albert Einstein in the early twentieth century, this cosmologist theorized that the universe is expanding. de Sitter found that this follows from Einstein's equations. It took time to convince Einstein of this, however. See *Big Bang* and *Einstein, Albert*.

**detection** 1. Sensing; indicating the presence of. 2. Extraction of the information from a modulated signal. See *detector*.

**detector** 1. A sensor; a device that indicates the presence of a certain phenomenon, such as ultraviolet radiation. 2. A device that removes the information from a modulated signal, making it suitable for listening, watching, or feeding to a computer or recorder. See *amplitude modulation, frequency modulation, pulse modulation*.

**detergent** A chemical that increases the wetting ability of water. Allows the water to penetrate where it otherwise would not. Used in washing and also in certain types of laxatives. Might be either natural (usually fat compounds or fatty acids) or artificially made.

**determinant** An indicator of whether or not a set of linear equations is consistent, that is, has a unique solution. If this is the case, the determinant is equal to zero. Consult a text on linear algebra for details.

**deuterium** An atom of "heavy hydrogen," with a nucleus having one proton and one neutron (see drawing). Normally, hydrogen nuclei consist of just a single proton. Deuterium is important in hydrogen-fusion reactions. See *fusion, hydrogen, tritium*.

DEUTERIUM

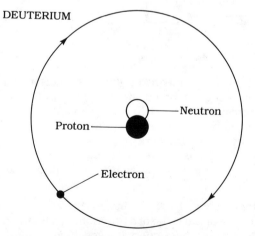

**deuteron** A subatomic particle consisting of a proton and a neutron; a deuterium nucleus. See *deuterium*.

**deviation** 1. A change from the normal or usual values. 2. An error or unscheduled change in the course of a ship, aircraft, or space vessel. 3. The maximum or instantaneous amount by which the carrier frequency swings away from the center frequency in a frequency-modulated signal. See *frequency modulation*.

**Devonian period** A period lasting from about 405 million to 350 million years ago. The sharks and insects appeared at this time. This is sometimes called the "age of the fishes." All fish today had ancestors in the Devonian period. See the appendix GEOLOGIC TIME.

**dew** Condensation that forms on surfaces when the air temperature drops below the value at which the relative humidity is 100 percent. The air can no longer hold all of the water vapor, so some of it condenses. See *dew point, relative humidity*.

**DEW line** See *distant early warning line*.

**dew point** The temperature at which dew forms. Usually it is lower than the air temperature. The closer the dew point to the air temperature, the higher the relative humidity. When the air temperature and the dew point are the same, the humidity is 100 percent. See *relative humidity*.

**diac** A semiconductor device with two terminals. A voltage is applied across these terminals. When the voltage reaches or exceeds a certain critical value, the device conducts.

**diabetes insipidus** 1. A disease in which the kidneys cannot concentrate the urine. The main symptom is excessive thirst and urination. 2. A disorder of the pituitary gland, causing excessive thirst and urination.

**diabetes mellitus** A disease that occurs when the pancreas does not produce enough insulin, or when the body does not properly use the insulin that the pancreas does produce. Blood glucose (sugar) cannot be efficiently metabolized. Various complications result. Treatment usually consists of injected insulin to properly metabolize the glucose in the blood, in conjunction with a medically prescribed diet.

**diagnosis** Determination of a specific illness that a patient has when he goes to a doctor or hospital. This allows proper treatment to be given.

**diagnostic medicine** A branch of medical practice, concerned with finding out the nature and causes of illnesses in individual cases. Once the cause of an illness is known, the right treatment can be given.

**diagram** An illustration that shows how a device looks or works. There are many different kinds of diagrams, used for various purposes. Some, such as schematic diagrams, show the wiring connections in electronic circuits. Others show the theory of operation of a machine. Still others might show such diverse things as the course of a ship or the strategy for a military operation. See *flowchart, isometric drawing, oblique drawing, schematic diagram*, and the appendix SCHEMATIC SYMBOLS IN ELECTRONICS.

**dial** 1. A rotary control, often with a visual display, in a machine or electronic device. 2. A rotary switch on an old-fashioned telephone that allows access to phone circuits. It has digits from 0 through 9. 3. To operate a rotary control as defined previously.

**dialing** Accessing a telephone circuit from a receiver, by means of the dial control. See *dial*.

**dialysis** A process in which the blood is detoxified by a machine. The machine does the work of the kidneys. Used when the kidneys are diseased or are not working, or have been removed or injured. See *kidney*.

**diameter** 1. The distance through a sphere or circle, as measured along a line through the center. 2. Inside diameter: The distance through a cross-sectional circle inside a tube or conduit, as measured along a line through the center. 3. Outside diameter: The same as the second definition, but measured between the outermost points of the cross section. 4. The distance through an irregular object, as measured along a line more or less through its center of mass or geometric center. See the drawing.

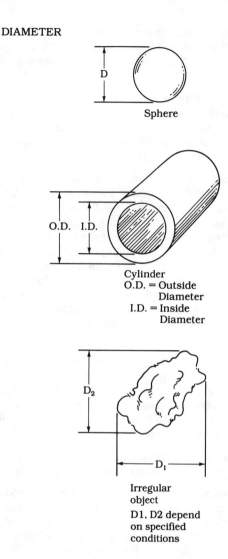

DIAMETER

Sphere

Cylinder
O.D. = Outside Diameter
I.D. = Inside Diameter

Irregular object
D1, D2 depend on specified conditions

**diametric** 1. Exactly the opposite of. 2. On the opposite side of. 3. Pertaining to diameter. See *diameter*.

**diamond** Carbon that has crystallized under pressure. It is the hardest known natural substance. It is cut into regular polyhedra shapes and has commercial value as a gemstone. In pure form it is clear, like glass, but it refracts light in a colorful and beautiful way. See *carbon*.

**diaphragm** 1. The vibrating part of a microphone, that moves when sound waves hit it, and whose energy is converted to electrical impulses by a crystal or a coil-and-magnet assembly. 2. A large muscle in the abdomen that makes the lungs expand and collapse, for breathing.

**diapir** Any object, such as a large blob of magma, that pierces the crust of the earth. See *crust, magma*.

**diarrhea** Abnormally rapid passage of feces through the intestines. Might cause water loss (dehydration) if severe. Often associated with improper diet, food poisoning, certain illnesses, or overuse of laxatives and cathartics.

**diastole** The portion of the heartbeat during which the ventricles are relaxed. This is when the blood pressure is lowest. See *diastolic, systole*.

**diastolic** Minimum pressure; pertaining to the blood pressure in between heartbeats, when it is lowest. Normal diastolic pressure is about 60-80 millimeters of mercury (mmHg) above atmospheric pressure. See *systolic*.

**diatom** A one-celled algae. Might exist in an infinite variety of shapes, and always leaves a fossil of silicate material, resembling glass (see drawing). When seen through a microscope, such a crystal is fascinating in its complexity and beauty.

DIATOM

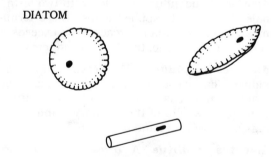

**dichlorodiphenyltrichloroethane** Abbreviation, DDT. A pesticide now banned in the U.S. Still used in some countries that export food. Known to cause harm to wildlife, it is suspected as a threat to human health also.

**dichotomy** 1. In plants, the splitting of the stem in two, forming branches. 2. A repeated branching by pairs. 3. A situation in which either of two conditions must hold; for example, true or false. 4. A process of dividing into two very different groups. 5. The moon at first quarter or third quarter. 6. Venus or Mercury illuminated so that we see a half disk.

**dieback** A population reduction following a rapid increase. Caused by the inability of the environment to support such large numbers of the species. It is thought that this will someday happen to humanity. Many experts think the world has far too many people already. See also *Malthusian model*.

**dielectric** A material that is an electrical insulator and has low loss when electric fields are placed across it. Some dielectrics cause a large increase in the energy that can be stored in an electric field. These materials especially include ceramics, mica, certain oxides, glass, some plastics, and even paper. See *dielectric breakdown, dielectric constant*.

**dielectric breakdown** A condition in which the voltage across a dielectric is so large that the dielectric conducts. This might cause temporary or permanent failure of the material. In air it is temporary; in paper it is often permanent.

**dielectric constant** For a dielectric substance, the ratio of the capacitance of a two-plate capacitor using the material, to the ratio of the same two-plate capacitor using dry air as the dielectric. Almost always larger than 1. Common values are 2 to 5. See *dielectric*.

**diesel engine** An internal combustion engine that works without the use of spark plugs. The heat for combustion is generated by the pressure in the cylinder as the piston moves up.

**diet** 1. The general pattern of eating, and types of food a species consumes. 2. A medically prescribed eating pattern, intended to alleviate or cure various disorders, or to cause a change in weight or water content of the body. 3. A restricted eating pattern intended to produce weight loss.

**dietary fiber** Undigestible carbohydrate, such as cellulose, found in some foods and serving to help the bowels function normally. Lack of fiber can cause constipation; too much causes bloating and gas. Wheat bran is the most common natural source. Also various powders and pills are available that contain fiber.

**dietary supplement** Any concentrated substances that help in ensuring adequate nutrition for the body. Vitamins, minerals, fiber powders and pills, and tonic liquids are examples. Some of these substances can be toxic in large amounts. It is always best to consult a doctor before taking these products. Ideally, a healthy diet should supply all the nutrients needed by a normal person. See the appendices RECOMMENDED DAILY ALLOWANCES and VITAMINS AND MINERALS.

**dieting** The practice of restricting one's eating in order to lose weight.

**differential** A very small difference in a variable. Used in calculus, it is a value that is made smaller and smaller, approaching zero but never quite getting there. As it does this, exact solutions are found to equations, because as the differential approaches zero, some unknowns approach these exact solutions. The solutions are found by observing this "approaching" process. See *differential triangle*.

**differential calculus** See *calculus*.

**differential equation** An equation in which derivatives appear. Solving these equations is a classical and sophisticated branch of mathematics. See also *derivative*.

**differential triangle** The triangle formed by the difference in y and the difference in x when evaluating the slope of the function $y = f(x)$ between two points P and Q on the function (see drawing). As P and Q are brought closer together, the triangle shrinks. The differences in y and x are called differentials when the triangle has shrunk down to arbitrarily small size. The slope of the line PQ then approaches the derivative of the function at the point P.

**differentiation** The process of finding the derivative either at a certain point, or in general. See *derivative*.

DIFFERENTIAL TRIANGLE

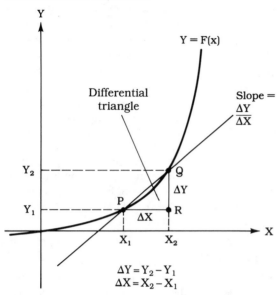

**differentiator circuit** An electronic circuit that produces an output that is equivalent to the rate of change of the input. It "takes the derivative" of the input waveform. See *derivative, integrator circuit*.

**diffraction** A phenomenon in which a wave disturbance "bends around" a barrer. This

DIFFRACTION

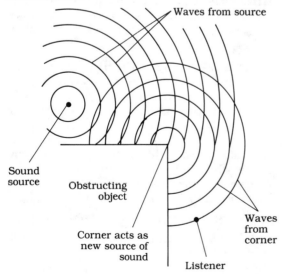

takes place especially if the barrier has a corner that is tiny compared to the wavelength. In this case the barrier acts as a new source of the wave disturbance (see drawing).

**diffraction grating** A clear film on which are drawn many fine black lines, extremely close together. When light passes through, the clear spaces between the lines cause diffraction. The result is an interference pattern or a spectrum. See *diffraction, interference, interference pattern, spectrum*.

**diffuse** Scattered over a region, rather than concentrated all in one spot.

**diffuse nebula** A cloud of interstellar dust and gas, scattered in space over a large region. See *nebula*.

**diffusion** 1. The scattering of light when it is reflected from a rough surface like a whitewashed wall, or when it passes through something translucent like frosted glass. 2. The scattering of any wave disturbance when it passes through an irregular medium or is reflected from an irregular surface.

**digestibility** The ease with which a food or chemical is digested. Some substances are not digestible at all, such as cellulose. See *dietary fiber*.

**digestion** The processing of food by the body into a form suitable for energy and tissue-building needs. It begins in the mouth and continues through the stomach and small intestine.

**digestive enzyme** A chemical that aids in digestion, and is secreted by the body. Sometimes digestive enzymes are given in tablet or syrup form when a person does not produce enough of them.

**digestive system** The system of the body responsible for digestion of food (see *digestion*).

**digit** 1. In the decimal numbering system, a numeral 0 through 9 in a specific position representing a power of 10. 2. In a binary numbering system, a numeral 0 or 1 in a specific position representing a power of 2. 3. In a base-n number system, a numeral 0 through n − 1, representing a power of n. See also *modulo*. 4. A small limb on the body, such as a finger or toe. 5. An element of a set.

**digital** Having either of two states, usually called "high" and "low." Might also be called "1" and "0," or "true" and "false," or "on" and "off." See the following definitions.

**digital communication** One-way or two-way message transfer, in which the signals actually sent are digital. Digital signals propagate more efficiently than analog ones. See *analog, digital*.

**digital enhancement** A method of improving the clarity of a satellite picture, or of a picture sent back by a space probe. Useful in mapping the earth, observing weather phenomena, and looking at features of distant planets.

**digitalis** A drug used to stimulate the heart. Also acts to remove excess water from the body. Useful in some heart ailments.

**digital meter** An indicating device that displays a value in numerals, usually 0 through 9. Many wristwatches are digital meters (chronometers). Digital meters are easier to read than analog ones. But for some purposes, analog meters are more useful, such as when it is necessary to adjust a control for a maximum level of current. See *analog meter*.

**digital signal** A means of sending information in discrete levels or steps, rather than as a smooth waveform. Might consist of pulses (on and off), or several different levels. The drawing shows how an analog waveform can be approximated in eight different levels. Digital signals propagate more efficiently than analog ones. See also *analog, pulse modulation*.

DIGITAL SIGNAL

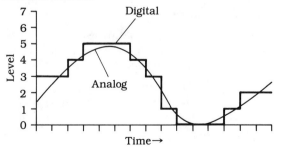

**digital-to-analog converter** Abbreviation, D/A or DAC. An electronic circuit that converts a digital signal (see *digital signal*) to smoothly varying form. Does the exact opposite of the function of an analog-to-digital converter (A/D or ADC). In some communications circuits, an A/D converter is used at the transmitting end, and a D/A converter at the receiving end. This allows the digital signals to be sent over the air or cable. See also *analog, analog-to-digital converter, digital signal*.

**dihedral wing** In an aircraft, a slanting of the wings slightly upward, rather than having them be exactly level (see drawing). This improves stability, so the plane is less likely to roll out of control.

DIHEDRAL WING

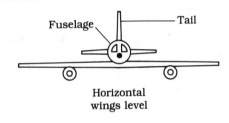

Horizontal wings level

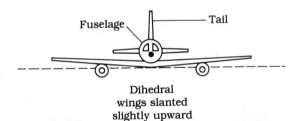

Dihedral wings slanted slightly upward

**dike** 1. A barrier that protects low-lying land against flooding. One of the largest such barriers in the world is in the Netherlands. Another system of dikes protects New Orleans in the U.S. 2. In geology, a tabular igneous intrusive rock that cuts across overlying rock strata.

**dimension** 1. A coordinate axis, used in a system to specify uniquely the position of a point. See *coordinate system*. 2. Height, width, or depth. 3. Any phenomenon that can vary in a continuous manner. 4. The number of coordinates needed to uniquely define a point. Also called dimensionality.

**dimorphism** A significant difference between two categories. An example is the greater height of males than females.

**dinosaur** A diverse species of giant lizard that is believed to have lived on the earth millions of years ago. The drawing shows some of the types. Some lived near water; some roamed the landscape. Some ate only plants, and others ate animals. Some could even fly. Some stood 50 feet tall. Then they suddenly became extinct. It is thought that a worldwide climate change was caused by some disaster such as a meteorite impact, and that cooler climate killed off the cold-blooded lizards.

**diode** A two-terminal electronic device, usually made from semiconductor materials, that conducts current in one direction but not in the other. Used as a rectifier, and sometimes as a detector. See *detection, rectification*.

**Diophantus** Also known as Diophantus of Alexandria. Lived in ancient Greece in the third and fourth centuries A.D. Diophantus was to algebra what Euclid was to geometry some centuries earlier (see *Euclid*). He is sometimes called the "father of algebra."

**Diophantine equation** Named after Diophantus, who enjoyed working with equations having exponents. An equation of the form $a + b = c$, where a, b and c represent the variables x, y and z all raised to the same integer power. See also *Diophantus, Fermat's Last Theorem*.

**diopter** For a lens having focal length d in meters, the number 1/d. The longer the focal length, the smaller the diopter, and vice-versa. The diopter increases as the magnification, or "power," of a convex lens increases. See *convex lens*.

**dioxin** Also spelled dioxan. A solvent, found in the ash from some incinerators, and known to cause cancer in animals. Also might be a threat to humans.

**DIP** Abbreviation for *dual inline package*.

**diphtheria** A contagious bacterial disease of the respiratory system. Also might cause neurological damage. Symptoms are flu-like, with sore throat and fever. If severe, the disease has complications affecting the heart, kidneys, and other organs. Immunization has largely eliminated this disease in the U.S.

**diplex** 1. The use of two transmitters with the same antenna. 2. Multiplex in which there are two signals. See *multiplex*.

**dipole** A pair of electric or magnetic poles, of opposite sense or charge, separated by a space. A characteristic pattern of flux lines exists between the pole centers (see drawing).

DINOSAUR

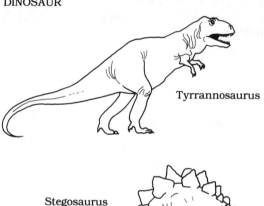
Tyrannosaurus

Stegosaurus

Brontosaurus

Triceratops

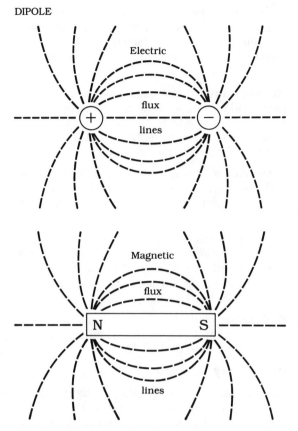

DIPOLE

**dipole antenna** A radio antenna, measuring 1/2 electrical wavelength, and fed at the center with coaxial cable or parallel-wire line. A simple and efficient antenna. Also called a doublet antenna.

**direct current** An electrical current that does not reverse direction. The intensity of the current might change, perhaps rapidly; but the polarity always stays the same. See *alternating current*.

**direction** 1. The coordinates or values that uniquely define the orientation of a line or ray in space. 2. See *azimuth*. 3. A general, imprecise expression of the orientation of a line or ray in space, such as "upwards" or "towards the constellation Orion."

**directional antenna** A radio antenna that works best in certain directions, chosen deliberately. A unidirectional antenna has just one favored direction; a bidirectional antenna has two (usually exactly opposite each other). Some antennas work well in three or even four different directions.

**direction angles** A set of three angles that uniquely define the direction of a line in Cartesian three-space. See also *Cartesian coordinates*.

**direction numbers** A set of three numbers that uniquely define the direction of a line in Cartesian three-space. See *Cartesian coordinates*.

**direct memory access** Abbreviation, DMA. In a computer system, the transfer of memory from one location to another, without having to go through the central processing unit (CPU). See *central processing unit, computer, memory*.

**direct motion** The apparent motion of a planet with respect to the distant stars, most of the time. This is west to east. Sometimes, the direction reverses; this is called retrograde motion. See *retrograde motion*.

**directrix** For a symmetrical curve or surface, an axis of symmetry. For example, a parabolic reflector has a directrix that denotes the direction in which it is aimed.

**dirigible** Also called a blimp. A large, helium-filled flying machine. It flies because it is lighter than air. Propulsion systems move it along. At one time it was thought that these would be the flying machines of the future; they were called airships. Then faster flying machines (airplanes) were developed.

**dis-** A synonym for non-. Means not.

**disaccharide** A sugar consisting of two simple sugar molecules joined together. Table sugar, or sucrose, is the most common example. It consists of one molecule of glucose and one of fructose. Milk sugar, or lactose, is another example. Disaccharides are split by digestive enzymes into their simple-sugar constituents.

**discharge** 1. In a battery, the gradual conversion of electrochemical energy into electricity. See *battery*. 2. The equalization of charge between the plates of a capacitor. See *capacitor*. 3. A sudden equalization of a charge difference between two poles; often takes the form of a spark. Lightning is the most dramatic example. 4. To cause any of the foregoing things to happen.

**discontinuity** 1. A break in an electrical circuit. 2. An abrupt change in the value of a function. See *discontinuous function*. 3. In geology, an interruption in sedimentation, producing a gap in a sample. 4. A zone of abrupt change between layers in the earth.

**discontinuous** Not continuous; that is, having one or more discontinuities. See *discontinuity, discontinuous function*.

**discontinuous function** A mathematical function that is not continuous; that is, it makes one or more "jumps," as shown in the drawing. See also *continuous function*.

DISCONTINUOUS FUNCTION

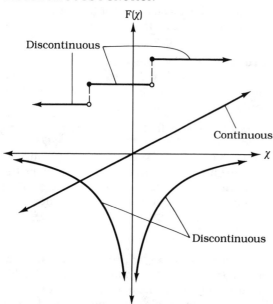

**discrete** Having definite, identifiable, individual parts, rather than being smooth and continuous. The set of integers is a discrete set; the set of real numbers is not.

**discriminant** A quantity that indicates the nature of the solutions of a quadratic or cubic equation. It allows one to tell whether the solutions are real or complex. Simple for quadratics and more complicated for cubics. For details, consult a text on intermediate or advanced algebra.

**disinfectant** A chemical used to clean surfaces and eliminate bacteria and viruses. Usually contains alcohol, phenols, or other substances that kill practically all living organisms without being overly toxic.

**disjunction** A sentence containing an OR logical function. Usually an inclusive OR, but sometimes an exclusive OR. See *logic, logic function*.

**disk** 1. Any circular data storage medium, such as a phonograph record, a laser phonograph record, or similar. 2. A circular magnetic data storage medium used by computers. See *floppy disk, hard disk*.

**disk drive** A device for recording and retrieving data from a magnetic computer disk. See *floppy disk, hard disk*.

**disk galaxy** See *spiral galaxy*.

**dispersion** 1. The splitting of light into its constituent colors or wavelengths, by means of a prism (see drawing). 2. Scattering of light or other energy by reflection from an irregular surface, or by passing through a translucent medium.

DISPERSION

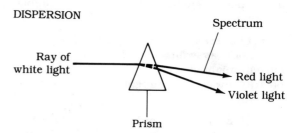

**displacement** 1. Distance covered by a moving object in a certain length of time. 2. The volume taken up by a submerged object, or the part of a floating object that is beneath the surface of a liquid. 3. A condition in which one substance takes the place of another. For example, chlorine might displace oxygen in the blood, causing unconsciousness.

**dissection** A process of taking apart and examining the parts of something. In particular, pertains to examining the bodies of animals or humans to learn about anatomy or to discover the cause of death, or to observe the effects of disease, poisoning, radiation, or other harmful agents.

**dissipation** 1. Conversion of energy or power in one form to energy or power in some other form, usually heat or electromagnetic radiation. 2. Power that is wasted in the form of heat, and that cannot go towards the intended purpose. There is always some of this, because no power conversion device is 100-percent efficient.

**dissociation** A process where the molecules of a substance split into their constituent atoms or into smaller molecules. An example is the electrolysis of water, where an electric current causes hydrogen and oxygen atoms to form at the electrodes.

**dissociative** 1. Tending to break apart. 2. Pertaining to dissociation. See *dissociation*.

**dissociative reaction** A chemical reaction in which molecules are split into their constituent atoms (elements) or into smaller molecules. See *dissociation*.

**dissonance** A combination of sounds that is unpleasant.

**distal** Farther away. This term is used especially in medicine. We might speak, for example, of the distal part of the small intestine, referring to that part nearer to the colon and farther away from the stomach. See also *proximal*.

**distant early warning line** Abbreviation, DEW line. A network of radar stations in northern Canada intended to detect bombers coming over the pole from Russia. Defenses of this kind are still in use, even though relations between the U.S. and the former European communist countries are much better than they were a few decades ago when the stations were first set up. See *Ballistic Missile Early Warning System*.

**distillation** 1. A process for removing impurities. The substance is boiled and the vapor collected by condensation (see drawing). The condensed vapor does not contain any impurities that did not boil out along with the desired substance. This is sometimes done with water to remove minerals like calcium and magnesium, and also various heavy metal impurities, bacteria and contaminants or poisons. 2. A process of refining. See *fractional distillation*.

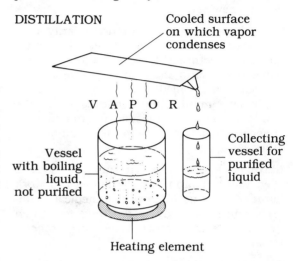

DISTILLATION

**distilled water** Water that has been purified by boiling and recondensing. See *distillation*.

**distortion** 1. In an amplifier, nonlinear operation. A condition where the output does not change in exact proportion to changes in the input. 2. Unacceptable sound quality or signal quality.

**distribution** 1. The arrangement of events in a probability function. 2. A probability function. See, for example, *normal distribution*. 3. Execution of the distributive law for mathematical operations. See *distributive law*. 4. The general arrangement of items, events, or phenomena. For example, an outbreak of measles might be confined to one town or spread out all over the whole state or region of the country.

**distribution frame** In a telephone switching system, the point where all lines come together for interconnection. A computer completes the calls by connecting the appropriate lines.

**distributive law** In mathematics, a property that two operations might have with respect to each other. Suppose we have two operations, + and *. Further suppose that for any three numbers x, y and z, it is always true that $x*(y+z)=(x*y)+(x*z)$. Then we say that * is distributive with respect to +. You might recognize this as true if * is multiplication and + is addition. It is not in general true the other way around, however.

**disturbance** Any change or irregularity in an otherwise stable or regular pattern. For example, the smooth east-to-west flow of the trade winds might spiral into a low-pressure system, also sometimes called a tropical disturbance. It might later mature into a hurricane. On the sun, disturbances are visible as prominences, sunspots, and flares. There are innumerable other examples.

**diuretic** A substance that causes the body to shed water; it promotes the kidneys to work harder. This is useful in certain medical conditions when the body is retaining too much water; however, overuse can cause mineral imbalances.

**diurnal** 1. Taking place once a day. 2. Active mainly in daylight.

**divergence** 1. A tendency to get farther apart. 2. A function in vector calculus. See a text on advanced calculus for details.

**divergent plate boundary** A boundary between two sections of the earth's crust, where the sections (plates) are moving apart. Generally, these places are at the midocean ridges, where new crust is being formed by the solidification of molten rock coming up from below. See *midocean ridge*.

**divergent series** In mathematics, a series that keeps growing, either positively or negatively, without limit as terms are added. It never comes to a definite numerical value; sometimes it is said to "approach infinity." See *convergent series, series*.

**diverging lens** See *concave lens*.

**diverging mirror** See *convex mirror*.

**diverticulum** In the large intestine, a protruding sac or pouch that often develops as a person ages (see drawing). To some extent these are normal. But they can become inflamed and this causes considerable discomfort and possibly illness. The condition of having diverticuli is called diverticulosis. When they get inflamed, the person has diverticulitis.

DIVERTICULUM

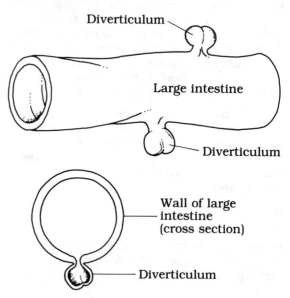

**divider** 1. A circuit that divides frequency by some integer number. 2. A device that can be adjusted to a certain spacing for reference measurement. 3. A partition.

**division** 1. Splitting into two or more parts. 2. A mathematical operation. The expression x divided by y is written $x/y$. If $x/y=z$, $x=yz$. Division by zero is not defined. 3. A part or segment of a larger whole. 4. To take a whole-number fractional part of a quantity. For example, we might put a 100-megahertz (MHz) signal through a divide-by-10 circuit to get a 10-MHz signal.

**Dixon, Bob** A professor at Ohio State University who observed signals from space in 1977 using a radio telescope. The signals were coming from the direction of the constellation Sagittarius, toward the center of our galaxy. He was involved in a search for signals from other worlds. The origin of the signal he heard in 1977 has never been determined.

**D layer** See *ionosphere*.

**DMA** Abbreviation for *direct memory access*.

**DNA** Abbreviation for *deoxyribonucleic acid*.

**dodecahedron** A 12-sided polyhedron. It is possible to have a regular polyhedron with this number of faces. In other words, we can have a 12-sided polyhedron with all the faces identical. See *polyhedron*.

**Dodgson, C. L.** A mathematician from Oxford University in England, better known as Lewis Carroll. He composed stories for children, in which logical games were played. *Alice in Wonderland* is the best known of his works.

**doldrums** A region near the equator where there is very little wind. Sailing ships had problems because of this. See *horse latitudes*.

**dolomite** A mixture of calcium carbonate and magnesium carbonate. Sometimes used as a mineral supplement to supply calcium and magnesium in the diet. Also called lime.

**domain** 1. In mathematics, the range of values for the independent variable for which a function is defined. See *function, independent variable, range*. 2. A territory in which a certain species lives. 3. A region in which a species, or certain members of a species, show territorial behavior.

**domestication** The taming of wild animals so that they can serve various purposes. We might tame wild horses so that we can use them for transportation. Some animals are easy to tame and others are more difficult.

**dominant trait** A reproductive trait that tends to appear more often than would be dictated by pure chance. Curly hair and brown eyes are dominant traits in humans, for example. See *recessive trait*.

**donor** 1. A person who gives blood or a body organ so that someone else who needs it in an emergency can get it. 2. A semiconductor dopant that adds electrons. See *acceptor, dopant*.

**dopamine** A chemical that assists in nerve function, especially in the brain; also has effects on the heart and kidneys. Used in the treatment of shock, congestive heart failure, and kidney failure.

**dopant** An impurity added to a semiconductor to make it behave a certain way. This makes it possible to design transistors, diodes, and integrated circuits for various purposes. See *doping, semiconductor*.

**doping** The addition of an impurity to a semiconductor to produce certain properties for manufacture of diodes, transistors, integrated circuits, and other semiconductor devices. See *acceptor, donor, semiconductor*.

**Doppler effect** The change in wavelength that occurs when an object moves. The wavefronts are shortened in the direction of motion, and lengthened in the opposite direction. See the drawing. See *blue shift, Doppler radar, Doppler shift, red shift*.

DOPPLER EFFECT

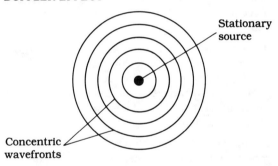

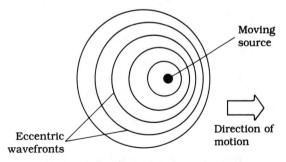

**Doppler radar** A radar instrument that detects not only the echoes, but also whether the object is getting closer or farther away with time (radial motion). Especially useful in locating severe weather, and in particular, tornado-producing thunderstorms. See *Doppler effect, Doppler shift*.

**Doppler shift**  The change in frequency or wavelength observed when an object has radial motion. If an object is getting more distant with time, wavelength is increased and frequency is decreased. If the object is getting closer with time, the wavelength is decreased and the frequency is increased. See *blue shift, Doppler effect, Doppler radar, red shift*.

**dorsal**  On the back or backside. For example, a shark has a prominent fin on its back, called a dorsal fin. See *ventral*.

**dose**  1. The amount of drug administered or needed to produce a certain effect in the body. 2. The amount of radiation to which a person, animal, or test sample has been exposed in a given period of time. Measured in rads. A dose of 100 rads within a few hours usually produces illness. A dose of about 500 rads is generally fatal if received in a short period of time. See *dose rate, rad*.

**dose rate**  The relative intensity of radiation. It is measured in rads per unit time, such as per hour, per day, or per year. A person normally gets about 10 rads in a lifetime. The dose rate might go up by a huge factor near the site of a nuclear accident. See *rad*.

**dosimeter**  A device that measures the dose of radiation to which a person, animal, or test sample has been exposed in a given period of time. Calibrated in rads. See *dose, dose rate, rad*.

**dosimetry**  The science of determining the amount of ionizing radiation, including alpha rays, beta rays, gamma rays, and X rays, to which a person, animal, or test sample has been subjected. See *alpha rays, beta rays, dose, dose rate, dosimeter, gamma rays, gray, rad, X rays*.

**dot product**  Also called the scalar product of two vectors. Let A = (v,w) and B = (x,y). Then the dot product A.B = vx + wy. The ordered pairs represent the end points of A and B in the Cartesian plane. The formula is similar for more than two dimensions. Consult a vector analysis text for further details. See *vector*.

**double helix**  Two intertwined helixes. This is the structure of the DNA molecule (see *deoxyribonucleic acid*). The strands of these molecules can reach lengths of several feet, although they are tightly curled up in cells. See the drawing.

DOUBLE HELIX

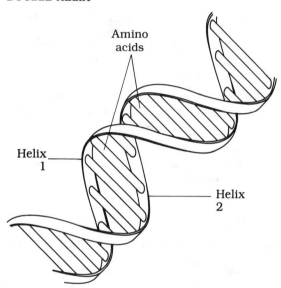

**doublet**  1. A type of antenna in which two identical sections are fed at the ends and placed end to end. 2. A half-wave length of wire or tubing fed at the center and used as an antenna. Also called a *dipole antenna*.

**double-vortex thunderstorm**  A thunderstorm in which there is an updraft, usually on the southwest side in the Northern Hemisphere, and a downdraft, usually on the northeast side (see drawing). The updraft spins counterclockwise and the downdraft spins clockwise in the Northern Hemisphere. This pattern is characteristic of severe thunderstorms. The updraft can develop into a tornado. See *thunderstorm, tornado*.

**Douglas, Donald**  An engineer who designed and built the DC-3, a passenger plane and the first of its kind, in 1935. It proved an exceptionally reliable aircraft, with two propeller-type

DOUBLE-VORTEX THUNDERSTORM

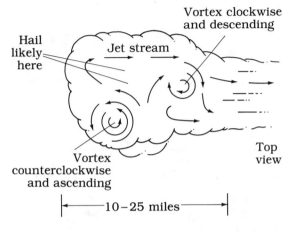

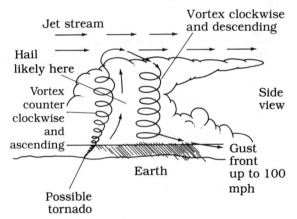

engines. Also designed the "gooney bird," an important cargo plane in World War II.

**downburst** 1. The downward-rushing air in the clockwise vortex of a severe thunderstorm in the Northern Hemisphere. This air causes high winds that blow outward in front of the storm. 2. A violent downdraft in a severe thunderstorm, also called a microburst. Can produce gusts of wind in excess of 100 miles per hour. See *double-vortex thunderstorm*.

**downlink** The signal sent down from a communications satellite, containing the information transmitted from the earth by the uplink. See *uplink*.

**Down's syndrome** Also called mongolism. A congenital defect usually caused by an extra chromosome. More common among older mothers. Mental and physical retardation are the main symptoms. As the person ages, heart disease and leukemia are more likely than with normal people; this reduces the life expectancy.

**downwelling** Downward movement of a fluid in the earth, such as magma in the mantle or surface water in the oceans. See *convection, upwelling*.

**dowsing** An ancient, unscientific method of searching for subsurface water. Sticks, called divining rods, are held out over land. When the sticks dip, there is supposed to be water under the ground. Of course, there is water underground almost anywhere, if one digs down far enough.

**draft** 1. The depth to which the hull of a boat goes when it is sitting in the water normally. 2. A copy of a file or document in progress, subject to revision. 3. An air current.

**drag** 1. The air resistance produced when an object moves in the atmosphere. 2. Unwanted air resistance against an airfoil. See *airfoil*. 2. Unwanted water resistance against a hydrofoil. See *hydrofoil*.

**drain** 1. The current drawn by a circuit. 2. The electrode in a field-effect transistor toward which charge carriers flow, and from which the output is usually taken. See *field-effect transistor*. 3. To get rid of excess water, especially in a parcel of land such as a field or a development area.

**drainage** 1. The runoff of excess water from a parcel of land. 2. The pattern followed by runoff water from a region.

**drainage basin** A region where all of the excess water flows into a single river. First, it forms small creeks or streams, then larger streams, and finally it flows into a major river.

**Drake, Frank** A radio astronomer who conducted the first serious search for signals from extraterrestrial beings. In 1959 he used the Green Bank, West Virginia facility for this purpose, calling his project Ozma (after the land of Oz). He also developed the Drake Equation for speculating about the number of planets in the galaxy likely to have intelligent life. See *Ozma Project*.

**drawdown** A process in which a species overuses resources. This is currently being done by humans with fossil fuels. It is predicted by some scientists that reserves of oil will be gone by about the year 2050, if current rates of consumption continue. These reserves took millions of years to be formed.

**dredging** The clearing of a channel in a river or bay or inlet. This is done by literally removing material from the bottom and putting it somewhere else. It allows larger boats to pass through than would otherwise be possible.

**drift** 1. A slow change in frequency, output voltage, or other characteristic of an electronic circuit, that ideally should remain constant. 2. To slowly change position or geographic location. 3. A formation of snow produced by blowing.

**drift ice** Ice in polar regions, especially in the Arctic, that drifts as winds push it. This has been a problem to explorers who didn't know the direction or speed at which the ice was drifting. In some cases, they were actually going south even though they were walking north, or vice-versa. See *ice pack*.

**drive** 1. The input to an electronic amplifier circuit. 2. Any force that propels a moving object.

**dropsy** See *edema*.

**drought** A prolonged shortage of rainfall below the normal or typical amounts for a region.

**drought cycle** A periodic recurrence of droughts, thought by some scientists to be related to the 22-year sunspot cycle. Some scientists, on the other hand, don't think there is any pattern or cycle, and that droughts are just chance happenings.

**drumlin** A long hill or irregularity in the terrain, formed by the pressure of moving glaciers long ago.

**dry adiabatic lapse rate** The rate at which dry air cools when it rises. As the air rises, the pressure goes down. The temperature in the atmosphere normally drops by 5.4 degrees Fahrenheit for every 1000 feet increase in altitude. This is about 1 degree centigrade per 100 meters. As dry air sinks it warms at this same rate.

**dry cell** An electrochemical cell, also called a battery, in which the chemicals are sealed in and resemble a paste or wet powder. This makes it possible to jostle or tumble the battery without upsetting its contents. See also *battery*.

**dry ice** Frozen carbon dioxide. At atmospheric pressure it evaporates directly to a gas, bypassing the liquid phase. It is called dry because it doesn't melt and get things wet. Used for food storage in transport.

**dual inline package** A method of manufacturing integrated circuits. The pins are in rows on either side of the molded epoxy or plastic case (see drawing), making it easy to insert and remove the device from a socket designed for it. See *integrated circuit*.

DUAL INLINE PACKAGE

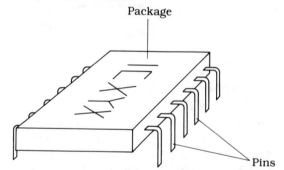

**duality** 1. A condition of having two different or contrasting states, seemingly at the same time. A good example is the particle/wave nature of light. 2. In geometry, the notion that the words "point" and "line" can be interchanged in a theorem about points and lines, and the result is still a valid theorem. Logically justified by Julius Plucker in the early 19th Century.

**duct** 1. In animals, a canal through which fluid is passed. For example, we have tear ducts in our eyes. 2. A device designed to guide matter or energy from one place to another. 3. In the atmosphere, a phenomenon where layers of air, having different temperatures, cause radio waves of very high or ultrahigh frequencies to travel long distances because of internal refraction or reflection at the boundaries between the air masses.

**ductile** Flexible, bendable, or malleable. Lead and copper have this property, whereas calcium, in pure form, does not. The opposite of brittle.

**dune** 1. A windblown pileup of sand, dust, dirt, or snow. Especially pertains to sand. In the case of snow, it is called a drift. 2. Sand embankments on an ocean shore, often with plants such as sea oats that help keep the embankments in place and discourage erosion. A fragile environment with a variety of species.

**duo-** 1. Pertaining to pairs; two. 2. Two phenomena or objects that work together to produce a certain effect.

**duodenum** The place where the stomach empties into the small intestine. This is where absorption of food and nutrients really begins.

**duralumin** An alloy of aluminum used in the construction of aircraft. Has excellent ability to withstand stresses without fatiguing. Also is light in weight, which is important for aircraft.

**dura mater** A protective covering over the central nervous system. It helps to cushion the brain from shock in the event of blows to the head.

**durum** A hard form of wheat used to make pasta such as macaroni. It has been used as food since about the First Century B.C.

**Dust Bowl** Primarily, the decade of the 1930s when much of the central United States was struck with drought. Soil erosion occurred on a vast scale. Sometimes we hear about the possibility of having another occurrence of this kind, especially in times of drought. See *drought*.

**dust storm** A condition that develops when high winds blow over dry land with loose topsoil, or over a desert. Large amounts of dirt are carried aloft and blown for miles. In a severe storm of this kind, the whole sky darkens and dust gets into everything. Might last for hours or even days.

**duty cycle** The percentage of the time that a machine or electonic circuit is required to be in the "on" or "high" condition. Can vary from zero (off all the time) to 100 percent (on all the time). See the drawing.

DUTY CYCLE

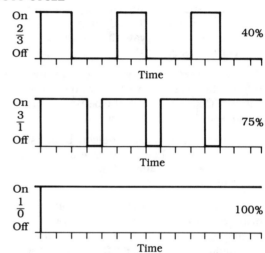

**dwarf** 1. A person who is far shorter than normal. Pituitary dwarfism: Small body but normal proportions, caused by underactive pituitary hormones (growth hormones). Hypothyroidism: Small body with the proportions similar to those of an infant. 2. Anything small by comparison with the normal or average type.

**dwarf star** 1. A star with a mass of only about $1/10$ that of the sun or less. Such stars tend to last a long time because they use up their fuel slowly. But they are very dim compared with the sun. 2. A star that is dead or almost dead. Very dim and extremely dense, consisting of heavy elements. See *black dwarf, white dwarf*.

**Dy** Chemical symbol for dysprosium.

**dye** A chemical compound commercially used to give artificial color. There are various different chemical classifications, used for such purposes as coloring food, clothing, and paper. They are usually organic substances.

**dyn** Abbreviation for dyne.

**dynamic** 1. Pertaining to the behavior of a system, machine or circuit when conditions are constantly changing. See *dynamic behavior, dynamic meteorology, static.* 2. Moving or in motion.

**dynamic behavior** The way that a system operates when conditions are constantly changing. In an electronic amplifier, for example, the input is usually an alternating or fluctuating current or voltage. The dynamic behavior of this circuit is the way that it operates with respect to this input signal.

**dynamic meteorology** A field in meteorology, or the science of weather and weather forecasting. Concerned with the behavior of moving air masses, and the way they interact.

**dynamics** The study of objects or substances in motion and how they interact. Subfields include aerodynamics, fluid dynamics and mechanics.

**dynamite** An explosive based on nitroglycerin. See *nitroglycerin.* Nowadays, the substance is pressed into sticks like large firecrackers, and is used in mining and construction.

**dynamo** See *generator.*

**dynamometer** 1. A machine used to evaluate the performance of automobile engines and other internal combustion engines. 2. A device that measures force.

**dyne** A unit of force, equivalent to 0.00001 newton. The amount of force needed to accelerate a mass of one gram at one centimeter per second per second. See also *newton.*

**dysentery** An intestinal ailment caused by a bacterium. Symptoms include diarrhea, bloating, and gas. When severe, there might be bleeding and mucus in the stools, with general discomfort. Might spread to the liver and right lung if left untreated. In exceptionally serious cases, widespread infection occurs.

**dysfunction** A malfunction, especially pertaining to processes in the body.

**dyslexia** An apparently hereditary problem that some people have, often mistaken in childhood for retardation or "learning disability." Words appear mixed up on the page, or letters appear backward or switched around. There have been major advances recently in the diagnosis and treatment of this problem.

**dysprosium** An element with atomic number 66. The most common isotope has atomic weight 164. In pure form it is a silver-colored metal. It is used in various alloys for nuclear research.

**dystrophy** 1. Wasting of body tissue, or retarded development of body tissue, especially muscle. 2. See *muscular dystrophy.*

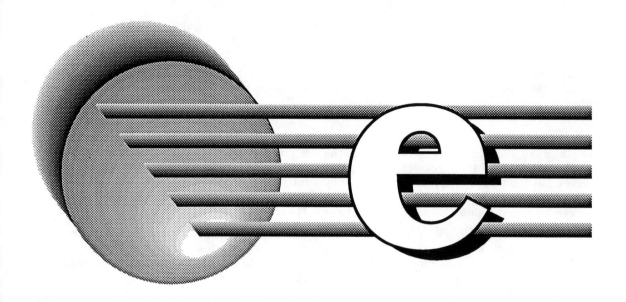

**E** Abbreviation for electric field, exa- (see appendix PREFIX MULTIPLIERS), voltage.

**e** 1. Abbreviation for electron charge. 2. The transcendental, irrational number whose first few digits are 2.7182818. This is the natural logarithm base. See *logarithm*.

**Eads, James** A civil engineer of the 19th Century. He is noted for his work with bridges. He invented and developed a new means of setting bridge foundations under river bottoms.

**ear** The organ of the body that converts sound waves into nerve impulses. These impulses are sent to the brain for processing. See the illustration for a diagram of the human ear.

**eardrum** The part of the ear that vibrates as air molecules strike it. The compression waves in the air are thus changed into mechanical movement. This movement is transferred to the middle and inner ear, where it is converted into nerve impulses. See the illustration for ear.

**Earle, Wilton** A physiologist who first succeeded in growing live human tissue outside of the body. He did this in containers, in which conditions could be controlled.

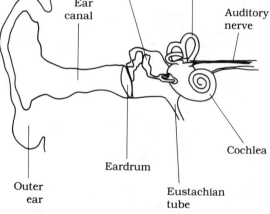

**earphone** A miniature speaker that can be placed on or in the ear.

**Earth** 1. The third planet in the Solar System of our parent star, the sun. About 8,000 miles in diameter and 93 million miles from the sun. It is the only planet in the Solar System known to harbor life. 2. Ground or soil.

**earthflow** A slow movement of soil downslope, caused by saturation with water, in combination with gravity. Slower than a mudslide, but faster than a creep. Only grass can grow on the slope.

**earthquake** The sudden breaking of the earth's crust. The fracture sends out shock waves that can be detected many miles away by seismic instruments. Near the epicenter (see *epicenter*), there might be serious structural damage caused by the vibrations.

**earthquake lights** Illumination of the sky, visible at night, that sometimes accompanies earthquake activity.

**earth science** A large field of science involved with the physical study of the earth. Subfields include such disciplines as geology, oceanography, and meteorology. See the table in the front of this book for subfields of earth science that are defined here.

**earthshine** The reflected light from the earth that produces some illumination on that part of the moon visible from the earth, except when the moon is full or nearly full. This glow is greatest when the earth looks full from the moon, that is, when the moon is new or nearly new (see the drawing). You might see this glow on the moon when it is in its crescent phase.

EARTHSHINE

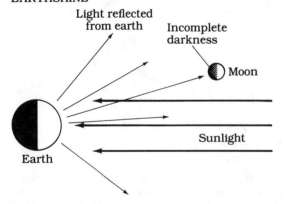

**Eastern Daylight Time** Eastern Standard Time plus one hour, or Coordinated Universal Time minus four hours. Generally used from early April till the last weekend in October in the U.S. See *Eastern Standard Time, Coordinated Universal Time*.

**Eastern Standard Time** Coordinated Universal Time minus five hours. Used along and somewhat either side of the longitude line 75 degrees West. This runs through the eastern U.S. Used from the end of October till the beginning of April in most applicable places. See *Eastern Daylight Time, Coordinated Universal Time*.

**Eastman, George** One of the people who made early movies. Thomas Edison convinced Eastman to make rolls of film that were run as the first movies.

**East Pacific rise** A mountain range 6000 miles long, in the eastern part of the Pacific ocean. It is the counterpart of the mid-Atlantic ridge. Some of the strangest life forms on the earth live on the floor of the ocean in this region. See *midocean ridge*.

**Eaton, Amos** A historian who traveled around the northeastern U.S. in the early nineteenth century teaching people about science. In particular, he gave lectures on earth science and life science.

**ebb tide** An outgoing tide. That is, the period of time after high tide during which currents tend to go out to sea. See *tide, tidal bore, tidal current*.

**EBS** Abbreviation for Emergency Broadcast System.

**eccentricity** 1. Degree of being off center. 2. The extent to which an ellipse differs from a circle. The more elongated the ellipse, the greater the eccentricity. See *ellipse*.

**ecdysis** See *moulting*.

**ECG** Abbreviation for electrocardiogram.

**echelon** A device for splitting light into its constituent wavelengths. Has extremely high resolution, and can examine spectral absorption and emission lines in great detail. See also *absorption line, emission line, spectroscope*.

**echinoderm** A type of marine life. Examples are starfish and sea urchins. They all have exoskeletons. They are an ancient life form. Some existed as long ago as the Ordovician period. See the appendices ANIMAL CLASSIFICATION and GEOLOGIC TIME.

**echo** 1. In sound, a reflected compression wave that arrives with a delay. 2. The reflected electromagnetic wave from a radar target or obstruction. 3. The reflected sonic or ultrasonic wave underwater from a sonar target, obstruction, or the bottom. See *depth sounding, sonar*.

**echolocation** A means of finding one's position by sounding from at least two different directions. Two circles can then be drawn from the points where the echoes occur. The point where the circles intersect is the position of the vessel.

**echo sounder** A device for determining the distance to an object by sending out sound waves that are reflected from the object. The time delay for arrival of the echo is used to calculate the distance to the object, when the speed of sound in the medium (air or water, usually) is known. See *depth sounding, sonar*.

**eclipse** 1. Lunar eclipse: The moon passes through the earth's shadow. If the umbra (darkest part) of the earth's shadow completely covers the moon, the eclipse is total. If not, it is partial. See drawing. 2. Solar eclipse: The moon's shadow falls somewhere on the earth. The moon and the sun are both about 1/2 degree in diameter as seen from the earth; therefore the area of totality for a solar eclipse is small. A total solar eclipse is dramatic. A partial solar eclipse might pass unnoticed, especially on a cloudy day.

**eclipse season** A short time during the year when eclipses are likely. Every year has two such times, six months apart, when lunar or solar eclipses are most likely, because the moon passes through the ecliptic when it is on a line with the earth and the sun. See *eclipse, ecliptic*.

**eclipsing binary star** A binary star (see *binary star*) in which one member is dim and large, and the other is bright and small, and the plane of their orbit is edgewise to us. Periodically, the big, dim star eclipses (passes in front

ECLIPSE

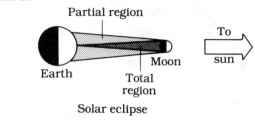

Solar eclipse

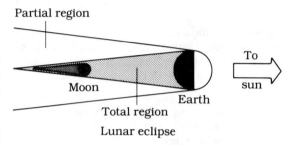

Lunar eclipse

of) the small, bright star, and the image we see is dimmed at regular intervals.

**ecliptic** 1. The plane of the earth's orbit around the sun. 2. The line on the celestial sphere representing the plane of the earth's orbit around the sun. The sun seems to move slowly along this line from west to east, making one complete cycle in a year. See *celestial sphere*.

**Ecole Normale** A French school at which some of the most famous mathematicians and scientists learned and taught. The students were less carefully selected than at the other major school opened in the late Eighteenth Century, the Ecole Polytechnique. See *Ecole Polytechnique*.

**Ecole Polytechnique** A select military school in France opened in the late 18th Century. The equivalent of modern-day California or Massachusetts Institutes of Technology. Elite engineers, scientists, and mathematicians studied and taught here.

**ecological balance** A condition in which nature is allowed to take its course, with minimal interference from human activities. Some human activities, such as farming and fishing, do not normally unbalance the ecology; others, such as extensive burning of fossil fuels or use of chlorofluorocarbons, do.

**ecology** 1. The environment in a specific area or region. 2. The study of the environment and the factors that affect it. Especially pertaining to living things.

**economics** A branch of social science that deals with large-scale market phenomena (macroeconomics) and small-scale phenomena (microeconomics). Economists try to find the causes of various problems, and to make forecasts and recommendations, in the budgets of whole countries as well as for individual people. This often involves sophisticated mathematics.

**ecosystem** A region with naturally interacting life forms and physical phenomena. An example is a lake, with fish, weeds, perhaps streams draining into it, and trees around it. This system could be upset in various ways, such as from a change in the course of a river leading into it (a natural event) or acid rain (could be natural or human-made).

**ecto-** A prefix meaning outside of.

**ectoderm** In an embryo, the layer of tissue that eventually becomes the skin.

**ectomorph** A person who is usually thin, introverted and high-strung. See also *endomorph, mesomorph*.

**ectoplasm** See *cytoplasm*.

**eczema** 1. Any inflammation of the skin. 2. A skin problem with the hands, often seen in people who wash dishes often or who wear rubber gloves. Probably caused by simple irritation, or by bacteria, viruses, or fungi that thrive in warm, moist environments.

**Eddington, Sir Arthur** A noted English astronomer and cosmologist who was a professor at the University of Cambridge in the early twentieth century. He did much of his work during the time that Einstein was gaining fame for his relativity theory. He saw inertia as a natural consequence of the total gravitational effect of everything in the universe. See *inertia, relativity*.

**eddy** A whirlpool in a liquid, often with a "dimple" or depression at the center (see drawing). Caused by drainage from underneath, or by turbulence in the liquid.

EDDY

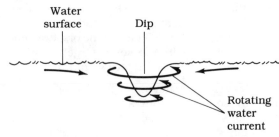

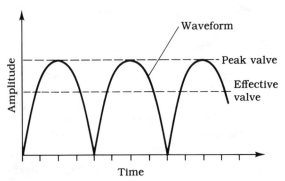

**eddy current** In a ferromagnetic material, a current that tends to develop when the substance is in a strong, alternating magnetic field. This can cause heating of the material and inefficiency in devices like transformers. The core material in transformers is usually broken into thin laminations, or else powdered, to reduce this loss, known as eddy-current loss.

**edema** An abnormal retention of water by the body. Occurs in some women before their menstrual cycles. Also can occur in various disease states or with mineral or hormone imbalances. The main symptoms are bloating and pain in the feet when standing.

**Edison, Thomas** Perhaps the most prolific inventor of all time. Born in 1847 in Ohio. Died in 1931. Known for such things as the electric light, the phonograph ("gramophone"), and many other devices we use today. It is said that he worked up to 20 hours a day and never got tired of his work.

**EDT** Abbreviation for Eastern Daylight Time.

**EDTA** Abbreviation for ethylenediaminetetracetic acid.

**EEG** Abbreviation for electroencephalogram.

**effective value** For a fluctuating phenomenon, the value of the same phenomenon, unchanging, that produces the same effect averaged over time. The drawing shows a full-wave rectified current waveform, and the equivalent direct-current value, which is the effective value of the waveform.

**efficiency** 1. A general expression for how much power or energy is used for the intended purpose in a system. 2. For a device with input power P and output power Q, both measured in the same units (such as watts), the fraction $Q/P$, or the percentage $100(Q/P)$.

**egg** The female cell that, when combined with a sperm cell, produces a zygote that develops into an embryo and finally into an animal. See *sperm, zygote*.

**EHF** Abbreviation for extremely high frequency.

**EIA** Abbreviation for Electronic Industries Association.

**eigenvalue** 1. A solution to a characteristic equation of a matrix. See a text on linear algebra or advanced algebra. 2. A fixed energy level for a wave function. See a text on advanced physics.

**Eightfold Way** A theory, depicted in a paper, *The Eightfold Way*, published by Murray Gell-Mann at California Institute of Technology in 1961. The paper dealt with a theory that predicted a nuclear particle called Omega Minus. There were eight properties describing the interaction among subatomic particles. This provided a neat and elegant mathematical solution to a problem that up till then was baffling. See a text on advanced physics for details.

**Einstein, Albert** A German scientist, born in 1879. He was never satisfied with ordinary, humdrum life. He became one of the most brilliant theoretical physicists of all time. While working days in the Swiss Patent Office, he spent his evenings and weekends formulating his relativity theory (see *relativity*) in the early years of the twentieth century. Always modest, Einstein disliked snobs and did not especially care for all the fuss that was made about him in his later life. He said that he never lost his childish sense of wonder—he never "grew up"—and this helped him achieve the mental intensity needed to make his discoveries.

**Einstein equation** The famous expression that energy is equivalent to mass times the speed of light squared ($E = mc^2$).

**einsteinium** Chemical symbol, Es. An element with atomic number 99. Does not occur in nature. Occurs as a by-product of nuclear weapons explosions. The most common isotope has atomic weight 254. It is a metal in pure form. It loses half of its radioactivity in about nine months.

**EKG** Abbreviation for *electrocardiogram*.

**elastance** An electrical quantity that is the reciprocal of capacitance; an expression of the ease with which a component loses its electrical charge. If C is the capacitance in farads, the elastance S, in darafs (farad spelled backwards), is 1/C. Elastance should not be confused with conductance, also often symbolized S. See also *capacitance*.

**elastic collision** See *collision*.

**elasticity** The ease with which an object can be stretched and then return to its original shape or length when the stretching force is removed. Rubber has good elasticity. Copper wire does not; it can be stretched, but once this has been done, it stays stretched.

**elastic rebound theory** A theory that earthquakes depend on the stretchability of rocks in the earth.

**E layer** See *ionosphere*.

**electret** A device that has a permanent electric field; the electrical counterpart of a permanent magnet. Certain waxes, ceramics and plastics can be made into electrets by heating them, and then letting them cool in a strong electric field.

**electrical conductivity** The ease with which a substance lets charge carriers, usually electrons, move. Substances with very poor conductivity are called insulators. Those with high conductivity are conductors; those with medium conductivity are semiconductors. Can be quantitatively expressed in siemens per meter. The reciprocal of resistivity. See *siemens*.

**electrical distribution system** The network of power lines and transformers that carries current from the generating plants to the users. See the diagram.

ELECTRICAL DISTRIBUTION SYSTEM

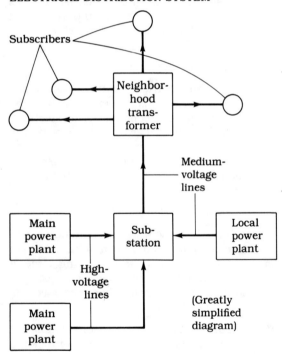

(Greatly simplified diagram)

**electrical energy** Energy expressed in kilowatt hours or watt hours. Electrical power expended over a period of time. See *kilowatt hour, watt hour*.

**electrical engineering** A branch of engineering concerned with the design and construction of electrical systems, such as power plants, transmission lines, and transformers. Sometimes includes the field of electronic engineering. See also *electronic engineering*.

**electrical ground** An earth connection that presents a low resistance for direct current and 60-hertz alternating current. Used primarily for safety purposes. The third prong (center) of a three-wire appliance cord is connected to electrical ground.

**electrical polarity** 1. The orientation of the positive and negative charges in an electric circuit. See *electric charge*. 2. The direction in which current flows in a circuit.

**electrical resistivity** The extent to which a substance resists the flow of charge carriers, usually electrons. Substances with high resistivity are called insulators. If a material has moderate resistivity, it is a semiconductor. If the resistivity is low, the material is a conductor. Can be quantitatively expressed in ohms per meter. See *ohm*.

**electric charge** An excess or deficiency of electrons in a substance. An excess produces a negative charge; a deficiency produces a positive charge. Measured in coulombs. One coulomb of charge is about $6.24 \times 10^{18}$ electrons. See *coulomb*.

**electric engine** A motor, used to drive a mechanical device such as a car or boat. See *motor*.

**electric eye** A device that detects the passage of people, cars or other objects. Usually works by means of a beam of light or infrared that is interrupted by the passing object(s), as shown in the illustration. Used for counting and also for intrusion detection.

ELECTRIC EYE

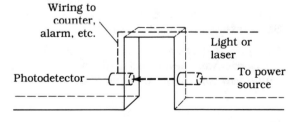

**electric field** A region surrounding an electrically charged object, in which other charged objects are attracted or repelled. Negative repels negative, and positive repels positive; negative attracts positive. The field intensity is measured in terms of the flux per unit cross-section (see *electric flux*). It varies inversely with the square of the distance from the charged object.

**electric flux** Imaginary "lines of force" surrounding a charged object. The lines radiate inward towards, or outward from, the object in straight paths. When other charged objects are brought near, the flux lines bend in complex ways depending on the relative polarities and intensities of the charges. See also *electric charge, electric field*.

**electric heating** Heating accomplished by passing electricity through resistive wires. This causes the wires to become hot. The heat is transferred to the environment by radiation, convection and/or conduction.

**electricity** 1. An excess or deficiency of electrons in a substance causing a force field to develop around the object. Can be quantitatively expressed in coulombs or volts. This is static electricity. See *coulomb, volt*. 2. The flow of charge carriers between poles of electron excess and deficiency. Can be quantitatively expressed in amperes. This is dynamic electricity. See *ampere*. 3. The branch of science concerned with charge, current, and resistance.

**electric light** 1. Any device that converts electric current into visible light. 2. A device, invented and perfected by Thomas Edison in the Ninteenth Century, that creates visible light by intense heating of a resistance wire. The wire glows white hot when current passes through it. Modern lights use tungsten wire. See *arc lamp, fluorescent lamp, incandescent lamp, mercury-vapor lamp, neon lamp, sodium-vapor lamp*.

**electric motor** See *motor*.

**electric potential** The excess or deficiency of electrons that produces an electric field, and results in a flow of charge carriers if there is a conductive or semiconductive path between poles. Measured in volts. See *volt*.

**electric power** The product of voltage and current in an electric circuit. If the voltage E is given in volts and the current I is given in amperes, then the product of these, EI, is the power P in watts. The power also can be calculated as the resistance R times the square of the current ($P = I^2R$), or as the square of the voltage divided by the resistance ($P = E^2/R$).

**electric shock** The effects of electric current when it passes through human tissues. A brief shock of a few milliamperes through the heart can cause heart fibrillation (see *fibrillation*). Extremely high-current shocks cause deep burns. Permanent tissue damage can result. As little as 20 to 40 volts can produce a fatal shock under some conditions.

**electrocardiogram** A device for recording or displaying the electrical impulses of the heartbeat. Different parts of the wave are shown in the diagram; they are called P, Q, R, S, and T. The recorded waveshape is called an electrocardiograph.

ELECTROCARDIOGRAM

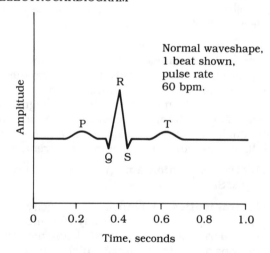

Normal waveshape, 1 beat shown, pulse rate 60 bpm.

**electrochemical** Pertaining to devices that convert chemical energy to electrical energy or vice-versa. Examples are most batteries (chemical to electrical) and the electrolysis of water to yield hydrogen and oxygen (electrical-to-chemical).

**electrochemical equivalent** Usually abbreviated Z. For a metal, the mass in grams that is deposited by one coulomb of charge in an electrolysis reaction. See *electrolysis*.

**electrochemistry** The science dealing with electrical effects of chemical reactions, and chemical phenomena produced by electricity. A cross between the sciences of electricity and chemistry.

**electrode** A conducting bar or plate used in electrochemical devices. Might serve a variety of purposes. Usually it establishes electrical contact between a conducting wire and a chemical medium.

**electrodeposition** The use of electricity to cause a substance to be deposited on another substance. Electrolysis is a common method of doing this (see *electrolysis*).

**electrodialysis** A means of getting liquids to flow by applying an electric field across a porous barrier. Ions in the liquid are attracted and repelled by the electric charge poles causing the liquid to flow through the holes.

**electrodynamics** A branch of electricity and physics involved with the interaction between electrical and mechanical effects. See *electrokinetics*.

**electrodynamometer** A meter for measuring electric current. Instead of a permanent magnet that interacts with the magnetic field produced by the meter coil, an electromagnet is used. This electromagnet operates from the current to be measured, so the device can measure either direct current or alternating current.

**electroencephalogram** A plot of brain waves. See *brain waves*.

**electrokinetics** A branch of electricity and physics, concerned with the way that moving, charged particles behave. Also involved with the behavior of moving materials in electric fields. See also *electrodynamics*.

**electroluminescence** 1. The property of certain materials that enable them to glow when subjected to an electric field. 2. The glowing of such a substance when an electric field is applied across it.

**electrolysis** 1. The effect in which a current, passing through a conducting solution (electrolyte), causes a chemical change in the solution and/or in the electrodes. Usually, elements or compounds from the solution are deposited on the electrodes or bubble off of them. 2. An electrical method of destroying hair roots.

**electrolyte** 1. A substance that ionizes in solution. The solution then conducts electricity. Salt is an example; when placed in water it causes the solution to become a fair conductor. 2. A mineral needed by the body, in particular sodium or potassium. See *electrolyte balance*.

**electrolyte balance** In the body, a delicate interaction between such elements (minerals) as sodium and potassium. Fluid losses, stress, and certain illnesses can cause an imbalance and unpleasant symptoms. A medical doctor's evaluation is needed to determine if electrolytes are out of balance.

**electrolytic capacitor** A capacitor in which one plate is an aluminum can and the other electrode is an electrolyte solution. The dielectric is an oxide film that forms on the can. Provides a large capacitance in a small volume. See *capacitor, dielectric*.

**electrolytic cell** A cell containing an electrolyte and two or more electrodes. Examples are most batteries, electrolytic capacitors, and devices for electrolysis. See *battery, electrolysis, electrolytic capacitor*.

**electromagnet** A coil of wire with an iron or nickel core that becomes highly magnetized when a current passes through the wire (see drawing). The magnetic field disappears when the current is shut off, except possibly for a small residual magnetism in the core.

**electromagnetic field** The type of force field responsible for radio waves, infrared, visible light, ultraviolet, X rays, and gamma rays. Consists of an electric field at right angles to a magnetic field. Both fields alternate in direction at a certain frequency that might vary from a cycle every few years to trillions or more cycles per second. See *electromagnetic spectrum*.

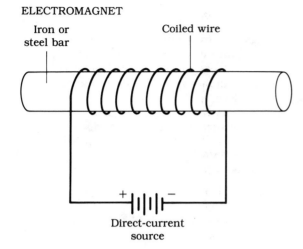

ELECTROMAGNET
Iron or steel bar
Coiled wire
Direct-current source

**electromagnetic induction** A phenomenon in which an alternating current in one conductor causes alternating currents of the same frequency in nearby conductors. This is the principle by which a transformer works. See *transformer*.

**electromagnetic particle** See *photon*.

**electromagnetic pulse** A strong electromagnetic field produced for a fraction of a second by lightning, nuclear explosions, and other phenomena in which many electrons are caused to move suddenly. This pulse can cause damage to electrical appliances; the electromagnetic induction results in large currents in anything that conducts. See *electromagnetic induction*.

**electromagnetic radiation** The propagation of an electromagnetic field. See *electromagnetic field*.

**electromagnetic spectrum** The range of possible frequencies or wavelengths of electromagnetic fields. Theoretically there are no limits to how long or short the wavelength can get, and to how low or high the frequency can be. But in practice, it is rare to observe wavelengths longer than a few hundred miles or shorter than the microscopic size of gamma rays. See the appendices ELECTROMAGNETIC SPECTRUM, RADIO SPECTRUM.

**electromagnetic units** A system of electrical units now replaced by the units derived from the Standard International System. See the appendix STANDARD INTERNATIONAL SYSTEM OF UNITS.

**electromagnetic wave** The observed effect of an electromagnetic field that seems wave-like. A particle, called a photon, having a certain energy, produces a certain electromagnetic wavelength. See *particle/wave dichotomy, photon*.

**electromagnetism** 1. A magnetic field caused by the flow of an electric current. 2. The science dealing with the relationship between electric currents and fields, and magnetic currents and fields. 3. See *electromagnetic wave*.

**electromechanical** Pertaining to any device that converts energy from electrical form to mechanical form, or vice-versa, such as a motor or generator. See *generator, motor, motor/generator*.

**electrometer** A voltmeter that measures extremely low voltages. This can be done in a variety of ways. See *voltmeter*.

**electromotive force** Abbreviation, EMF or emf. The potential, or voltage, that moves charge carriers, causing a flow of electric current. This term is generally used to refer to voltage produced by an electric generator, thermocouple, solar cell, or battery. Measured in volts. See *volt, voltage*.

**electron** A negatively charged subatomic particle. Can be either free or bound to the nucleus of an atom. Electrons in atoms exist in spherical shells of various radii, representing energy levels. The larger the shell, the higher the energy in the electron. See *electron charge, neutron, proton*.

**electron charge** Also called an elementary charge. The charge on a single electron; this is the smallest possible quantity of electric charge. The charge on a proton is equal but of opposite polarity. See *coulomb, proton*, and the appendix CONSTANTS.

**electronegative** Having a negative electric charge or negative electric polarity; this is caused by a surplus of electrons. See *electropositive*.

**electron-hole pair** In a semiconductor substance, an electron and a related hole (electron deficiency). When an electron gains energy within an atom and moves to a different shell, it leaves a vacancy in the shell it has left; this is a hole. See *electron, hole*.

**electronic** Pertaining to devices that use electrical amplification, signal generation, signal processing, control, or switching. Examples are diodes, transistors, and integrated circuits.

**electronic engineering** A branch of engineering concerned with the design and manufacture of electronic devices. Sometimes this includes computers and robots. See *electronic*.

**electronic game** An entertainment device, usually with a microcomputer, that allows the user to manipulate controls and sometimes develop certain skills. Sophisticated electronic games are used as simulators in flight training, military operations, and other important jobs.

**Electronic Industries Association** Abbreviation, EIA. An American group of electronics manufacturers and scholars that sets standards, distributes information, provides interaction between industry and government, and maintains public relations.

**electronic intelligence** See *artificial intelligence*.

**electronic music** 1. Music in which electronic devices are used to enhance the quality, characteristics or loudness of the sound. 2. Music that is actually written by a computer and played via a set of oscillators.

**electronic navigation** Radionavigation, satellite navigation, echo sounding and other electronic means for determining the position of an aircraft, ship, or space vessel. This has largely replaced celestial navigation. See *celestial navigation*.

**electronics** A field of engineering and science concerned with complex current-carrying circuits that perform a tremendous variety of functions. See the preceding several definitions.

**electron microscope** A type of microscope that has greater magnifying power than an ordinary visible-light instrument. The wavelengths of visible light limit the usefulness of optical microscopes. It is impossible to render detail smaller than the size of a wavelength (approximately). The electrons can be made to show far greater detail, because they can be accelerated to energy levels so high that the wavelengths are much shorter than they are with visible light. A simplified functional diagram is shown.

ELECTRON MICROSCOPE

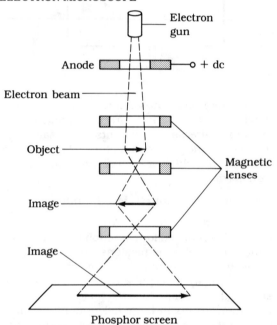

**electron rest mass** See the appendix CONSTANTS.

**electron tube** See *vacuum tube*.

**electronvolt** Abbreviation, eV. The energy acquired by a unit charge (such as an electron or a proton) moving through a potential difference of one volt. This is a tiny amount of energy compared with more familiar units such as watt hours or joules. See the appendix ENERGY UNITS.

**electrophile** A substance that tends to attract electrons.

**electrophoresis** 1. The movement of dielectric particles through a liquid in which they are suspended, when charged electrodes are placed in the liquid. 2. A process of coating materials by attracting particles in a suspension using charged electrodes.

**electroplating** A process in which one metal is deposited on the surface of another by means of electrolysis. The metal to be plated is charged electrically and placed in a solution containing ions of the substance that forms the plating. See *electrolysis*.

**electropositive** A condition of having positive electric charge or polarity; a deficiency of electrons. See *electronegative*.

**electroscope** A simple instrument for detecting the presence of electric charge. When a charge is present, the leaves repel each other because they have the same charge polarity. See the drawing.

ELECTROSCOPE

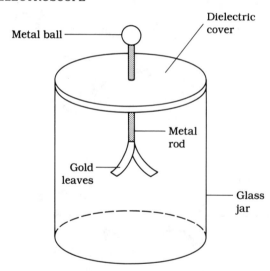

**electroshock therapy** A method of treating certain mental disorders. Electrodes are placed on the scalp and a current is sent through the brain, causing a seizure and unconsciousness. Some patients are improved for a time after this treatment.

**electrostatic** Pertaining to stationary electric charges and nonfluctuating electric fields, their effects and their behavior.

**electrostatic charge** See *electric charge*.

**electrostatic field** See *electric field*.

**electrostatic generator** A device for producing high-voltage electric charges. These operate in various ways, such as with coils and/or capacitors, or with moving belts. See *van de Graaff generator*.

**electrostatic precipitation** The attraction of particles from a liquid or gas onto charged electrodes. A dust precipitator works by this process.

**electrostatic shield** See *Faraday shield*.

**electrostatic units** A generally outmoded system of units based on the centimeter, gram, and second. Today we use the Standard International system. See the appendix STANDARD INTERNATIONAL SYSTEM OF UNITS.

**electrostriction** The contraction of certain substances, such as ceramics, with the application of a voltage. The electrical counterpart of magnetostriction. See *magnetostriction*.

**element** 1. A chemical substance that cannot be reduced into component substances. There are 92 chemical elements that occur in nature; several more can be human-made. 2. A part of or component of some larger device or system. For example an element of an antenna, or of a circuit, or of a vacuum tube. 3. A member of a set in mathematical set theory. 4. In computer practice, a subunit that cannot be further categorized. A word element, for example, is a bit, and a file element is a record.

**elementary particle** 1. A subatomic particle considered as small in scale as we need to go for the purpose at hand; for example, an electron or a proton. 2. A particle that cannot be broken down into smaller particles. Some scientists doubt whether we will ever truly find such a particle. Others think that quarks, photons, and electrons are elementary particles. See *electron, photon, quark*.

**elevation** 1. Altitude above sea level for a certain location on land, or on an artificial structure on land. 2. The angle, usually in degrees, that an object subtends relative to the horizon.

**elevator** A device on the rear horizontal stabilizer of an aircraft (on the tail wing) that controls the angle at which the plane meets the air. It is an up/down-movable device that can be controlled by the pilot.

**ELF** Abbreviation for extremely low frequency.

**elimination** 1. The process whereby the body gets rid of waste. 2. A logical process of deduction when the cause of something is found by proving all other possible causes false or implausible. 3. A process of cancelling out a certain variable or factor in a mathematical equation.

**ellipse** The curve formed on a plane when a cone intersects the plane in a certain way. If you shine a flashlight onto a flat surface, the edge of the large ring of light forms an ellipse if it is not a perfect circle and if all the light actually lands on the surface. The Cartesian graph and equation for an ellipse are shown in the illustration.

ELLIPSE

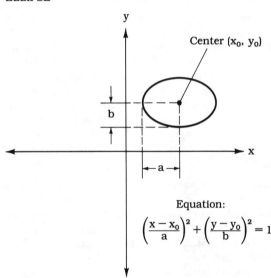

Center $(x_0, y_0)$

Equation:
$$\left(\frac{x-x_0}{a}\right)^2 + \left(\frac{y-y_0}{b}\right)^2 = 1$$

**ellipsoid** An ellipse (see *ellipse*) rotated completely around one of its axes (either the major axis or the minor axis), forming a three-dimensional figure shaped like a football or a flattened sphere. The latter figure is sometimes called an oblate spheroid.

**elliptical galaxy** A galaxy in the shape of an ellipsoid. Classified from E-0 (perfectly spherical) to E-7 (cigar-shaped).

**elliptical orbit** The shape of most orbits. The center of gravity is one focus of the ellipse. The orbiting object follows an elliptical path through space according to classical (Newtonian) physics. See *ellipse, Kepler's Laws*.

**elliptical polarization** Electromagnetic waves in which the lines of flux rotate rather than staying in the same plane. The rate of polarization change increases and decreases somewhat with each rotation, so the polarization appears strongest in one plane and weaker in the others. See *circular polarization*.

**El Niño** A periodic warming of the water near the equator in the Pacific ocean off the west coast of South America. Literal translation: "The Child," after the Christ, because the effect often occurs around Christmas time. See the drawing. An intense, prolonged El Niño event has effects on weather throughout a large part of the world.

EL NIÑO

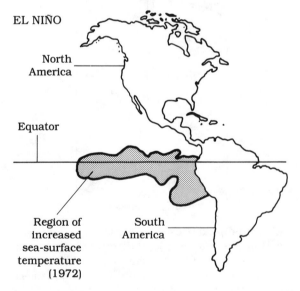

From Jon Erickson, THE MYSTERIOUS OCEANS (TAB Books, Inc., 1988).

**embalming** A method of preserving a dead organism, protecting it against decay. This allows it to be used later for laboratory research.

**embolism** A blood clot in the circulatory system that breaks off and lodges somewhere such as the heart, kidney, lung, or brain where it causes tissue death. Can be permanently crippling and possibly fatal.

**embryo** 1. In mammals, the early developmental stage of a fetus. 2. A part of the seed in certain plants that later develops into a new plant.

**embryology** The science concerned with animal reproduction after the egg has been fertilized, and until birth.

**EME** Abbreviation for Earth-Moon-Earth. See *moonbounce*.

**emerald** A form of beryl (see *beryl*) with a characteristic green color, used as a gemstone. It is found in mines and also can be artificially made.

**Emergency Broadcast System** A scheme for using the broadcast airwaves to alert the public in case of a national emergency, such as a nuclear war. All normal broadcasting would cease and every station would broadcast an attention tone, followed by specific instructions.

**emery** A hard mineral, aluminum oxide with various impurities. When ground into powder, it is useful as an abrasive agent. The powder can be glued to paper for use as a durable form of sandpaper. The powdered substance is also added to concrete for heavy-duty construction purposes.

**emesis** 1. Vomiting. 2. The administration of an emetic. See *emetic*.

**emetic** A substance given to induce vomiting. Used sometimes in cases of poisoning. With certain poisons, however, an emetic should not be given. It is important to know basic First Aid and to determine the type of poison that has been consumed before giving an emetic. If possible, a physician should always be consulted before giving an emetic. Information can be obtained by dialing 911.

**EMF** Abbreviation for electromotive force.

**emission** 1. Radiation of energy in some form. 2. The output of a device intended to produce a signal. 3. The type of output from a radio transmitter. There are various different classifications according to the method of modulation. See *modulation*. 4. A discharge from some part of the body.

**emission line** In a spectrum, a bright band caused by excitation of atoms. The drawing shows a hypothetical spectrum with emission lines in the visible range. Each element or compound has a unique "signature" of emission lines. This lets astronomers tell what substances are in glowing nebulae.

EMISSION LINE

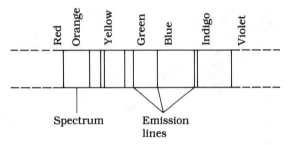

**emitter** 1. In a transistor, the semiconductor electrode that sends off the charge carriers (electrons or holes). 2. A device that sends off energy in a certain form, such as a radio antenna or a laser.

**emotion** A complex form of intelligence combined with some reasoning and consciousness. Anger, fear, and love are examples. Animals show some emotions; there is debate as to whether or not they have all of the emotions that humans have.

**EMP** Abbreviation for electromagnetic pulse.

**emphysema** A degenerative disease of the lungs. Symptoms are like those of a chest cold. Affects many people who breathe polluted air for much of their lives. Might occur along with lung cancer.

**empirical** Determined on the basis of experiments and laboratory data.

**empirical engineering** A method of developing designs by repeated testing to see what works and what does not. This is a trial-and-error process. Usually, theoretical work ("the drawing board") is done first, so empirical engineers don't just start from scratch without guidelines.

**Empiricism** A philosophy concerning the nature of reality. An empiricist believes that ideas can exist only on the basis of observable facts. Aristotle was a philosopher/scientist who held this view of the universe. See *rationalism*.

**empty set** See *null set*.

**emulsion** A suspension of tiny liquid droplets in another liquid substance. An example is emulsified castor oil. Normally, castor oil and water would separate; however, by breaking up the oil into fine droplets, the two substances can be made to mix.

**enable** 1. A computer command or operation actuating a system. 2. A digital pulse that actuates a system. 3. To make possible.

**enamel** 1. The hard outer coating of a tooth. Consists mainly of calcium compounds. 2. A hard substance used to protect and decorate various machine parts or objects. Basically, paint. 3. A thin, hard coating sometimes applied to wire, especially copper, to prevent corrosion or short-circuiting.

**encephalitis** An acute inflammation of the brain. Can be caused by viruses or a reaction to a toxic substance. This is a serious illness and requires a doctor's care. See *encephalomyelitis*.

**encephalomyelitis** An acute inflammation of the brain and the spinal cord. Causes and treatment are similar to encephalitis. See *encephalitis*.

**encoder** 1. An electronic circuit that converts an intelligible signal into coded form. The opposite of a decoder (see *decoder*). 2. See *analog-to-digital converter, digital-to-analog converter*. 3. A keyboard machine on which an operator can type and some form of digital code is sent out. 4. A tone generator, used to actuate the receiver in a private-line or telephone communications system.

**endocrine** Pertaining to hormones or to the system of glands that produce and regulate body hormones. See *endocrine glands*.

**endocrine glands** The ductless glands that secrete hormones which regulate various body functions. See the illustration. The pituitary controls growth and serves as a control center for the other endocrine glands. The thyroid regulates metabolism. The parathyroids control calcium metabolism. The pancreas produces insulin to metabolize blood glucose. The adrenals aid in stress and also interact to regulate blood glucose levels. The ovaries (in the female) and the testicles (in the male) are the sex glands. See also *adrenal gland, ovary, pancreas, parathyroid, pituitary gland, testicles, thyroid*.

ENDOCRINE GLANDS

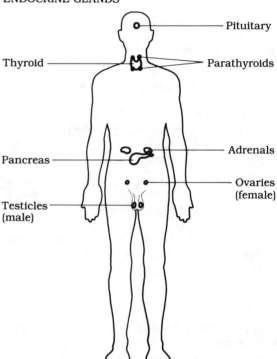

**endocrinology** A branch of science, and in particular of medicine, concerned with the endocrine glands. See *endocrine glands*.

**endoderm** In a fetus, the interior part of the body that will eventually become the digestive system.

**endomorph** A type of person and personality. Generally heavy-set, outgoing, and carefree. See *ectomorph, mesomorph*.

**endorphins** Substances produced by the body that act as a pain reliever, anxiety reducer, and mood-elevating drug. Some drugs stimulate the pituitary gland to produce these substances. Hard physical exercise also causes increased production of endorphins. This is why you often are in a good mood after a workout.

**endoscope** A device that doctors use to look inside the body, especially in the lungs and the digestive system. Optical fibers transmit laser light and also send images back to be viewed on a monitor screen. Often there are cables to control devices for cutting, or for closing wounds or ulcers. See the drawing.

ENDOSCOPE

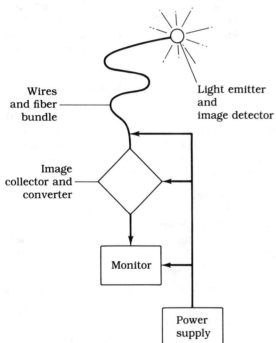

**endoskeleton** A skeleton inside an animal. Humans have such skeletons. See *exoskeleton*.

**endosperm** In a seed, the tissue that surrounds the germ or embryo. This is the "starchy" part of the rice, wheat or corn kernel. But there is also significant protein and other nutrients there.

**endpoint** 1. A point representing an extreme value on a line segment or ray. 2. A transition point. 3. The completion of a chemical reaction as displayed by an indication such as a color change in a solution.

**energy** 1. Power expended over a period of time. See *power*. 2. The capacity to do work. See *potential energy*. 3. The movement of molecules or particles. See *kinetic energy*. 4. Reserves of materials commonly used to produce energy for society, such as fossil fuels.

**energy band** 1. In an atom, an electron shell representing a specific energy level that the electron might have. 2. The difference between the maximum and minimum possible or allowable energy level for a particle.

**energy conservation** Maximizing efficiency and minimizing consumption of energy resources so that they will last longer. This reduces the chances that we will eventually face a crisis because of a shortage of resources. See *energy crisis*.

**energy conversion** Changing energy from one form to another, such as potential energy to kinetic energy. For example, gasoline (potential) to car moving down highway (kinetic). See *energy units*.

**energy crisis** Insufficient energy production for the needs or demands of a society. This might take place because of overpopulation, poverty, war, or a shortage of natural resources.

**energy, kinetic** See *kinetic energy*.

**energy, potential** See *potential energy*.

**energy storage** 1. A reserve of potential energy that can be used later in kinetic form. An example is the electricity in a battery stored as chemical energy. 2. The conversion of energy to some form where it can be kept for later use.

**energy transformation** See *energy conversion*.

**energy units** The various different means by which energy can be measured. See the appendix ENERGY UNITS. The most common unit employed by scientists is the joule, or watt sec-

**engine** 1. Any device that converts potential energy into mechanical energy. 2. Any device that converts potential energy into motion; a propulsion unit. 3. A motor used for propulsion.

**engineer** 1. A person who designs or supervises the manufacture and testing of various devices. 2. Someone who supervises the operation of a system.

**engineering** A branch of science involved with practical matters in design, construction, and operation of hardware and software.

**engineering seismology** The application of earthquake studies to engineering, and the development of construction techniques to make buildings and other structures earthquake-resistant.

**entomology** A branch of biology or life science concerned with insect anatomy and behavior.

**entropy** 1. The flow of heat energy from regions of more energy to regions of less. An equalization process. 2. The gradual and, some scientists think, irreversible and universal process whereby temperature differences tend to even out. The final result would be equal thermal energy everywhere, and all life processes would grind to a halt. This might not actually be true; the universe might collapse in a few hundred billion years. See *Big Bang, Big Bounce, Big Crunch*.

**environment** 1. The earth, the balance of climate, and the interaction of species. Generally, our surroundings. 2. A set of conditions for the operation and functioning of a system. 3. The climate and surroundings on some planet other than the earth.

**environmental action** Political and social activity directed toward preservation and responsible management of the environment. See *environmentalism*.

**environmental engineering** 1. A branch of engineering concerned with making machines and other devices as harmless to the environment as possible. 2. The branch of engineering that deals with artificial modification of the environment. See *environment*.

**environmental hazard** Any activity or substance that presents a danger to the environment or any species living in a given region.

**environmental impact** The effect, or supposed effect, that a given human activity has or will have on the environment.

**environmentalism** A philosophy concerned with humanity and our responsible use and management of the earth's natural resources. Environmentalists realize that resources are finite, and wastefulness can threaten humanity as well as other species.

**Environmental Protection Agency** An agency of the Federal Government responsible for making and enforcing regulations intended to protect the environment from damage by human activity.

**enzyme** A substance important in certain chemical reactions within the body, especially digestion. Various enzymes in the stomach and intestines break down proteins, carbohydrates, and fats so that they are easily absorbed. Some enzymes operate within individual cells, assisting in their digestion and metabolism.

**Eocene epoch** A time during the Cenozoic era, or the era of recent life, starting about 54 million years ago and lasting for about 17 million years. During this time, some believe such species as whales and horses evolved. See the appendix GEOLOGIC TIME.

**eolian** 1. Pertaining to the wind. 2. Pertaining to anything carried by the wind, such as dust or radioactive fallout or certain plant seeds.

**eon** A very long period of time; the longest time unit generally used in defining geologic time. Encompasses many millions of years, or even as long as 1,000,000,000 years or more.

**EPA** Abbreviation for Environmental Protection Agency.

**ephedrine** A drug similar to epinephrine (see *epinephrine*). But it acts more slowly, and for a longer time. It can produce extreme reactions such as sleeplessness and tremors. It is one of various compounds related to adrenaline.

**ephemeral plant** A plant that completes its life cycle in just a fraction of one year. Dandelions are a good example. They might appear two or three times during a growing season, such as in May and again in August.

**Ephemeris** An almanac that gives future positions for various celestial objects. Provides information on such things as eclipses of the sun and moon, planetary alignments, conjunctions, and oppositions.

**Ephemeris Time** Abbreviation, ET. A means of defining time according to the heavens, independently of the motions of the earth. Started in the year 1900. Nowadays, atomic standards are used to measure time, and the solar day is the most common unit used.

**epicenter** The point where an earthquake occurs, or the point on the surface beneath which the quake effects are centered. Can be located by triangulation as shown in the drawing. Most of the damage normally occurs near this point. But there are exceptions when the quake is centered in an uninhabited region or under the sea.

EPICENTER

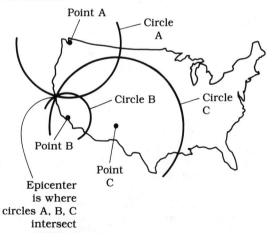

**epicycle** A planetary suborbit according to the theory of Ptolemy, who taught that all of the objects in the sky revolve around the earth. To explain retrograde motion of certain planets, Ptolemy proposed that they have complex orbits with a main, large orbit (deferent) and a smaller orbit around a point on the deferent (see drawing). This model explained observed planetary motions remarkably well, and was accepted for centuries. See *deferent, Ptolemaic model*.

EPICYCLE

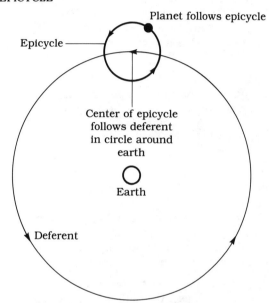

**epidemic** A widespread occurrence of a contagious disease. Can be spread by droplets in the air, by water, by insects, or by sexual contact, contaminated blood, and shared hypodermic needles, such as the case for acquired immune deficiency syndrome (AIDS).

**epidermis** The outermost layer of skin, consisting of dead cells. These cells wear off regularly. See *dermis* for illustration.

**epiglottis** A flap in the throat that guides food into the esophagus when you swallow, keeping it out of the windpipe.

**epilepsy** An inherited seizure disorder. In "petit mal" the seizures involve only a loss of memory for a few seconds. In "grand mal" they are severe, and often violent. Some of history's greatest people have had this problem; it need not be crippling. See *seizure*.

**epinephrine** A hormone released from the adrenal glands under conditions of stress. It assists in nerve function and causes general stimulation, increased alertness, and can also cause tremors and excitement.

**epistemology** A branch of philosophy from which scientific method is derived. Literally means "theory of knowledge." In epistemology, attempts are made to answer questions such as, "What does knowledge consist of? Where does it come from? How do we get it? How can we increase it? Can we ever know everything there is to know?"

**epitaxy** The "growth" of one chemical solid on another. Atoms accumulate and maintain their alignment. This is the method by which many semiconductor substances are made. In particular, crystals of silicon are "grown" in this way. The result is an essentially pure material. See *crystal*.

**epithelium** Tissue that makes up the skin and the linings of the digestive system, and various blood vessels and membranes. Keeps substances in their proper places; for example, blood in the arteries and veins, and food in the intestine to be gradually absorbed.

**epoch** A time interval, on the order of a million to perhaps 10 or 15 million years, that makes up part of the age of recent life, the Cenozoic era. See the appendix GEOLOGIC TIME. We now live in the Holocene epoch.

**epoxy** A compound produced artificially and used as a filler or glue. Has a consistency like plastic. It is an insulator and a dielectric. Some epoxy is flexible, and some is fairly rigid. It is durable and not easily damaged by solvents or other corrosive substances. See *epoxy glue*.

**epoxy glue** A glue made using epoxy. The most common form comes in packages of two tubes, one of "resin" and the other of "hardener." When mixed they form a tough, adhesive, nonconducting material in about $1/2$ hour to 4 hours, depending on the exact chemical composition. See *epoxy*.

**equation** 1. In all sciences except logic or chemistry, a statement of the form $A = B$, where A and B are expressions and $=$ means that they are identical. 2. In logic, a statement of the form $A = B$, where A and B are sentences, and $=$ means "if and only if." 3. In chemistry, a statement of a reaction in symbolic form, such as $C + O = CO$ (carbon plus oxygen yields carbon monoxide).

**equator** 1. The line running around the earth midway between the poles, and assigned zero degrees latitude. 2. The line running around a star or planet, midway between its poles.

**equatorial** 1. Pertaining to the equator. 2. Aligned with the celestial equator; see *equatorial mount*. 3. Pertaining to regions of the earth, or of some other planet, near the equator.

**equatorial mount** A method of mounting a telescope so that the movements of the sun, moon, or stars caused by the earth's rotation can be followed by moving only one bearing. See the drawing. The mount is adjusted according to the latitude at which the telescope is used. See also *azimuth-elevation mount*.

EQUATORIAL MOUNT

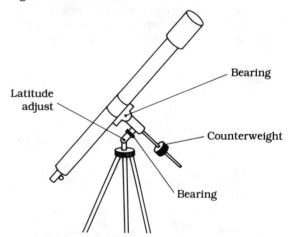

**equilibrium** Balance, or a state of balance. We might speak of the sense of balance that a tightrope walker has. We also might consider chemical balance, hormone balance, thermal balance, and electrical balance, where two or more interacting effects operate in a stable way.

**equinoctial** Pertaining to the equator or to the regions on the earth near the equator.

**equinox** Those times of the year when daylight and darkness are equally long. The dates are usually March 20 or 21, and September 22 or 23. The sun is directly over the equator at these times. See the illustration. In the Northern Hemisphere, the March equinox is called vernal, and the September equinox is called autumnal.

EQUINOX

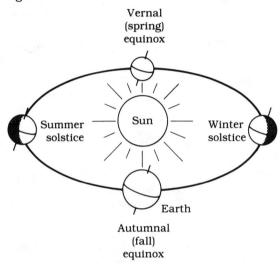

**equivalence** 1. Logical equivalence: Sentences A and B are logically equivalent if and only if A is true whenever B is true, and vice-versa. 2. The property of having equal value or equal effect. 3. In mathematics: a condition where two equations have the same solution set. Also, existing in a one-to-one correspondence (homomorphism).

**equivalence principle** A postulate made by Einstein in his theory of relativity. In general, acceleration-produced force is just the same, in the way it affects matter and energy, as gravitational force.

**equivalence relation** In mathematics, a relation * with three properties. Reflexive: For all x, x*x. Symmetric: For all x and y, if x*y then y*x. Transitive: For all x, y and z, if x*y and y*z, then x*z. Equality (=) is an equivalence relation. But "less than" (<) and "greater than" (>) are not; they are transitive, but they are not reflexive or symmetric.

**Er** Chemical symbol for erbium.

**era** A time interval of tens or hundreds of millions of years. The most recent era, and the one in which we live now, the Cenozoic, began about 65 million years ago. See the appendix GEOLOGIC TIME.

**Eratosthenes** The first person to measure the size of the earth. He lived more than 2000 years ago. By measuring the north-south distance between two places, and the difference in sun angle, he calculated the earth's circumference to be 25,000 miles. This is accurate to within a few percent.

**erbium** Chemical symbol, Er. An element with atomic number 68. The most common isotope has atomic weight 166. It is a soft metal in pure form. Scientists use it in nuclear research because it absorbs neutrons.

**erg** A unit of energy. Roughly equivalent to the work done by a mosquito taking off. The energy of 1 erg = 0.0000001 joule (one ten-millionth of a joule). Not often used; the joule is more commonly employed as an energy unit. See the appendix ENERGY UNITS.

**ergocalciferol** A form of vitamin D, or calciferol. See the appendix VITAMINS AND MINERALS.

**ergonomics** A branch of human engineering. Deals with optimizing worker performance by means of changing, or adjusting, the conditions in which they do their jobs.

**ergosterol** Also called provitamin D. The substance that is converted into vitamin D by ultraviolet radiation, such as from the sun. See the appendix VITAMINS AND MINERALS.

**erosion** 1. Loss of soil or rock, caused by the action of the wind, or of windblown particles of dust and sand over periods of years, decades or longer. 2. Loss of soil or rock washed away by water. Might take place over hundreds, thousands, or millions of years to sculpt canyons and riverbeds.

**erratic** 1. Not taking place at regular or predictable intervals. 2. Not following any definite pattern. 3. Unnatural. 4. A large boulder dropped by a glacier.

**erythrocyte** See *red blood cell.*

**Es** Chemical symbol for einsteinium.

**escapement** A part of a mechanical "wind-up" clock, or pendulum clock. It transfers the tension in the spring, or the force of gravity on the pendulum, into a constant, precise movement of the second hand. The hour hand is geared at a ratio of 1:60 relative to the minute hand; the minute hand is geared at this same ratio relative to the second hand.

**escape velocity** The speed with which an object must be hurled to get completely away from the gravitational field of a planet or star. For the earth, this is about 25,000 miles per hour, or 7 miles a second.

**escarpment** A steep face (like a cliff) at the edge of a land region having high elevation.

**esker** An ancient riverbed, consisting of deposited material that has settled from, or been eroded by, glacial water.

**esophagus** The path that food and liquids take from the mouth to the stomach. See the illustration. Matter is pushed along by squeezing contractions called peristalsis. See *peristalsis.*

ESOPHAGUS

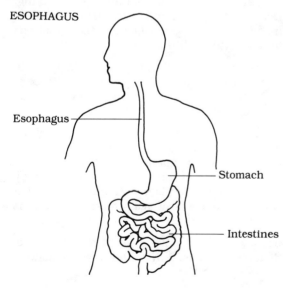

**ESP** Abbreviation for extrasensory perception.

**essential amino acids** The amino acids (see *amino acid*) that must be obtained from food, because they cannot be made from other substances in the body. These are cystine, histidine, isoleucine, leucine, lysine, methionine, phenylalanine, threonine, tryptophan, tyrosine, and valine. In civilized countries, people eating a reasonably balanced diet with sufficient calories will always get enough amino acids in their food. But bad eating habits, or starvation, can produce deficiencies.

**essential fatty acids** Two types of substances, called linoleic and linolenic acid, needed by the body in small amounts. In civilized countries, deficiencies are rare; in fact, the problem is more likely to be overconsumption of fats in general.

**essential nutrients** Protein, vitamins, minerals, essential fatty acids, and dietary fiber, along with a certain amount of carbohydrate needed for the body to be healthy. We probably haven't identified all the essential nutrients. To get them all, you must have good eating habits. Some of the more important ones are listed in the appendices RECOMMENDED DAILY ALLOWANCES and VITAMINS AND MINERALS.

**EST** Abbreviation for *Eastern Standard Time, electroshock therapy.*

**ester** A compound that results from the combination of an acid and an alcohol. Resembles a fat or fatty acid, often with a distinctive flavor.

**estrogen** A female sex hormone. Can be administered in certain hormone imbalance conditions. Normally produced to some extent in both males and females, but more in females.

**estuary** A small inlet or bay where a river flows into the ocean. Caused by interaction between the river current and the ocean tides.

**ET** Abbreviation for *Ephemeris Time, extraterrestrial.*

**ethanal** See *acetaldehyde.*

**ethane** A major constituent of natural gas, along with methane (see *methane*). Contains two carbon atoms and six hydrogen atoms, linked as shown in the drawing. It is highly flammable, and odorless, although a scent is often added so that gas leaks can be easily detected.

ETHANE

$C_2H_6$

○ Carbon atoms

• Hydrogen atoms

— Bonds (CH)

**ethanoic acid**  Fermented alcohol present in vinegar and largely responsible for the taste of vinegar.

**ethanol**  Also known as ethyl alcohol. An alcohol distilled from plants, such as corn, rice, and potatoes, or produced by fermentation as in wine and beer. The type used in liquors and alcoholic drinks, and also added to some types of gasoline. See *alcohol*.

**ethene**  Also called ethylene. A gas consisting of two carbon and four hydrogen atoms, highly flammable, colorless, and odorless. It is probably the most versatile hydrocarbon; when polymerized, it forms polyethylene (see *polyethylene*), a familiar plastic. Ethene is also a constituent of antifreeze, solvents, drugs, and explosive chemicals, as well as paints and synthetic rubber.

**ether**  1. A solvent with a peculiar odor and capable of causing drowsiness or giddiness. Was once used as a general anesthetic before surgery. Not often used nowadays. 2. A substance that was supposed to exist in order to transmit light through space, like air serves to transmit sound on the surface of the earth. In the early part of the 20th Century, Einstein rejected the idea of this so-called luminiferous ether, and today scientists do not usually consider it to exist.

**ethology**  A branch of biology and life science concerned with the behavior of animals (and also of humans). The cornerstone of this science is the belief that behavior evolves along with the biological characteristics of species, with certain types of behavior tending to promote survival and further evolution.

**ethyl**  1. See *ethanol*. 2. A group of hydrocarbons consisting of two carbon atoms and five hydrogen atoms. 3. Slang for high-octane gasoline. 4. Slang for gasoline to which ethanol has been added.

**ethyl alcohol**  See *ethanol*.

**ethylene**  See *ethene*.

**ethylenediaminetetracetic acid**  Abbreviation, EDTA. A compound used in the manufacture of detergents.

**ethyne**  Also called acetylene. A colorless gas with two carbon atoms and two hydrogen atoms. It is highly flammable and has a characteristic odor. Used in welding and cutting because it burns with a hot flame.

**etiology**  1. Origin. 2. In medicine, the cause(s) and history leading to disease. 3. The science of cause-effect.

**Eu**  Chemical symbol for europium.

**Euclid**  Also known as Euclid of Alexandria, or "the father of geometry." The author of the book *Elements*, a mathematics text written about 300 B.C. He also wrote books on other subjects in science. He was one of the most respected scholars of his time. In Alexandria (Egypt), he taught at the school called the Museum, an ancient equivalent of Oxford or Harvard.

**Euclidean**  Pertaining to the theories and principles of Euclid.

**Euclidean geometry**  Geometry with points, lines, and planes usually on a "flat" surface but also sometimes in three dimensions. The simple geometry taught in junior high school. See *non-Euclidean geometry*.

**Euclidean space**  A space in which every plane triangle, no matter how large or small, has interior angles that add up to a total measure of 180 degrees (see illustration). A "flat" space. Can have any number of dimensions, although two-space and three-space (of two and three dimensions) are most common.

EUCLIDEAN SPACE

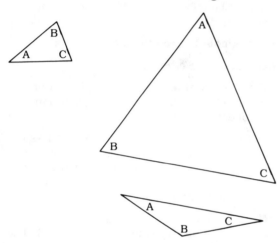

Sum of measures of angles A+B+C = 180 degrees for all triangles

**eudiometer** A device that measures volumes of gases at constant pressure. Useful in certain chemical tests.

**Eudoxus** An ancient mathematician who discovered the theory of proportion, or ratios, around 360 B.C. In modified form we recognize it as the rule of cross multiplication: for numbers a, b, c, and d, we say a/b = c/d if and only if ad = bc, with the constraint that neither b nor d may be equal to zero.

**eugenics** A controversial branch of life science involving the possibilities of improving species characteristics (including of humans) by means of applied genetics. This might include "selective breeding" as well as actually modifying genes in a lab, or cloning. See *genetic engineering*.

**eugenol** A food preservative used in olden times made naturally by putting evergreen or clove into supplies of food. Not used much nowadays, because there are chemicals that are more effective and reliable.

**Euler, Leonhard** An eighteenth-century Swiss mathematician. Studied under Jean Bernoulli of the well-known Bernoulli family (see *Bernoulli family*). Added to knowledge in almost every branch of mathematics, pure as well as applied. He worked with such esoteric ideas as the square root of −1.

**europium** Chemical symbol, Eu. An element with atomic number 63. The most common isotope has atomic weight 153. In its pure form it is a soft metal. It is useful in nuclear lab work because it absorbs neutrons.

**eustachian tube** An opening leading from the middle ear to the throat. Equalizes pressure. Might become congested, resulting in middle-ear inflammation (otitis). See the drawing.

EUSTACHIAN TUBE

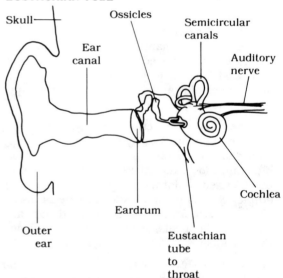

**eutectic** Pertaining to a mixture of solids, in which the freezing point (and thus the melting point) is made as low as can be obtained with any mixture of the same substances. This might be done with tin and lead, for example. We would then have a eutectic tin/lead solder alloy.

**eutectic salt** A material that is being examined as a possible heat storage medium for solar energy collectors.

**euthanasia** The killing of a terminally ill person, based on the rationale that it will save him/her unnecessary pain. A controversial practice. Most doctors refuse to do this in the United States.

**eutrophication** A condition in which a body of water has too many nutrients. This causes certain organisms to multiply too fast. The result is an imbalance with such effects as changed levels of dissolved oxygen in the water. Some life-forms then die off, and others thrive and dominate.

**eV** Abbreviation for *electronvolt*.

**evaporation** The process of conversion from liquid or solid form to gaseous form. This happens naturally with most liquids and solids, but is greatly speeded up by heating, and especially by boiling.

**evaporation rate** The speed with which a substance changes from liquid or solid form to a gas or vapor. Can be measured in grams per second, or kilograms per hour, or some other units of mass per unit time. Can also sometimes be given in terms of volume per unit time.

**evaporite deposit** Minerals left when a large body of water evaporates. Examples are the salt and gypsum deposits left after the inland seas retreated at the beginning of the Permian period about 290 million years ago.

**evapotranspiration** The transfer of water from the soil by means of capillary action, and from plants and animals to the atmosphere as a byproduct of respiration. This water returns to the atmosphere as vapor when the sun shines on wet ground, or when you exhale.

**even-even** Pertaining to an atomic nucleus having an even number of protons and an even number of neutrons. The first designator refers to the proton count and the second to the neutron count. See *even-odd, odd-even, odd-odd*.

**even number** An integer divisible by two. See *integer, odd number*.

**even-odd** Pertaining to an atomic nucleus having an even number of protons and an odd number of neutrons. The first designator refers to the proton count and the second to the neutron count. See *even-even, odd-even, odd-odd*.

**event horizon** As an object collapses under the influence of its own gravitation, the point where it seems (as seen from the outside) to slow to a stop forever. This occurs because time is "stretched" infinitely, as is space (see drawing), when the object gets so dense that even light cannot escape from it. See *black hole, gravitational collapse, Schwarzchild radius*.

EVENT HORIZON

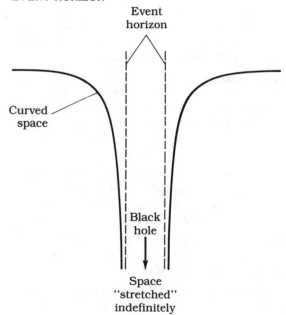

**evergreen** 1. A plant that stays green all year round, such as a pine tree. Generally applies to plants in temperate climates, where deciduous trees lose their leaves in the winter. 2. Pertaining to a plant that stays green all year round. See *deciduous*.

**evolution** 1. A natural process in which species become adapted to their environment over many generations. 2. The development of new species based on adaptations to the environment. 3. See *Evolutionism*. 4. A process of change, directed by environmental factors. 5.

Gradual perfection or improvement of the design of a machine.

**Evolutionism** The theory that all species of living things, including human beings, evolved from lower species. Thus we would all have, as common ancestors, the first cells in the primeval oceans. Although this theory is popular among scientists today, a logically rigorous proof of it has not been found. See *Creationism*.

**exa-** See the appendix PREFIX MULTIPLIERS.

**exact science** A science in which there is no error. Strictly speaking, there is no such science; even logic, the foundation for all reasoning, has been found "incomplete," so that there are statements that cannot be proven nor disproven. Thus, "exact" is a relative term. Some sciences, such as Euclidean geometry, are more exact than others, such as medicine.

**excitation** 1. An increase in the energy level of an atom. 2. Power delivered to an amplifier or load, such as a transistor or an antenna. 3. The application of current to a circuit. 4. The activation of an electromagnet by passing a current through the coil.

**exciter** 1. A device that supplies a signal to a load, such as an amplifier or an antenna. 2. The circuit in a radio transmitter that supplies power to the final amplifier.

**exciton** In a semiconductor material, an electron and its associated hole. See *electron, electron-hole pair, hole, semiconductor*.

**exclusion principle** A principle in atomic theory. Given two particles in an atom that are of the same kind (such as electrons), they must have different sets of quantum numbers. See *quantum number*.

**excretion** 1. Elimination of waste, resulting from metabolism and digestion. 2. The passage of urine or feces.

**exercise** 1. Physical activity, done mainly for the purpose of maintaining or improving body health. 2. A practice run or test run for a device or machine. 3. To carry out.

**exfoliation** 1. Peeling or sloughing off of dead skin. 2. Shedding of a coat of dead skin; see *moulting*. 3. Separation of the outer layers of a rock.

**exitance** For a glowing object, the brightness per unit area. Usually measured in watts per square meter or lumens per square meter. For an object giving off a constant amount of light, the exitance would decrease if the surface area increased.

**exo-** A prefix meaning "outside" or "outside of."

**exocrine gland** A gland that secretes either onto the skin or into a specific place inside the body, rather than into the bloodstream or lymph stream. The salivary glands are a good example; saliva goes into the mouth to aid digestion. See *endocrine glands*.

**exoergic process** Any nuclear reaction that produces energy output. Uranium fission and hydrogen fusion are two examples. These processes can be used to produce energy or to make bombs. Hydrogen fusion has yet to be "tamed" for use as a source of energy.

**exogamy** Breeding between different species or between species that are only somewhat related to each other.

**exogenous** Pertaining to outside substances put into the body, and that cannot be made in the body. An example (in the case of humans) is vitamin C. Most drugs are also substances of this type, but not all; insulin and thyroxin are made in the normal human body, but they also can be taken as drugs.

**exoskeleton** A skeleton that is outside the body. It is like a hard shell. This is the type of skeleton that arthropods have. See also *endoskeleton*.

**exosphere** The extreme upper part of the atmosphere, above the ionosphere, starting at about 300 miles altitude. This is the beginning of space. It literally means "outside of the sphere," beyond the environment of the earth.

**exothermic** Pertaining to any process or reaction in which heat is given off. A special form of exoergic, because heat is a special form of energy. See *exoergic*.

**expand** To grow larger yet remaining a single unit, either because of heating or increased pressure, or because of some other process,

**expanding-earth hypothesis** A theory that the primordial earth expanded as internal radioactive decay heated it up. This expansion could explain why the continents separated, but it is not a generally accepted theory, because if it were true, the outlines of the continents would not "fit" together so well. See also *continental drift*.

**expanding universe** A theory accepted today by most astronomers. The red shift (see *red shift*) of distant galaxies suggests they are all moving away from us. The farther away they are, the faster they are receding, so the whole universe is expanding. A popular theory has space shaped like a four-dimensional sphere that is getting bigger like a balloon being inflated (see drawing). See also *Big Bang*.

EXPANDING UNIVERSE

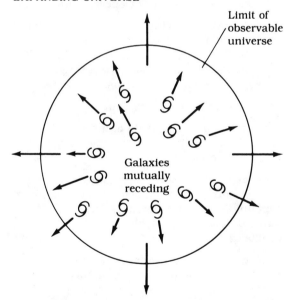

**expansion** Increasing in size; getting physically larger with the passage of time. Can occur because of heating, pressure, or as a result of unknown causes.

**expansion steam engine** An older form of steam engine in which the hot vapor expanded in a cylinder driving a piston. More sophisticated machines of this kind had multiple cylinders. The first such engine was used in 1854. Later the steam turbine made this type of engine obsolete. See *steam engine, steam turbine*.

**expansivity** The extent to which an object increases in size because of heating. Can be measured in terms of increased length, surface area, or volume per unit temperature rise. A steel bar might get a millimeter longer for every 10 degrees Fahrenheit the temperature goes up, for example.

**experiment** An activity carried out to see what happens under certain controlled conditions. Sometimes the general nature of the outcome is known, but measurements must be made to see how great the effect will be. Sometimes the outcome is unknown, and the result is a surprise. See *qualitative, quantitative*.

**experimental** 1. Determined on the basis of experiments. 2. Under test. We might speak of an experimental flu vaccine, for example.

**experimentalist** A scientist who makes most of his/her discoveries by conducting experiments. The experiments often yield results that would not be expected on the basis of theory alone. Hence the saying, "One experimentalist can keep a dozen theorists busy." See also *theorist*.

**explosion** A rapid chemical or nuclear reaction with the release of energy in a short time. See *explosive*.

**explosive** Any chemical combination that reacts suddenly and violently under certain conditions. Gunpowder must be ignited; nitroglycerin must receive a blow or shock; uranium-235 must be pressed together in an amount exceeding the critical mass. In each of these cases, energy will be released in great amounts in a very short time.

**exponent** A number or variable written as a superscript indicating that a quantity is to be raised to the power given by the value of the superscript; for example, $2^2 = 4$. The superscript "2" means that a number is to be multiplied by itself (squared); a superscript "3" means that the base number is to be cubed, that is, squared and then multiplied by itself again. In higher mathematics, exponents may have fractional, irrational or even complex-number values.

**exponential function** A function that can be expressed in terms of a base number raised to an exponent value, with the exponent being the independent variable. Examples are shown in the illustration, with the base being the irrational-number constant e, approximately equal to 2.718. Such functions grow larger at a rate that increases, so they have a characteristic "inflationary" shape. See *exponent, exponential growth*.

EXPONENTIAL FUNCTION

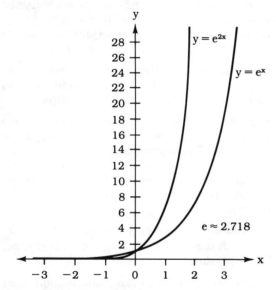

**exponential growth** An increase that occurs as an exponential function. A good example is the increase in the world population over the past several decades. The growth rate increases with time so that as the function gets larger, its value goes up faster and faster. See *exponential function*.

**exposure** 1. A condition of being subjected to radiation, especially at infrared, visible, ultraviolet, or shorter wavelengths. 2. The amount of radiation received by a sample over a period of time. 3. A stress condition caused by being unprotected against environmental extremes such as heat or cold.

**exposure meter** 1. A device that measures the amount of a certain kind of energy radiation that a sample has received. 2. A light meter for determining how to adjust a camera to get a good photograph.

**extender** A substance added to a chemical, so that the chemical is diluted but still effective for its intended purpose. For example, adding water to window cleaner. The water would be an extender in this case.

**extensometer** An instrument that precisely measures the extent to which an object is stretched. This can be done electronically, or by means of lasers to actually count the number of visible-light wavelengths by which the object stretches.

**extinct** No longer in existence as a species. The situation (for a species) that occurs when the last living member of a species dies.

**extinction** The dying out of an entire species. This occurs naturally as part of the evolution process. But recently, some species have died out because of human activities, causing concern among environmentalists.

**extra-** A prefix meaning "outside of."

**extracellular** 1. Outside of a cell or cells. 2. In the fluid or medium containing cells, but not in the cells themselves.

**extract** 1. To remove, especially pertaining to a single substance to be taken out of a medium containing many different substances. 2. A purified form of a substance, for example, vanilla extract.

**extraction** 1. The process of removing one substance from a medium containing many different substances. 2. A step-by-step process whereby a square root can be obtained. See a text on algebra for further details. 3. Purification of a metal from crude ore. 4. Refining of oil from crude deposits or from coal.

**extrapolation** A form of educated guesswork, sometimes done with the help of computers, to extend or predict values of a function beyond the known domain. An example is shown in the drawing, where population is graphed versus time. Extrapolation for the near future (in this case) is fairly easy. But for the long term (in this case) it is difficult because the world cannot continue to support population growth at an exponential rate. See *Malthusian model*.

EXTRAPOLATION

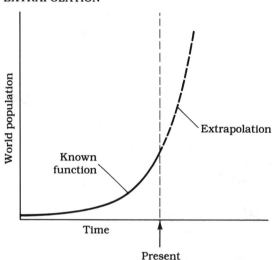

**extrasensory perception** Abbreviation, ESP. The ability to sense things by other means than with the five usual senses of sight, hearing, touch, smell, and taste. Scientific research in this field is hard to do, and some scientists doubt if ESP exists.

**extraterrestrial** 1. On a planet or moon other than the earth. 2. Above the earth, in the atmosphere or in space. 3. Pertaining to something that comes from outer space.

**extratropical** 1. Pertaining to tropical storms that have moved out of the tropics and lost their tropical characteristics. 2. Pertaining to weather or climate in the temperate zone, outside of the tropics.

**extreme** 1. Unusually displaced from the normal levels, values or range. We might have temperature extremes during a heat wave or cold wave, for example. 2. The largest or smallest value of a function within a certain domain. 3. Much more severe or intense than normal.

**extremely high frequency** Abbreviation, EHF. The band of the radio spectrum extending from 30 gigahertz (GHz) to 300 GHz. These are wavelengths from 10 millimeters (mm) to 1 mm. They are sometimes called millimetric waves. See the appendix RADIO SPECTRUM.

**extremely low frequency** Abbreviation, ELF. The band of electromagnetic radiation from 30 hertz (Hz) to 300 Hz. These are in the low end of the audio-frequency band. They are not useful for radio, but underground currents at these frequencies can be used for communication with submarines. The standard utility line frequency (60 Hz) is within this range.

**extrinsic** 1. Pertaining to a semiconductor with impurities added. The impurities make the material conduct better, and cause it to have properties useful in making diodes, transistors, and integrated circuits. See *intrinsic, semiconductor*. 2. Originating outside of a substance or environment.

**extrusion** 1. A process of manufacturing in which a substance is forced through a hole or die. A good example is wire, made by passing molten metal through a hole in a diamond or other hard substance. The size of the hole determines the gauge of the wire. 2. A natural process where molten rock or other material is ejected from within the earth.

**extrusive** Pertaining to rocks or other materials that have come out from within the earth. Volcanic rocks are a good example.

**eye** An organ that converts light into nerve images and nerve impulses. The drawing shows the human eye, responsive to light in the range from about 7500 Angstroms to 3900 Angstroms wavelength (see *Angstrom unit, visible spectrum*). Other species have different eye structures, and some species, notably insects, can see in the ultraviolet range.

**eye of storm** The central part of an intense, revolving storm, especially a well-developed hurricane. The winds are light and the barometric pressure is lower than anywhere else in

EYE

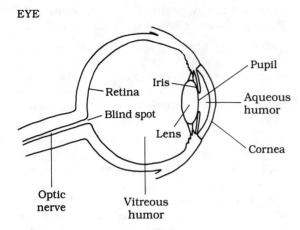

EYE OF STORM

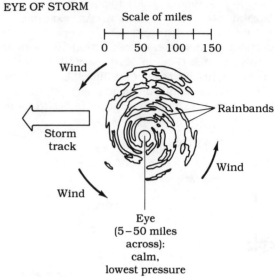

the storm. The region is several miles wide. On radar, it shows up as as a circular, dark region without the echoes that come from rainbands or the eyewall (see drawing). Some intense winter storms have central regions similar to the eyes of hurricanes. See also *hurricane*.

**eyepiece**  A lens or set of lenses, used to magnify the image from the objective of a telescope or microscope, and used for direct viewing with the eye. Magnification varies inversely with the focal length of the assembly.

**eyewall**  The intense band of severe weather than surrounds the eye of a hurricane or tropical storm. See *hurricane*.

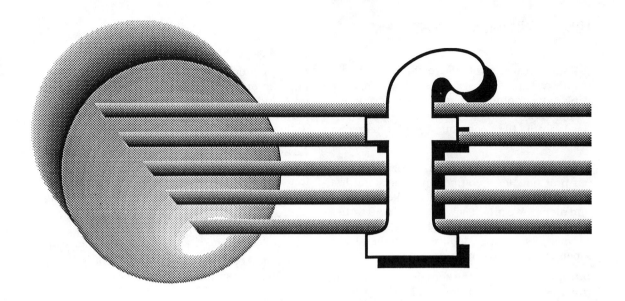

**F** Abbreviation for farad, Faraday constant; chemical symbol for fluorine.

**f** Abbreviation for femto- (see PREFIX MULTIPLIERS appendix), frequency.

**face** 1. A flat surface on a crystal, where cleavage occurs (see *cleavage*). 2. One of the flat surfaces on a polyhedron (see *polyhedron*).

**facsimile** 1. A method of sending still pictures over the telephone. Often used to send letters, especially when signatures are needed. It is like a quick postal service for short documents. The image is converted into electronic pulses that can be transmitted in a narrow-band signal, right through telephone circuits. 2. A near-replica of something; an imitation.

**fact** 1. A statement or hypothesis that has been proven true. 2. A truthful statement in arithmetic, involving numbers and operations.

**factor** 1. A cause in a cause-effect relation. 2. One of two or more discrete expressions that are multiplied together to obtain a more complicated expression. 3. To break a complicated mathematical expression down into simpler form.

**factorial** A function of a nonnegative integer. If the integer is n, then the factorial is written $n! = n(n-1)(n-2) \ldots (3)(2)$. That is, n! is the product of all the integers starting with 2, up to and including n. By default, $0! = 1$ and $1! = 1$.

**FAD** Abbreviation for *flavin adenine dinucleotide*.

**fading** 1. In radio communication, a changing of the strength of a received signal. Can be caused by atmospheric conditions, by phase interaction, or by objects getting in the way. 2. A gradual washing-out of color from an object, caused by exposure to light and other radiation.

**Fahrenheit temperature scale** A means of measuring temperature, generally used by nonscientists in the United States and some other Western countries. Pure water at one atmosphere pressure freezes at $+32$ degrees Fahrenheit (F); pure water at one atmosphere boils at $+212$ degrees F. Absolute zero is about $-459$ degrees F. One Fahrenheit degree is $5/9$ the size of one centigrade degree. If C is the centigrade temperature, the Fahrenheit temperature F is given by $F = (9/5)C + 32$. The Fahrenheit and centigrade scales agree exactly at $-40$ degrees. See *centigrade temperature scale*.

**failsafe** A means of failure protection in electronic systems. The system shuts down to prevent catastrophic damage in the event of a component failure. A backup circuit is then used until the component is repaired or replaced.

**failure rate** The average number of failures per unit time for a component or system. Usually given for individual devices like transistors and diodes. We might say that one in 50 transistors will fail within a year of being placed into service. The one-year failure rate is then 1/50, or 2 percent. See *hazard rate*.

**fallopian tube** In the female, the ducts between the ovaries and the womb. This is where the egg and the sperm join to form the zygote. Then the zygote moves into the uterus, where it develops into an embryo, and finally a fetus. See *embryo, fetus, uterus, zygote*.

**fallout** See *radioactive fallout*.

**familial trait** An inherited characteristic tending to run in families. There are innumerable examples: Curve of the smile, mannerisms, habits. Such traits can be physiological or psychological.

**family** 1. A male, female, and all their descendants. Especially pertains to animals who raise their young, and to human beings. 2. A set of similar things, used for similar purposes. We might have a family of logic integrated circuits, for example. 3. A group of related organisms, ranking above a genus and below an order.

**family of curves** A set of curves, all with similar functions and all having roughly the same shape. Used sometimes in two-dimensional graphs, to give a rough idea of the nature of a function that requires three dimensions to depict exactly.

**famine** Widespread starvation in a geographic area over a certain period of time. In recent decades, large portions of Africa, bordering on the Sahara desert, have been subjected to famine, because the weather is getting drier, and because there are too many people and no effective contraception.

**farad** The standard unit of capacitance. When the voltage changes at one volt per second, and there is a current of one ampere, the capacitance is one farad. This is an extremely large unit, and in practice we usually see microfarads (a millionth of a farad) or picofarads (a million-millionth of a farad). See *capacitance*.

**faraday** A unit of electrical charge quantity. The amount of charge needed in electrolysis to free one gram of an element. Not often used; equal to about 96,500 coulombs. See *coulomb*.

**Faraday effect** When radio waves bounce (refract or reflect) from the ionosphere, the change in polarization that occurs. See the drawing. At the receiving station, the polarization fluctuates randomly. See *polarization*.

FARADAY EFFECT

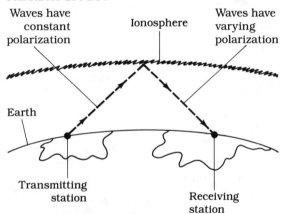

**Faraday, Michael** One of two independent inventors of the electric motor. This was in the early nineteenth century. Faraday is known for his research in electric and magnetic fields. Several effects and devices bear his name.

**Faraday shield** A device that allows magnetic fields to pass through, but blocks electric fields. Used in some antennas for direction-finding, and also in radio-frequency tuned circuits. Also sometimes called an electrostatic shield.

**Faraday's Laws** 1. The law of magnetic induction. When a conductor moves with respect to a magnetic field, a current is induced in the conductor. 2. In an electrolytic cell, the more charge that passes through, the more material is deposited on the electrodes. See *electrolysis*.

**far infrared** See *infrared*.

**Farnsworth, Philo** A communications and electronics engineer, born in 1906. When he was just a teenager, he invented an important scanner that was to lead eventually to the development of television cameras.

**farsightedness** A condition in which objects at great distances appear in good focus, but objects close up are blurred. This can occur either because the lens is inflexible or because the eyeball is "too short." See the drawing. See also *nearsightedness*.

FARSIGHTEDNESS

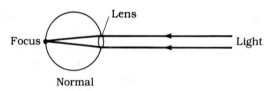

Normal

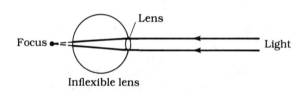

Inflexible lens

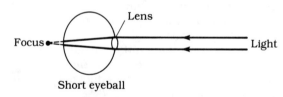

Short eyeball

**far ultraviolet** See *ultraviolet*.

**fat** 1. A hydrocarbon compound with concentrated energy, about 9 kilocalories (9000 heat calories) per gram. Found in many foodstuffs, especially dairy products, and meats. Also in certain vegetables. 2. A reserve source of fuel and insulation for the body, the same hydrocarbon compound stored in various body tissues. See *fat cell*.

**fat cell** A cell in which fat is stored in the body. This fat serves as an important fuel source when food is in short supply. Nowadays famines are rare in some parts of the world (such as the U.S.) and these energy stores are not often needed. See *fat*.

**fathom** A unit of water depth used by sailors. Equal to six feet (two yards), or 1.83 meters.

**fathometer** An instrument that measures the depth of a body of water in a certain location. Uses sonar principles. See *depth sounding, sonar*.

**fatigue** 1. A condition in which the body is overstressed and exhaustion occurs. 2. Loss of mechanical strength in a substance, especially a metal, because of prolonged stress. This might lead to catastrophic breakdown of the material.

**fatty acid** Any of various hydrocarbon compounds. A few of these are needed in the diet of humans, but the essential amount is just a few drops a day. Some fatty acids have short carbon chains and some have long chains. Some can be metabolized by the body for use as fuel. See *essential fatty acids*.

**fault** 1. A place where rocks in the earth's crust have broken because of an earthquake. 2. A boundary between slowly moving masses of crustal rock. Occasionally, these rocks slide with jerking movements, causing earthquakes. 3. A defect in a component or device.

**fax** See *facsimile*.

**Fe** Chemical symbol for iron.

**fecundity** An expression of fertility, usually of the female. The ability of the female to conceive and bear offspring. See *fertility*.

**feedback** A process where some output gets back to the input of a circuit. This can be facili-

FEEDBACK

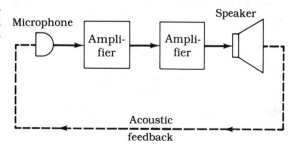

tated on purpose, or it might occur accidentally. The drawing shows undesirable feedback from a speaker to the microphone of an audio amplifier. This would probably result in "howling." Feedback can be positive (regenerative) or negative (degenerative). Positive feedback increases amplifier gain and might cause oscillation. Negative feedback reduces amplifier gain.

**feeding center**  A nerve center in the brain that controls appetite. See *appestat*.

**Fehling's solution**  A chemical used to test liquids for sugar content. Can be used, for example, to test urine for glucose. A positive indication would mean that the patient could have diabetes mellitus. Sugar causes the solution to turn red when Fehling's is added.

**feldspars**  The most common minerals in crustal rocks. They contain silicon, and also calcium, aluminum, oxygen, sodium, and other common elements.

**femoral**  1. Pertaining to the femur or thigh bone. 2. Pertaining to the upper leg; for example, the femoral artery.

**femto-**  See the PREFIX MULTIPLIERS appendix.

**femur**  1. The thigh bone. 2. The upper leg, especially of an arthropod or arachnid.

**Fermat, Pierre de**  A mathematician known for his work with probability, especially predicting the outcome of rolling dice. He is said to have proved a theorem about Diophantine equations, and then died before publishing it. See *Fermat's Last Theorem*.

**Fermat's Last Theorem**  A theorem that Pierre de Fermat supposedly proved. He made a note in a margin of a paper: "I have found a truly marvelous proof!" Then he died and no one ever saw the demonstration. The statement of the theorem is: Suppose we have an equation $a+b=c$, where a, b and c represent variables x, y and z all raised to the integer power n; then if n > 2, there are no positive integers x, y and z such that $a+b=c$. See *Diophantine equation*.

**Fermat's Law**  When a ray of light travels from one point to another, it always takes the path requiring the smallest possible amount of time. That is, because light has constant speed, it always follows the shortest distance (geodesic) between two points in space.

**fermentation**  A process in which bacteria metabolize a substance, such as a sugar or alcohol, changing it to another form. Intestinal bacteria can ferment certain foods, causing gas. Sugar in fruit can be fermented to make wine. Does not require the presence of oxygen.

**fermi**  A standard unit of length in nuclear physics, equal to a quadrillionth of a meter (0.000,000,000,000,001 m) or a trillionth of a millimeter.

**Fermi, Enrico**  A physicist who was a refugee from Fascist Italy during World War II. First used neutrons to bombard uranium in an attempt to produce nuclear fission. The result was not a huge explosion, but the transformation of a small percentage of uranium nuclei into other substances.

**fermion**  A type of subatomic particle with spin $1/2$. This means that, in order for it to look the same, it must be spun through two complete revolutions. The matter in the universe is made up of these kinds of particles. See *spin*.

**fermium**  Chemical symbol, Fm. An element with atomic number 100. The most common isotope has atomic weight 257. All isotopes of this heavy element are rather unstable; they are radioactive and they decay within hours or days. This element is a by-product of hydrogen-bomb detonations.

**fern**  A fairly sophisticated plant with characteristic leaf patterns, usually growing close to the ground in moderate or subdued light, such as on forest floors. They reproduce by means of spores. See the PLANT CLASSIFICATION appendix for the place of the ferns in the evolutionary process.

**Ferrel's Rule**  In the Northern Hemisphere, a moving object is pushed towards the right; in the Southern Hemisphere, towards the left. This explains why the prevailing winds generally blow east-west instead of north-south.

**ferric** Containing iron. Especially, pertaining to an iron compound.

**ferrite** A form of powdered iron with very high permeability. Used as a core for inductors at low, medium, and high frequencies. It is inefficient at very-high and ultra-high frequencies. Other substances are added to obtain the desired permeability. See *permeability*.

**ferroalloy** Any of various metal alloys containing iron.

**ferroelectric** A dielectric substance that can be electrically polarized when there is an electric field present. The electrical equivalent of ferromagnetic (see *ferromagnetic*). Examples of such materials are barium titanate and Rochelle salt. These substances tend to concentrate lines of electric flux.

**ferromagnetic** A material that conducts a magnetic field easily. Iron is a good example; nickel is another. Such materials tend to concentrate magnetic flux; they have high permeability. So they make good cores for transformers and electromagnets. There are various commercially made ferromagnetic substances for many different applications. See *electromagnet, ferrite, permeability, transformer*.

**fertile** 1. Capable of producing offspring (in animals). 2. Capable of supporting plant growth (in soil).

**fertility** 1. The ability of a species to produce offspring. The opposite of sterility. In the male, fertility is shown by a high sperm count. In the female, the ability to regularly produce eggs is the measure of fertility. 2. The ability of soil to support plant growth, especially as pertains to certain crops, such as corn, oats, wheat, soybeans, and the like.

**fertilization** 1. The union of a sperm with an egg to form a zygote. See *zygote*. 2. The addition of fertilizer to soil to enhance plant growth. See *fertilizer*.

**fertilizer** A natural or synthetic, nitrogen-containing compound or compounds, that can be added to soil to cause enhanced plant growth. Natural fertilizers include manure and compost (basically rotten leaves and mulch). Synthetic fertilizers are available in various forms that keep well for fairly long periods of time.

**FET** Abbreviation for field-effect transistor.

**fetal alcohol syndrome** A condition that occurs when a pregnant woman drinks alcohol excessively and for prolonged periods. The fetus becomes addicted to alcohol; when born, the baby might go into alcohol withdrawal. Other problems, such as stunted growth and mental retardation, also can occur.

**fetch** The stretch along which wind gets a chance to make waves on water. Generally, the larger the body of water, the longer the fetch, and the bigger the waves on the downwind shore. Over the open sea, we might have a large fetch at certain latitudes such as 40–50 degrees south (the "roaring forties"), with huge swells as a result.

**fetus** A developing baby in the uterus. Recognizable because it has the beginnings of a face, hands, feet, and body organs. Might even suck its thumb, toss, and kick.

**Feynman, Richard** A pioneer in the field of microminiaturization, and a physics professor at California Institute of Technology. Won a Nobel prize for his work in electrodynamics.

**fiber** 1. A filament of tissue, sometimes capable of expanding or contracting to cause movement (muscles in animals). 2. A filament of material used for reinforcement; for example, spun glass used to make fiberglass. 3. The tough, stringy structure of a plant stem. 4. An optical fiber. See *fiberoptics*. 5. See *dietary fiber*.

**fiberoptics** 1. The transmission of light through clear glass or plastic filaments (see drawing) for any of various purposes. Modulated light can be used for communications; thousands or millions of signals can be sent on a single beam of light in this way. Many optical fibers bundled together can be used to transmit a video image, such as from deep inside the body where there isn't enough space for even the smallest television camera. 2. The science of transmission of light through clear glass or plastic filaments for any of various purposes. There are two categories. In *incoherent* fiberoptics, light of varying intensity, such as a communications signal, is transmitted. In *coherent* fiberoptics, light images are transmitted through bundles of fibers.

FIBEROPTICS

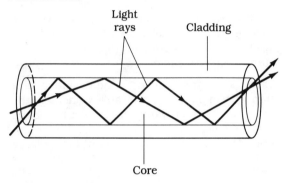

**Fibonacci sequence** A sequence of numbers, where the first two terms are 1, 1 and each term after that is the sum of the two just before it. Thus the series is 1, 1, 2, 3, 5, 8, 13, 21, 34 and so on. This series has been found to occur often in nature, in diverse places.

**fibrillation** An abnormal heart condition in which the muscles do not contract rhythmically, but twitch at random instead. Can occur as a result of extreme stress, electrolyte imbalance, or electric shock. If normal heartbeat is not restored, the victim will die because the tissues will not get oxygenated blood.

**fibrin** A form of protein that causes red blood cells to get tangled up to create a blood clot. Fibrin is the result of action between fibrinogen and thrombin. See *fibrinogen, thrombin*.

**fibrinogen** A blood enzyme, important in the process of clotting. It combines with thrombin in the blood, along with calcium, to form fibrin. Acts most effectively in the presence of air, and also when blood flow is sluggish or obstructed. See *fibrin, thrombin*.

**fibroblast** In the body, a cell that helps in the manufacture of fibers of connective tissue. See *connective tissue, fiber*.

**fibrosis** A condition in which there is excessive fiber in a part of the body. This might occur in scarring, for example. It also sometimes takes place when organs are injured slowly over long periods.

**fibula** The thinner of the two shin bones. Runs parallel with the tibia. In animals, the thinner of the two bones between the ankle and the knee. See *tibia*.

**field** 1. A pattern of electric, magnetic, or other flux surrounding an object that is electrically charged, magnetically polarized or having other force-producing properties. 2. Practical use. We might say that a circuit is to be tested "in the field," for example.

**field coil** 1. A coil that produces a constant or fluctuating magnetic field for any of various purposes, such as the operation of a motor or speaker. 2. The main coil of a relay. 3. The fixed coil in an electric meter.

**field-effect transistor** Abbreviation, FET. An electronic semiconductor device. It consists of a channel through which charge carriers flow from the source to the drain. The gate regulates the flow of charge carriers. When a varying signal is applied to the gate, the flow of charge carriers from the source to the drain varies also, and usually to a much larger extent, so the device amplifies. See *semiconductor, transistor*.

**field ion microscope** A device similar to the electron microscope (see *electron microscope*), except that helium ions are used instead. These particles are far heavier than electrons, so they have shorter wavelengths for a given speed. This helps to increase the resolving power, because this depends directly on the wavelength.

**field lens** In an eyepiece using multiple lenses, the lens that the light from the objective strikes first. This lens determines how wide the viewing angle (field of view) will be.

**field magnet** 1. A permanent magnet in a dynamic speaker or dynamic microphone. 2. A permanent magnet used in some electric motors or generators.

**field strength** 1. A measure of the intensity of an electric or magnetic field. Can be given in flux lines per unit cross-sectional area, or in teslas or webers (see *tesla, weber*), or in volts per meter. 2. A measure of the intensity of an electromagnetic field. Usually given in microvolts per meter.

**FIFO** Abbreviation for first-in/first out.

**filament** 1. The hot, glowing part of an incandescent lamp. See *incandescent lamp*. 2. The thin resistance wire that heats up the cathode of a vacuum tube. See *vacuum tube*. 3. Any threadlike structure, such as might be found in some body tissues or plant tissues.

**file** 1. A self-contained set of data, especially if stored in a computer or in computer memory, such as magnetic disk. 2. The action of putting data into storage for later use.

**filter** 1. A device that purifies water or air, removing pollutants, toxins, and/or dust particles. 2. In electronics, a circuit that lets some signals or currents pass, but blocks others. There are various different types. 3. To cause to be passed through a purification device or selective circuit.

**filtration** 1. The process of being passed through a filter. 2. Purification of water by means of a filter.

**fin** 1. In many water-dwelling animals, any of various appendages that help in stabilization and propulsion. 2. In aerodynamics or hydrodynamics, a device that directs the flow of the gas or liquid so as to control direction and help with stabilization. A good example is the rudder on a boat.

**finite** 1. Having some natural number of elements, where the set of natural numbers is $N = \{0,1,2,3,...\}$. 2. Having a measurable length, area, or volume. 3. Representable by a real or complex number.

**finite mathematics** Mathematics confined to finite values or objects. See *finite, finite number, finite set*.

**finite number** 1. A natural number. 2. An integer. 3. A real number. 4. A complex number.

**finite sequence** 1. A sequence with a defined start and end. 2. A sequence whose elements can be named without missing any, when given enough time.

**finite series** The sum of the members of a finite sequence. See *finite sequence*.

**finite set** A set with a finite number of elements. That is, given enough time, you could name all the members of the set.

**fire** 1. Rapid oxidation, resulting in heat and flame. 2. To process at high temperature, for the purpose of causing a certain reaction (especially hardening) to occur rapidly. 3. To operate a weapon; especially, to launch a projectile such as a bullet, shell or missile.

**fireball** 1. A piece of oxidizing material in a projectile, such as a meteor or a piece of space junk during reentry. 2. Any hot, glowing sphere of material. 3. The original Big Bang from which the universe is thought to have formed. See *Big Bang*. 4. The bright cloud in a nuclear explosion.

**fireproofing** The treating of material to make it resistant to rapid oxidation. Especially pertains to building construction. Asbestos was once used widely for this, but recently it has become suspect as a possible cause of cancer, so other methods are being developed.

**firmware** In a computer, software that has been programmed in permanently, and is not easily altered. See *software*.

**first-in/first-out** A form of buffer, in which storage occurs as shown in the accompanying illustration. The output is at a constant speed, regardless of changes in the speed of the input, as long as there are some elements stored in the buffer. Elements come out in the same order that they go in. See also *buffer, pushdown stack*.

**first-in/last-out** See *pushdown stack*.

**First Law of Motion** See *Newton's Laws of Motion*.

**First Law of Thermodynamics** See *thermodynamics*.

FIRST-IN/FIRST-OUT

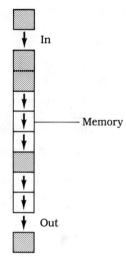

**first-surface mirror** A mirror in which the reflective material is facing the oncoming light particles. The light therefore does not pass through the glass at all. This prevents distortion and double-imaging. See the illustration.

FIRST-SURFACE MIRROR

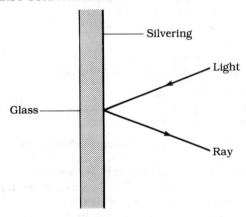

**fish** Any of various water-dwelling animals that extract their oxygen from the dissolved air in water, using gills. First appeared about 500 million years ago during the Ordovician period. See the ANIMAL CLASSIFICATION appendix for the position of the fishes on the evolutionary scale. See also the GEOLOGIC TIME appendix.

**fissile** A material whose atomic nuclei will break up under certain conditions, such as bombardment by high-speed particles. See *fission*.

**fission** 1. The splitting of an atomic nucleus. This is how the first nuclear weapon worked (see drawing). In this case uranium-235 was brought together into a lump larger than the critical mass, and fission took place as a result. 2. The splitting of a cell. 3. In general, breaking up of particles.

FISSION

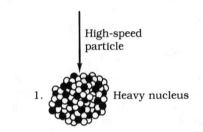

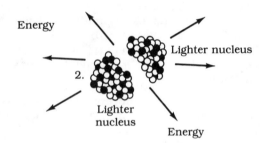

**fissionable** See *fissile*.

**fissure** 1. A crack in the surface of the earth, often with steam or other subsurface material escaping. 2. A crack in the ocean floor from which hot material often escapes. 3. A break in the mucous membrane, such as in the large intestine.

**fissure eruption** A crack in the crust of the earth, through which magma flows or spews. The most common type of volcanic eruption. See *fissure, magma, volcano*.

**fistula** A surgically made opening in a hollow organ such as the stomach, leading to some other organ or to the skin.

**fitness** 1. An expression of the level or extent to which the body can withstand exertion without becoming tired. 2. A state of mind in which high priority is given to good health habits, especially with respect to eating, smoking, alcohol and drug consumption, and physical exercise. 3. The extent to which a life form has adapted to its surroundings.

**fixation** A means of preserving biological specimens for later observation under a microscope.

**Fizeau, Armand** A scientist who developed a unique method of measuring the speed of light. He used rapidly rotating mirrors to create a sweeping beam that was transmitted over a distance. Thus, even in a very short instant of time, the beam direction would change considerably, and this could be observed as an eclipsing effect.

**flaggelum** A whiplike appendage on certain cells, that allows the cell to move around under its own power. See the drawing.

FLAGELLUM

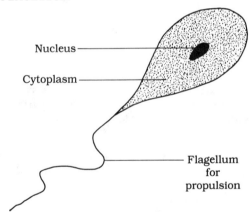

**flame** The glowing gases around a region where fire is taking place. The color depends on the gases being oxidized. The familiar orange of the candle flame or wood fire is largely caused by sodium vapor. See also *fire*.

**flameproofing** See *fireproofing*.

**flap** On the wing of an aircraft, a device that deflects air downward to increase lift. Flaps also increase drag, thereby slowing the aircraft down. Used during landing.

**flash distillation** A method of desalination for converting seawater to fresh water. The seawater is boiled and the steam collected by condensation. See *distillation, distilled water*.

**flash flood** A sudden, often catastrophic flood that occurs when localized, heavy rainfall causes a stream to overflow its banks. The water might rise several feet within a few hours.

**flash point** For a combustible liquid, the lowest temperature at which the vapor will burn. This tends to happen rapidly, causing an explosion or "flash."

**flatworm** A primitive type of worm, comparatively low on the evolutionary scale. See the ANIMAL CLASSIFICATION appendix. Many of these worms, whose bodies are generally flat (hence their name), are parasites. The tapeworm is a good example. See also *fluke, tapeworm*.

**flavin adenine dinucleotide** Abbreviation, FAD. A substance that helps the body to metabolize certain nutrients.

**flavonoid** A substance commonly found in plants. Some of these substances are responsible for the colors of the plants. Some might have nutritional benefits.

**F layer** See *ionosphere*.

**Fleming, Alexander** The scientist who discovered penicillin. This made it possible to cure infectious diseases, such as bacterial pneumonia, that previously had high mortality rates.

**Fleming, J. A.** An English scientist and engineer who devised the first vacuum tubes to rectify alternating currents. He got his idea mainly from the work of Thomas Edison in conjunction with the incandescent lamp.

**Fleming's Rules** Simple rules involving the relationship between electric fields, magnetic fields, voltages, currents, and forces. These rules are demonstrated using the hands. The relationships are rather complicated three-dimensional rules. See an electronics text for details.

**Fletcher, Horace** A man who lost considerable weight by chewing every bite of food for a very long time. Lived around the beginning of the twentieth century. A fad, called "fletcherization," developed when he made his findings public.

**flight control** 1. Any of the devices used to guide an aircraft in its flight. 2. The process of guiding an aircraft along a prescribed flight course. Nowadays this can be done by a computer, but there are always human pilots, copilots, and flight engineers.

**flight engineering** In an aircraft crew, the duties of maintenance and the keeping of the aircraft's technical diary.

**flight instruments** The dials, alarms, and video equipment that display important factors during aircraft flight. For example, an altimeter shows the height of the aircraft above sea level and/or above actual terrain.

**flint** Any of various hard rocks, comprised of black or dark gray quartz, used for cutting and to produce sparks for ignition.

**flip-flop** A simple digital electronic circuit with two stable states. The circuit is changed from one state to the other by a pulse or input signal. Then the circuit maintains this state until another change signal is received. There are several different types of flip-flops, used in various applications in digital technology.

**flocculation** 1. A process for removing the mud and other particle impurities from water. 2. A process where particles stick together, making larger clumps. This might have taken place during the birth of the Solar System, resulting in the formation of the planets and moons.

**floe** A floating, drifting chunk of ice in the ocean. Floes can break apart or crash together, so they are in a constant state of change. Polar explorers have drifted on them, sometimes for hundreds or even thousands of miles.

**flood** 1. The overflowing of a stream or river, causing land to be submerged that is normally above the water level. 2. To saturate. 3. To bombard with radiant energy, especially visible light.

**flood control** Any of various means by which property is protected against the dangers of floods. This might include dikes, landscaping, and damming of streams and rivers. However, the simplest method of avoiding flood-induced catastrophe is to avoid building on the floodplain. See *floodplain*.

**floodplain** The lowlands around a river or stream that are submerged during floods as a normal natural occurrence (see drawing). Because people have built cities on these lands, disasters are frequent. The land on a floodplain might be revitalized by the silt from the floods. This was the case for centuries along the Nile River in Egypt, until people built dams to control the floods.

FLOODPLAIN

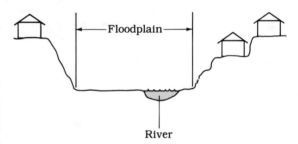

**floodwall** An artificial bank for a river or stream that keeps it confined during times of flooding. Usually made with reinforced concrete.

**floppy disk** A magnetic storage medium used with computers. Usually measures 3.5 or 5.25 inches in diameter. Might store up to more than a megabyte (about a million bytes) of data. Gets its name from the fact that it is flexible, or "floppy," although the 3.5-inch disks are encased in a rigid plastic frame.

**flowchart** A diagram that depicts a process or algorithm logically, in a way that is easily followed. Different shaped boxes are used for different types of functions, such as decisions (yes/no) and commands. See *algorithm*.

**flower** In some plants, the apparatus for sexual reproduction. The bright colors attract pollinating insects. They have evolved in many different ways, so that they can be pollinated by the insects that best ensure efficient reproduction. Some flowers are self-pollinating.

**flowering plant** A plant that makes use of flowers for reproduction. See *flower*; see also the PLANT CLASSIFICATION appendix for the place of these plants on the evolutionary scale.

**flowmeter** A device that indicates how fast water is being pumped through a pipeline. Usually gives its indication in liters or gallons, per second or minute.

**fluid** 1. Generally, any liquid or gas that conforms to the shape of a container in which it is held. Some substances that seem solid are actually fluids over long periods of time; an example is the earth's crust. 2. Having the properties usually attributed to a fluid. 3. Continuous and uninterrupted.

**fluid dynamics** A field of physics concerned with the behavior of fluids (liquids and gases) in motion. See *fluid mechanics, fluid statics*.

**fluidics** A means of circuit design in which fluids are used rather than electricity. Switching functions and other simple operations can be performed. Such devices are almost totally immune to the effects of strong magnetic or electric fields, but they work slowly compared with electric circuits.

**fluid mechanics** A field of physics concerned with the behavior of fluids, either in motion or stationary. See *fluid dynamics, fluid statics*.

**fluid statics** A field of physics concerned with the behavior of fluids when they are not moving. See *fluid dynamics, fluid mechanics*.

**fluke** 1. An unusual occurrence; a coincidence. Might be responsible for unexpected events or experimental errors. 2. A type of flatworm (see *flatworm*), some of which are parasites and invade body organs such as the liver.

**fluorescence** The emission of radiation, usually visible light, by a substance when that substance absorbs radiation from someplace else.

**fluorescent lamp** A source of visible light, powered by electric current. More efficient than incandescent lamps of the same wattage (see *incandescent lamp*). A glass tube is filled with argon gas and mercury vapor (see drawing). Electricity causes the gases to ionize, producing ultraviolet that strikes a phosphor on the inside of the glass tube. The phosphor then glows, or fluoresces, in the visible range.

FLUORESCENT LAMP

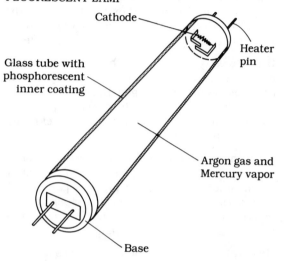

**fluoridation** The addition of fluoride compounds, such as sodium fluoride or calcium fluoride, to water. This might help to reduce tooth decay in people who drink the water. See *fluoride*.

**fluoride** A compound of fluorine and a mineral element, usually sodium. Sodium fluoride is the substance added to many toothpastes to help prevent tooth decay. Other common fluorides include potassium fluoride, calcium fluoride, and magnesium fluoride.

**fluorine** Chemical symbol, F. An element with atomic number 9; the most common isotope has atomic weight of 19. Highly reactive, it is a member of the so-called halogen family of elements. See *halogen*.

**fluorite** Another name for calcium fluoride. It is a crystal solid and has various industrial applications. Usually the crystals are bluish, although impurities cause some variability of the color.

**fluorocarbon** See *chlorofluorocarbon*.

**flux** 1. The presence of an electric, magnetic, or electromagnetic field. 2. A measure of the intensity of an electric, magnetic, or electromagnetic field. Flux lines depict the orientation of the field. Flux density is a direct expression of the intensity of the field at a given location. Flux is normally the most dense near the poles, such as positive and negative, or north and south. See also *electric field, electric flux, flux density, flux lines, magnetic field, magnetic flux*.

**flux density** The strength of an electric or magnetic field, as given in flux lines per unit cross-sectional area. In general, the more flux lines that pass through a given area perpendicular to the lines, the more intense the field. See *flux, flux lines*.

**flux lines** Imaginary lines in space surrounding electric or magnetic poles. When you put a bar magnet underneath a piece of paper, and then sprinkle iron filings on the paper, the filings tend to show the position of these imaginary lines. See *flux, flux density*.

**fluxmeter** An instrument that gives an indication of flux density, usually of a magnetic field, in teslas or webers. See *flux, flux density*.

**Fm** Chemical symbol for fermium.

**FM** Abbreviation for frequency modulation.

**F = ma** See *force/mass/acceleration equation*.

**focal length** 1. In a convex lens, the distance between the center of the lens and the focal point. See *focal point*. 2. In a concave mirror, the distance between the center of the mirror surface and the focal point. See the illustrations.

**focal point** The point to which incoming parallel rays are focused by a convex lens or a concave mirror. See the drawings. If a point source of light is placed at this point, the outgoing rays passing through the lens or reflected from the mirror are parallel. See also *focal length*.

**foci** Plural of focus. Generally used in reference to a mathematical relation or function having more than one focus, such as an ellipse. See *focus*.

**focus** 1. A point of symmetry in a mathematical relation or function. 2. The point for a convex lens or a concave mirror where incoming parallel rays of light converge. See *focal point*. 3. The control in an optical instrument that allows for obtaining a sharp image. 4. To adjust an optical instrument for a sharp image. 5. The central point of a theory or logical argument.

**foehn** A warm, dry wind on the leeward side of a mountain range. For example, in the northern Rockies this would occur on the eastern slopes. See *chinook*.

**fog** Clouds that form at ground level. This requires that the relative humidity be 100 percent, and that condensation occur on particles suspended in the air.

**folacin** Another name for folic acid. See *vitamins* and the VITAMINS AND MINERALS appendix.

FOCAL LENGTH
FOCAL POINT

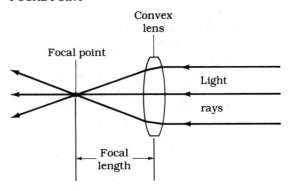

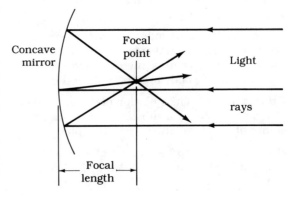

**fold** In the earth's crust, distortions in the rock layers, usually observed as dips or upward thrusts. A dip-shaped distortion is called a syncline, and an upward, peak-shaped distortion is an anticlyne.

**folic acid** See *vitamins* and the VITAMINS AND MINERALS appendix.

**follicle** In the skin, the place where a hair shaft has its root. The hair shaft grows from the follicle. Oil glands (sebaceous glands) lubricate the hair shaft. See the drawing.

FOLLICLE

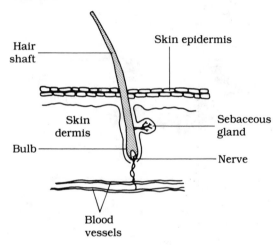

**food additive** Any substance that is added to food to preserve it, to alter or enhance its nutritional value, or to change its flavor or color. All additives used in the United States must be approved by a federal agency called the Food and Drug Administration.

**food chain** The sequence of organisms through which energy passes, usually from plants to lower animals, and finally to higher animals. An example is plankton, eaten by small fish, eaten by larger fish, eaten by humans. There are many "branches" to this chain. That is, animals and people can get their food in various different ways.

**food fad** A nutritional belief not based on scientific fact that leads people to follow bizarre and often unhealthful eating habits. Much of this faddism occurs with weight-loss diets. It is always best to consult a doctor if you want to lose weight. Some food fads develop when certain foods are thought to have properties that they do not have.

**food processing** The refining of foodstuffs to make them last longer in storage, or to eliminate harmful bacteria, toxins, or parasites. Also can be done to enhance the flavor, improve the texture, or otherwise make food easier to prepare and to eat. Cooking was one of the earliest of these procedures.

**food supplement** Vitamins, minerals, fiber, or enzymes that are sometimes taken with food in an attempt to get all the necessary nutrients even if the diet by itself is deficient. Some food supplements can be toxic in large amounts; this is especially true of vitamin A, vitamin D, and some minerals. See the RECOMMENDED DAILY ALLOWANCES and VITAMINS AND MINERALS appendices. It is best to consult a doctor if you plan to take supplements in excess of the recommended daily allowances.

**foot** 1. A unit of length equal to 12 inches or about 30.5 centimeters. 2. The appendage at the end of the leg in terrestrial animals and birds.

**foot-lambert** The average brightness of a surface that emits light of one lumen per square foot. More often, this is expressed in candelas per square meter. See *candela, lumen*.

**foot-pound** A unit of work or energy equivalent to that needed to raise a weight of one pound up by one foot. This is about 1.356 joules. See the ENERGY UNITS appendix.

**foramen** Any of various small openings in the skull (see drawing).

**foraminifera** Microorganisms that lived in the ocean during the last ice age, but became extinct when the ice sheets melted between 16,000 and 13,000 years ago. This put a cold, fresh-water lid on the ocean. Under these changed conditions, most of the microorganisms perished; some survived to the present.

**force** 1. The influence that causes a mass to be accelerated. 2. Attraction or repulsion,

FORAMEN

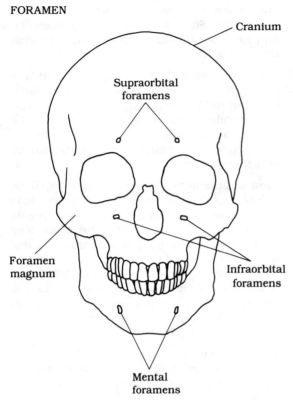

caused by electric fields, magnetic fields, gravitation, or subatomic effects. Measured in dynes or newtons. See *dyne, newton*.

**forced extinction** A species extinction caused by the activities of humans rather than as a natural process.

**force/mass/acceleration equation** The equation $F = ma$, or force equals mass times acceleration. The units must all be in the same system. We might specify acceleration in meters per second per second ($m/s^2$), and mass in kilograms. Then force would be in kilogram meters per second per second ($m/s^2$), or newtons.

**Ford, Henry** An American engineer of the early twentieth century. He is best known for his work in the manufacture of affordable and efficient cars. He introduced standardization and the assembly line.

**forebrain** In the embryo, the group of cells that eventually matures into the cerebrum, thalamus, and hypothalamus. See *brain, cerebrum, hypothalamus, thalamus*.

**forecastle** An elevated, forward living space or forward deck in a ship.

**foreshock** An earth tremor that precedes an earthquake. Can provide some warning, but the main shock often follows rapidly.

**forest** A region in which trees naturally grow over 100 percent or near 100 percent of the terrain. The drawing shows the parts of the world naturally covered by forest. This drawing does not take deforestation into account. See *deforestation*.

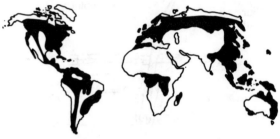

SHADED AREAS SHOW FORESTED REGIONS OF THE WORLD.

From Erickson, THE LIVING EARTH (TAB Books, Inc., 1989).

**forestation** 1. The natural growth of forests. 2. The deliberate cultivation of forest by planting trees. This might be done to produce lumber for future generations, or to slow down or prevent unwanted global warming. See *global warming*.

**formaldehyde** A gas, formed by oxidation of methanol (wood alcohol). When dissolved in water, it makes a preservative solution technically known as formalin, but also sometimes called formaldehyde. This solution has a characteristic odor. Biological samples are often kept in this solution to slow the process of decay.

**formalin** See *formaldehyde*.

**Formalism** A philosophy of knowledge and especially of mathematics. Pure formalists believe that mathematics is nothing but a game or an art form played with symbols according to agreed-on rules. The mathematician Hilbert was among those who subscribed to this philosophy. See *Intuitionism, Logicism.*

**formants** See *speech formants.*

**form drag** Air resistance or water/fluid resistance, caused by the simple fact that a moving object presents a certain surface area (see drawing).

FORM DRAG

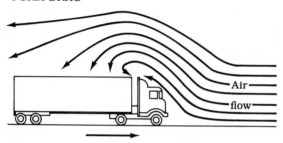

**formic acid** A liquid, clear and with a characteristic odor, used in the manufacture of dye. In nature, this substance is found in ants. It also can be chemically produced.

**formula** 1. A combination of chemicals that results in a compound, mixture, or suspension for a certain purpose. 2. An equation or process that gives the solution to a problem having variables. 3. The chemical abbreviation for a substance or reaction. For example, the formula for the reaction of hydrogen and chlorine, yielding hydrochloric acid, is $H + Cl = HCl$.

**FORTRAN** A computer language that is used for science and engineering. The word is a contraction of "FORmula and TRANslation." It is an easy-to-learn, high-level language similar to BASIC but with more adaptility to scientific and engineering problems.

**fossil** A preserved imprint or footprint of a life form that existed long ago. Usually found in rocks. Sometimes bones and teeth are preserved along with the imprint. Occasionally whole organisms are preserved intact in ice. See *petrification.*

**fossil fuel** Any combustible substance in common use, and obtained from drilling or mining. Oil, coal, natural gas, oil shale, and tar sands are examples. These can be refined for various purposes, such as running cars, planes, and trains. They are also used for heating homes and making certain medicines. Natural stores of these fuels are finite.

**fossilization** The process whereby fossils are formed (see *fossil*).

**Foucault pendulum** A weight suspended from a long string or wire, used to prove that the earth rotates. Once it is started, the plane of its swing will slowly rotate unless the pendulum is located at the equator. The higher the latitude, the faster it rotates, approaching one complete revolution every 23 hours and 56 minutes for 90 degrees north or south latitude (either geographic pole).

**four-channel stereo** See *quadrophonics.*

**four-dimensional** 1. Pertaining to a space in which four coordinates, such as w, x, y and z, are necessary to uniquely define a point. 2. Pertaining to three-space with the addition of time as a dimension. See *fourth dimension, space-time.*

**Fourier series** A series that shows any periodic (repeating) function as a combination of sine and cosine terms, added up. The series is usually infinite, allowing an approximation of the actual value. Theoretically, any waveform, such as square, sawtooth, sine-squared, or cycloid, can be expressed in this way. The theory of Fourier series is called Fourier analysis and is quite sophisticated.

**fourth dimension** 1. A dimension defined by a coordinate axis, usually denoted the t axis or the w axis, that runs through the origin of Cartesian three-space in such a way that it is perpendicular to all three axes x, y and z. It cannot be envisioned, but it is mathematically workable. See *Cartesian coordinates.* 2. Time, in the context of space-time. See *space-time.*

**fovea** In the eye, the region on the retina where the greatest sensitivity exists. It is where the light falls from an object that you are looking directly toward.

**FPS system** A system of units based on the foot, pound, and second. The meter, kilogram, and second are used more often in science and engineering. See the STANDARD INTERNATIONAL SYSTEM OF UNITS appendix.

**Fr** Chemical symbol for francium.

**fractal** A theoretical object that seems complicated no matter how closely one examines it. A hypothetical example is shown in the drawing. In recent years, the idea has developed that our universe might actually be like this. For example, we might keep on finding tinier and tinier subatomic particles, without limit, except for the amount of energy we can concentrate for that purpose. See *elementary particle, Mandelbrot set.*

FRACTAL

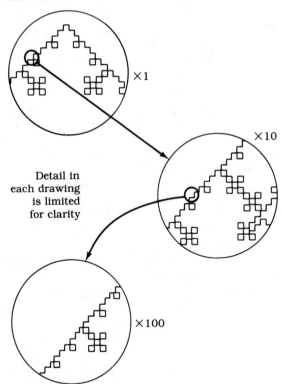

Detail in each drawing is limited for clarity

**fraction** 1. A rational number between 0 and 1, written as a quotient of two integers. 2. A quotient of any two integers. 3. A part of.

**fractional crystallization** A means of separating the dissolved solids in a solution. The temperature is slowly lowered. The dissolved materials then crystallize in order of solubility from least to most. See also *fractional distillation.*

**fractional dimensions** Pertaining to a space that has, for mathematical purposes, a fractional (not whole) number of dimensions. The object in the drawing for *fractal* might be thought of as taking up two dimensions, because it is a plane figure, or one dimension, because it is, in a sense, a fantastically twisted line. Yet it cannot be followed like a line; we could say it has 1.5 dimensions. Some mathematicians have developed a sophisticated method of calculating the exact number of dimensions for an object such as this.

**fractional distillation** A method of separating liquids in a mixture. Different substances in the mixture vaporize at different temperatures. This is effectively put to use by drawing vapor off at different levels in a vertical tube. The substances that vaporize at the lowest temperatures tend to "boil off" first and are therefore at the highest level in the tube. This process is extensively employed in oil refineries. See *fractional crystallization.*

**fractionalization** 1. The rendering of a substance into its various component chemicals. 2. The breaking-up of a complex into its constituent parts. 3. The writing of a number in fractional form. With rational numbers it can be done precisely; with irrational numbers, approximately.

**fractocumulus clouds** Also called scud. The fast-moving, shredded clouds that often accompany storms.

**fracture** 1. To break or cause to be broken. 2. A line or plane along which an object is broken. Might occur in the earth as a fault (see *fault*) or in a bone as a result of injury.

**fragmentation** 1. The breaking-up of an object or substance into smaller parts or pieces. 2. A form of reproduction similar to budding where pieces of an organism break off and evolve into new organisms. See *budding.*

**frame** 1. A single, complete television picture, repeated 25 times per second (usually), with 625 horizontal lines and a 4-to-3 horizontal-to-vertical ratio. 2. See *frame of reference*. 3. The skeleton of the body in vertebrates. 4. See *distribution frame*.

**frame of reference** In physics, and especially in relativity theory, the point of view. Must always be specified with respect to some other point of view, such as standing still on the surface of the earth. Can be stationary, moving, or accelerating.

**francium** Chemical symbol, Fr. An element with atomic number 87. The most common isotope has atomic weight 223. It occurs naturally as a radioactive substance along with uranium.

**Franklin, Benjamin** An American statesman, philosopher and inventor who lived in the eighteenth century. He is best remembered for demonstrating that lightning is a large-scale phenomenon just like static electricity. Founded a scientific organization in 1727 in the English colonies (later to become the United States).

**fraternal twins** Twins that develop from two separately fertilized eggs, but in the same womb and alongside each other. They might be of the same sex or of opposite sexes. They do not usually look any more alike than would be the case with siblings born separately. See *identical twins*.

**Fraunhofer letters** Designators for dark lines in the spectrum of the sun. The lines are called Fraunhofer lines, after the German optician who first noticed them. The letters are assigned with A for the longest wavelength (reddest light), and progressing in the alphabet as wavelengths get shorter. See *absorption line*.

**free electron** An electron that is not bound to any particular atomic nucleus, but instead, wanders among the nuclei in a substance (see drawing). This causes the substance to conduct electricity easily. See *bound electron*.

**free energy** The capacity for a system to do work. Also called available energy. See *energy*.

**free fall** The condition of existing in the earth's gravitational field without any means of support, and neglecting air resistance. A lead shot dropped from a hovering balloon is essentially in free fall until it strikes the ground. See *free-fall acceleration*.

**free-fall acceleration** The acceleration of an object in a vacuum if it is dropped near the surface of the earth. This is about 9.807 meters per second per second (m/sec/sec). It is sometimes rounded off to 10 m/sec/sec or 32 feet per second per second (ft/sec/sec). This means that, each second, the speed of the object increases by 10 m/sec or 32 ft/sec. Referred to as one "gravity," or 1 g. See *free fall*.

**free radical** A molecular fragment that readily combines with other substances. It is an atom with an unpaired electron, tending to capture electrons in an "attempt" to become complete. One free radical can cause a chain reaction affecting hundreds or even thousands of molecules. In the body, these substances are suspected as being harmful.

FREE ELECTRON

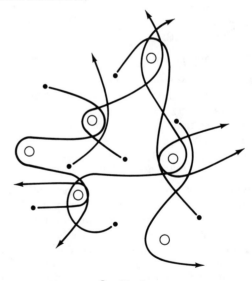

◯ = Nuclei
• = Electrons

Electrons "wander" among nuclei

**free space** 1. Interplanetary or interstellar space, well away from the gravitational influence of planets or stars. 2. In electronics, a vacuum with no nearby objects or fields that would influence the propagation of electromagnetic waves.

**freeze drying** A method of preserving organic substances. The material is frozen and then subjected to extremely low pressure. This causes the ice to evaporate directly into vapor. The vapor is removed, and the material warmed back up. It has thus been dehydrated with minimal effects on its properties.

**freezing point** As a liquid sample is cooled, the temperature at which it begins to solidify. For pure water at sea level and normal atmospheric pressure, this temperature is 0 degrees centigrade or 32 degrees Fahrenheit.

**frequency** For an oscillating or varying quantity, the number of complete cycles per unit time. Usually this is given in cycles per second, or hertz (Hz). Also can be given in thousands of cycles per second, or kilohertz (kHz), or in millions of cycles per second, or megahertz (MHz). Frequencies can become so high that large-magnitude prefix multipliers must be used. See the PREFIX MULTIPLIERS appendix.

**frequency counter** See *frequency meter*.

**frequency meter** A device for measuring or counting frequency. Modern digital frequency meters simply count the number of pulses per unit time, usually per second. This can be done inexpensively using integrated circuits, to an accuracy of six or seven digits. More expensive meters can read out to eight or more digits.

**frequency modulation** Abbreviation, FM. A means of conveying information by varying the frequency of a signal. This method has certain advantages. It is less susceptible to interference from static and ignition noise than amplitude modulation (AM). See the drawing. See also *amplitude modulation, modulation, pulse modulation*.

**fresh water** Water that is found in most aquifers, containing essentially no salt.

FREQUENCY MODULATION

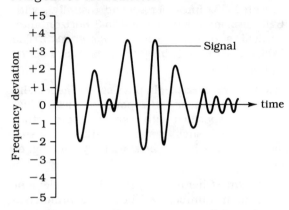

**fresnel** A frequency of one million megahertz, or one terahertz (1,000,000 MHz = 1 THz). Used to express frequencies in the visible, infrared, and ultraviolet ranges. The visible spectrum extends from about 400 to 770 fresnel.

**Fresnel lens** A type of lens in which the refraction takes place because the plastic or glass is cut in rings with slanted surfaces. This allows large lens diameter without great bulk (see drawing). Such lenses do not have good resolv-

FRESNEL LENS

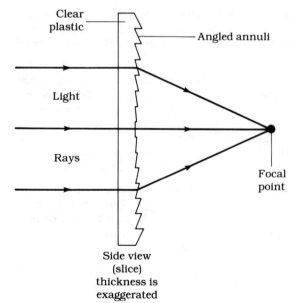

ing power and are not suitable for use in precision optical instruments. They are used in lighthouses and in overhead projectors.

**Freud, Sigmund** The pioneer of psychoanalysis. He did most of his work in the late nineteenth century, during a time when theoretical science enjoyed rapid advances. He is best known for his theories concerning the subconscious mind and its influence on our behavior. See *psychoanalysis*.

**friction** A force that acts contrary to an applied force, especially when two substances come into contact. Air resistance, water resistance, or rolling resistance are forms of friction. There is no such thing as a truly frictionless system, although a spacecraft can coast almost unaffected by friction.

**friction drag** 1. A measure of the extent to which friction affects a system (see *friction*). 2. See *form drag*.

**Friedman, Herbert** An astronomer who, with others, has theorized that X-ray stars are binary stars orbiting very close to each other. See *binary star, X-ray star*.

**Friedmann, Alexander** A cosmologist who discovered that Einstein's equations imply an expanding universe. Willem deSitter also found this at about the same time, in the early part of the twentieth century. The theory of the expanding universe is now generally accepted. See *Big Bang*.

**frigidity** In the female, inability to have orgasm; lack of sexual drive or interest.

**frigid zone** The regions of the earth near the poles. Specifically, that part of the earth inside the Arctic Circle, and that part inside the Antarctic Circle. See *Antarctic Circle, Arctic Circle*.

**front** See *weather front*.

**frontal bone** A part of the cranium. The bone in the skull that covers the front of the brain.

**frontalis** The muscles in the forehead. These contribute to facial expression, especially movement of the eyebrows.

**frontal lobe** The part of the cerebrum toward the front. In the 1930s a form of surgery, called a prefrontal lobotomy, was tried in an attempt to cure violent mental illness. See *brain*.

**frost** 1. Water that has condensed on a surface whose temperature is below the freezing point. The vapor crystallizes directly on the surface. Sometimes the crystals form intricate and fantastic patterns. 2. Water in topsoil that has frozen.

**frost heaving** Buckling of a road surface caused by repeated freezing and thawing of the ground (see drawing). This phenomenon is most noticeable in the spring, as the frost in the earth is melting. It is caused by the repeated expansion and contraction of water as it freezes and thaws. It is a major road-maintenance problem in northern parts of the United States. It also brings up rocks in Northern farmers' fields.

FROST HEAVING

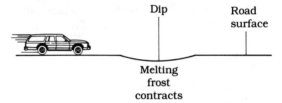

**fructose** Also mistakenly called "fruit sugar." Part of the sucrose (table sugar) molecule, along with glucose. Fructose is two-thirds sweeter than ordinary sugar and can be used as a substitute to save calories. Some people cannot tolerate the taste. See *glucose, sucrose*.

**fruit** A fertilized flower that has matured. Contains seeds that can be released when the fruit breaks open or when it rots on the ground after having fallen off the plant. Some fruits are edible and are sources of carbohydrate, fiber, vitamin C, and various minerals that depend on the richness of the soil in the region where the plant grows.

**frustum** A pyramid, tetrahedron or cone whose top has been truncated (cut off) in a plane parallel to the plane of the base.

**ft** Abbreviation for foot.

**fuel** 1. A source of energy, especially for conversion to heat or to mechanical work. 2. Oil and its products, including gasoline, kerosene, and similar combustible substances. 3. Coal and its derivatives. 4. Natural gas. 5. Rocket propellant, such as liquid hydrogen and oxygen. 6. Radioactive isotopes for nuclear power. 7. Food calories, especially fat and carbohydrate, that are converted by the body into useful energy to carry out body functions.

**fuel cell** An electric cell or battery that produces its current by storing energy in the form of fuel, particularly hydrogen and oxygen. This type of device can be recharged by electrolyzing water into its components, hydrogen and oxygen, and when used, its only byproducts are heat and water.

**Fujita scale** A means of categorizing tornadoes according to destructive power. See the illustration and table. Devised by T. Fujita of the University of Chicago. The most powerful F-5 (intensity 5 on the Fujita scale) tornadoes cause almost complete destruction to trees and frame buildings.

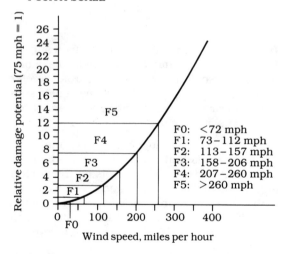

**Fuller, R. Buckminster** A twentieth-century scientist, inventor, and philosopher. Devised the geodesic dome, a lightweight and strong structure. See *geodesic dome*.

**fumarole** A hole in the ground through which hot gas, usually water vapor, escapes. A geyser is a good example.

**function** 1. In mathematics, a relation such that the dependent variable has at most one value for any value of the independent variable. Graphically this means that the curve never "folds back" on itself. See *relation*. A function can be abbreviated by a single small letter. We might write $y = f(x)$, saying that "y equals f of x," and meaning that the variable y is a function of x, and we call the function f. Suppose $f(x) = 3x$. Then, $y = 3x$, and f can be graphed as a line in Cartesian coordinates, passing through the origin and having a slope of 3. See *Cartesian coordinates*. Some functions operate on more than one variable; we might have, for example, $w = f(x,y,z)$. 2. A cause-effect relationship. See *cause-effect*.

**Functionalism** The theory that something must perform a specific function in order to be beautiful, and the belief that there is beauty in the orderliness of things. This philosophy was popular during the eighteenth century.

**function theory** The theory of mathematical functions. See *function*.

**fundamental** 1. A basic principle. 2. A simplified rule of operation. 3. See *fundamental frequency*.

**fundamental frequency** 1. The main or intended frequency of oscillation or communication in a radio system. 2. The lowest frequency in a radio signal. 3. The carrier wave in a modulated signal. See *harmonic, sideband*.

**fundamental interaction** Gravitation, electromagnetic interaction, subatomic, or nuclear force. Subatomic force can be either *weak* or *strong*. Gravitation is what pulls us to earth and keeps planets in orbit. Electromagnetism is the attractive or repulsive force that we see with charged objects and magnets. The weak nuclear force is a complicated concept in physics, as is the strong nuclear force. They affect particle decay and the "sticking together" of atomic nuclei. All these types of interaction can explain the behavior of all particles and energy in the universe; hence, we call them fundamental.

**fundamental number** Constants in nature that seem universal and that can be expressed as numbers. The speed of light in a vacuum is an example (about 186,000 miles or 300,000 kilometers per second). See the CONSTANTS appendix.

**fungicide** 1. A substance that kills fungus in lawns or soil; a form of weed killer. 2. A substance that kills fungus in the body or on the skin; a drug.

**fungus** A primitive plant, lacking the ability to synthesize energy from light (photosynthesis) and thus existing as a parasite. Mushrooms are a good example. Might live in the body and cause considerable distress if it spreads too much. Live in warm, damp environments and can thrive even in total darkness.

**Funk, Casimir** A Polish chemist who first discovered the substance that cures beriberi (see *beriberi*). Now we know this as vitamin B-1 or thiamine. In Funk's time, the early 1900s, the vitamins had not yet been categorized, and the causes of most deficiency diseases were not accurately known. See the VITAMINS AND MINERALS appendix.

**funnel** 1. A truncated, inverted cone. 2. A device that makes it easier to pour liquids into a container having a small mouth. Shaped basically like an inverted, truncated cone, usually with a tube attached at the bottom. 3. A smokestack, especially in a ship.

**funnel cloud** A tornado that has not yet reached the ground (see drawing). So called because it often is funnel-shaped, although it might also look like a snake or a rope. Water droplets cause the vortex to be visible. See also *tornado*.

**funny bone** A sensitive nerve in the body. The most well-known such nerve is in the elbow.

**fuse** 1. A device that limits the current that might flow in a circuit. It usually consists of a thin wire made from a metal with a low melting temperature. If the current exceeds the maximum, the wire melts, breaking the circuit. Some fuses blow almost instantly; they are called fast-break or fast-blow. Some take longer to break the circuit and are called slow-blow. See also *circuit breaker*. 2. A device for detonating an explosive.

**fuselage** The long, hollow, cylindrical part of an aircraft to which the wings and tail are attached. The freight or passengers ride in cabins in this part of an aircraft. It is usually pressurized to make breathing easy even at high altitudes.

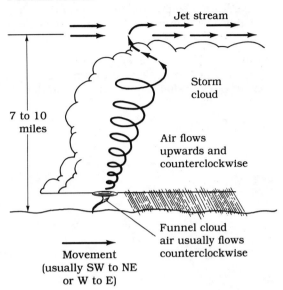

FUNNEL CLOUD

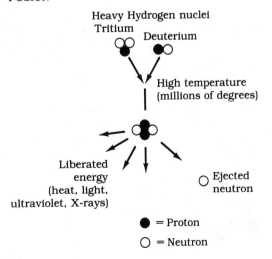

FUSION

**fusible alloy** Alloys that melt at fairly low temperatures. Solder, made of tin and lead, is probably the commonest example. See *alloy, solder*.

**fusion** 1. The joining of two atomic nuclei, either identical or different, to make a single, heavier nucleus. Takes place in the interiors of stars. This is how all heavy elements were originally formed. 2. The combination of hydrogen nuclei to make helium nuclei (see drawing). This is the process that makes the sun shine, and also that gives hydrogen bombs their power. It has yet to be harnessed as a steady, reliable source of energy.

**G** Abbreviation for conductance, gauss, giga- (see the appendix PREFIX MULTIPLIERS), gravitational constant.

**g** Abbreviation for free-fall acceleration, gram.

**Ga** Chemical symbol for gallium.

**gabbro** A type of rock that forms when magma crystallizes in the crust of the earth. It resembles basalt, but has a coarser grain. See *basalt, igneous rock, magma*.

**gadolinium** Chemical symbol, Gd. An element with atomic number 64. The most common isotope has atomic weight of 158. It is a metal in pure form. It is used as a neutron-absorbing substance in nuclear physics experiments.

**Gagarin, Yuri** The first man to orbit the earth in a spacecraft. He was a Russian cosmonaut, who went once around the earth in the capsule *Vostok I* on April 12, 1961. The craft orbited at an altitude of 100 to 200 miles.

**Gaia Hypothesis** A theory in which the earth is regarded as one large organism, like a huge cell in which all life-forms interact, depending on the health of the cell for their existence. Many environmentalists subscribe to this theory nowadays. If we hurt the earth, we threaten our survival. Conversely, if we have an interest in the health of the earth, we ensure that we will have a good place to thrive.

**gain** 1. An increase in a quantity, such as your weight or the level of atmospheric noise. 2. Amplification in an electronic circuit. 3. A quantitative statement of the amplification in an electronic circuit, usually given in decibels (dB). Can be specified for current, voltage, or power. See *decibel*.

**gal** Abbreviation for gallon.

**galactic halo** A spherical "shell" of older stars and globular clusters that surrounds our galaxy. This sphere of stars and clusters has a radius of 65,000 light years. See *galaxy*.

**galactic noise** Radio-frequency energy that comes from the plane of our galaxy, the Milky Way. Most of this noise is observed in the direction of the constellation Sagittarius, because the center of our galaxy lies in this direction. See *galaxy*.

**galactose** A simple sugar. The lactose molecule contains one molecule of glucose and one of galactose. In the digestive systems of some people, lactose is split into glucose and galactose; in others it is not. Galactose is metabolized like other carbohydrates. See *glucose, lactose*.

**galaxy** A congregation of stars held together by their mutual gravitation, and existing in space apart from other similar groups of stars. Our galaxy, including the halo (see *galactic halo*), has a mass of about 300 billion (300,000,000,000) suns. Some galaxies might have as many as a trillion stars. Galaxies might be spiral-shaped, spinning like huge hurricanes in space; they might be elliptical or spherical. They might also be irregular. The drawing shows elliptical and spiral galaxy classifications.

GALAXY

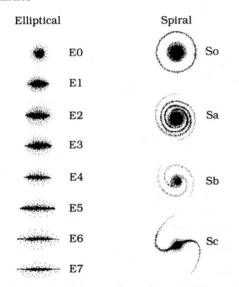

**gale** A strong wind, having a speed of at least 39 miles per hour and up to 46 miles per hour. A wind of 47–54 miles per hour is called a whole gale. Sometimes the term is used imprecisely to refer to any strong or damaging wind. See the appendix BEAUFORT SCALE.

**Galen** A physician of ancient times whose ideas influenced medicine for 18 centuries. Lived during the Second Century A.D. He wrote more than 100 books about medicine. Some of his terminology is still used today.

**galena** Lead sulfide, and the ore from which lead is obtained. It appears as a shiny metal-like rock. It has a characteristic cube-shaped crystal. See *lead*.

**Galilei, Galileo** A seventeenth-century Italian astronomer who built and used one of the first astronomical telescopes. He discovered moons orbiting Jupiter, and found that our own moon has craters. He also observed sunspots, and developed theories of the solar system similar to those we believe today. But in his time, the Church dictated what people were to believe. They did not agree with all of Galileo's theories, so he was punished.

**gallbladder** A small cavity where bile is stored and released to help digest food, particularly fats. Located underneath the liver in the right part of the abdomen. See also *bile*.

**gallium** Chemical symbol, Ga. An element with atomic number 31. There are various isotopes; the most abundant isotope has atomic weight 69. It is most often used in the manufacture of semiconductors. See *gallium arsenide*.

**gallium arsenide** Abbreviation and chemical formula, GaAs. Usually pronounced gas. A semiconductor substance used in high-technology circuits, especially amplifiers where sensitivity and low noise are required.

**gallon** A British unit of fluid measure. Used commonly in England and the U.S. by lay people. One gallon is a little less than four liters. It is equal to four quarts, eight pints, 16 cups, or 128 fluid ounces. See *liter*.

**gallstone** An accumulation of minerals or salts in the gallbladder. More common in older people and in women. Symptoms include indigestion and abdominal pain, although many patients don't have symptoms at all. See *gallbladder*.

**galvanism** The generation of an electric current using chemical reactions. A simple example is a dry cell. The word comes from the name of an eighteenth-century scientist, Luigi Galvani. Besides being responsible for the functioning of electrochemical cells, galvanism can cause corrosion. See *electrochemistry*.

**galvanizing** A process where iron or steel is plated with zinc. This helps to prevent rust when the metal must be exposed outdoors. Electrical conduits and sheet metal are often plated with zinc in this way.

**galvanometer** A device for detecting the flow of an electric current. The simplest device can be made by wrapping wire in a coil around a compass as shown in the drawing. When a current flows, the compass needle will be deflected either to the left or to the right, depending on the direction of current flow.

GALVANOMETER

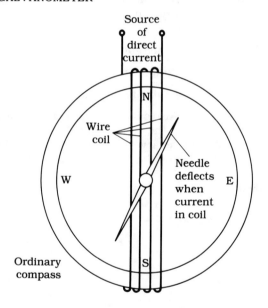

**gamete** Cells that form a zygote when they combine. In humans these are the egg and sperm. See *egg, sperm, zygote*.

**gamma globulin** A substance that can be isolated from human placenta and used to combat infectious diseases. See *placenta*.

**gamma rays** Electromagnetic fields with very short wavelengths and therefore high particle energy. The wavelength ranges from 1 Angstrom unit down to about 0.001 Angstrom (see *Angstrom unit*). This is a few millionths of the wavelength of visible light. Gamma rays are highly penetrating and dangerous in high doses. See *alpha rays, beta rays, X rays*.

**Gamow, George** A contemporary cosmologist, who has referred to the primordial fireball as "ylem." He is a proponent of the Big Bang theory. See *Big Bang*.

**ganglion** A region in the body where nerves converge. They are "branch points" in the circuit of the nervous system.

**gangrene** Tissue death. Can occur because of infection, or because the blood circulation has been cut off. It might spread and can be life-threatening. When this process affects a part of the body, amputation is usually necessary.

**Ganymede** One of the moons of Jupiter, about the size of the planet Mercury. One of the four moons visible through binoculars or a small telescope when observing Jupiter from the earth.

**garnet** A mineral containing silicon and various other elements, forming a hard crystalline material. This can be ground up and employed in sandpaper and grinding apparatus. Sometimes larger crystals are used in jewelry.

**gas** 1. One of the three fundamental states of matter, the others being liquid and solid (see *liquid, solid*). A gas will fill up a container almost immediately, so that the density is uniform throughout the container. 2. Slang for *gasoline*.

**gas constant** A universal quantity that defines the behavior of gases. It is given in units of joules per kilomole per degree Kelvin. See the appendix CONSTANTS.

**gasoline** A refined liquid fossil fuel used in internal combustion engines. Complete combustion (burning) yields only heat, mechanical energy, carbon dioxide, and water. However, the combustion is usually incomplete. This produces carbon monoxide, a dangerous gas.

**gasoline engine** A device that uses gasoline to produce mechanical energy by means of internal combustion (burning). Operates by exploding small amounts of gasoline vapor and air-suspended droplets within a confined space at regular intervals. Various machines convert this energy into useful torque.

**gastric** Pertaining to the stomach. See *stomach*.

**gastric juice** Liquids in the stomach, produced by the body to help digest food that has been taken in. Consists of hydrochloric acid and various enzymes. See *stomach*.

**gastritis** An inflammation of the stomach lining. Might occur with viral or bacterial infections ("stomach flu") or as a result of poisoning or excessive consumption of irritating foods or drinks. Symptoms range from dull aching to violent vomiting.

**gastroenteritis** Inflammation of the digestive system. Might occur for the same reasons as gastritis (see *gastritis*). Symptoms include diarrhea and abdominal pain, along with general fatigue.

**gastrointestinal** Pertaining to the digestive system, that is, to the stomach and intestines. See *intestine, stomach*.

**gastroscope** An endoscope for observing the esophagus and stomach. See *endoscope*.

**gas turbine** A large fan that develops torque as hot gases are forced through it. It is a sort of windmill under pressure. The gases are heated by combustion of fuels such as gasoline or kerosene. See *steam turbine*.

**gate** 1. A logic circuit that performs a specific logic function, such as AND, NOT, OR, or NOR. 2. The control electrode in a field-effect transistor. See *field-effect transistor*. 3. A barrier installed in a dam for the purpose of flood control. Also called a floodgate. 4. Any device that controls the flow of a current or fluid in an electronic or fluidic system.

**gauss** A unit of magnetic field intensity. One gauss is one "line of flux" passing at a right angle through a square centimeter of cross-sectional area. Equal to 0.0001 tesla. See *tesla*.

**Gauss, Karl Friedrich** A mathematician who worked in the early nineteenth century. He is best known for his work with analysis and functions. Some of his mathematics was extremely sophisticated; it was mostly applied, especially in electricity and magnetism.

**Gaussian function** A mathematical function used by engineers to design bandpass filters having certain characteristics. Such a filter is called a Gaussian filter. It allows a pulse to pass with very little distortion.

**gaussmeter** A meter that indicates the strength of a magnetic field. Reads out in gauss or in teslas. See *gauss, tesla*.

**Gauss' Theorem** An expression for determining the strength of an electric field. In general terms, the theorem is stated: For any closed surface in an electric field, the electric flux passing through the surface is directly proportional to the charge quantity in coulombs. See *electric field, electric flux*.

**Gb** Abbreviation for gilbert.

**Gd** Chemical symbol for gadolinium.

**Ge** Chemical symbol for germanium.

**gegenschein** Also called counterglow. A faint counter-image of the sun, reflected off of particles in interplanetary space. Appears exactly opposite the sun on the celestial sphere. Caused

GEGENSCHEIN

by the fact that the reflected images of particles are brightest when they are in "full phase" as shown in the drawing.

**Geiger counter** A device for measuring radiation intensity. The ionizing particles or photons cause momentary conduction in a tube. The pulses are counted and read out on a meter that indicates the relative strength of the radiation. The device might also produce audible clicks.

**Gell-Mann, Murray** A scientist who wrote a paper in 1961 called *The Eightfold Way*, a mathematical work that proposed a new theory of subatomic particles. Eight particles were needed to explain the behavior of the forces acting within atoms. Gell-Mann worked with other atomic physicists to predict the existence of a particle they called omega minus. See *Eightfold Way*.

**gem** Any hard, clear substance regarded as precious, and occurring in nature. Some gems, such as emeralds, can be artificially made. Gems are generally rare, and they often refract light in beautiful patterns because of their high indexes of refraction.

**Gemini** 1. A constellation, meaning "The Twins," seen in the night sky at some times of the year. 2. A space program of the 1960s, leading to the Apollo project and the later moon landings. Gemini capsules carried two astronauts. Their missions included space walks and docking maneuvers. See *Apollo*.

**gender** An expression of sex, either to male or female.

**gene** A part of a chromosome, in which traits are encoded. Each gene is responsible for a certain feature of the organism, for example, brown eyes. See *chromosome*.

**gene pool** The set of all genes that exist for a certain species of organism. This set might change as mutations and natural selection proceed and the species evolves. See *mutation, natural selection*.

**general anesthesia** A medical procedure in which the patient is "put to sleep." This is done during major surgery, when local anesthesia would not be effective. See *local anesthesia*.

**generator** A device that produces electric current from mechanical energy, usually rotation. A magnet can be rotated within a coil, or a coil rotated in a magnetic field (see drawing). This results in alternating current in the coil. A generator is built much like a motor. See *motor, motor/generator*.

GENERATOR

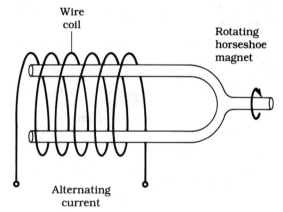

(Greatly simplified)

**genetic code** Information that is encoded in deoxyribonucleic acid (DNA), and that determines characteristics of cells as they reproduce. See *deoxyribonucleic acid*.

**genetic engineering** A field of biological science concerned with controlling the characteristics that a cell or life-form will have, by changing its genes. See *genetics*.

**genetics** A branch of biological science involving the way that genes affect evolution. Scientists in this field study the behavior of organisms as relates to their genes. They sometimes try to predict the course of evolution in the future. See *genetic engineering*.

**genital** 1. Pertaining to the genitalia. 2. The genitalia. See *genitalia*.

**genitalia** The sex glands. In the female these are the ovaries; in the male they are the testicles. The ovaries produce eggs, and the testicles make sperm cells. The genitalia are also responsible for hormone production. If they are removed, hormone disturbances occur. See *ovary, testicles.*

**genitourinary system** The system of the body involving the genitalia, kidneys, bladder, ureters, and in the male, the penis. See *bladder, genitalia, kidney, penis, ureter.*

**genome** All of the genes in an organism. Each species has a certain combination of genes that is unique to that species. See *gene, genetic code.*

**genotype** An expression of an organism's type, by specifying its genes.

**genus** 1. In topology, a term that defines the nature of an object according to the number of holes it has. We might say a light bulb has genus zero; a coffee cup has genus one; a piece of Swiss cheese might have genus six or seven. 2. A class of organisms, ranking above a species and below a family.

**geocentric theory** An ancient cosmological theory in which the earth was believed to be the center of the universe (see drawing). All the other planets, and the moon and sun, were thought to orbit the earth. See *heliocentric theory, Ptolemaic model.*

**geochemistry** A branch of chemistry, or of geology, concerned with the nature of the substances that make up our planet. For example, we know the earth has an iron core, and that silicon is abundant in the crust. We know there is plenty of nitrogen in the atmosphere. But there is still much about the earth that we do not know.

**geodesic** 1. On a sphere, especially the earth, the shortest distance between two points. 2. In a sphere of more than three dimensions (such as a four-sphere universe), the shortest distance between two points, as defined according to the path followed by a photon of light.

**geodesic dome** A spherical shell constructed using segments that are all identical polygons. The segments are usually triangular, small compared to the size of the sphere, and flat. Invented by R. Buckminster Fuller, there are numerous such domes in existence today. They can enclose large spaces, and they will withstand high winds without damage. There is one at Epcot Center in Orlando, Florida, as a part of a scientific exhibit. See *Fuller, R. Buckminster.*

**geodesy** 1. Geometry on the surface of the earth. 2. The ancient science of locating points on the earth, and of learning the shapes of the continents and other characteristics of the earth.

**geodetic** See *geodesic.*

**geographic north** See *true north.*

**geography** The study of the earth, the continents, the climate, the terrain, and the different types of life that exist in various places. There is a social branch to this, concerned with the customs and history of peoples in different parts of the world.

**geologic time** Time on a scale of thousands, millions, or billions of years. Divided into various eons, eras, periods, and epochs. See the appendix GEOLOGIC TIME. See *eon, epoch, era, period.*

GEOCENTRIC THEORY

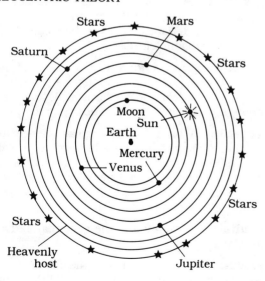

**geology** A branch of earth science or physical science, concerned with the structure of our planet, its behavior, and what we can expect to happen to it in the future. There are numerous subspecialties within this field.

**geomagnetic field** The magnetic field that surrounds the earth. The lines of flux come together at the geomagnetic poles, which do not exactly coincide with the geographic poles (see drawing). The solar wind distorts these flux lines at great distances from the earth. See *geomagnetic field reversal, geomagnetic pole, magnetic field, magnetic flux, solar wind.*

GEOMAGNETIC FIELD

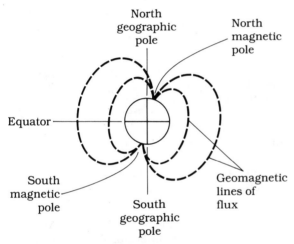

**geomagnetic field reversal** A periodic switch of the earth's north and south magnetic poles. This has occurred about 300 times in the past 170 million years. We can expect that it will occur again. The effects it can have on us are not fully known. See *geomagnetic field, geomagnetic pole.*

**geomagnetic pole** On the earth, either of two points where the lines of flux from the geomagnetic field converge. The north magnetic pole is located near Bethurst Island in extreme northern Canada. The south magnetic pole is just off the Adelie Coast of Antarctica. See *geomagnetic field, geomagnetic field reversal.*

**geomagnetic storm** A disturbance in the geomagnetic field that takes place a few hours after a solar flare. A barrage of charged particles reaches the earth, and is accelerated by the geomagnetic field. The moving particles in turn produce magnetic effects of their own. One effect is the aurora. See *aurora, geomagnetic field.*

**geomagnetism** The magnetism caused by the geomagnetic field. See *geomagnetic field.*

**geometric mean** See *mean.*

**geometric progression** A sequence of numbers, in which each term is obtained by multiplying the previous term by a constant. If we call the constant k, and the starting number x, then the sequence is (x, kx, kkx, kkkx, ...). As an example, let k = 2 and x = 1. Then the sequence is (1, 2, 4, 8, 16, ...).

**geometric** 1. Pertaining to geometry. See *geometry.* 2. Increasing in geometric progression. See *geometric progression.* 3. Having a pattern or structure that can be represented by simple mathematical functions.

**geometry** A branch of mathematics concerned with ideal physical objects and their relationships. Undefined terms are *point, line* and *plane.* Taught in junior high or high school as an axiomatic system of postulates, definitions, and proofs. See *Euclidean geometry, non-Euclidean geometry.*

**geomorphology** A branch of geology concerned with the evolution of features of the landscape. See *geology.*

**geophysics** A mathematical branch of geology. An example is the study of the waveforms and propagation of seismic disturbances caused by the shifting of the crust. Another example is the resonant behavior of the earth for low-frequency currents passing through it. See *geology.*

**geosphere** The earth, its atmosphere and its oceans.

**Geostationary Operational Environmental Satellite** Abbreviation, GOES. Any of various satellites in geostationary orbit. Used for weather forecasting and storm tracking. Several of these satellites have been launched by the

National Oceanic and Atmospheric Administration (NOAA) since 1974. See *geostationary orbit, geostationary satellite.*

**geostationary orbit**   An orbit such that the orbiting satellite always stays over the same spot on the earth's equator. See *Geostationary Operational Environmental Satellite, geostationary satellite.*

**geostationary satellite**   A satellite in an orbit such that it stays above the same place on the earth all the time. This requires an altitude of about 22,500 miles over some point on the equator. Such a satellite has the advantage that ground-based antennas can be fixed, always pointing towards the satellite. Thus the satellites are useful for television broadcasting and other communications purposes. See *Geostationary Operational Environmental Satellite.*

**geostrophic wind**   A horizontal wind resulting from atmospheric pressure differences and Coriolis effects. See *Coriolis effect, pressure gradient.*

**geosyncline**   See *syncline.*

**geothermal**   Pertaining to heat energy within the earth. This heat comes from the decay of radioactive materials deep inside the planet.

**geothermal energy**   1. Heat stored inside the earth. 2. The use of heat stored within the earth for human needs.

**geothermal power plant**   An installation for converting geothermal energy into electric power. In the drawing, a scheme is shown where water is pumped deep into the earth. The water is heated to boiling by the subsurface rocks. The steam drives turbines that turn generators to make electricity. If seawater is used, the steam can be condensed to provide drinkable water. See *desalination, geothermal energy.*

**geotropism**   The tendency of plants to grow either with gravity (positive geotropism, or downward) or against gravity (negative geotropism, upward). This would be upset in a spacecraft environment, for example, with zero gravity.

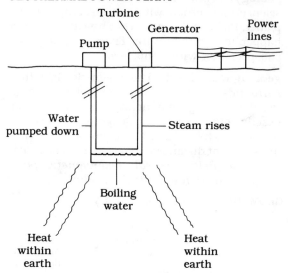

GEOTHERMAL POWER PLANT

**germ**   1. The embryo in a seed that forms a new plant when conditions are right (the seed is planted or embedded in the soil). This part contains many of the vitamins and minerals; for example, wheat germ is a nutritional supplement rich in B complex and other nutrients. 2. Slang for a bacterium or virus responsible for causing an illness in animals or people. 3. An initial concept from which a new theory or procedure develops.

**germanium**   Chemical symbol, Ge. An element with atomic number 32. The most common isotope has atomic weight 74. This substance is a semiconductor, and was used in diodes and transistors extensively until silicon largely replaced it. But it is still sometimes seen. See *semiconductor, silicon.*

**German measles**   See *rubella.*

**germ cell**   A cell that becomes either a sperm or egg (in humans and other mammals), or that contributes to reproduction.

**germicide**   A chemical that kills harmful bacteria and/or viruses, usually on physical objects or in the environment.

**germination**   The beginnings of the process in which a seed becomes a plant. Observable in many plants as "sprouting."

**Gernsback, Hugo** The founder of numerous science-related magazines in the early part of the 20th Century. His publications contained much science fiction, and we still see Gernsback Publications today. Gernsback wanted science fiction to seem real and believable.

**gestation** Also called pregnancy, especially in humans. The time from conception until birth, while the fetus develops in the womb.

**getter** A device for getting rid of the last little bit of gas in a picture tube or other device that needs a vacuum in order to work. Usually a metal such as magnesium or cesium, energized with high-frequency current that causes it to react with any remaining gas in the device.

**GeV** Abbreviation for gigaelectronvolt.

**geyser** A spring that ejects hot water and/or steam, often at regular intervals. "Old Faithful" is a good example of a periodic geyser in Yellowstone National Park.

**g force** 1. The force of gravitation on the earth, equivalent to an acceleration of 9.8 meters, or 32 feet, per second per second. 2. The force of gravitation on some other planet or moon. 3. Acceleration force produced when a rocket takes off, or during flight, especially of high-speed jet or rocket planes. Might be 10 or more times the force at the earth's surface. Measured in g's, where 1 g is the force of gravitation experienced at the surface of the earth. 4. See *artificial gravity*.

**Gibbs, Josiah W.** A physicist who worked at Yale during the nineteenth century. His specialty was thermodynamics. He is known for his role in aviation, explosives and refrigeration. In particular, he is largely responsible for inventing the process that makes refrigerators work today.

**giga-** See the appendix PREFIX MULTIPLIERS.

**gigaelectronvolt** Abbreviation, GeV. A unit of energy equal to 1,000,000,000 electronvolts. See *electronvolt*.

**gigahertz** Abbreviation, GHz. A unit of frequency equal to 1000 megahertz (MHz), or 1,000,000,000 hertz (Hz). See *hertz*.

**gigantism** A growth abnormality resulting from the pituitary gland producing too much growth hormone. The result is great height, sometimes more than eight feet, and shortened lifespan. See *pituitary gland, pituitary hormone*.

**gilbert** A unit of magnetomotive force. Equivalent to 0.796 ampere of current. More commonly, magnetomotive force is expressed in terms of the current in a coil, multiplied by the number of turns in the coil. See *magnetomotive force*.

**gill** 1. The organ that a fish uses to extract dissolved oxygen from the water in which it swims. 2. A spore-producing fold on the underside of the cap of a mushroom.

**gimbal** A device that allows rotation in two planes, and provides for mounting of a globe or gyroscope. See the drawing.

GIMBAL

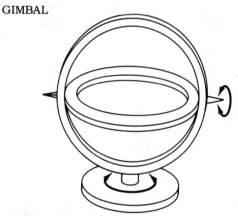

**gingiva** See *gum*.

**gingivitis** An inflammation of the gums. Can be largely prevented by adequate dental care including brushing, flossing, and regular checkups. See *gum*.

**gizzard** In certain animals, an organ that helps to digest food by grinding it up. The animal might swallow sand or gravel, where it collects in the gizzard and serves to pulverize the food.

**glacial lake** A lake formed by glaciers from the last ice age, or as a result of glacial melting in recent times.

**glacial surge** A possible result of global warming. Large ice sheets might become unstable and break off into the ocean. This would cause a rise in sea level and more glacial surges. It would become a vicious circle. See *global warming*.

**glacial till** Rocks, sand, and clay that have been carried by the movement of glaciers. Evidence of this has been found on every continent except Antarctica. See *glacier, ice age*.

**glaciation** 1. A condition of being covered by glaciers. 2. The extent to which a landmass is covered by glaciers. 3. The formation or advance of glaciers. See *glacier, ice age*.

**glacier** An enormous sheet of ice that forms especially in mountains where the climate is cold. Can also form over a period of centuries when the summers are not warm enough to completely melt winter snows. During the ice ages, glaciers covered much of North America and Europe. Glaciers cause significant changes in the landscape because of their great eroding power. See *ice age*.

**gland** An organ that secretes chemicals or hormones into body cavities or into the bloodstream to serve various body functions. Glands might have ducts, such as the tear and salivary glands and the gallbladder; or they might be ductless, secreting hormones into the bloodstream.

**glass** A clear, hard substance made from sand, calcium oxide, and sodium carbonate. Can also occur naturally. Other chemicals can be added to produce extra strength, resistance to temperature changes, or coloring.

**glass fiber** A fine thread of glass. Many such fibers can be pressed together to make a hard, durable, corrosion-resistant material (fiberglass). A single glass fiber can carry light for communications purposes. See *fiberoptics*.

**glaucoma** A disease in which the pressure inside the eyeball is greater than normal. If not treated, this can cause blindness because the abnormal pressure injures the retina. See *eye*.

**glaze** 1. A coating of ice that sticks to roads, trees and utility wires, and to other objects after an ice storm. See *ice storm*. 2. A coating of hard material, such as glass or ceramic, that serves to protect a material from weathering or wear.

**Glenn, John** The first American to orbit the earth in the Mercury capsule, on February 20, 1962. He went around the planet three times, taking photographs and communicating with ground-based personnel via radio. Later became a senator.

**glider** An aircraft without an engine, flying only by means of its lift. The wings are longer than they are on a typical airplane, and the craft itself is light, usually carrying just one or two people. Must be launched from a standard aircraft, from which it might glide a fairly long distance.

**Global Positioning System** A group of navigation satellites that provides guidance for aircraft, missiles, and ships. Distances are measured with extreme accuracy from the aircraft, missile, or ship to three or more satellites whose positions are known. See *radiolocation, radionavigation*.

**global warming** An increase in the average temperature of the earth caused by human activities. Mainly, the burning of fossil fuels causes more carbon dioxide to be put into the air than normal, increasing greenhouse effect. There might be a warming of several degrees in just the next 100 years or so, and this could cause climate changes, a rise in sea level, and other effects that would inconvenience or endanger the human race. See *greenhouse effect*.

**globular star cluster** A group of stars held together by their mutual gravity, and appearing as a blob near the center, where the stars seem to run together (see drawing). There are sometimes more than a million stars in a single cluster. These clusters tend to be arranged in a halo around our galaxy. See *galactic halo, open star cluster*.

**globulin** A protein found in plants and animals, occurring in various forms such as antibodies. See *antibody*.

**glossitis** An inflamed or swollen tongue. Occurs as a symptom of many different diseases, some serious and some not.

GLOBULAR STAR CLUSTER

**glossopteris** A fern that existed late in the Paleozoic era. Its fossils are found in the Southern Hemisphere but not in the Northern. This suggests that the continents broke up fairly late in the history of our planet, rather than early, as was originally thought. See *continental drift*.

**glottis** In humans and other mammals, the space where the vocal cords are located, next to the windpipe. See *epiglottis*.

**glove box** A chamber in which dangerous materials, such as contaminated specimens or radioactive substances, are safely handled. It is a metal box, shielded if necessary against radiation, with flexible gloves inserted into a side wall. This allows a person to "handle" the material without touching it or being exposed to its harmful effects.

**glow discharge** A luminous electrical effect, that takes place when current flows in an ionized gas at low pressure. The color of the glow depends on the gas or gases present.

**glucagon** A hormone that raises the level of the blood sugar (glucose) when energy is needed by the body. It does this by breaking down glycogen stored in the liver and muscles. See *glucose, glycogen*.

**glucose** A simple sugar found in the blood and used as a major source of fuel for the body. It is found in many natural sources, especially in fruits. It is obtained from various other sugars and starches when these are broken down in the intestine.

**glucose tolerance** 1. The ability of a person to metabolize glucose efficiently. See *glucose*. 2. A test in which a sample of glucose, usually 100 grams, is administered, and blood glucose is monitored for several hours afterwards. The level is plotted as a curve versus time. This allows a doctor to tell if a person is able to metabolize glucose normally.

**glutamic acid** An amino acid. Can be manufactured in the human body from other amino acids, so it need not actually be present in the diet. See *amino acid, essential amino acids*.

**glutamine** An amino acid. Can be manufactured in the human body from other amino acids, so it need not actually be present in the diet. See *amino acid, essential amino acids*.

**gluten** A protein found in wheat. Some wheat contains more of this than other forms of wheat. Mainly found in the endosperm (see *endosperm*).

**glyceride** A form of fatty acid. The best known is probably triglyceride, found in the blood. The level of triglyceride depends on various factors, including diet. There are other glycerides, all of which are esters of glycerol. See *glycerol*.

**glycerin** See *glycerol*.

**glycerol** A form of alcohol, a clear liquid in pure form. It is a building-block for various substances found in foods and in the body, especially fatty substances. Used industrially as a solvent.

**glycine** An amino acid. It is manufactured from other amino acids in the body, so it is not needed in the human diet. See *amino acid, essential amino acids*.

**glycogen** Also sometimes called "animal starch." A complex carbohydrate stored in the liver and muscles of mammals. Glucose from the blood is stored as glycogen, which can be used when needed by the body. See *glucagon, glucose*.

**glycolysis** The metabolism of glucose in the body to yield energy. There are by-products of this process, including adenosine triphosphate (ATP). See *adenosine triphosphate, metabolism*.

**GMT** Abbreviation for Greenwich Mean Time.

**Gnab Gib** See *Big Crunch*.

**gneiss** A coarse-grained metamorphic rock, with alternating layers of different minerals, such as quartz and some other dark rock. It is similar to granite. See *granite*.

**gnomon** 1. The device that casts the shadow in a sundial (see drawing). It is aligned depending on latitude. It always points to the North Star, or Polaris. This makes the sundial as accurate as possible. 2. What is left over of a parallelogram after a similar parallelogram, containing one of the corners of the original, has been taken away.

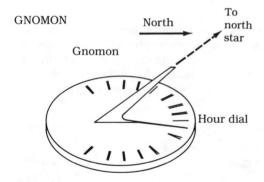

GNOMON

**Goddard, Robert** The engineer who pioneered rockets. He built model rockets similar to those children play with today. He developed the first liquid-fuel rocket in 1908. See *rocket, rocket engine*.

**Godel, Kurt** The mathematician who proved the Incompleteness Theorem and other theorems of mathematical logic. His discoveries revolutionized mathematical thought. He demonstrated the incompleteness of first-order logic, probably the greatest work of his life, in 1930, when he was just 24 years old. See *Incompleteness Theorem*.

**GOES** Abbreviation for Geostationary Operational Environmental Satellite.

**goiter** An enlarged thyroid gland, resulting from an insufficient amount of iodine in the diet. The person with this problem has a swollen throat, as if they had swallowed a baseball.

**goiter belt** Parts of the country or the world where there is little or no iodine in the soil. People living in such regions must get iodine from imported food, or from supplements such as iodized salt. Otherwise they are likely to have goiter. See *goiter*.

**gold** Chemical symbol, Au. An element with atomic number 79. The most common isotope has atomic weight 197. A precious, yellow-colored, rather heavy metal used in various industrial applications and as a standard of money exchange. It is malleable and is a good conductor of heat and electricity. It is corrosion-resistant and is sometimes used to plate electrical contacts.

**golden section** A ratio of two lengths that occurs frequently in nature and is held by some to be special, hence the term "golden." The drawing shows the manner in which the ratio is constructed. The ratios of the lengths AQ to AB, and AP to AQ, are identical inside the regular pentagon. We say that AQ is a golden section of AB, and that AP is a golden section of AQ. The ratio is approximately equal to $5/8$.

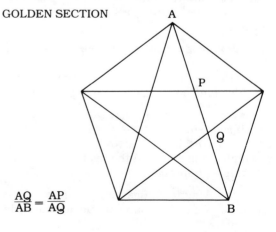

GOLDEN SECTION

**Gold, Thomas** A student at Cambridge University who, along with his professor Herman Bondi, proposed the Steady State theory of the universe. See *Steady State theory.*

**Golgi apparatus** A complicated set of membranes in the cytoplasm of cells. Important for many of the cell functions and metabolism. See *cell, cytoplasm.*

**gonad** In the male, the testis or testicle; in the female, the ovary. See *ovary, testicles.*

**gonadotropin** A hormone that regulates various endocrine functions involving the sex hormones. Secreted by the pituitary gland. This hormone is largely responsible for the changes that occur during puberty and menopause, for example. See *menopause, puberty.*

**Gondwanaland** A major supercontinent that existed before the landmasses broke up. See the drawing. Gondwanaland contained present-day South America, Africa, Antarctica, and Australia. The outlines have changed somewhat since the supercontinent first broke apart. See *Laurasia.*

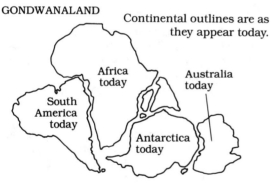

GONDWANALAND Continental outlines are as they appear today.

From Erickson, THE LIVING EARTH (TAB Books, 1989).

**goniometer** See *radio direction finder.*

**goniometry** The engineering practice of locating radio transmitters. This might be used to find a pirate radio station or to find a lifeboat lost at sea. See *radio direction finder.*

**gonorrhea** A venereal disease caused by bacteria and spread through sexual contact. It is the most prevalent venereal disease in the civilized world. It can be treated with antibiotics. See *venereal disease.*

**googol** The number resulting when you write 1 followed by 100 zeroes. This number is vastly larger than the number of atoms in the known universe. It is sometimes mentioned in discussions about very large finite numbers, or when trying to gain some concept of infinity.

**googolplex** The number resulting when 10 is raised to the googolth power. That is, you would have to write a 1 followed by a googol zeroes. Even this is small, however, compared to some finite numbers, and it is nothing compared to infinity. See *googol.*

**gout** A chronic disease caused by accumulation of certain substances (monosodium urate) in the joints. It causes joint pain, especially in extremities such as the toes. It is associated with excessive levels of uric acid in the blood. See *arthritis.*

**governor** 1. A device in a motor or engine that keeps it from running too fast and damaging itself. 2. A device that keeps the speed of a direct-current (dc) motor constant. 3. A device that prevents a motor or engine from slowing down even when the turning resistance (load) changes.

**graben** A valley, formed by a down-dropped fault block. See *fault.*

**gradient** 1. Rate of change with distance for a physical quantity such as charge or air pressure. The drawing shows the pressure gradient between low-pressure and high-pressure weather systems. 2. The slope of a function at a certain point or within a defined small interval.

**graft** 1. The process of attaching healthy skin onto a severe wound or burn, so that it will grow and heal. 2. A section of healthy skin that has been attached to a wound or burn for healing. 3. An artificial means of cultivating plants. Live stems or shoots are attached to roots; they grow together.

**gram** Abbreviation, g or gm. A unit of mass equal to 0.001 kilogram (kg). See *kilogram* and the STANDARD INTERNATIONAL SYSTEM OF UNITS appendix.

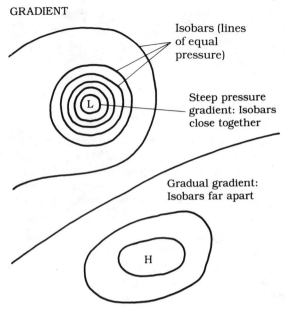

GRADIENT

Isobars (lines of equal pressure)

Steep pressure gradient: Isobars close together

Gradual gradient: Isobars far apart

**gram-molecular weight** The weight (or mass), in grams, of one mole of a compound. See *mole*.

**Gram-negative** Bacteria that do not retain the chemical stain in Gram's method of treatment. A classification of bacteria. Examples are the bacteria that cause typhoid fever and salmonella.

**Gram-positive** Bacteria that retain the chemical stain in Gram's method of treatment. A classification of bacteria. Staphylococci are an example. They cause infections such as pneumonia and gastroenteritis.

**Gram stain** Also called Gram's method. A technique for staining bacteria in order to classify them. See *Gram negative, Gram positive*.

**grandfather well** An ancient type of well, driven into the ground by the Chinese and taking several generations to complete. They were made with bamboo and sometimes were hundreds of feet deep.

**grand mal** 1. A violent, dramatic epileptic seizure. The patient loses consciousness, has convulsions, and might vomit. 2. A form of epilepsy in which violent seizures occur. See *epilepsy, seizure*.

**granite** A rock with a coarse grain made up mostly of quartz and feldspar. It is thought to have originated as a molten material deep within the earth. It is the main rock that makes up the continents. See *feldspar, quartz*.

**graph** A pictorial way of showing a relation or function. There are various different coordinate systems for this. A graph might be simple, needing just two dimensions, or it might be in three or more dimensions. See *Cartesian coordinates, coordinate system, polar coordinates, spherical coordinates*.

**graphics** 1. A set of drawings, showing the construction and functioning of a system, or accompanying a technical work. 2. In computer science, a branch dealing with the generation of illustrations.

**graphite** A form of carbon that occurs as a soft, black powder. It can be pressed into pencil leads and electrodes. It is a good conductor of electricity. It is sometimes used as a lubricant.

**graph theory** A branch of mathematics, dealing with graphs of functions. See *Cartesian coordinates, coordinate system, function, polar coordinates, spherical coordinates*.

**grass** A perennial plant that grows in various familiar forms. May attain a height of more than six feet, or two meters, in some cases. For the place of the grasses on the evolutionary scale, see the PLANT CLASSIFICATION appendix.

**gravitation** See *gravity*.

**gravitational collapse** A situation where an object becomes so dense that its own gravity overpowers everything. This is thought to happen to heavy stars when they die. Without outward heat pressure, gravity makes the star shrink. This increases the gravity at the surface, causing further shrinkage. The vicious circle ends in a black hole. See *black hole, event horizon, Schwarzchild radius*.

**gravitational constant** Abbreviation, G. A fundamental constant of the universe, believed to be the same everywhere in the universe, or at least within the local group of galaxies. Has to do with the intensity of the force of gravity. See *gravity* and the CONSTANTS appendix.

GRAVITATIONAL COLLAPSE

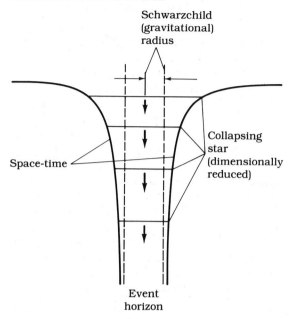

**gravitational field** The gravity and associated curvature of space that surrounds every object that has mass. The intensity of the field at any point is proportional to the mass of the object, and inversely proportional to the square of the distance from the object.

**gravitational radius** See *event horizon, Schwarzchild radius.*

**gravitational red shift** A phenomenon seen with light coming from an object having intense gravity. The wavelength is made longer, and the spectral lines appear shifted toward the red end of the spectrum because of this. This is explained according to the principles of Einstein's general theory of relativity. See *red shift.*

**gravitational waves** See *gravity waves.*

**graviton** A hypothetical particle that carries the force of gravity; thought to be responsible for gravity. See *gravity.*

**gravity** 1. The force of attraction that a particle produces on another when both particles have mass. The force is proportional to the product of the masses of the particles, and inversely proportional to the square of the distance between them. 2. The pull of the earth, a planet, or any other large celestial object on masses near it. 3. The mutual attraction of all the objects in the universe. 4. Any of the effects that accompany the aforementioned force.

**gravity dam** The most common type of dam. It holds back the water because of its mass. It is so constructed that gravity adds to its strength. Might be made of rock or earth, often along with concrete; also can be made only from concrete.

**gravity waves** Disturbances in space-time that occur when an object undergoes gravitational collapse, or when other events take place that cause an object to suddenly change its mass. These waves are thought to travel at the speed of light, and are like "ripples" in the space-time continuum. See *gravity.*

**gray** A unit of radiation. Expresses the amount of alpha rays, beta rays, gamma rays, or X rays that a sample has received over an interval of time. For a mass of one kilogram, absorbing one joule of energy from the radiation, the absorbed dose is one gray (1 Gy). This is equivalent to 100 rads. See *alpha rays, beta rays, gamma rays, rad, X rays.*

**gray line** See *terminator.*

**gray matter** The control center for the nervous system, located in the brain. So called because of its color. See *brain, central nervous system.*

**great circle** A geodesic on a sphere. See *geodesic.* This is an arc that forms part or all of a circle, whose center is the same as the center of the sphere. On the earth, ships and planes follow such paths when possible, because they represent the shortest distance between two points on the surface.

**Green Bank Observatory** A radio telescope in West Virginia, one of the most important for the observation of space at radio wavelengths. Frank Drake, who worked at this observatory, was largely responsible for a search for extraterrestrial intelligence that was carried out using radio telescopes to listen for signals from other worlds. No signals were ever heard that could be definitely attributed to aliens. See *Ozma Project*.

**greenhouse** A glass or plastic enclosure used for keeping plants warm in all seasons. The device traps heat (see *greenhouse effect*) while allowing the needed sunlight to enter. Tropical plants can be grown in such enclosures, even at temperate latitudes.

**greenhouse effect** The principle by which a greenhouse stays warmer than the surroundings. Light and short-wavelength infrared enter through the glass, heating objects inside. The objects give off long-wavelength infrared, and the glass acts like a mirror for this, keeping it in. In the atmosphere, carbon dioxide has a similar effect (see drawing). This is why an increase in the earth's carbon dioxide, caused recently by human activities, is such a concern. See *global warming*.

GREENHOUSE EFFECT

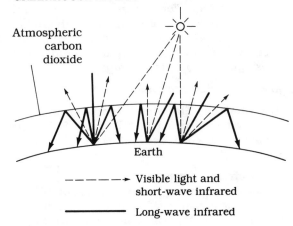

- - - - → Visible light and short-wave infrared
———— Long-wave infrared

**Greenough, Horatio** A sculptor who put forth a theory that has been followed by architects. This theory states that form must change according to changes in function. The result is that building architecture has tended to be based on the needs of the people who use the buildings.

**greenstone belts** Lava flows and sediments, possibly from volcanic islands caught in a squeeze between drifting continents. Many gold mines are found in these regions.

**green sulfur bacteria** A predecessor to blue-green algae, believed to be one of the earliest forms of life on the earth. A photosynthetic bacteria, it derived energy from sunlight and first appeared about 3.5 billion (3,500,000,000) years ago.

**Greenwich Mean Time** Standard time at the Greenwich meridian (zero degrees longitude), more often called the prime meridian. This is the time zone that passes through Greenwich, England. See *coordinated universal time*, *prime meridian*.

**Greenwich meridian** See *prime meridian*.

**grey matter** See *gray matter*.

**grid** 1. A pattern or arrangement; a matrix, whose points are at the intersections of criss-crossing sets of parallel lines. 2. A criss-crossing set of parallel lines or wires. 3. A control electrode in a vacuum tube. See *vacuum tube*.

**Grimm brothers** Famous for writing fairy tales, but also for theorizing in general about the nature of languages. They did their work in the nineteenth century.

**grooming** A form of animal behavior. Cats are especially known for this; they lick their fur to keep it clean. It is also a form of social behavior in some animals.

**gross** 1. A quantity of 12 dozen, or 144. 2. The weight of an object, especially of packed material, or vehicles, ships or planes, including everything. The gross weight of a crate of apples, therefore, would include the crate itself as well as the apples.

**ground shaking** Motion of the ground during an earthquake. See *earthquake*.

**ground speed** The speed of an aircraft relative to the surface of the earth. This is usually different from airspeed because of winds aloft. See *airspeed*.

**ground state** For an atom, the condition of having the least possible amount of energy. This means that the electrons are in the lowest orbits. Such an atom might absorb radiant energy, and one or more electrons will move into higher orbits. See *atom, electron*.

**groundwater** Pockets of water beneath the surface of the earth. In many parts of the world this is the main supply of water for drinking, bathing, and other use by the population. The depth of the pockets varies; they might be only a few feet under the surface, or many hundreds of feet down. See *aquifer*.

**groundwater pollution** Contamination of the groundwater by seepage, such as from leaky gasoline tanks or broken sewage pipes. Anything that contaminates the soil may also pollute the water.

**groundwater table** The general level of the top of the groundwater pocket in a given region (see drawing). This is the minimum depth that one must drill to get a reliable supply of well water. Lakes tend to form if the ground dips below the level of the water table anyplace.

GROUNDWATER TABLE

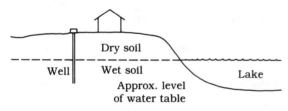

**ground wave** In a radio signal, the combination of the line-of-sight wave and the surface wave. See *line of sight, surface wave*.

**ground zero** 1. For an air-burst nuclear detonation, the point on the earth directly underneath the place where the bomb explodes. 2. For an underground or surface nuclear bomb, the point on the surface above the blast, or at the location of the blast. 3. The epicenter of an earthquake. See *epicenter*.

**group** 1. A mathematical set with certain properties relative to a certain operation *, such that: (a) For x and y in the set, x*y is also in the set; (b) there is an element i in the set such that x*i = x for all x in the set; and (c) for every x in the set, there exists a y in the set such that x*y = i. 2. A set of related elements in the periodic table. See *group 0 element* through *group VIII element*. 3. A set of related organisms. 4. A combination of atoms forming a molecule.

**group theory** A branch of mathematics dealing with the behavior of sets of objects having certain properties. See *group*.

**group therapy** In psychiatry and drug rehabilitation treatment, a method in which several people share their problems and experiences. There are various different methods by which this can be done. A counselor or psychologist or psychiatrist, called the facilitator, leads the group.

**group 0 element** Any of the elements helium (He), neon (Ne), argon (Ar), krypton (Kr), xenon (Xe), and radon (Rn). These are sometimes called the inert gases. See the PERIODIC TABLE OF THE ELEMENTS appendix.

**group 1 element** A classification of elements in the periodic table. Divided into two subgroups, 1-A and 1-B. In subgroup 1-A are the elements hydrogen (H), lithium (Li), sodium (Na), potassium (K), rubidium (Rb), cesium (Cs), and francium (Fr). In subgroup 1-B are copper (Cu), silver (Ag), gold (Au), terbium (Tb), and berkelium (Bk). See the PERIODIC TABLE OF THE ELEMENTS appendix.

**group 2 element** A classification of elements in one form of the periodic table. Divided into two subgroups, 2-A and 2-B. In subgroup 2-A are the elements beryllium (Be), magnesium (Mg), calcium (Ca), strontium (Sr), barium (Ba), and radium (Ra). In subgroup 2-B are zinc (Zn), cadmium (Cd), and mercury (Hg). See the PERIODIC TABLE OF THE ELEMENTS appendix.

**group 3 element** A classification of elements in one form of the periodic table. Divided into two subgroups, 3-A and 3-B. In subgroup 3-A are the elements boron (B), aluminum (Al), gallium (Ga), indium (In), and thallium (Tl). In subgroup 3-B are scandium (Sc), yttrium (Y), lanthanum (La), and actinium (Ac). See the PERIODIC TABLE OF THE ELEMENTS appendix.

**group 4 element** A classification of elements in one form of the periodic table. Divided into two subgroups, 4-A and 4-B. In subgroup 4-A are the elements carbon (C), silicon (Si), germanium (Ge), tin (Sn), and lead (Pb). In subgroup 4-B are titanium (Ti), zirconium (Zr), hafnium

(Hf), cerium (Ce), and thorium (Th). See the PERIODIC TABLE OF THE ELEMENTS appendix.

**group 5 element** A classification of the elements in one form of the periodic table. Divided into two subgroups, 5-A and 5-B. In subgroup 5-A are the elements nitrogen (N), phosphorus (P), arsenic (As), antimony (Sb), and bismuth (Bi). In subgroup 5-B are vanadium (V), niobium (Nb), and tantalum (Ta). See the PERIODIC TABLE OF THE ELEMENTS appendix.

**group 6 element** A classification of the elements in one form of the periodic table. Divided into two subgroups, 6-A and 6-B. In subgroup 6-A are the elements oxygen (O), sulfur (S), selenium (Se), tellurium (Te), and polonium (Po). In subgroup 6-B are chromium (Cr), molybdenum (Mo), and tungsten (W). See the PERIODIC TABLE OF THE ELEMENTS appendix.

**group 7 element** A classification of the elements in one form of the periodic table. Divided into two subgroups, 7-A and 7-B. In subgroup 7-A are the elements fluorine (F), chlorine (Cl), bromine (Br), iodine (I) and astatine (At). In subgroup 7-B are manganese (Mn), technetium (Tc), and rhenium (Re). See the PERIODIC TABLE OF THE ELEMENTS appendix.

**group 8 element** A classification of the elements in one form of the periodic table. Includes iron (Fe), ruthenium (Ru), osmium (Os), cobalt (Co), rhodium (Rh), iridium (Ir), nickel (Ni), palladium (Pd), and platinum (Pt). See the PERIODIC TABLE OF THE ELEMENTS appendix.

**Groves, General Leslie** The general in charge of the Los Alamos nuclear project in 1943. This led ultimately to the development of the atomic bombs that were dropped on Hiroshima and Nagasaki, Japan in 1945, ending World War II.

**growing season** The time from the last frost in spring till the first frost in the autumn. Ranges from zero in regions of permafrost to all year round in the tropics. In temperate latitudes it can range from about three months to eight or nine months. There is one growing season each year.

**growth** 1. A biological process in which cells reproduce, thereby increasing the size of an organism up to maturity. 2. Accumulation of atoms on a substance. See *epitaxy*. 3. A cyst or tumor. See *cyst, tumor*.

**growth hormone** The pituitary hormone that regulates growth. Secretion of this hormone is determined according to genetic codes present from birth. Too little growth hormone causes dwarfism; too much causes gigantism. See *pituitary gland, pituitary hormone*.

**growth ring** In the cross section of a tree or other woody plant, one of several or many concentric circles. Each circle indicates the growth for one season (see *growing season*). The width of the ring is an indicator of how good or bad the weather was during the given year. The age of a tree can be determined by counting the growth rings in its trunk.

**guanine** A major part of deoxyribonucleic acid (DNA) and ribonucleic acid (RNA). Joins with cytosine to hold the double helix together. See *deoxyribonucleic acid, ribonucleic acid*.

**gulf stream** Part of a large current circulating in the Atlantic ocean. It is a warm current, spilling eastwards and northwards from the Gulf of Mexico, around the tip of Florida, and near the East Coast of the U.S. (see map). Then it flows across the North Atlantic to England and Northern Europe. The warm waters cause the weather in regions near the stream to be warmer than it would otherwise be. The weather in England is

GULF STREAM

therefore much more moderate than at the same inland latitudes in Manitoba, Canada.

**gullet** See *esophagus*.

**gum** 1. The soft tissue around the roots of the teeth in most mammals. 2. A material in plants, useful in various industrial applications. It is a pliable, sticky substance. Some gums are used in food processing.

**Gunn diode** A special type of semiconductor diode that oscillates under certain conditions, at ultrahigh frequencies.

**gunpowder** A powdered mixture of sulfur, carbon, and potassium nitrate. When ignited it burns extremely fast. If there is very much of it packed together, an explosion results. This is how firecrackers are made. It was first invented by the Chinese in the eleventh century A.D.

**Gutenberg, Beno** A seismologist, known for his discussion with Albert Einstein in which he elaborated on earthquakes as Einstein told of his relativity work. While they were walking and talking, the famous 1933 Los Angeles earthquake occurred. They didn't even notice till they looked around and saw people panicking.

**guyot** Pronounced "*gee*-oh." An undersea volcano with a flat top. Evidently, wave action wears the peak down over time. The peak almost seems as if it were sawed off. These formations are used as proof that the seafloor is spreading. See *continental drift, plate tectonics*.

**Gy** Abbreviation for gray.

**gymnosperm** Plants that evolved in the Permian period, about 290 million years ago. They bore seeds with no fruit covering. A good example of these today are the conifers, or cone-bearing trees.

**gynecology** A branch of medicine, dealing with the physiology of the human female. Especially concerned with the female reproductive system.

**gypsum** Any of various forms of calcium sulfate. It might range in hardness from rock-like to chalk-like. It is used in the manufacture of many different industrial substances.

**gyre** Also called a ring. A vortex, or large swirling current, in the ocean that shows up in radar images. Displaces water from one part of the ocean to another.

**gyrocompass** A gyroscope that indicates compass direction. It is set using a compass, and its gyroscope effect then keeps it oriented just as the earth's magnetic field keeps the needle of a compass oriented. Used in navigation, especially on aircraft. See *gyropilot, gyroscope, gyroscope effect*.

**gyropilot** A pair of gyroscopes in an aircraft, used for automatic piloting. The gyroscopes detect any change in horizontal or vertical heading, and motors operate the controls to correct the error. See *gyrocompass, gyroscope*.

**gyroscope** A device that has a heavy, rapidly rotating disk in a gimbal (see drawing). The rotating mass tends to stay in the same plane, resisting changes in its orientation. Used for navigation in aircraft, and also to stabilize spacecraft.

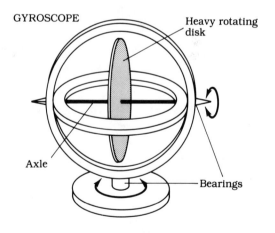

**gyroscope effect** The tendency for a rapidly rotating, heavy disk or wheel to stay in the same plane. See *gyroscope*.

**H**  Abbreviation for henry, magnetic flux; chemical symbol for hydrogen.

**h**  Abbreviation for hecto- (see PREFIX MULTIPLIERS appendix), hour, Planck's constant.

**habitat**  The location and environment for a certain plant or animal where it tends to thrive and where it prefers to be. The term is more often used for animals than for plants.

**habituation**  A condition of being used to something, such as thumb sucking. A psychological dependence, but not severe enough to be called addiction. Might occur with certain mild medications such as aspirin, that are not truly addictive substances. See *addiction, addictive substance*.

**Hadean eon**  See *Archean eon*.

**Hadley cell**  A sustained, large-scale convection pattern in the atmosphere. Gives rise to the prevailing winds. See *convection*.

**Hadley, George**  An English scientist who proposed a theory of heat convection in 1735. He thought that air moved between the poles and the equator in gigantic *cells*, or convection currents. We now know that this is true, although the pattern is somewhat more complicated than Hadley originally thought. See *convection, Hadley cell*.

**hadron**  A matter particle made up of quarks. The most common are neutrons and protons. See *neutron, proton, quark*.

**hafnium**  Chemical symbol, Hf. An element with atomic number 72. The most common isotope has atomic weight of 180. It is a metal in pure form. It has a variety of industrial uses, especially in the filaments of incandescent lamps.

**hair**  1. A body covering, resembling fur but longer, that serves to keep an animal warm. 2. An individual fiber of interconnected cells, that grows from the skin in many warm-blooded animals, especially mammals. See *follicle*. 3. A silk-like growth found in some plants.

**hair follicle**  See *follicle*.

**Haldane, J.B.S.**  An English scientist who is reputed to have said that the universe is not only stranger than we think it is, but stranger than we *can* think it is.

**half-life**  In radioactive decay, the time required for the radiation intensity of a sample to decrease to half the starting value (see drawing). This is a constant time, no matter what the starting value, but differs among various radioactive materials. Some substances have half-lives of hours or days; others might take centuries or eons.

HALF-LIFE

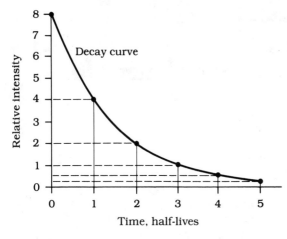

**halite** Sodium chloride (ordinary table salt) that is found naturally in the earth. It remains from ancient seas.

**Hall Effect** When a current-carrying strip is placed in a magnetic field, with the magnetic flux perpendicular to the strip, a voltage develops across the strip. This is especially true of certain materials.

**Halley, Edmund** An astronomer who studied comets and published a book about them in 1705. He discovered that a certain comet had a recurring history of apparitions, or appearances, every 76 years. The comet was later named for him. See *Halley's Comet*.

**Halley's Comet** A well-known comet with a period of 76 years. Its most recent appearances were in 1910 and 1986. See *comet* and *Halley, Edmund*.

**hallucination** A malfunction of the brain and senses, causing a person to see, hear, feel, or smell things that aren't there. These are called visual, auditory, tactile, and olfactory hallucinations, respectively.

**hallucinogen** A chemical substance that, when taken into the body, causes a person to have hallucinations. See *hallucination*.

**halo** 1. A visible, glowing ring that surrounds the sun or moon when its light shines through high-altitude ice crystals. 2. A glowing sphere that might appear around a star as the radiant energy, especially ultraviolet and X rays, causes gas clouds to fluoresce. Visible through large telescopes as a ring. 3. Any spherical, glowing region of energized gas surrounding a radiation source.

**halocarbon** A compound containing carbon and a halogen, especially chlorine or fluorine. See *chlorofluorocarbon*, *halogen*.

**halogen** Any of the elements fluorine, chlorine, bromine, iodine, and astatine found in group 7 of the periodic table. They readily form compounds. They are used to kill bacteria and viruses, thus preventing infection. They are also used for various industrial purposes.

**halogenation** 1. The addition of a halogen to a compound, resulting in a reaction. See *halogen*. 2. Addition of a halogen to a substance, such as chlorine to water, for a specific purpose, such as killing harmful bacteria and viruses.

**harbor** 1. A natural protected area, such as a bay or large inlet, where ships can be safely kept, loaded and unloaded. 2. A place that has been built up for use as a port for ships, because of its natural characteristics that provide protection from storm effects.

**hard disk** A magnetic disk in a computer, so called because it is rigid rather than flexible (see *floppy disk*). The hard disk is usually installed in the computer permanently; floppy disks can be exchanged by the operator easily.

**hardness** An expression of durability or cutting ability, especially for mineral rocks. Usually it is given on a scale of 1 to 10. Diamond has a hardness of 10. Chalk, talc, and gypsum are soft, at the lower range of this scale. Hardness can be determined by testing a substance to see what will scratch it and what it will scratch.

**hard radiation** Referring to the shorter wavelengths for a given type of radiation. Therefore, hard ultraviolet would be almost in the X-ray range; hard X rays would be almost in the gamma-ray range. The shorter the wavelength, the more penetrating and energetic the rays.

**hardware** 1. Heavy industrial equipment. 2. In a computer, the physical components, such as integrated circuits, keyswitches, disks, and disk drives. See *firmware, software*. 3. Tools for maintaining and repairing a system.

**hardware engineering** A branch of engineering, and especially of computer engineering, concerned with the design and construction of electronic circuits.

**harmonic** A multiple of the frequency of a sound note or radio wave. The fundamental frequency might be called f, and given in hertz (Hz). Then the second harmonic has frequency 2f, the third harmonic 3f, and so on. Most waves are not pure, but contain harmonic energy as well as energy at the fundamental frequency. See *fundamental frequency, octave*.

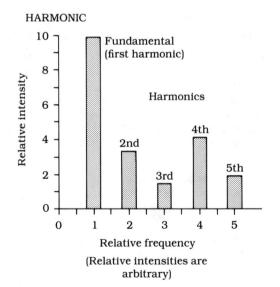

HARMONIC

(Relative intensities are arbitrary)

**harmonic division** A sophisticated mathematical technique that allows one to draw exact tangent lines to conic sections. See *conic section*.

**harmonic motion** A special type of oscillating motion. Imagine a friend swinging a ball on a string around in a perfect circle. When you look at this from a distance, in the plane of the ball's orbit, the ball seems to oscillate back and forth. This to-and-fro movement is the simplest form of harmonic motion. More complicated forms of this motion are possible, when smaller orbits are superimposed on the main orbit.

**harmonic series** The series $1/2 + 1/3 + 1/4 + 1/5 + ...$, going on forever. This is a divergent series. That is, its sum does not converge to any finite number. This was proved by Nicole Oresme in the fourteenth century A.D. See *convergent series, divergent series*.

**Harvey, William** The medical scientist who first showed, about 400 years ago, how blood circulates in the body: from the heart, through the arteries, then into the veins, and then back to the heart. This is an oversimplification, but he had the general idea.

**Hawking, Stephen** A noted cosmologist and Lucasian Professor of Mathematics at Cambridge University in England. His position is the same as that once held by Isaac Newton. Hawking is paralyzed, but has written a top-selling book, *A Brief History of Time*, in which he describes the nature of the universe and attempts to formulate a unified field theory. See *unified field theory*.

**Hawkins, Gerald S.** An American astronomer who showed that the stones in Stonehenge are placed so that observers can mark the beginnings of the seasons, predict eclipses, and keep track of other events in the sky. See *Stonehenge*.

**hazard rate** For a given component, the probability that it will fail immediately, as soon as it is put to use. This is lower than the failure rate as given for a certain period of time. Applies especially to electronic devices. See *failure rate*.

**HDL** Abbreviation for high-density lipoprotein.

**He** Chemical symbol for helium.

**headphone** A pair of small speakers worn on or in the ears; used for private listening or for communications.

**headwater** The source of a river, usually a lake or underground spring.

**"health" food** Any of various foods that are claimed by food faddists to have special properties. The foods that get this label vary with time, as fads come and go. There is no special food with magical powers, although there certainly are "junk" foods with little or no nutritional value.

**health physics** The physics of keeping scientific and medical personnel safe from radioactive materials. Also concerned with the safe handling of atomic waste.

**heart** The organ of the animal body that pumps blood through the circulatory system, to nourish the tissues of the body. The drawing shows the way in which a human heart functions. (It is not an exact physical representation.) From the right ventricle, blood is pumped to the lungs, where it is replenished. Then it goes to the left auricle, and a valve regulates its flow into the left ventricle. The left ventricle pumps the blood to the body tissues via the arteries. Blood comes back to the right auricle through veins, and a valve regulates its flow once again to the right ventricle. See the following several definitions. See also *artery, auricle, circulatory system, pulmonary artery, pulmonary vein, vein, ventricle*.

**heart arrhythmia** An irregular heartbeat. To some extent this is normal, but sometimes it indicates a disease or other problem. It is best evaluated by a physician.

**heart attack** Death of some part of the heart muscle caused by a lack of oxygen. This is in turn caused by a shortage of blood to the affected muscle. Most often the blood shortage occurs because there is a partial blockage in a coronary artery.

**heart disease** 1. A progressive, degenerative disease in which the coronary arteries become narrowed. Aggravating factors seem to include smoking, lack of exercise, high blood pressure, and obesity. A high level of cholesterol in the blood is associated with increased incidence of this disease. 2. Any abnormality of the heart that threatens the health of the person.

**heart failure** Also called cardiac arrest. A stoppage of the heart for any reason.

**heart fibrillation** See *fibrillation*.

**heart-lung machine** A mechanical device that performs the functions of the heart and lungs; that is, it pumps blood through the circulatory system and provides oxygen to the blood, and also removes carbon dioxide from the blood. Used as a life-support system when the heart and/or lungs are not working for a short time, such as during certain types of heart surgery.

**heart murmur** An abnormal noise in the heartbeat, heard with a stethoscope. It can be identified by a physician. It may or may not indicate a serious condition.

**heart pacemaker** An electronic device that regulates the heartbeat by stimulating the heart muscle at timed intervals. Used in patients with certain forms of heart disease.

**heart palpitation** A sudden series of rapid heartbeats that takes place for no apparent reason. Might be the result of excessive dose of cer-

HEART

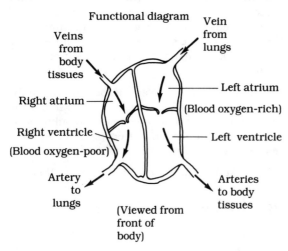

tain drugs; it can also indicate some serious illness or mineral imbalance.

**heart rate** The number of heartbeats per minute. In a well-trained, male adult athlete, the resting heartbeat might be as low as about 45 beats per minute. Normal adult rates range from 65 to 80 beats per minute at rest. In children the rate is higher. In poorly conditioned adults it might be as high as 85 to 100 beats per minute at rest. The rate increases under stress and during exercise.

**heart transplant** The surgical removal of a human heart, and replacement with a heart from another person or even from an animal such as a baboon. This is a major surgical procedure.

**heart valve** A valve between the auricle, or atrium, of the heart, and the ventricle. Keeps blood from flowing back into the auricle when the ventricle contracts. See *auricle, heart, ventricle*.

**heartwood** Dead cells near the center of a woody plant stem, branch, or trunk. This wood is comparatively hard, and does not contribute to the flow of nutrients in the plant.

**heat** 1. A transfer of energy because of a temperature difference. 2. A condition of being at a temperature relatively higher than the surroundings. 3. Warm or hot weather.

**heat budget** The mechanism that maintains the earth's temperature and climate within ranges suitable for life. It is a balance between energy received from the sun, and energy radiated back into space. Atmospheric changes might alter this mechanism. See *global warming*.

**heat capacity** 1. The amount of energy needed to raise a sample by a certain temperature. Measured in joules per degree Kelvin. 2. The amount of heat needed to raise one kilogram of a substance by one degree centigrade or Kelvin. Measured in joules per kilogram per degree Kelvin. 3. The amount of heat needed to raise one mole of a substance by one degree centigrade or Kelvin. Measured in joules per mole per degree Kelvin.

**heat conductivity** The extent to which a substance will transfer heat. Usually, but not always, electrical conductors are good heat conductors, and vice-versa. Silver and copper are among the best heat conductors. Fiberglass is among the worst, and is used as an insulator in buildings for this reason.

**heat exhaustion** A condition in which body functions slow down because of overheating. There is a slight fever, sweating, and there might be muscle spasms and dehydration. See *heat stroke*.

**heat flow** 1. Heat transfer. 2. The rate of heat transfer, in joules per second or in energy per unit time. See *heat transfer*.

**heat lightning** Lightning seen at night, from such a great distance that its actual source cannot be identified. Incorrectly attributed to "heat," it actually takes place in distant thundershowers.

**heat loss** 1. A loss of energy in a system that occurs as heat dissipation. For example, a radio transmitter might have a power input of 1000 watts, and an output of 600 watts; the other 400 watts is heat loss in the amplifier transistor. 2. Heat energy that escapes when it is not supposed to. Might occur, for example, from a badly insulated house in the winter. Can take place as conduction, convection, radiation, or a combination of these. See *conduction, convection, radiation*.

**heat mapping** See *thermal mapping*.

**heat of atomization** For a certain compound, the energy that is necessary to break up one mole of it into its individual elements. See *mole*.

**heat of combustion** For a given substance, the amount of energy produced by the combustion of one mole of it. See *mole*.

**heat of fusion** The amount of heat needed to melt a unit mass of a solid that has reached its melting point, without raising its temperature. Depends on the substance. Usually given in joules per gram or joules per kilogram.

**heat of reaction** For chemicals in a reaction, the extent of energy exchanged when one mole of the substances undergoes the reaction. This can be either a combination of elements into a compound, or the splitting of a compound into elements. See *mole*.

**heat of vaporization** The amount of heat needed to convert a unit mass of liquid to vapor, once it has reached its boiling point, without raising its temperature. Depends on the substance. Usually given in joules per gram or joules per kilogram.

**heat pump** A machine that transfers heat energy by condensing vapor into liquid. Essentially works like an air conditioner. When the vapor condenses it gives off heat; this makes it possible to take heat energy from a cooler place and transfer it to a warmer place.

**heat radiation** 1. The transfer of heat energy in the form of electromagnetic fields. This occurs mostly in that part of the spectrum called infrared (see *infrared*). 2. A misnomer for infrared.

**heat shield** A device that protected a reentering spacecraft of the Mercury, Gemini and Apollo types. Friction with the atmosphere caused intense heat during reentry; the shield kept the space capsule from being destroyed by this (ablation). See the drawing. See *Apollo, Gemini, Mercury*.

HEAT SHIELD

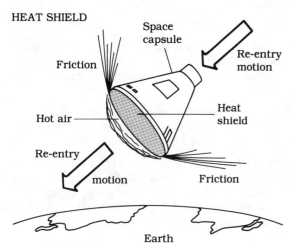

**heatsink** A means for getting rid of extra heat in an electronic component. This increases the current-carrying or power-handling capacity of the device. A finned piece of metal conducts heat away, dissipating it by convection and radiation.

**heat stroke** The most severe reaction to body overheating. The skin is hot, pale or grayish, and dry, and the victim is often unconscious. The body temperature rises to extreme levels. A medical emergency. Without immediate hospitalization the victim might die.

**heat transfer** The movement of energy from one place to another in the form of heat; that is, because they are at different temperatures. See *conduction, convection, radiation*.

**heavy water** Water in which the hydrogen atom is in the form of deuterium. See *deuterium*.

**hecto-** See the PREFIX MULTIPLIERS appendix.

**Heimlich maneuver** A method of helping someone who is choking to expel the object on which he or she is choking. This technique is taught by the American Red Cross in conjunction with emergency First Aid.

**Heinlein, Robert** A science-fiction writer of the early twentieth century. Some of his work was politically controversial.

**Heisenberg Principle** See *uncertainty principle*.

**Heisenberg, Werner** The quantum physicist remembered for his uncertainty principle. Certain things happen in a way that cannot be predicted on a particle-for-particle basis, but only as a probability or average. His ideas suggest that our fate is not really predestined, but that there are certain random factors involved. See *uncertainty principle*.

**helicable** A spring-like wire used in heart pacemakers. It is tiny, just half the diameter of a human hair, but is constructed in such a way that it will not break with prolonged use.

**helicopter** An aircraft that works by means of a rotating airfoil and stabilizers. The rotating airfoil looks like a propeller. Helicopters can land straight up-and-down, and can hover, so that they are more maneuverable than airplanes. But they cannot go as fast.

**heliocentric theory** The theory put forth by Copernicus and others, that the earth and other planets revolve around the sun (see drawing). Before this theory was accepted, it was thought that all of the objects in the heavens revolve around the earth. See *geocentric theory, Ptolemaic model.*

HELIOCENTRIC THEORY

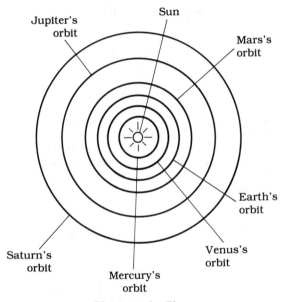

(Not to scale. The asteroids, Uranus, Neptune and Pluto were unknown when this theory was first proposed.)

**helium** Chemical symbol, He. An element with atomic number 2, and atomic mass almost always 4. It is a gas except at extremely cold temperatures. In stars, most of the energy is given off when hydrogen nuclei fuse to form helium nuclei. Helium is the second most abundant element in the universe, after hydrogen. See *hydrogen, fusion.*

**helix** A three-dimensional geometric figure with a shape resembling a spring ("Slinky"). The *pitch* of the helix is given by the number of revolutions per unit length. A helix with a high pitch is like a compressed spring; one with a low pitch is like a stretched-out spring.

**hema-** 1. Pertaining to the blood. 2. Pertaining to the element iron.

**hematite** High-grade iron ore. Originally abundant in such places as the Mesabi Iron Range in northeastern Minnesota. Nowadays, lower-grade ores must often be used because the supply of hematite has been used up. See *taconite.*

**hematology** 1. The branch of medicine concerned with the study of the blood. 2. The collection and evaluation of blood specimens in medicine.

**hematoma** The spilling of blood into body tissues. Caused by the breakage of an artery, capillaries, or vein. Might be a simple "blood blister," bruise, or a more severe case of internal bleeding. If it occurs in certain places on a large scale, it can be life threatening. See *hemorrhage.*

**heme** A compound that contains iron and that is an essential component of hemoglobin and other oxygen-containing substances in the body. See *hemoglobin.*

**hemi-** 1. A prefix meaning half of. Basically the same as semi-. 2. A prefix meaning similar to.

**hemicellulose** A carbohydrate similar to cellulose, found in the bran of grains such as wheat, oats, and rye. Serves as dietary fiber. See *cellulose, dietary fiber.*

**hemichordate** A primitive, water-dwelling animal that preceded more sophisticated vertebrates. For place in the animal family tree, see the ANIMAL CLASSIFICATION appendix.

**hemisphere** 1. One-half of a sphere; a bowl-shaped object. 2. One-half of the surface of the earth, usually given either as the Northern, Southern, Eastern or Western. See the drawing.

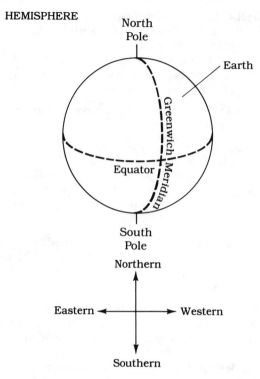

HEMISPHERE

**hemoglobin** In red blood cells, the iron-containing substance that transports oxygen to the body cells. It is red in color when rich in oxygen, after coming from the lungs. This is why oxygen-rich blood in the arteries of the body is bright red.

**hemophilia** An inherited disorder in which blood does not clot normally. A minor injury can cause a victim to bleed to death. Medical treatment is now available for this problem.

**hemorrhage** Internal bleeding. A severe form of hematoma (See *hematoma*). In a hemorrhage, blood spills into the tissues where it can interfere with the absorption of nutrients and oxygen. Massive tissue death can result. In some types of hemorrhage, a patient may actually bleed to death, or drown in his/her own blood.

**henry** Abbreviation, H. The unit of inductance. The value of inductance when one volt is produced by a current that is changing at one ampere per second. It is a large unit, and we more often see values in the range of millihenries (mH), equal to 0.001 H, or microhenries ($\mu$H), equal to 0.001 mH. See *inductance*.

**Henry Draper Catalogue** See *Cannon, Annie*.

**Henry, Joseph** The first American to design a working electric motor. See *motor*.

**hepatic** Pertaining to the liver, or to functions of the liver.

**hepatic coma** A state of unconsciousness caused by the effects of a severely deranged liver. Might occur in advanced alcoholism along with cirrhosis of the liver. See *cirrhosis*.

**hepatitis** Inflammation of the liver. An acute illness. Might be caused by various things, such as foreign substances or microorganisms. Symptoms include nausea and pain on the right side of the abdomen. When severe, it can be life threatening.

**heptane** A hydrocarbon and a member of the alkane series, with seven carbon atoms and 16 hydrogen atoms, arranged as shown in the drawing. It is a flammable liquid at room temperature and at atmospheric pressure.

HEPTANE

**herbaceous** Pertaining to any plant such as corn or wheat or sumac, whose tissue is not wood.

**herbivore** An animal that eats only plants. See *carnivore, omnivore.*

**hereditary** Inherited. We might speak of a hereditary tendency to develop diabetes, for example, or a hereditary trait such as curly hair. See *heredity.*

**heredity** 1. A person's ancestors. 2. The characteristics of one's ancestors. 3. The characteristics of a race of people, such as Italians or northern Europeans, especially with regard to functions of the body, such as ability to digest lactose. 4. The traits inherited from one's parents. We might speak of "heredity versus environment" when discussing certain types of illnesses or personality characteristics.

**Herman, Robert** One of the scientists who first theorized that we might observe, using radio telescopes, the red-shifted radiation from the Big Bang. He thought of this in 1948; it was finally observed in 1965 by Arno Penzias and Robert Wilson of the Bell Laboratories. See *Big Bang, red shift.*

**hermetic seal** An airtight seal that protects materials from damage by corrosion from the air. Can be done by welding a metal case, by soldering or the use of special glues. The sealed chamber can be filled with some nonreactive gas, such as helium or xenon.

**hernia** A condition in which a small part of the intestine pushes through the abdominal muscle wall. There might be few symptoms, but if it is severe it can cause acute inflammation, intestinal obstruction, and resulting serious illness.

**Herodotus** A Greek historian, who lived in the fifth century B.C. He recorded facts about science and medicine. For example, according to his records, ancient Egyptians thought that all diseases came from effects of food eaten.

**Herophilus** An ancient scientist who dissected the bodies of humans and learned about the internal organs and the circulatory system. This was done nearly 2000 years ago, when medicine was extremely primitive by modern standards.

**herpes** A chronic venereal disease caused by a virus (See *herpes virus*). The main symptom is a recurring irritation of the genitalia, such as burning and itching. It is spread through sexual contact.

**herpes virus** The virus that causes herpes. There are various forms of the virus, and it can infect other parts of the body besides the genitals. For example, cold sores are often caused by this virus.

**Herschel, William** An astronomer of the late eighteenth century, noted for his 48-inch telescope. It allowed people to see celestial objects never before observed. Through it, one could see nebulae and other galaxies, although at the time, it was not known that there were galaxies outside our own.

**hertz** Abbreviation, Hz. The standard unit of frequency. Equivalent to one complete cycle per second. The alternating-current electricity in the utility lines has a frequency of 60 Hz. Higher frequencies are given in kilohertz (kHz), where 1 kHz = 1000 Hz; also in megahertz (MHz), where 1 MHz = 1000 kHz. Extreme frequencies may be specified in gigahertz (GHz), where 1 GHz = 1000 MHz; and even in terahertz (THz), where 1 THz = 1000 GHz. See *frequency.*

**Hertz, Heinrich** One of the first scientists to discover and work with radio waves. Various effects and phenomena in electronics are named after him. See *hertz.*

**Hertzsprung, Ejnar** A Danish astronomer who, early in the twentieth century, found a correlation between the color of a star and its absolute brightness. See *Hertzsprung-Russell diagram.*

**Hertzsprung-Russell diagram** A graph-type way of showing different types of stars. See the drawing. Spectral type indicates color, with O being hottest and bluest, and M being coolest and reddest. The *absolute visual magnitude* indicates brightness, with lower numbers being brighter. When stars are graphed as points on this system, they tend to fall mostly into a curved, but defined, region called the *main*

## HERTZSPRUNG-RUSSELL DIAGRAM

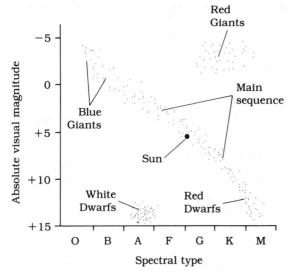

sequence. The exceptions are found in clumps, the *white dwarfs* and *red giants*.

**hetero-** A prefix that indicates a combination of various different sorts of things. See, for example, *heterogeneous*. The opposite is *homo-*.

**heterodyne** 1. For radio signals, to mix together, resulting in beat frequencies. See *beat frequency*. 2. One of the beat signals produced when radio signals are mixed. 3. A radio carrier wave that is heard as a steady tone in a receiver equipped to listen to code or sideband signals. See *carrier*.

**heterogeneous** 1. Consisting of different kinds of things in a combination, working together. 2. Containing various different substances, species, or ingredients.

**heuristic method** A "shotgun" approach to problem solving or troubleshooting. When there is no definite procedure that can be used, one must simply try various things, narrowing down gradually to the solution. This method can sometimes be used in equipment design: One simply tries different things, making educated guesses, till something works.

**Hewish, Anthony** An astronomer who, along with Jocelyn Bell, discovered pulsars (see *pulsar*). They made their discovery using the radio telescope at Cambridge University, England in the mid-1960s.

**Hewlitt, William** An engineer and a 1934 graduate of Stanford University, who founded, along with David Packard, the Hewlitt-Packard company. The company manufactures electronic devices, especially test equipment, and also makes calculators.

**hex-** A prefix meaning "six-." For example, hexane has six carbon atoms; a hexagon has six sides.

**hexachlorophene** A chemical used in hygienic cleansers. It kills germs. A compound of carbon, chlorine, hydrogen and oxygen, it is a powder in pure form.

**hexagon** 1. A six-sided polygon, lying in a single plane. 2. A regular six-sided polygon, lying in a single plane. This is one of the figures that can be repeated in a pattern, with all the individual polygons fitting exactly together. It is seen often in nature. An example is a honeycomb.

**hexane** A hydrocarbon and a member of the alkane series, with six carbon atoms and 14 hydrogen atoms, arranged as in the drawing. It is a liquid at room temperature.

**Hf** Chemical symbol for hafnium.

**HF** Abbreviation for high frequency.

**Hg** Chemical symbol for mercury.

**hibernation** A behavior pattern of certain mammals, such as bears, for the purpose of conserving energy when food is scarce. This is usually in wintertime. The animals find a warm hole to curl up in and fall asleep continuously or almost continuously, for weeks or months.

**hieroglyphics** 1. An ancient form of writing. It is at least 5000 years old, dating from the time of the building of the Pyramids. Probably derived from cuneiform (see *cuneiform*), hieroglyphics were written on paper and had a base-10 system of numbers as well as other mathematical symbols.

**high** 1. A weather system in which atmospheric pressure is higher than the surrounding pressure. Might be hundreds of miles across. Air circulates around it (see drawing). Usually brings good weather. 2. A condition of being elevated in level, intensity or physical position. 3. See *high-altitude*. 4. In digital electronics, the more positive of the two voltages. See *low*. 5. Under the influence of drugs (slang).

HIGH

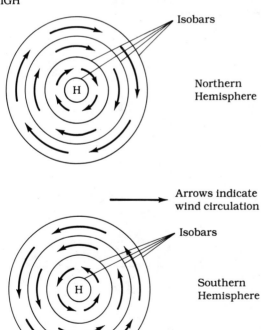

Isobars

Northern Hemisphere

Arrows indicate wind circulation

Isobars

Southern Hemisphere

**high-altitude** 1. In the upper atmosphere. 2. Relatively high above the surface of the earth or another planet.

**high-density lipoprotein** Abbreviation, HDL. The medium for so-called "good cholesterol." One of two cholesterol-carrying components in the blood. Recently, research suggests that the ratio of HDL to LDL (low-density lipoprotein) is more important to heart health than the total cholesterol level. See *cholesterol, low-density lipoprotein*.

**high-energy physics** The physics of subatomic particles, as explored using particle accelerators to smash atoms. High levels of energy are required in order to see tinier and tinier particles. See *particle accelerator*.

**higher-order language** Also called high-level language. A computer language that the operator uses directly, such as BASIC. See *BASIC, COBOL, FORTRAN, machine language*.

**high frequency** Abbreviation, HF. The range of radio frequencies from 3 megahertz (MHz) to 30 MHz. In the old days of radio these were called "short waves," even though their wavelengths are rather long, from 10 to 100 meters (about 33 to 330 feet). Signals at HF often propagate around the world via the ionosphere. See *ionosphere*.

**high-level language** See *higher-order language*.

**high-level radioactive waste** The by-product of reprocessing certain types of nuclear fuels. It stays radioactive for thousands of years; that is, it has a long half-life (see *half-life*). It is therefore especially dangerous.

**high-octane** Pertaining to fuel, especially gasoline, with an above-average octane content. This type of fuel is less likely than low-octane gasoline to cause knocking in the engine.

**high-test** 1. For gasoline, high-octane. 2. Fuel that yields much energy very fast, used for such purposes as auto racing or in fighter aircraft.

**high-tension line** A power line carrying extremely high voltage. Long-distance power lines are of this type. The voltage is stepped up using transformers. Such a line might carry 1 million volts. For a given amount of power, the higher the voltage, the less the current in the wires, and therefore, the lower the loss in the power line over a great distance.

**high-voltage**  1. Pertaining to circuits or devices operating at more than a certain level. Usually this is a few hundred volts, but the term is relative. 2. Pertaining to the plate power supply in a vacuum-tube circuit.

**hindbrain**  The cerebellum and medulla of the brain in the embryo. Develops to control basic body functions. See *forebrain, midbrain*.

**Hindemith, Paul**  A musician who became interested in the spectrum of emission lines from hydrogen. He had a scientist translate these spectral lines into musical notes. The idea was that hydrogen, being the most abundant element in the universe, ought to yield a beautiful sound. But the experiment failed; the resulting notes were boring. This refuted the idea that everything in nature must be "beautiful."

**Hindu-Arabic numerals**  See *Arabic numerals*.

**hipbone**  A large, somewhat bowl-shaped, strong bone that connects the thigh bone to the spine. There are two of these, sometimes also called pelvic bones, one for each leg.

**Hipparchus**  An ancient Greek astronomer who, in the Second Century B.C., observed the earth's shadow on the moon during a lunar eclipse, and determined from this that the distance to the moon is 30 earth diameters. He might also have been the first to use the astrolabe for finding the elevation of celestial objects above the horizon. See *astrolabe*.

**Hippocrates**  A Greek physician who lived in the fifth century B.C. He is remembered for a book called the *Hippocratic Corpus*. Many of his recommended practices are still followed today. The doctor's oath is named for him. See *Hippocratic Oath*.

**Hippocratic Oath**  An oath that a person must take in order to become a doctor. It is a pledge to use the skill for healing purposes only. This is why doctors are reluctant to perform certain tasks such as execution by means of lethal injection.

**hirsutism**  1. The growth of hair. 2. A condition of having more than the normal amount of hair.

**histamine**  A substance released in the body during allergic reactions. It causes the blood vessels to get larger in diameter (dilate), and generally increases blood circulation. This causes a characteristic flushing. Skin eruptions, such as hives, may occur when this substance is released in large amounts for a period of time. See *hives*.

**histidine**  An essential amino acid. This substance must be present in dietary protein, because it cannot be manufactured by humans from other substances. See *amino acid, essential amino acids*.

**histogram**  A type of graph, in which vertical bars are used and the independent variable (on the horizontal axis) is displayed in discrete values, rather than continuously. See the drawing. See *Cartesian coordinates*.

HISTOGRAM

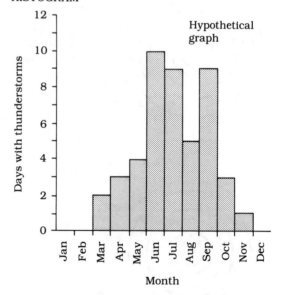

**histology**  Also called microscopic anatomy. The study of anatomy of cells and other things too small to be seen with the unaided eye. See *anatomy*.

**histone** A protein that plays a role in the activity of chromosomes and genes, especially when cells divide. See *cell division, chromosome, gene*.

**historical geology** A branch of geology, dealing with the evolution of landforms. Also concerned with the study of fossils, the evolution of rock strata (layers), plate tectonic motions, and paleoclimates.

**HIV** Abbreviation for human immunodeficiency virus.

**hives** Itchy bumps on the skin that form as an allergic reaction to something eaten or that has come in contact with the skin. Might also occur along with certain infectious diseases such as hepatitis.

**Ho** Chemical symbol for holmium.

**Hodgkin, Alan** A well-known medical scientist who did research in various fields, particularly involving the nervous system. He shared the Nobel Prize for medicine in 1963 for his work.

**Hodgkin's disease** A disease of the lymph system. Tumors occur along with many different symptoms. The most common is pain in the affected areas after drinking anything alcoholic. The cause is still somewhat of a mystery.

**holding pattern** The circling of airplanes waiting to land. The pattern is watched over by air-traffic controllers. Different planes are assigned different altitudes and different patterns to avoid collisions.

**hole** In electronics, the absence of a single electron. This carries a unit positive charge, because electrons carry negative charge. In some semiconductors, it is easier to think of current flow in terms of holes, rather than in terms of electron movement. See the drawing for an illustration of hole flow. See *semiconductor*.

**holistic** Pertaining to the treatment of the body as a whole. In holistic medicine, for example, the aim is to maintain overall body health, because a well-treated body is less likely to get sick than a poorly treated body.

HOLE

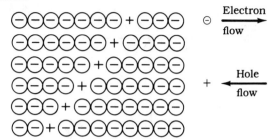

**Hollerith code** A code for punched cards used with early computers. Not used nowadays; magnetic media such as disks have taken the place of punched cards. See *floppy disk, hard disk*.

**holmium** Chemical symbol, Ho. An element with atomic number 67. The mean atomic weight is 165. In its pure form it is a metal.

**Holocene epoch** Also called the time of recent life. It began about 10,000 years ago, after the glaciers of the Pleistocene epoch retreated. This is the time during which humans became civilized. Written records go back a little more than half of this time. See the GEOLOGIC TIME appendix.

**hologram** A photograph in which three dimensions are portrayed. The image actually has perspective, and changes when the observer moves. Thus it is possible to render objects more or less "true to life," whereas in a normal photograph, the image is the same no matter what the viewing angle. Some holograms need special lighting to appear most vivid; others can be viewed in ordinary light. Sometimes holograms are projected into the air using lasers. See *holography*.

**holography** The making of holograms. One method is shown in the drawing. The best holograms are made using lasers. There are other methods besides the one shown. See *hologram*.

## HOLOGRAPHY

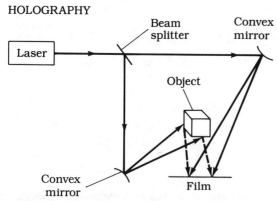

**homeomorphism** A property of objects in topology. Two objects, or *topological spaces*, are homeomorphic if and only if it is possible to map the points continuously, in a one-to-one correspondence, from one to the other, and from the other to the one. This can be visualized as twisting, stretching, or compressing but never tearing, cutting, or overlapping, the one object to get to the other.

**homeopathy** A form of medicine, in which extremely diluted medicinal substances are used. Some of this theory is criticized by conventional medical people. Nevertheless, much of homeopathic medical belief is common sense.

**hominid** The species including humans and our ancestors who walked upright on two legs. See *Homo erectus, Homo habilis, Homo sapiens*.

**homo-** 1. A prefix meaning "all of the same type" or "uniform" or "continuous." 2. Pertaining to human beings or their ancestors.

**Homo erectus** An early human-like species who walked upright and obtained food by hunting, rather than by gathering as had been the main way for earlier species. See *Homo habilis*.

**homogeneous** Pertaining to a substance that is uniform throughout; that is, it is the same every place. An example is pure liquid water.

**Homo habilis** A human-like species that first appeared about 2,000,000 years ago. It resembled the Australopithecus, but had a larger brain, and also had bones in the limbs that were more human-like than those of the earlier Australopithecus. The brain was roughly half the size of that of modern humans. See *Australopithecus*.

**homomorphism** 1. A type of mapping or function between sets, in which the elements of the sets can be paired off in a one-to-one correspondence. See *function*. 2. In botany, a situation in which a plant has perfect flowers all of which are of the same kind.

**Homo sapiens** The species of modern humans, including Cro-Magnon and Neanderthal.

**Hooke, Robert** A seventeenth century physicist. Published the first book about microscopic things, called *Micrographia*, in 1665. He observed things and drew detailed diagrams of them.

**Hooke's Law** A rule in physics concerning strain on objects. In a solid object, stress is directly proportional to strain. This is true as long as the stress is not so great that the solid is permanently deformed.

**horizon** 1. The line midway between the nadir and the zenith on the celestial sphere. See *celestial sphere, nadir, zenith*. 2. The farthest point that can be seen on the surface of the earth, from a given vantage point in a given direction. 3. The set of all points for a given location, as defined in (2). 4. The farthest point that a radio wave can be received by a station on the surface of the earth, for a given transmitter location, at frequencies where there is no ionospheric propagation. 5. The set of all points for a given location, as defined in (4).

**horizontal oil drilling** A method of getting at oil deposits by drilling horizontally under the ground once a certain depth has been reached (see drawing). This increases the amount of oil that can be gotten from a well, so it is necessary to drill fewer wells.

**hormone** A substance secreted by any of the endocrine glands (See *endocrine glands*). The substance regulates certain body functions, such as metabolism and sexual development.

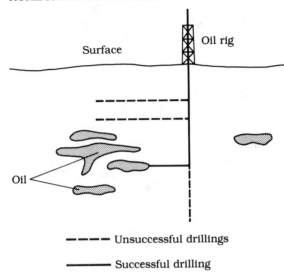

HORIZONTAL OIL DRILLING

- - - - - Unsuccessful drillings
———— Successful drilling

**hornblende** A dark mineral common in igneous rock and metamorphic rock, made up of silicates of various elements, especially magnesium, aluminum, sodium, iron, and calcium. See *igneous rock, metamorphic rock.*

**hornfels** A hard, slate-like, silicate metamorphic rock with a fine grain. Can contain various minerals. See *metamorphic rock.*

**Horsehead Nebula** A gas-and-dust nebula in the constellation of Orion. So called because of its shape. Dark dust blocks the light from glowing gases behind; the dark region has a shape something like a head.

**horse latitudes** The region around 30 degrees north latitude, where a semipermanent high-pressure region exists. Sailing ships were becalmed there for days; the crews ran short of food, so they ate food meant for their horses. The horses were thrown overboard, and could be seen floating in the sea by other ships' crews.

**horsepower** Abbreviation, hp. A unit of power equal to 746 watts. Often used to specify the power that an engine or electric motor can deliver. When the term was coined, it was thought that this amount of power is roughly the power provided by one healthy horse pulling a certain weight on a level surface.

**horsetail** A fern-like, perennial plant. Reproduces without flowers. For its place in the plant kingdom, see the PLANT CLASSIFICATION appendix.

**horst** A block of the earth's crust that has been raised up by faults on both sides.

**hot-air balloon** A large balloon, several meters in radius, that is filled with air heated by a controlled flame at the base. The lifting power is sufficient to loft one or two people in a basket. Hot-air ballooning is a popular sport.

**hot spot** A small region of isolated volcanic activity. There are more than 100 of these in various places around the world. The Hawaiian Islands are a good example. The lava from a hot-spot volcano differs from that of a more typical volcano; there are more alkali minerals. See *volcano.*

**hot spring** An artesian spring that has been heated because it is extremely deep. Geothermal heat causes the rise in water temperature with increasing depth. See *artesian spring, geothermal energy.*

**hot-wire device** A meter that works by measuring the heat that a current produces in a wire. The meter can measure current, either alternating or direct. Or it can measure some other quantity. For example, we might measure wind speed with an anemometer that spins a generator, that in turn causes current to flow in a wire, heating the wire.

**hour** Abbreviation, hr. A unit of time equal to $1/24$ mean solar day. The earth rotates about 15 degrees of arc on its axis in this length of time. Equal to 60 minutes or 3600 seconds.

**hovercraft** An all-terrain vehicle that rides on a cushion of air. The air pressure can be developed in a variety of different ways. In the drawing, one method is shown for a small, one-person craft. The craft is steered by leaning in the desired direction.

**Hoyle, Fred** One of the astronomers who formulated the Steady State Theory of the evolution and structure of the universe. Also known for his theory that radio galaxies may contain

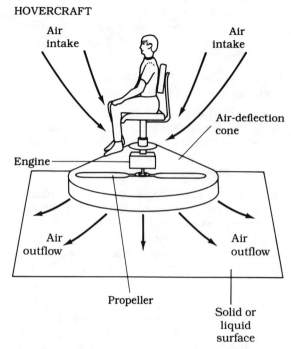

HOVERCRAFT

massive explosions in their centers. See *radio galaxy, Steady State Theory*.

**hp** Abbreviation for horsepower.

**hr** Abbreviation for hour.

**Hubble age** An estimate of the age of the universe, calculated according to the value of the Hubble constant. According to this constant, galaxies 15 billion (15,000,000,000) light years away are retreating at the speed of light. Nothing can go faster than this. This has led some astronomers to suggest that the Big Bang occurred 15 billion years ago. This is called the Hubble age of the universe. See *Big Bang, Hubble constant* and *Hubble, Edwin*.

**Hubble constant** A constant first pointed out by the astronomer Edwin Hubble. The apparent speed of retreat of distant galaxies is directly proportional to their distance. The constant is approximately one-tenth of the speed of light (0.1c) for every 1.5 billion (1,500,000,000) light years of distance from us. This constant has been revised periodically as more accurate measurements of distance have been made. See *Big Bang* and *Hubble, Edwin*. See also *red shift*.

**Hubble, Edwin** An early twentieth century astronomer noted especially for his research into measuring great distances in the universe. Noticed a linear relation between the distance to a galaxy and the extent to which its light has been red-shifted. This helped to verify the notion that the whole universe is expanding as if it blew up about 15 billion (15,000,000,000) years ago. See *Big Bang, Hubble constant, red shift*.

**Hubble space telescope** An orbiting telescope launched by the United States using the Space Shuttle on April 25, 1990. When deployed it proved to have a flaw in its alignment. It will be repaired eventually. Then it should allow astronomers to see with clarity they have never known before, because there is no atmosphere to blur the images.

**hue** An expression of color based on the peak wavelength. Can actually be given in Angstrom units, where red light is about 7000–7500 Angstroms, violet light about 3900–4200 Angstroms, and the other colors are in between. Can also be specified in general terms, such as "blue-green." See *Angstrom unit, saturation*.

**hum** The noise produced in speakers or headphones when utility-line currents are picked up in an audio amplifier. The sound has a characteristic low pitch (60 hertz), corresponding to the power-line frequency.

**human engineering** A branch of engineering concerned with the "user-friendliness" of equipment. Convenience and efficiency are among the main objectives. Safety and fatigue reduction are also important.

**human immunodeficiency virus** Abbreviation, HIV. The virus that causes AIDS. See *acquired immune deficiency syndrome*.

**Humason, Milton** An astronomer who worked with Edwin Hubble in measuring the distances to other galaxies early in the twentieth century. See *Hubble constant* and *Hubble, Edwin*.

**Hume, David** An Empiricist philosopher who believed that ideas exist only as they are shown to us by things that we can observe. See *Empiricism*.

**humerus** The bone of the upper arm, between the elbow and the shoulder.

**humidifier** A device for increasing the relative humidity in the indoor air. Especially useful in dry climates or during cold winters when indoor humidity tends to be low. The most common method is simply to boil water and combine the vapor with the air in the ventilating system of the building.

**humidity** Water vapor in the air. The air is capable of containing a certain amount of water vapor. The higher the temperature, the more water vapor that can exist in a given amount of air. See *relative humidity*.

**humor** 1. Any of four different types of body fluids according to ancient medicine: blood, phlegm (mucus), black bile and yellow bile. 2. Any body fluid.

**humus** Partially decomposed remains of plants and/or animals. This is the part of the soil that provides much of the nutrient value for growth of new plants.

**hurricane** A tropical cyclone with maximum sustained winds of 74 miles per hour or more. Circulates counterclockwise in the Northern Hemisphere, and clockwise in the Southern Hemisphere. They do not occur near the equator because of the lack of Coriolis effect there. The storms are actually gigantic heat engines (see drawing). The most intense storm activity is in a ring-shaped wall of clouds, the *eyewall* or *wall cloud*, immediately surrounding the *eye*, or calm center. See *cyclone, eye of storm*.

**Hurricane Forecast Centers** Part of the National Oceanic and Atmospheric Administration (NOAA) in the United States, responsible for tracking and predicting the paths of hurricanes, especially when they threaten land. See *hurricane*.

**Hutton, James** A scientist noted for his work in bringing together the sciences of geography, paleontology, and mineralogy into the broader science of geology. He is sometimes called the "Father of Geology." See *geography, geology, mineralogy, paleontology*.

**Huxley, Thomas** A scientist who traveled around England in the middle and late nineteenth century, giving lectures on Darwin's theory of evolution. He is credited with making this theory more widely known, sooner, than it might have been otherwise. See *Evolutionism*.

**hybrid** 1. Consisting of two or more different types of components. For example, we might have a radio transmitter that uses integrated circuits in the oscillator, transistors in the intermediate amplifiers and modulator, and a tube in the final amplifier stage. 2. Produced by breeding plants or animals having different traits. This often results in hardier plants or animals.

**hybrid circuit** See *hybrid*.

**Hydra** A constellation in the southern celestial hemisphere, generally south of Cancer and Virgo.

HURRICANE

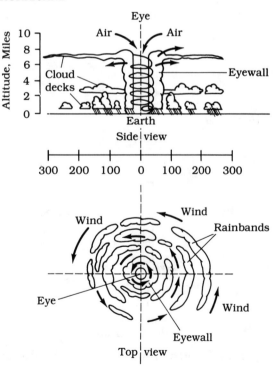

**hydra** A small animal that lives in fresh water. It is cylinder-shaped with tentacles around the mouth at one end.

**hydrate** 1. To add water to a substance. 2. A substance that is made up of something else with water added. 3. A compound containing hydrogen and oxygen.

**hydration** A condition in which rocks contain water molecules in their atomic arrangements. Silicate rocks in the earth's crust are hydrated; this has led to a theory that the world's seas developed as this water was gradually liberated over time.

**hydraulic** Making use of water to perform work, especially to provide a greatly increased force per unit area. See *hydraulic lift, hydraulic press, hydraulics.*

**hydraulic lift** A device that makes use of the fact that water in liquid form is practically non-compressible. See the drawing. A piston pushes water into a container, where a much larger force, although over a smaller displacement, is developed. See *hydraulics.*

HYDRAULIC LIFT

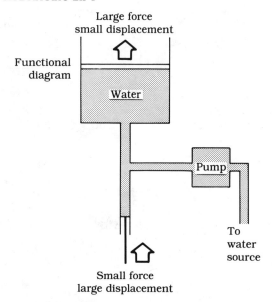

**hydraulic press** An industrial tool used for forming metal and for other processes requiring extreme pressure. Operates using hydraulic principles, in a manner similar to the hydraulic lift. See *hydraulic lift, hydraulics.*

**hydraulics** A branch of engineering, concerned with the use of water to obtain great force or pressure. Water cannot be compressed in its liquid form, and this is a useful property. It allows for the redistribution of forces, something like the way a set of pulleys works. A small force over a large displacement can be converted to a large force over a small displacement, or vice-versa. See *hydraulic lift, hydraulic press.*

**hydrocarbon** Any compound consisting only of carbon and hydrogen. See *alkane, alkene* for examples. These compounds are often useful as fuels, because they readily combine with oxygen, and this generally occurs rapidly as burning. The compounds are found in fossil fuels such as oil. They also occur naturally in gaseous form. Fuel compounds can be synthesized.

**hydrochloric acid** A powerful acid formed by the combination of hydrogen and chlorine (HCl). These elements react readily with each other. This is the acid found in your stomach. See *hydrochloride.*

**hydrochloride** A compound produced when hydrochloric acid reacts with certain chemicals, especially organic chemicals. This facilitates putting the chemical in liquid form.

**hydrocortisone** An anti-inflammatory drug, available in creams and ointments, and used for minor skin irritations. If used over large areas of skin, it should be done under the supervision of a doctor, because this drug has hormone-like properties in large amounts.

**hydroculture** The commercial raising of fish; fish farming. In saltwater this is called *mariculture.*

**hydrodynamics** The study of the behavior of fluids, especially water, in motion. Interrelated with hydraulics (See *hydraulics*).

**hydroelectric energy** Electric energy derived from falling water. See *hydroelectric power plant.*

**hydroelectric power plant** A power plant that converts the energy from falling water into electrical energy. This can be done at natural waterfalls, but more often, dams are built to provide artificial, controlled waterfalls. The force created by gravity on the mass of the water drives turbines (see drawing). Hydroelectric power plants do not pollute, but they do sometimes cause controversy when the dams create reservoirs that flood people's land.

HYDROELECTRIC POWER PLANT

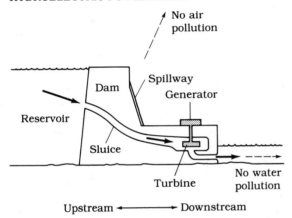

**hydrofoil** A lifting device that works with water, in a manner similar to the way an airfoil works in the air. It allows a boat to run mostly above water, minimizing friction.

**hydrofoil craft** A high-speed boat that rides mostly or entirely above the water, except for a hydrofoil underneath that provides stabilization with a minimum of friction. See *hydrofoil*.

**hydrogen** Chemical symbol, H. The most abundant element in the universe, with atomic number 1. The most common isotope has atomic weight 1; there are heavier isotopes (see *deuterium, tritium*). Hydrogen is a light gas at room temperature. It is the fuel that makes stars shine. See *fusion*.

**hydrogenation** 1. The addition of hydrogen to a substance, especially to an unsaturated fat, to retard spoilage. The process causes these fats, usually liquid at room temperature (such as corn oil), to become semisolid or solid. 2. A process in which coal is converted to oil or natural gas by adding hydrogen to the carbon to make a hydrocarbon. See *hydrocarbon*.

**hydrogen bomb** An explosive that makes use of the energy liberated by the fusion of hydrogen. To provide the temperatures needed to cause this fusion reaction, atomic bombs are used. A hydrogen bomb might liberate as much energy as there would be in up to 100 million tons of TNT. See *fusion, atomic bomb*.

**hydrogen bond** A bond between hydrogen and another atom in a compound. The hydrogen atom might either share its electron with another atom, or accept an electron from another atom (see drawing). Hydrogen can therefore combine with many different elements to make various compounds.

HYDROGEN BOND

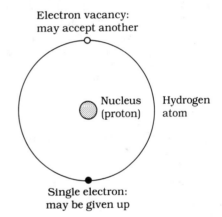

**hydrogen chloride** The gas that produces hydrochloric acid when dissolved in water. See *hydrochloric acid*.

**hydrogen fluoride** A highly corrosive compound formed by the union of hydrogen and fluorine. Also sometimes called hydrofluoric acid when dissolved in water. It is so caustic that it will "eat at" things that other substances cannot. It is often used for etching glass.

**hydrogen fuel cell** A source of energy using hydrogen. When it combines with oxygen by burning, the result is water. This water is easily converted back to hydrogen and oxygen by electrolysis in a manner similar to charging a battery.

**hydrogen fusion** The fusion of hydrogen to get helium and energy. This is the process that makes most stars shine. It is also responsible for the enormous destructive power of modern nuclear weapons. See *fusion, hydrogen, hydrogen bomb*.

**hydrogen iodide** A compound formed by one molecule of hydrogen and one of iodine. It is a corrosive gas at room temperature.

**hydrogen lines** The absorption lines or emission lines caused by the element hydrogen. One of these lines is at a radio wavelength of 21 centimeters. This has been postulated as a possible "marker" signal, near which alien civilizations might send signals. There are other lines, some in the visible spectrum. See *absorption line, emission line*.

**hydrogen peroxide** A compound in which two oxygen atoms combine with two hydrogen atoms. It is like water with an extra oxygen atom in each molecule. This clear, odorless liquid is used to sterilize cuts, and can also be used as a bleach or general-purpose cleaner and germ killer.

**hydrogen spectrum** See *hydrogen lines*.

**hydrogen sulfide** A gas formed by the reaction of hydrogen with sulfur. Has a rotten-egg smell. This gas is formed by the burning of sulfur-containing fossil fuels; it is a pollutant.

**hydrologic cycle** See *water cycle*.

**hydrological mapping** The making of maps that show the distribution and flow of water on the surface of the earth. This is done largely with the help of environmental satellites.

**hydrology** The study of the distribution of water on the earth, especially in solid form and liquid form. Snow cover, rivers, lakes, and glaciers are all studied from the ground, from the air, and from space.

**hydrolysis** 1. The reaction of a compound with water creating new chemicals. 2. The addition of water to a substance to alter its nature or to create new substances.

**hydrolyzation** The splitting of starches to form easily absorbed sugars. This occurs in the digestive system when you eat such foods as potatoes or rice.

**hydrophilic** Tending to absorb water. An example is the psyllium that you might have used to aid digestion and elimination. The psyllium granules absorb and hold water, creating bulk in the intestines.

**hydrophily** The fertilization of a flower as a result of either the flower or the pollen being carried by water.

**hydrophobia** Fear of water or a violent and unpleasant reaction to water. This term is sometimes used to describe rabies, because choking on water is a symptom of the disease. See *rabies*.

**hydroplaning** 1. Riding on water at high speed, for example, water skiing. 2. An undesirable tendency for tires to ride up on a film of water on wet pavement at highway speeds (see drawing). This can be minimized by a deep tire tread, and also by roughening the highway surface.

HYDROPLANING

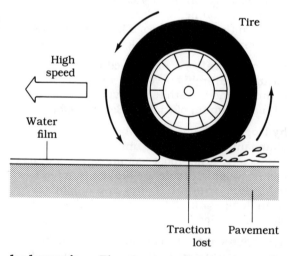

**hydroponics** The growing of plants in artificial substances, rather than in real soil. The plants are sprouted, and then, in effect, are left in the sprouting medium instead of being planted.

**hydrosol** A suspension of tiny particles in water. Essentially a colloid in which the liquid, or medium, is water. See *colloid*.

**hydrosphere** The earth's surface water, including that in all the oceans, rivers, and lakes, and also in glaciers, snow cover, ground water, and water vapor.

**hydrostatics** A branch of physics involving water not moving, especially when it has been confined for human needs. Interrelated with hydraulics. See *hydraulics, hydrodynamics*.

**hydrothermal** Pertaining to heated ground water. For example, we might find a mineral ore deposit that has been put into place by hot underground water.

**hydrotropism** The tendency of plants to grow toward water. In the soil, this is usually downwards, but it can also occur sideways when water is unevenly distributed under the ground.

**hydroxide** Any metal compound containing oxygen and hydrogen atoms bonded together, and whose chemical symbol therefore ends in -OH (for oxygen and hydrogen). Usually, these compounds are bases, such as sodium hydroxide. But some can under certain conditions, act either as an acid or as a base.

**hydroxyl** The -OH portion of a hydroxide. See *hydroxide*.

**hygrometer** A device for measuring relative humidity. Might have a dry-bulb thermometer and a wet-bulb thermometer, whose readings are compared, and the relative humidity is read from a table. Or it might be a meter-type instrument. The latter type is less precise, using a hair or other filament that expands or contracts as the atmospheric water-vapor content changes. See *relative humidity*.

**hygroscopic** Pertaining to materials that absorb water vapor. Cobalt chloride crystals are a good example. Small bags of this are often put into bottles of pills to keep the pills from absorbing water vapor. See *hydrophilic*.

**hyper-** A prefix meaning over-, overly, high, or excessive.

**hyperactivity** A disease often associated with children. It is frequently misdiagnosed. The main symptoms are restlessness and irritability. Weight loss and insomnia might also occur.

**hyperbola** The curve formed on a plane when a cone intersects the plane in a certain way. If you shine a flashlight over a flat surface, with the axis of the beam aimed horizontally so that the edge of the light ring only partially lands on the surface, you obtain half of a hyperbola. The shape is similar to that of a parabola. The Cartesian graph and equation for a hyperbola are shown in the drawing.

HYPERBOLA

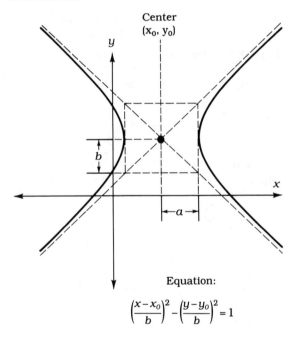

Equation:

$$\left(\frac{x-x_0}{b}\right)^2 - \left(\frac{y-y_0}{b}\right)^2 = 1$$

**hyperbolic cosecant** Abbreviation, csch. A special form of trigonometric function, in which a unit hyperbola, rather than a unit circle, is used. Also definable as csch

$$x = \frac{2}{(e^x - e^{-x})}$$

where e is approximately equal to 2.718. See *trigonometric function, unit hyperbola*.

**hyperbolic cosine** Abbreviation, cosh. A special form of trigonometric function, in which a unit hyperbola, rather than a unit circle, is used. Also definable as cosh

$$x = \frac{(e^x + e^{-x})}{2}$$

where e is approximately equal to 2.718. See *trigonometric function, unit hyperbola.*

**hyperbolic cotangent** Abbreviation, coth. A special form of trigonometric function, in which a unit hyperbola, rather than a unit circle, is used. Also definable as coth

$$x = \frac{(e^x + e^{-x})}{(e^x - e^{-x})}$$

where e is approximately equal to 2.718. See *trigonometric function, unit hyperbola.*

**hyperbolic function** A trigonometric function that is based on the unit hyperbola rather than on the unit circle. See *hyperbolic cosecant, hyperbolic cosine, hyperbolic cotangent, hyperbolic secant, hyperbolic sine, hyperbolic tangent, trigonometric function, unit circle, unit hyperbola.*

**hyperbolic geometry** A form of non-Euclidean geometry, in which the surface is not a plane but is a hyperboloid instead. See *hyperboloid, non-Euclidean geometry, spherical geometry.*

**hyperbolic secant** Abbreviation, sech. A special form of trigonometric function, in which a unit hyperbola, rather than a unit circle, is used. Also definable as sech

$$x = \frac{2}{(e^x + e^{-x})}$$

where e is approximately equal to 2.718. See *trigonometric function, unit hyperbola.*

**hyperbolic sine** Abbreviation, sinh. A special form of trigonometric function, in which a unit hyperbola, rather than a unit circle, is used. Also definable as sinh

$$x = \frac{(e^x - e^{-x})}{2}$$

where e is approximately equal to 2.718. See *trigonometric function, unit hyperbola.*

**hyperbolic tangent** Abbreviation, tanh. A special form of trigonometric function, in which a unit hyperbola, rather than a unit circle, is used. Also definable as tanh

$$x = \frac{(e^x - e^{-x})}{(e^x + e^{-x})}$$

where e is approximately equal to 2.718. See *trigonometric function, unit hyperbola.*

**hyperboloid** The three-dimensional figure obtained when a hyperbola is rotated on its axis. Shaped basically like two infinitely extended concave bowls facing away from each other.

**hypercharge** A property of certain subatomic particles, similar to electric charge in some ways. It is a theoretical quantity, developed to explain particle decay.

**hyperon** A subatomic particle that decays very fast. It is a form of hadron and is comprised of three quarks. See *hadron, quark.*

**hyperplasia** An enlargement of a body organ or a thickening of skin or connective tissue or mucous membrane, caused by an increase in the number of cells. See *hypertrophy.*

**hypersonic** See *supersonic.*

**hypertension** High blood pressure. A problem can exist if diastolic blood pressure is consistently above about 90 millimeters of mercury (90 mmHg). A doctor should evaluate blood pressure regularly. See *blood pressure, diastolic.*

**hypertonic** Pertaining to a solution having a higher osmotic pressure of certain dissolved substances, as compared with other solutions. Osmosis of the dissolved substances takes place into those other solutions. See *hypotonic, osmosis, osmotic pressure.*

**hypertrophy** An enlargement of a body organ, or a thickening of skin or connective tissue or mucous membrane, caused by enlargement of the individual cells. See *hyperplasia.*

**hypo-** A prefix meaning under, low, or deficient.

**hypocenter** The point of origin for an earthquake. Sometimes called the focus or focal point of the quake.

**hypochondriac** A person who thinks that he or she has illnesses of various sorts, often without any objective basis. Such a person might diagnose himself or herself as having cancer or heart disease, when in reality he or she is physically healthy.

**hypocycloid** Any inwardly curved figure as graphed in Cartesian or polar coordinates. In the drawing, a figure is shown with four cusps. There may be as few as three; there is no upper limit to the number of cusps possible.

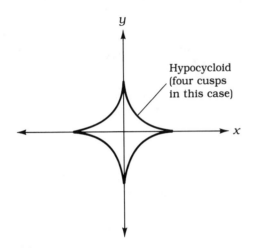

**hypothalamus** An area at the base of the brain below the thalamus. This center regulates many different autonomic body functions, such as sleeping, appetite, and metabolism. See *appestat, brain, thalamus.*

**hypotension** Consistently low blood pressure. This is best evaluated by a doctor.

**hypothermia** A condition in which body temperature is abnormally or dangerously low, usually because of exposure to cold weather. In extreme cases, emergency medical treatment might be needed.

**hypothesis** 1. A conjecture; an idea or theory that has not been proven, but which scientists are attempting to prove. 2. The antecedent in a logical "if/then" sentence; that is, the phrase or part after the "if" and before the "then."

**hypothetical** Imaginary; set up for the purposes of verbal demonstration or simulation. We might make up a hypothetical hurricane, for example, to show its effects on a large town. This can even be done by computers to simulate the effects an actual phenomenon would have.

**hypotonic** Pertaining to a solution with a comparatively low osmotic pressure, so that dissolved chemicals tend to osmose into it. See *hypertonic, osmosis, osmotic pressure.*

**hypoxia** A condition in which body tissues do not get sufficient oxygen. Can occur because of low air pressure, oxygen displacement by carbon monoxide or other gases, lung congestion, or various other reasons.

**hysterectomy** A surgical method of sterilizing a female by removing the uterus. See *uterus.*

**hysteresis** A tendency for electronic devices or machines to act "sluggish." For example, a thermostat set for 70 degrees Fahrenheit might not switch on the heating till it gets down to 68 degrees, and then might not switch it off again till it rises to 73 degrees. This is desirable in the

HYSTERESIS LOOP

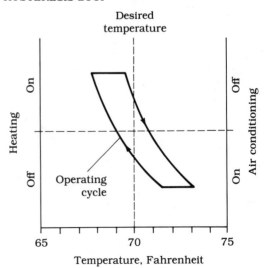

case of the heating unit (see diagram for hysteresis loop), because it keeps the machine from constantly switching on and off. But hysteresis can cause problems in some cases, such as in the core of a transformer. See *hysteresis loop, hysteresis loss.*

**hysteresis loop** A graph of a hysteresis process. The drawing shows an example for either a heating unit or an air conditioner.

**hysteresis loss** Power that is wasted, or lost, as heat in a ferromagnetic substance because of hysteresis in the substance. This causes loss in the cores of transformers. It becomes worse as the frequency increases, for a given core material.

**Hz** Abbreviation for hertz.

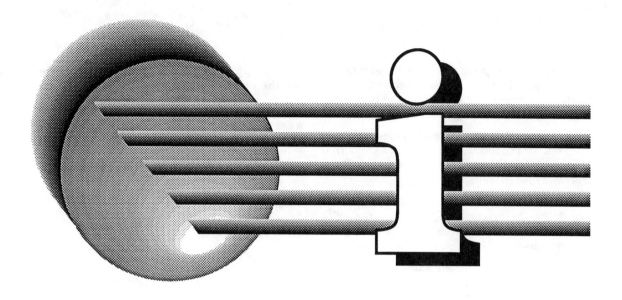

**I** Abbreviation for current; chemical symbol for iodine.

**i** Abbreviation for the square root of −1 as mathematicians use it. See *imaginary number*.

**ibuprofen** An aspirin substitute. For some people it is more effective than aspirin, especially for pain associated with muscle spasms, such as menstrual cramping. See *acetaminophen, acetylsalicylic acid*.

**Icarus** 1. The name of a man-powered, ultralight aircraft that can fly up to several miles at low altitude. 2. In ancient mythology, a person who flew with artificial wings that melted when he went too high and got too close to the sun. Today we know he would have frozen to death instead.

**IC** Abbreviation for integrated circuit.

**ICBM** Abbreviation for intercontinental ballistic missile.

**ice** Water in frozen form. Normally, pure water freezes at 32 degrees Fahrenheit or 0 degrees centigrade. Ice is slightly less dense than water in liquid form. This is why it floats on lakes, rivers, and the sea.

**ice age** Any of various periods in geologic history dominated by glaciation over large portions of the land mass of the earth (see illustration). The most recent, the Pleistocene epoch, ended about 10,000 years ago. There were other, longer ice ages about 2,000,000,000 and 700,000,000 years ago. Many scientists think another one is coming in a few thousand years. See the GEOLOGIC TIME appendix.

ICE AGE

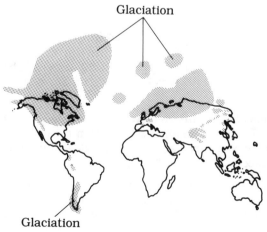

From Erickson, THE LIVING EARTH (TAB Books, Inc., 1989).

**iceberg** A large piece of a glacier that has broken off and floats in the sea. Sometimes icebergs are carried well down into the temperate latitudes if they are large enough.

**icebreaker** A boat designed to clear the way through frozen lakes, rivers or seas, so that another boat can pass. Used in polar exploration and also in transport.

**ice cap** 1. The ice over the North Pole or the South Pole of the earth. 2. The ice in the polar regions of Mars. 3. The ice in the polar regions of any planet that has polar ice. See *ice sheet*.

**ice crystal** A tiny, frozen water droplet in the atmosphere or on a surface. When water solidifies, it does so in patterns, just as do other substances. The patterns can build on themselves, forming fantastic shapes if conditions are right. This is how snowflakes form and get their interesting shapes.

**ice erosion** Erosion of the land caused by ice, especially glacial action. The landscape over much of the United States and over almost all of Canada, was carved by glaciers during the last ice age. On a smaller scale, ice can gradually alter the shoreline of a lake as the lake freezes and thaws over many seasons.

**ice floe** See *floe*.

**ice pack** The ice covering, or almost covering, the sea in the polar regions, or on a frozen lake.

**ice point** The freezing point of water, and especially, the temperature at which ice and water are in equilibrium, when the pressure is normal (about 14.7 pounds per square inch or 30 inches of mercury). This is 32 degrees Fahrenheit or 0 degrees centigrade or 273 degrees Kelvin.

**ice sheet** An extensive, nearly flat layer of ice that spreads in all directions from the center of a glacier. Greenland and Antarctica support ice sheets.

**ice storm** A rainstorm that occurs when the surface is below freezing. The rain thus freezes when it lands, giving everything a coat of ice up to several inches thick. This is often beautiful, and just as often destructive.

**iconoscope** A form of television camera tube. See the illustration. See *image orthicon, vidicon*.

ICONOSCOPE

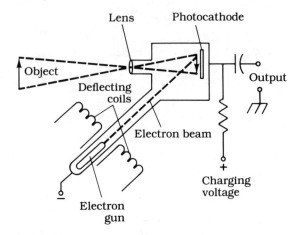

**icosahedron** A polyhedron with 20 faces. More commonly, a regular polyhedron with 20 faces that are all identical.

**ID** Abbreviation for identify, identification, identity.

**i.d.** Abbreviation for inside diameter (see *diameter*).

**ideal crystal** A crystal without any imperfections. It is built up with no flaws at all in its lattice structure.

**ideal gas** A gas that acts perfect; that is, it acts just exactly the way that the laws of physics would predict. Such a gas would have to be made of continuous matter, rather than of particles, or else the molecules would have to be geometric points. Real gases are never quite like this.

**identical twins** Two fetuses or people who have formed from a single egg. They look identical when they are born, and they are always of the same sex. Even when they grow up, they might look much the same.

**identification** 1. The process whereby the exact chemical composition of matter is determined. 2. The process in which a substance is categorized by name. 3. The broadcasting of the call sign of a radio or television station, as required by law.

**identify** 1. Determine the characteristics of. 2. Determine according to name, type, or species. 3. Broadcast the call sign or other assigned name or mark.

**identity** 1. In mathematics, for an operation *, the number i, in the system, such that for any x, x*i=x. For addition, i=0; for multiplication, i=1. 2. A set of characteristics that makes something unique and easy to recognize.

**IEEE** Abbreviation for Institute of Electrical and Electronic Engineers.

**igneous rock** Rock that has solidified from a molten state, such as volcanic lava. See *metamorphic rock, sedimentary rock*.

**ignimbrite** Volcanic deposits, created when hot, glowing solid particles are ejected from within the earth.

**ignition** 1. See *combustion*. 2. The starting of a combustion process, especially with rocket engines. 3. The starting of an internal combustion engine.

**ignition system** In a vehicle propelled by an internal combustion engine (such as a car or truck), the electrical and mechanical devices that provide the sparks in the pistons. The exception is a diesel engine that does not have spark plugs.

**IGY** Abbreviation for International Geophysical Year.

**ileum** The last section of the small intestine before the colon begins. Much of the digestive process occurs here.

**iliac artery** Either of two large arteries branching off from the aorta (main torso artery) and leading into the legs.

**iliac vein** Either of two large veins leading from the legs, and coming together to form the inferior vena cava (main vein in the torso).

**illuminance** The amount of light that lands on a surface, per unit surface area. Measured in lumens per square meter, also called lux. See *lumen, lux*.

**image** 1. The focused rays of light from a convex lens or concave mirror. See *real image*. 2. A scene as observed with the eyes. See *virtual image*. 3. A focused scene on the retina of the eye or the film of a camera. 4. A picture on a TV screen. 5. A pattern on a cathode-ray tube. 6. A picture focused on the mosaic of a TV camera tube. 7. A duplicated item in computer storage. 8. A false signal in a superheterodyne radio receiver.

**image orthicon** A television camera tube, with fast response and high sensitivity. It works well in dim light and can follow fast-changing scenes. It has a rather high noise output, however. See *iconoscope, vidicon*.

**imaginary number** A real-number multiple of the square root of $-1$. The unit imaginary number is called i by mathematicians, and is called j by engineers. It is the number that, when multiplied by itself, gives $-1$. See *complex number*.

**immune system** That system of the body responsible for protecting against disease. Includes antibodies and the hormones that stimulate the body to make them, and the glands responsible for producing the hormones.

**immunity** 1. Resistance to infection. 2. Not being affected by something that normally has an effect on most others. One who has had mumps, for example, does not catch it from someone else who has it; they have developed immunity. 3. Impervious to undesirable or degrading phenomena.

**immunization** A drug or vaccine that protects a person or animal against disease. There are different immunization formulas for different diseases.

**immunoglobin** See *antibody*.

**immunodeficiency** A malfunction of the immune system, so that a person or animal does not have normal resistance to infection and disease. Can be inherited, or be caused by some harmful substance, or be caused by viruses. See *acquired immune deficiency syndrome*.

**immunology** The branch of medicine concerned with the body's immune system and its behavior, and with substances or influences that can interfere with the functioning of this system.

**impact** 1. Collision, or the effect of a collision. See *collision*. 2. Overall effect of. We might speak of the impact of a chemical war, for example. 3. Crash landing. 4. Pack together tightly.

**impedance** An expression of the opposition that something offers to alternating current. Made up of resistance and reactance. The resistance is never negative; reactance can be either positive or negative. Reactance is often multiplied by the square root of $-1$, called the *j operator*, to express impedance as a complex number $R + jX$, where $R$ is the resistance and $X$ is the reactance. See *capacitive reactance, complex number, inductive reactance, j operator, resistance*.

**Imperial System of Units** An old-fashioned system of units based on the foot, pound, and second. Nowadays the Standard International System is preferred. See the STANDARD INTERNATIONAL SYSTEM OF UNITS appendix.

**impetigo** An infection of the skin, most often occurring in infants and children. Caused by bacteria, usually streptococcus or staphylococcus, which are common bacteria. Treated with antibiotics.

**implant** 1. To install in the body, as with an artificial heart valve. 2. A part installed in the body, as defined in (1). 3. See *implantation*.

**implantation** The placing of an egg in the uterus by artificial means, after the egg has been fertilized outside the uterus. Used in humans when a woman is unable to get pregnant by means of normal sexual intercourse.

**implosion** Falling inward, usually because of a lack of pressure within, or an excess of pressure outside. Can also take place under the influence of gravity.

**impotence** In the male, inability to maintain an erection. This makes it difficult or impossible to have normal sexual intercourse.

**imprinting** The tendency for a young animal to regard the first large, moving object it sees as its mother. In this way, researchers have convinced newly born animals that they (the scientists) were the mothers.

**impulse** 1. A short pulse of high voltage or current. 2. The electromagnetic pulse produced by a sudden current pulse. 3. The pulse or signal traveling along a nerve. 4. A force acting for a given length of time. 5. The product of force and time, given in kilogram-meters per second, or newton-seconds. See *force*.

**In** Chemical symbol for indium.

**in-** A prefix meaning not or non.

**inbreeding** Mating between animals or humans very similar to each other. In the case of human beings, this might mean that Northern Europeans would mate only with other Northern Europeans. Over a long time, this tends to weaken the species.

**incandescence** A phenomenon of glowing, as a result of high temperature. There is so much energy in the substance that some of it is given off in the visible-light range of the electromagnetic spectrum.

**incandescent lamp** A device for producing artificial light, by passing electric current

INCANDESCENT LAMP

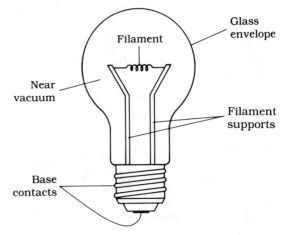

through a resistance wire till the wire glows white hot (see drawing). This was the earliest electric lamp, developed by Thomas Edison in the nineteenth century.

**incisor** One of the front teeth in mammals, responsible for cutting food when the first bite is taken.

**inclination** For a planet, the extent to which its orbit differs from the plane of the earth's orbit. Measured in degrees with respect to the plane of the ecliptic. See *ecliptic*.

**inclined plane** One of the six simple machines. A ramp or sloping horizontal surface. Makes it easier to get objects up, because they can be rolled or dragged, rather than lifted.

**incoherent** 1. Not aligned. See *incoherent light, incoherent radiation*. 2. Unclear; blurred, as in a visual image. 3. Not understandable; unintelligible, as in a radio signal.

**incoherent light** Light in which the wavefronts are not all aligned, and/or are not all of the same frequency. See *coherent light, coherent radiation, incoherent radiation*.

**incoherent radiation** Electromagnetic radiation in which the wavefronts are not all aligned, and/or have different frequencies (see drawing). This is the type of radiation given off by light bulbs, stars, or glowing coals. See *coherent light, coherent radiation, incandescence*.

**Incompleteness Theorem** Proven by Kurt Godel in 1930, when he was 24 years old. A revolutionary theorem in mathematics. It showed that there are statements in logic whose truth cannot be proven nor disproven. This seems to say that our universe is such that we can never know everything about it. See *Godel, Kurt*.

**incomplete protein** Protein that does not contain all of the essential amino acids. Plant proteins are usually incomplete when obtained from only one plant source; however, they can be combined to get complete protein. See *complete protein, essential amino acids*.

**increment** 1. A small change in a measurable quantity. We might say that an ammeter measures current in increments of one milliampere, for example. 2. A small change in a variable. See *variable*.

**incus** One of the three bones in the inner ear, responsible for transferring sound vibrations. It is also called the anvil.

**indefinite integral** See *integral*.

**independent variable** In a mathematical relation, the variable or variables that determine the value of the dependent variable. Graphed as the abscissa on the Cartesian plane. See *abscissa, dependent variable*.

**index of refraction** A measure of the extent to which a transparent substance causes light rays to be bent when entering or leaving at an oblique angle. A vacuum has an index of 1.0; substances such as water and glass have indexes higher than 1.0. Quartz and diamond have very high refractive indexes. See *refraction*.

**indicator** 1. An instrument that displays a specific quantity; a meter. 2. Some easily observed variable that tells us about the operation of a device or about the functioning of the body. 3. A chemical that shows us when a reac-

INCOHERENT RADIATION

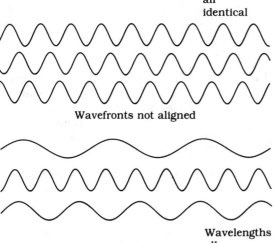

Wavelengths all identical

Wavefronts not aligned

Wavelengths all different

tion has taken place. 4. A chemical that shows us that some substance is present or absent.

**indigenous** Native to a certain region or location. Pertains to any species of plant or animal that naturally inhabits a place.

**indigo** 1. A deep purple color between blue and violet in the visible-light spectrum. 2. A plant dye of a dark purple color. 3. Any synthetic dye of a dark purple color.

**indium** Chemical symbol, In. An element with atomic number 49. The most common isotope has atomic weight 115. It is used to make certain semiconductors for electronic devices. In nature it is often found along with zinc.

**inductance** An electrical effect where energy is stored in the form of a magnetic field. The standard unit is the henry; more often it is given in millihenries (one thousandth of a henry) or microhenries (one millionth of a henry). See *henry*.

**inductive reactance** The opposition that an inductance offers to alternating current. Given in units of ohms (see *ohm*). In general, the higher the inductance, the higher the inductive reactance. If L is the inductance in henries, and f is the frequency in hertz, then the inductive reactance X is given by the formula $X = 6.28 fL$. Engineers consider inductive reactance to be positive. See *capacitance, capacitive reactance, impedance, inductance*.

**induction** 1. See *electromagnetic induction*. 2. See *inductive logic*. 3. In mathematics, a method of proof that something is true for a countably infinite (denumerable) set of numbers or objects. We assign each object a natural number in the set $N = \{0, 1, 2, 3, ...\}$. Then we prove the theorem for all n in the set N. This is done by first showing it true for $n = 0$, and then showing that if it is true for some n, it is true for $n + 1$.

**induction coil** A coil that generates a large voltage by storing up energy over a period of time in the form of a magnetic field, and then releasing this energy all at once. This is how the spark coil in a car engine works, for example.

**inductive logic** A method of showing that something is true most of the time, or is probably true in any given instance, but is not necessarily true all the time. It is more a philosophical argument than a scientific method.

**inductor** An electronic device that has inductive reactance. Usually a coil of wire, but also can be a length of transmission line, or an integrated circuit wired in a certain way. See *inductance, inductive reactance*.

**industrial engineering** A branch of engineering concerned with the efficient design and operation of factories.

**Industrial Revolution** A change in society from agriculture to industry, and especially the construction of buildings and heavy machines. Took place in the eighteenth and nineteenth centuries throughout Europe, and in the nineteenth century in America. Nowadays a transition is taking place from industry to information processing. See *Information Age*.

**inelastic collision** See *collision*.

**inertia** The tendency of a mass at rest to remain at rest, and for a mass in nonaccelerating motion to maintain that state of motion, unless acted upon by an outside force. See *Newton's Laws of Motion*.

**inertial frame of reference** A point of view in which the observer is not accelerating. This means that the observer feels zero gravitational force. See the drawing.

**inertial guidance system** A method of keeping a missile on its assigned path. Deviations from the assigned path are detected by sensitive instruments that can sense the forces on masses, caused by inertia in case the missile goes off course. See *inertia*.

**inertial navigation** A method of keeping an aircraft or space vessel on course, by means of devices that sense the forces on masses, caused by inertia. Computers aid in the exact course corrections to be made. This is especially important on spacecraft. See *inertia*.

INERTIAL FRAME OF REFERENCE

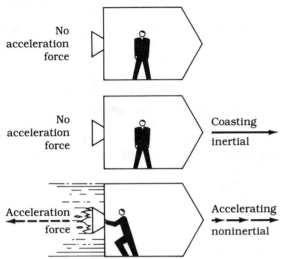

INFERIOR CONJUNCTION

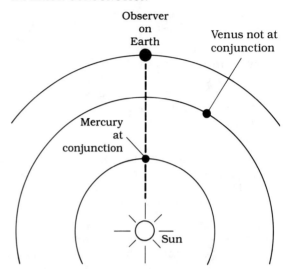

**inexact science** A science in which considerable error is to be expected, and occasional false or nonproductive results will take place. Medicine is a good example of this. See *exact science*.

**infarction** Death of body tissue, as a result of the blood circulation being cut off by a clot in an artery. Most often this term is used in reference to a heart attack. See *heart attack*.

**infection** Multiplication of harmful bacteria or viruses, with resulting damage to living tissues. The effect can be anything from a boil or abscess to a fatal disease.

**infectious disease** Any disease caused by infection. See *infection*.

**inferior** 1. Having the lower of two values, or the lowest value in a group. 2. Within; inside of. 3. Subordinate to; dependent on. 4. Of low quality or poor workmanship.

**inferior conjunction** For the planet Mercury or the planet Venus, the crossing of that planet in front of the sun, or the time when it is most nearly on a line between the earth and the sun (see drawing). See also *opposition, superior conjunction*.

**inferior vena cava** The main vein leading up the lower torso, having come from the two large veins in either leg (the iliac veins). Leads back to the heart, where blood is then pumped to the lungs to be replenished with oxygen.

**infertility** 1. Sterility in the female, and especially, lack of ability to produce eggs. 2. In agriculture, a lack of needed nutrients and minerals in the soil, or an excessively acid or alkaline soil, so that plants grow poorly.

**infiltration** The soaking of water into the ground through the soil. This occurs more easily in some soils than in others, and is an important factor, along with rainfall, in the water supply for a given region. It is also important in agriculture.

**infinite** 1. Having no end; continuing on forever. We might speak of an infinite sequence, for example. 2. Larger than any finite quantity. See *finite*. 3. Pertaining to an infinite ordinal or transfinite cardinal. See *infinite ordinal, transfinite cardinal*.

**infinite number** 1. See *infinite ordinal*. 2. See *transfinite cardinal*.

**infinite ordinal** A number that is larger than any finite number. The smallest infinite ordinal is the number you would get if you could count all the way to the end of the set of natural numbers. There are infinitely many larger ordinals. Infinite quantities are more often defined in terms of cardinals; see *transfinite cardinal*.

**infinite process** A process that goes on forever, occurring in steps that can be denoted, but that never come to an end. Many approximations are of this nature; the more steps that are carried out, the greater the accuracy, but the result is never exact. An example is the extraction of the square root of 2.

**infinite sequence** A list or set of numbers that can be denoted, or denumerated, but cannot all be listed. The best example is the set of natural or counting numbers $N = \{0, 1, 2, 3, ...\}$.

**infinite series** The sum of an infinite sequence. This can be a finite number, or it can be infinite. The sum of the sequence of natural numbers is infinite. But the sum $S = 1/2 + 1/4 + 1/8 + 1/16 + ... = 1$. See *infinite sequence*.

**infinite set** A set having an infinite number of elements. Such a set can be denumerably infinite, meaning that we can at least begin to list the elements. An example is the set $N = \{0, 1, 2, 3, ...\}$ of natural or counting numbers. Or an infinite set can be nondenumerable. The real and complex numbers comprise sets of this type.

**infinitesimal** 1. Of extremely small value, practically equal to zero. 2. An arbitrarily small quantity that can be made smaller and smaller for certain mathematical purposes (see drawing). 3. Any of various quantities that appear equal to zero in simple mathematics, but that are actually different from each other in the same sense that infinite ordinals or transfinite cardinals differ. This branch of mathematics remains largely unexplored. See *infinite ordinal, transfinite cardinal*.

**infinity** 1. Any number or quantity larger than any finite number. 2. See *infinite ordinal*. 3. See *transfinite cardinal*.

**influenza** A virus-caused illness, like a bad cold. Symptoms include aching joints, stuffy nose, sore throat, and fever. More severe illness might involve the lungs. In elderly or weak people, this disease can progress to pneumonia and become dangerous. See *pneumonia*.

**information** 1. Any material that can be written, or encoded, that tells us specific things about nature, people, or processes. 2. The transmission of material as defined in (1). 3. The storage of material as defined in (1).

**Information Age** The ongoing change in society from an industrial base to an information-processing base of employment. This has been largely the result of the development of high-speed computers with huge memory capacity and the ability to transfer data at high speed. See *Industrial Revolution*.

**information processing** See *data processing*.

**information theory** A branch of statistics and probability theory. Concerned with the processing and utilization of information to obtain certain results. It is used in the design of computers, communications systems, and even in the biological sciences and in economics.

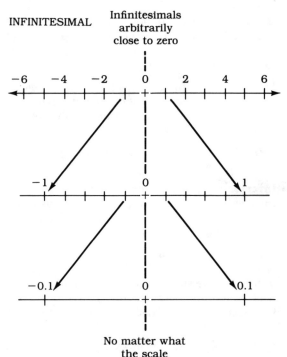

INFINITESIMAL Infinitesimals arbitrarily close to zero. No matter what the scale

**infrared** Electromagnetic radiation with wavelengths just longer than those of visible light. Near infrared extends from about 7500 Angstroms to 15,000 Angstroms. Middle infrared is from 15,000 to 60,000 Angstroms. Far infrared is from 60,000 to 400,000 Angstroms. Far-far infrared goes from 400,000 Angstroms to 1 millimeter. By comparison, visible light extends from about 3900 Angstroms (violet) to 7500 Angstroms (red). See *Angstrom unit* and the ELECTROMAGNETIC SPECTRUM appendix.

**infrared astronomy** The observation of space at infrared wavelengths. Special camera film is used with ordinary telescopes at near infrared. Special instruments are needed at longer infrared wavelengths. See *infrared*.

**infrared-emitting diode** An electronic device similar to a light-emitting diode, but that produces energy at infrared wavelengths. See *infrared, light-emitting diode*.

**infrared photography** The making of picture images from infrared, rather than from visible light. Ordinary cameras will work for near infrared (see *infrared*), but special film is needed, and the focusing is a little different.

**infrasound** Acoustic vibrations at frequencies lower than the lowest audible. This is from zero to about 20 hertz (cycles per second). In air, these acoustic waves have very long wavelength. When intense, the "rumbling" and vibrating can cause damage to buildings and can even shatter windows.

**in-line engine** An internal combustion engine, with all the cylinders in a line, one right behind the other. See *internal combustion engine*.

**innate** 1. Inherited; occurring naturally in an animal. Pertaining to a trait that does not need to be learned. 2. Instinctive. See *instinct*.

**inner ear** The part of the ear from which the auditory nerve leads directly to the brain. The cochlea and the semicircular canals are contained in this part of the ear. See *ear*.

**innervation** The nerves for a certain body part. These nerves send impulses for control, and send back impulses for sensations. See *nerve, nerve ending, nerve fiber*.

**inorganic** Pertaining to substances that are not living in any sense. These are the plain elements and compounds, such as sodium or carbon dioxide. Many mixtures and solutions are also included. See *organic*.

**inorganic chemistry** The branch of chemistry concerned only with elements, compounds, and reactions not involving biological processes in any way. See *biochemistry, organic chemistry*.

**inositol** A water-soluble nutrient, sometimes considered part of the B complex of vitamins. Assists in nerve function and metabolism. See *vitamins* and the VITAMINS AND MINERALS appendix.

**input** 1. The signal, voltage, or current supplied to an electronic circuit for processing. 2. The terminals in a circuit into which a signal, voltage, or current is fed in order to be processed. 3. To feed data or instructions into a computer.

**input/output module** A ready-made electronic circuit used to connect computers to their external devices such as keyboards, monitors, or robots. They are made in standardized, rectangular cases. See *interface*.

**insect** A large class of cold-blooded animal. They have exoskeletons and usually they have

INSECT

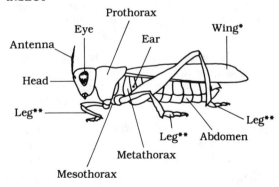

*Some insects have no wings

**Left leg only is shown

six legs. Some can fly. They are among the oldest animals on earth, and also are some of the hardiest, enduring all climates and catastrophes. A typical insect is shown in the drawing. Insects are important in the natural balance, although they are often regarded as pests, and sometimes carry harmful bacteria. For the place of the insects in evolution, see the ANIMAL CLASSIFICATION appendix.

**insecticide** A chemical that kills insects and other animals commonly thought of as "bugs" or "pests," such as spiders. Can be used for agriculture to keep certain harmful insects from breeding and eating the plants. Also can be used in buildings to keep pests from multiplying.

**inside diameter** See *diameter*.

**insolation** The rate of energy received from the sun over a unit surface area. Usually given in watts per square meter. This quantity is greatest when the sun shines straight down, and gets less and less as the angle of the sun decreases relative to the horizon.

**insomnia** Difficulty sleeping. Can be caused by various physical and psychological factors. Often the cause is drinking too much coffee and tea. Certain drugs also can cause this problem. It also can be caused by anxiety and by certain disease conditions.

**instant** 1. A very short period of time, approaching zero duration. 2. Taking place within a short interval of time.

**instantaneous** 1. Happening in an extremely short time. 2. Measured or determined over an arbitrarily small interval of time.

**instantaneous quantity** The value of a physical quantity as determined over a small interval of time. We might speak of instantaneous velocity, for example, in an accelerating car, as being 40 miles per hour at some point while going from zero to 60 miles per hour in 10 seconds. Often these quantities are found using differential calculus. See *calculus, derivative*.

**instinct** A form of behavior that is not learned, but that an animal is born with. For example, certain birds fly south in the winter and back north again in the summer. Bears hibernate in winter. Salmon go upstream to spawn. There are countless other examples. This behavior is programmed into the genes themselves for a particular species.

**Institute of Electrical and Electronic Engineers** Abbreviation, IEEE. A professional organization in the United States and in some other countries, devoted to advancing the state of the art in electronics.

**instrument** 1. An indicator or meter that displays a physical quantity, especially for control of a machine. An example is an altimeter in an aircraft, or a speedometer in a car. 2. A tool. 3. A surgical device.

**instrumentation** The set of instruments used in controlling a machine and in monitoring its operation. See *instrument*.

**instrumentation engineering** A branch of engineering involving the design of convenient, easy to use, and accurate instruments, especially for use with machines such as cars and aircraft. See *instrument, instrumentation*.

**insulator** 1. A material that is a poor conductor of electric current. 2. A substance that is a poor conductor of heat. 3. A device intended to keep electric current from flowing between two points. 4. A material used to keep buildings at a constant temperature regardless of the outside weather, and to reduce energy losses in heating or cooling.

**insulin** A hormone secreted by the pancreas that regulates the metabolism of glucose in the bloodstream. Sometimes this must be injected, if a person does not produce enough of it. See *diabetes mellitus, glucose, pancreas*.

**insulin shock** The effects of getting too much insulin, especially for a diabetic. The level of blood glucose falls and the person becomes shaky and dizzy, and might faint. In severe cases, death can occur. See *insulin*.

**integer** Any member of the set N of counting numbers {0, 1, 2, 3, ...} and their negatives. Generally, the set of integers is written Z = {..., −3, −2, −1, 0, 1, 2, 3, ...}.

**integral** 1. Definite integral: The area under a curve of a function, over a defined part of the domain. See the drawing in which the integral of function f is shown for the interval x = a to x = b. 2. Indefinite integral: Also sometimes called the antiderivative. A function that gives the original function when differentiated. Can be defined over an interval to obtain the definite integral. See *calculus, derivative*.

INTEGRAL

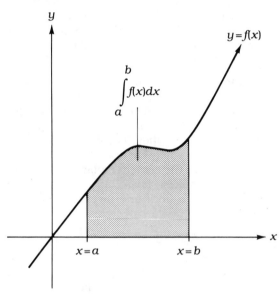

**integral calculus** See *calculus*.

**integrated circuit** Abbreviation, IC. A complete electronic circuit, factory-made to perform a certain function, etched onto a wafer of silicon and packaged in a standard case. Many electronic devices today consist mostly of ICs.

**integration** 1. The process of finding the integral of a function, either in general or over a certain interval of the domain of the function. See *integral*. 2. The process of making integrated circuits. See *integrated circuit*.

**integrator circuit** A device that produces the integral of a waveform. For example, a sine wave is shifted by 90 degrees to get what might be called a negative cosine wave. Basically the opposite of a differentiator circuit. See *differentiator circuit*.

**intelligence** 1. Information or data. 2. The ability to reason, and to learn from mistakes.

**intelligence quotient** Abbreviation, IQ. A measure of mathematical or verbal intelligence level. For a person who is x years old, who tests at an intelligence level of y years, the IQ is 100(y/x). If you are 13 and test as being 15 years old, for example, your IQ is $100(15/13) = 115$.

**intelligibility** In communications, the ease with which a signal, especially a voice signal, is understood. Can be given as a percentage of accurately received words or characters.

**Intelsat** Acronym for International Telecommunications Satellite Consortium. An organization for cooperation among nations in the launching and use of communications satellites. Formed in the middle 1960s.

**intensity** Relative or absolute level or strength, for a variable phenomenon such as sound, radio signals, ionizing radiation, electric fields, magnetic fields, or visible light. There are specific units of level for each phenomenon.

**intensity scale** A means of expressing the level of some phenomenon, such as sound, light, or signal strength. Often given in decibels (see *decibel*) for sound and for radio signals. More often given in lumens for visible light (see *lumen*).

**inter-** A prefix meaning between or among. Sometimes confused with intra-.

**interaction** 1. For two or more different effects, a tendency for a change in one to cause a change in some other or others. 2. A chemical reaction.

**interchangeability** In a machine, the ease with which one part can be replaced, either with an identical part or with a similar part.

**intercontinental ballistic missile** Abbreviation, ICBM. A rocket with one or more nuclear warheads attached, and capable of traveling thousands of miles over a preset course and accurately deliver the bombs to their targets.

**interface** 1. To provide a link between electronic devices that use different signal modes. 2. To make it possible for one type of circuit to work with accessories or with other circuits in a network. 3. A device that converts one type of signal to another to provide a link between communications circuits. 4. A boundary between two substances or media.

**interference** 1. In communications, any signal or noise that hinders the reception of a desired signal. 2. The interaction between or among multiple sources of wave disturbances. See *interference pattern*.

**interference pattern** 1. A pattern of lines or "fringes," alternating between bright and dark, that forms as a result of phase interaction between different light sources (see drawing). 2. A pattern of high-intensity and low-intensity bands or zones, caused by phase interaction between different sources of waves.

INTERFERENCE PATTERN

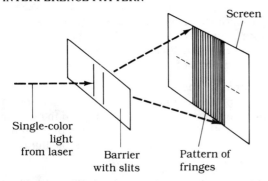

**interferometer** A radio telescope that uses multiple antennas to obtain a very sharp directional response. This allows radio astronomers to see the sky in great detail at radio wavelengths. See *radio astronomy, radio telescope*.

**interferon** A substance thought to inhibit the action of viruses. It might be a key to the development of a cure for cancer, as well as a vaccine against the common cold and flu.

**intergalactic** Pertaining to space, or to space travel, between or among galaxies. This involves distances of millions or billions of light years.

**interglacial** Pertaining to warm periods between ice ages. Most scientists agree that we are now in such a period. See *ice age*.

**intermediate frequency** Abbreviation, IF. In a superheterodyne radio receiver, the frequency obtained by mixing the incoming signal with the output of an oscillator. The resulting IF has a fixed frequency, so it is easy to process. See *superheterodyne radio receiver*.

**intermetallic** Pertaining to compounds of metals in an alloy. An example is aluminum hydroxide and magnesium hydroxide, commonly combined for use as an antacid.

**intermittent** 1. Taking place in an "on-again, off-again" manner. 2. Occasional. 3. An electronic component or circuit that fails some of the time, but not all of the time.

**intermodulation** The mixing of two or more radio signals, producing false or undesired signals at various frequencies. The drawing shows a spectral display of this effect. You might have noticed this effect when listening to a portable radio in a place where there are many transmitters, such as in the center of a large city. When severe, this causes "hash" all over the whole range of a radio unit, and reception is nearly useless.

INTERMODULATION

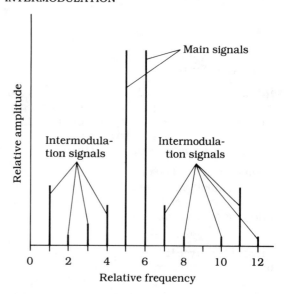

**internal combustion engine** A machine that converts chemical energy into mechanical energy or work. Works by burning fossil fuels or other flammable liquids or gases. Little explosions drive pistons outward in cylinders. These movements are harnessed mechanically to create rotational motion to drive wheels or propellers, or other mechanical devices.

**internal energy** The sum of the kinetic and potential energy of particles in a system. It is a theoretical quantity and cannot be directly measured. See *kinetic energy, potential energy*.

**internal resistance** For an electric or electronic device, the opposition that it offers to electric current. Given in ohms (see *ohm*), it is equivalent to the applied voltage divided by the current that flows through the device.

**International Date Line** A line running approximately along the meridian 180 degrees, opposite the prime meridian (see *prime meridian*). The line is a little crooked to avoid cutting through eastern Siberia (Russia) and certain Pacific islands. When you cross this line going west, you go ahead in time by one day; when you cross it going east, you go back in time by one day.

**International Geophysical Year** Abbreviation, IGY. A year-long period in which scientists from all over the world cooperated in research to learn more about our planet. Ran through the latter part of 1957 and into 1958, during a sunspot-cycle peak.

**International Morse code** See *Morse code*.

**International Telecommunication Union** Abbreviation, ITU. The worldwide organization responsible for setting standards for electromagnetic communication. This includes radio and television frequency assignments. Headquarters are in Geneva, Switzerland.

**International Unit** Abbreviation, IU. A unit of vitamin measure. Varies depending on the particular vitamin. Sometimes equivalent to a milligram (0.001 gram). See the RECOMMENDED DAILY ALLOWANCES and VITAMINS AND MINERALS appendices.

**interneuron** A connecting neuron (see *neuron*) that carries impulses throughout the body. There are many of these in an advanced animal or human connecting the brain, spinal cord, and body organs. See *central nervous system*.

**interplanetary** Pertaining to space or space travel between or among the planets. Distances range from 250,000 miles (the distance to the moon) up to the diameter of the whole Solar System, or about 7.4 billion (7,400,000,000) miles.

**interpolation** A process of "filling in" unknown or unmeasured data, either by educated guessing or by means of computers. An example is shown in the drawing. Although it is possible that a gigantic "spike" occurs in the gap, it is highly unlikely, so we can be fairly sure that the illustrated interpretation is correct. See *extrapolation*.

INTERPOLATION

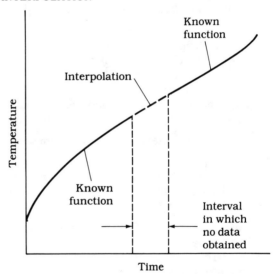

**interpreter** A computer program that translates commands from the operator into machine language. See *machine language*.

**interrogation** A question or set of questions, especially in computerized systems for the purpose of troubleshooting or routine checking of a system.

**interrogator** A device in air navigation. It "asks" a ground-based station a "question." The ground station "answers," and the time delay gives the distance between the aircraft and the station. Using several different ground stations, the plane's position is determined.

**intersex** 1. Having characteristics of both male and female, more or less equally. 2. An animal or human with hormone imbalances that give it characteristics of both sexes.

**interstellar** Pertaining to space or space travel between or among the stars. Distances range from a few light years up to 100,000 light years, the diameter of our galaxy.

**intertropical convergence zone** Abbreviation, ITCZ. The belt near the equator where the prevailing trade winds come together. It is a semipermanent low-pressure region with showers and thunderstorms much of the time. It moves slightly north of the equator in the northern summer, and slightly south of the equator (in most regions) during the northern winter.

**intestine** The part of the digestive system after the stomach. In humans this is a tubular organ about 20 feet long, ending with the colon and rectum. Nutrients are absorbed along its length. Food normally takes between 18 and 36 hours to completely move through the intestine.

**intra-** A prefix meaning within. Sometimes confused with the prefix inter-.

**intracellular** Pertaining to processes that take place within individual cells, or to components of a single cell.

**intrauterine device** A contraceptive device that a woman can place in her vagina. It prevents sperm from reaching the egg about nine times out of ten. Can be used along with other forms of contraception to minimize the chances of unwanted pregnancy.

**intravenous** Abbreviation, IV. Injected into the blood vessels. This is the fastest known way to administer medicine and nutrients when they cannot be given by mouth.

**intravenous feeding** Administration of essential nutrients, such as glucose, vitamins, and minerals by injecting directly into a vein. Used when a person cannot eat or cannot absorb food consumed by mouth.

**intrinsic** 1. Pure, especially in reference to semiconductor materials. 2. In semiconductors, having properties just like a pure material. Such semiconductors are almost insulators. See *extrinsic, semiconductor*. 3. Originating within an object or organism.

**intrusion** 1. A flow of magma into other rock or rock strata. See *magma*. 2. The movement of an undesirable substance, or of an impurity, into a medium. An example is the seeping of salt water into the drinking supply during and just after a coastal hurricane.

**intrusive** 1. Tending to flow into other substances. 2. Pertaining to rock that has flowed into other rock formations when it was molten, forming granite.

**Intuitionism** A school of mathematical thought, begun mainly by L. Brouwer around the beginning of the twentieth century. According to this philosophy, if we say an object exists, it means that the object can be mathematically constructed in a finite number of steps. See *Formalism, Logicism*.

**Invar** An alloy of iron, nickel and carbon that does not expand or contract very much when the temperature changes.

**invariant** 1. Never changing; having the same properties or values all the time and under all conditions. 2. In mathematics, something that does not change with certain operations.

**inverse** 1. Flowing in a direction opposite to normal or forward. 2. Pertaining to a function that "undoes" a given function. See *function, inverse function*. 3. The logical opposite. The inverse of high is low; the inverse of true is false. 4. For a number x, the value $-x$ (for addition), such that $x+(-x)=0$. 5. For a number y, the value $1/y$ (for multiplication), such that $y(1/y)=1$. 6. For a mathematical operation $*$, for a value z, that number $z'$ such that $z*z'$ gives the identity element.

**inverse function** A mathematical function that "undoes" a given function. For example, the square root is the inverse of the square function for nonnegative real numbers (see drawing). The function $y = x$ is its own inverse. In general, if we have a function f such that $f(x) = y$ for certain values of x, then the inverse is written $f^{-1}(y) = x$. Sometimes the inverse of a function is a relation, but not a function. See *function*.

INVERSE FUNCTION

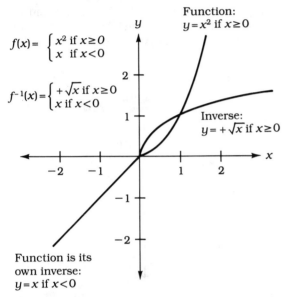

Function is its own inverse:
$y = x$ if $x < 0$

**inverse-square law** A rule concerning intensity of many types of radiation. If you multiply the distance from a point source by n, the intensity becomes less by a factor of n squared. An example is shown in the drawing for $n = 2$. If a source is very far away, it can be considered as a point source.

**inversion** 1. The process of turning upside-down, as in the image focused by a convex lens. 2. The action of an inverter circuit. See *inverter*. 3. A weather phenomenon in which the air temperature rises, rather than falls, with increasing altitude in a layer. This often causes pollutants to be trapped near the surface of the earth.

**invertebrate** 1. An animal having no backbone; these are primitive species. 2. Pertaining to an animal without a backbone. 3. An animal with an exoskeleton, such as an insect or shellfish. See *exoskeleton*.

**inverter** A logical circuit that negates a pulse. That is, it converts high to low, and vice-versa. Also called a NOT gate. See *logic gate*.

**in vitro** Pertaining to test-tube procedures, especially in the laboratory. We might speak of *in vitro* fertilization of an egg, resulting in a so-called "test-tube baby." See *in vivo*.

**in vivo** Pertaining to events within an organism, such as the body. Normally, this is how fertilization of an egg occurs, for example. See *in vitro*.

**involuntary** 1. Subconscious; not done or controlled by a conscious act of will. 2. Pertaining to a muscle that is not consciously moved or controlled. An example is the contractions of the stomach after eating a meal.

**involute** Spiraling inwards, or a curve that spirals inwards.

**iodide** Any compound containing iodine. Most often, the sodium iodide or potassium iodide that is added to table salt. See *iodized salt*.

INVERSE-SQUARE LAW

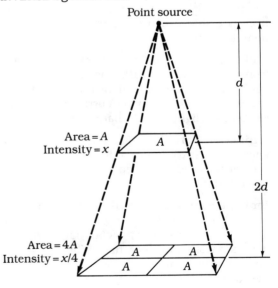

**iodine** Chemical symbol, I. An element with atomic number 53. The most common isotope has atomic weight 127. In pure form it is a metal and a poison. It is one of the halogens (see *halogen*). In tiny amounts it is a necessary nutrient. It serves as a germicide for minor cuts and scratches, although it stains fabrics.

**iodized salt** Table salt (sodium chloride) or a salt substitute (such as potassium chloride), to which a small amount of sodium iodide or potassium iodide has been added. This ensures that people using the salt get the iodine they need, preventing goiter. See *goiter, goiter belt*.

**ion** An atom with extra electrons or with a deficiency of electrons, resulting in its being electrically charged. An ion with extra electrons is negatively charged and is called an anion; an ion deficient in electrons is positively charged and is called a cation.

**ion engine** A device for obtaining propulsion in outer space. Not fully developed yet. It would accelerate ions (see *ion*) in a strong electric field. The ejected ions would serve to create the force for propulsion.

**ionic bond** An electrical attraction that holds elements together in certain compounds. Sodium chloride (salt) is a good example of a compound held together by an ionic bond. Electrons are shared between atoms of such compounds. See *ion*.

**ionization** The addition or removal of electrons from atoms, creating ions. This occurs with powerful electromagnetic radiation. It also can occur when a large current or voltage exists, as in lightning discharges. See *ion*.

**ionization potential** For a given atom, the energy necessary to ionize it; generally, this is the energy needed to strip the atom of at least one electron.

**ionizing radiation** Radiation commonly thought of as radioactivity, including high-speed alpha particles, beta particles and gamma rays, as well as X rays. When such rays strike matter, some of the atoms are ionized. See *alpha rays, beta rays, gamma rays, X rays*.

**ionosphere** The upper atmosphere of the earth, beginning at an altitude of about 35 miles. The thin air is ionized by ultraviolet from the sun. There are several layers, as shown in the drawing: the D layer (about 35–55 miles up), the E layer (about 60–70 miles up), the F1 layer (about 100 miles up) and the F2 layer (at an altitude of 260–280 miles). The D layer vanishes at night. The E layer usually does, too, but not always. The F1 and F2 layers merge into a single layer, called the F layer, about 200–250 miles up, during the night. The E, F1 and F2 layers are mainly responsible for radio communications at shortwave and medium frequencies.

IONOSPHERE

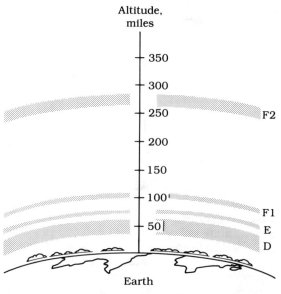

**ion pair** Two elements that are ionized equally and in opposite sense, so that they will readily combine to form a compound with an ionic bond. See *ion, ionic bond*.

**ipecac** A chemical used to cause vomiting. Might be used in certain cases of poisoning. However, there are some poisons for which ipecac should not be used. If the poison is known, instructions can be found on the container or by dialing 911. All cases of poisoning should be referred to a hospital emergency room as soon as possible.

**IQ** Abbreviation for intelligence quotient.

**iridescence** The effect produced when light shines through a very thin, transparent membrane. Some wavelengths of light get canceled out because of internal reflection. The result is a multicolored sheen. You might have noticed this when blowing soap bubbles.

**iridium** 1. Chemical symbol, Ir. An element with atomic number 77. The most common isotope has atomic weight 193. It is a shiny, heavy metal in its pure form. Often found naturally along with platinum. 2. An isotope of platinum that is abundant in some meteorites. It has been a tool for researchers who believe that mass extinctions, such as that of the dinosaurs, have been caused by meteorite bombardments.

**iris** The colored part of the eye, that dilates or constricts the pupil to adjust to dim or bright light. See *eye*.

**iron** Chemical symbol, Fe. An element with atomic number 26. The most common isotope has atomic weight 56. It is a gray, dull metal in its pure form. It is used in many industrial applications. Along with nickel, it is easily magnetized, and is used to make permanent magnets.

**iron chloride** A compound formed from iron and chlorine. The two elements react readily together, resulting in a yellowish or brownish substance with consistency similar to rust. This substance is found along with rust in untreated iron exposed to environments near the ocean, because of the chlorine that is a part of salt.

**iron ore** Usually iron oxide, found in rocks in various places on the earth. This ore is refined to obtain pure iron. High-grade ore is called hematite; low-grade ore is called taconite. To be minable, ore must be at least 25 percent iron. See *hematite, iron, taconite*.

**iron oxide** A compound formed from iron and oxygen. More often called rust. A reddish to maroon-colored to black, crumbly solid, it occurs inevitably when iron is exposed to the air. Rusting happens more quickly in moist environments. Iron can be protected against rusting by galvanizing. See *galvanizing*.

**irradiance** The radiant power that strikes a surface per unit area. Measured in watts per square meter. It can be measured for any type of electromagnetic radiation. In the visible-light range it is called illuminance. See *illuminance*.

**irradiated fuel** A form of nuclear waste. Also called spent fuel. It consists of uranium rods from the cores of nuclear reactors. These remain radioactive for thousands of years even after they are no longer useful in the reactor. This material has a high content of plutonium, a deadly radioactive element.

**irradiation** Deliberate exposure of materials to ultraviolet, X rays or ionizing radiation. Food can be irradiated to enhance its nutritional value. Sometimes the process is used for killing germs.

**irrational number** A real number that cannot be expressed as a ratio of two integers. When written in decimal form, such numbers are nonterminating and nonrepeating. See *rational number, real number*.

**irregular** 1. Not having any orderly pattern. 2. An exception to the rule. 3. Not having a defined shape.

**irregular galaxy** A galaxy that has no defined shape, but is more or less just a blob of stars. See *galaxy*.

**irreversible** Pertaining to a process that can go in only one direction. It cannot be "undone." An example is the burning of wood.

**irrigation** The use of groundwater, or water from lakes, reservoirs, or rivers, for the purpose of farming in regions where rainfall alone is not sufficient. See the illustration.

**irritability** 1. The relative sensitivity of an organism to environmental stimuli. 2. Increased sensitivity to pain and other unpleasant sensations.

**ischemia** A lack of blood to a certain part of the body, resulting in malnutrition and oxygen starvation of the tissues there.

**isentropic** Pertaining to a physical process that takes place without flow of heat energy from one place to another. See *entropy*.

IRRIGATION

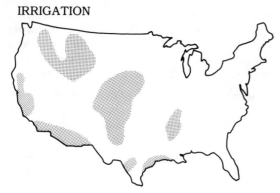

Shaded regions show heavily irrigated areas in the U.S.

From Erickson, EXPLORING EARTH FROM SPACE (TAB Books, Inc., 1989).

**island arc** A chain of islands caused by the crust on the ocean floor plunging underneath a continental plate (subduction), with resulting volcanic activity. Trenches are often found near these chains of islands. See *subduction*.

**Ismail, Abdel Aziz** An Egyptian biological scientist who has worked on finding treatments for ancient and modern diseases. Parasitic infections have been troublesome in Africa for centuries; atherosclerosis is just beginning to appear. His research involves both of these diseases, and others.

**iso-** A prefix meaning equal or equivalent.

**isobar** A line on a weather map denoting equal barometric pressures. An example is shown in the illustration. These lines denote systems of high and low pressure in the atmosphere.

**isochrone** Pertaining to the swing of an ideal pendulum, having the same period of cycle no matter how far back and forth it swings. This is what makes it possible to run a clock with a pendulum.

**isocline** A line on a map, representing points on the earth, all of which have the same angle of geomagnetic flux with respect to the surface. That is, the flux lines are not exactly horizontal, but are slanted by the same number of degrees everywhere on this line. See *geomagnetic field*.

**isodynamic line** A line on a map denoting points where the intensity of the geomagnetic field is the same. See *geomagnetic field*.

**isogonal line** A line on a map denoting points where the geomagnetic declination is the same. At all these points, a compass needle deviates from true North by the same amount and in the same direction. See *declination, geomagnetic field*.

**isoleucine** An essential amino acid. It must be obtained from protein in the diet, because the human body cannot make it from other amino acids. See *amino acid, essential amino acids*.

**isomer** 1. A substance with a unique arrangement of the atoms. Some substances, such as carbon, can exist in different arrangements that make a difference in how they appear (black powder or diamond in the case of carbon). 2. An atomic nucleus with a defined energy state.

**isometric** Having the same measure. Especially pertains to linear dimension, such as the lengths of lines.

ISOBAR

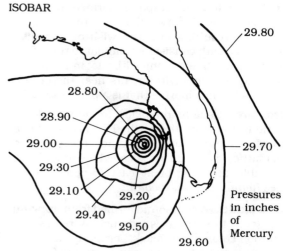

From Gibilisco, VIOLENT WEATHER: HURRICANES, TORNADOES AND STORMS (TAB Books, Inc., 1984).

**isometric drawing** A form of drawing in which equal distances have equal measures. Vertical lines are drawn vertically; horizontal lines recede to the right or the left at angles of 30 degrees (usually). See the illustration. See *oblique drawing, perspective.*

ISOMETRIC DRAWING

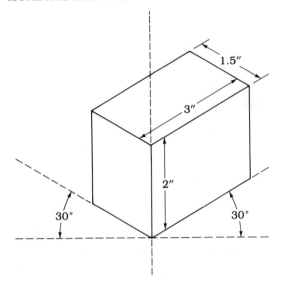

ISOTHERM

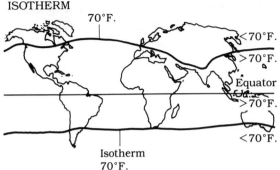

Isotherm 70°F.

**isomorphism** 1. Having the same general structure. 2. A relationship between or among objects, showing that they have the same structure. 3. A resemblance among different animals, caused by evolution. 4. A one-to-one correspondence between sets.

**isomorphism law** A rule in chemistry, that materials with similar or identical crystal lattice shapes are much alike chemically. See *crystal, crystal lattice.*

**isostasy** A tendency for the earth's crust to balance itself. Erosion, for example, reduces the weight of the crust, and as a result, it rises. Much of this crustal buoyancy is caused by deep roots into the mantle. See *mantle.*

**isotherm** A line on a weather map, denoting equal temperatures. An example is shown in the illustration.

**isothermal** Having the same temperature.

**isotonic** For a solution, the property of having identical osmotic pressure with one or more other solutions.

**isotope** For an element, the existence of a certain number of neutrons in the nucleus. The number of protons (atomic number) is always the same for a given element. But the number of neutrons can vary, resulting in various isotopes. Some isotopes can be radioactive for a certain element, while others are not. The isotope for an element is given by the sum of the numbers of protons and neutrons, after the name of the element; for example, carbon-12 (not radioactive) or carbon-14 (radioactive). See *atomic number, atomic weight.*

**isotropic** Having the property of being the same in all directions, or the same as seen from all directions. A sphere is an example of an object of this type, with respect to its center point.

**iteration** A single repetition in a process where the same thing is done over and over. For example, we might repeatedly multiply a number by 0.5, getting a sequence such as 16, 8, 4, 2, 1, 1/2, 1/4, ... . Each multiplication by 0.5 is one iteration. See *iterative process.*

**iterative process** Any process that can be broken down into one or more sets of steps that are all the same, and are repeated many times. See *iteration* for an example. Computers can perform such processes millions or even billions of times.

**ITU** Abbreviation for International Telecommunication Union.

**IU** Abbreviation for International Unit.

**IUD** Abbreviation for intrauterine device.

**IV** Abbreviation for intravenous.

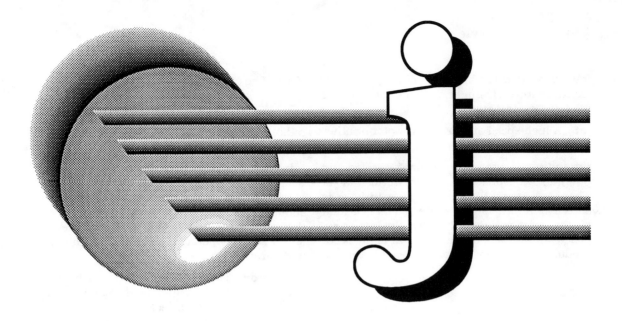

**J** Abbreviation for joule.

**j** Engineering symbol for the square root of −1. See *imaginary number, j operator*.

**jade** A hard, usually green stone regarded as semiprecious. It is a compound of aluminum, sodium, silicon, and oxygen. It is often carved into statuettes.

**jamming** 1. Deliberate interference to a radio signal or signals to prevent communication. 2. Deliberate interference to radar to cause false echoes. See *radar*.

**Jansky, Karl** A pioneer in radio astronomy. First noticed radio noise coming from the center of our galaxy, at a frequency of about 21 MHz or a wavelength of 15 meters, early in the twentieth century. He did not pursue radio astronomy further. See *radio astronomy, radio telescope*.

**jargon** Also called lingo. The language used in a certain field of science or engineering. For example, we have computer lingo, and amateur-radio lingo.

**jasper** Quartz in microcrystalline form with certain impurities, producing various colors. Used as a gemstone. See *quartz*.

**jaundice** The deposition of bile in the skin and mucous membranes, resulting from illness of the liver. See *bile*.

**jaw** 1. Either of two bones in the skull, to which teeth are attached. One or both of these bones is movable. In humans, only the lower jaw is movable. 2. See *mandible, maxilla*.

**Jeans, James** A British scientist who, early in the twentieth century, theorized that spiral nebulae were new matter, being introduced into the universe from some fourth dimension. This was before the spiral nebulae were found to be galaxies like ours. See *galaxy*.

**jejunum** The part of the small intestine that is mostly responsible for the absorption of nutrients from ingested food. It is the first part of the small intestine.

**jellyfish** A complex coelenterate with stinging tentacles and a floating sac from which the tentacles hang. The stingers paralyze the prey. Jellyfish stings are painful to people, but rarely harmful unless they occur in large numbers, or in the case of a man-o-war. See *man-o-war*.

**Jenney, William le Baron** An American architect and engineer who developed the steel frames to build tall buildings during the nineteenth century. He might be called the father of modern skyscrapers.

**jet** A high-speed stream of air, water or other gas or fluid.

**jet engine** A propulsion device perfected and put into widespread use after World War II. There are several different types. A turbojet is shown in the drawing. Compressed air is heated by the combustion of the fuel; this air is expelled out the back of the engine at high speed, producing thrust by the action-reaction principle.

JET ENGINE

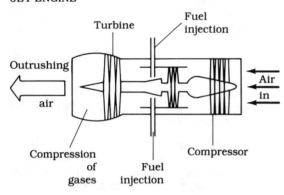

**jet propulsion** A means of obtaining high speed, especially for aircraft, using jet engines. Also can be used to propel boats. Some jet aircraft can attain speeds of up to several times the speed of sound. See *jet engine*.

**Jet Propulsion Laboratory** Abbreviation, JPL. A research-and-development group run by the United States government. It is involved with the study of the planets.

**jet stream** Any of several high-speed, high-altitude wind bands in the earth's atmosphere. These are only a few tens of miles wide but can blow at more than 100 miles per hour. They occur where major air masses meet. The most intense jet streams are at the fringes of the polar fronts. Weather systems, particularly low-pressure centers, tend to follow the jet streams, so they are important to the weather forecaster. See *polar front*.

**jiffy** See *chronon*.

**Jodrell Bank** A radio-telescope location in Cheshire, England. Much of our knowledge in radio astronomy has been gathered here. See *radio astronomy, radio telescope*.

**joint** 1. A place where two bones meet, and where movement in one or two planes is possible. Examples are the elbow (one plane) and the shoulder (two planes). 2. A mechanical device that allows rotation in one or two planes. 3. Mutual and cooperative, as between companies in different countries.

**j operator** The engineering expression for the square root of −1. Written as the small letter j. When multiplied by a real number, we write j followed by that real number; for example, j4.5. This unit imaginary number generally indicates reactance. See *capacitive reactance, complex number, inductive reactance*.

**Josephson Effect** When two superconductors are brought close together, a current flows across the narrow gap separating them. This causes an electromagnetic field to be generated.

**joule** The most common unit of energy. Equivalent to one watt expended for one second (one watt-second). Also equal to one newton of force moving over one meter of displacement (one newton-meter). See the ENERGY UNITS appendix. See *newton, watt*.

**joule heating** The heating of an electrical wire or other conductor when a current flows. This is inevitable because all conductors have some resistance. It is extremely small in superconductors, however. See *superconductor*.

**Joule-Kelvin Effect** An exception to Joule's Law for gases (See *Joule's Law*), caused by the fact that real gases are never ideal. When a gas is allowed to flow into a medium having reduced pressure, the temperature either rises or falls, depending on the starting temperature.

**Joule's Laws** 1. When an electric current flows through a resistive medium, heat is produced. 2. The heat, in watts, caused by a current in a resistance is equal to the square of the current in amperes, times the resistance in ohms. 3. For an ideal gas, the internal energy depends only on the mass of the gas and on its temperature.

**joystick** 1. A device that allows control in two or three dimensions. It can be manipulated forwards and backwards, to the right and left, and sometimes also by twisting. 2. A control in an aircraft that allows changing of altitude.

JOYSTICK

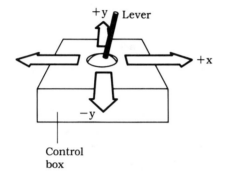

**JPL** Abbreviation for Jet Propulsion Laboratory.

**jugular vein** Either of two large veins in the neck that carry blood from the brain back to the superior vena cava, and then to the heart.

**junction** 1. A point at which two or more electrical conductors come together. 2. A point at which two or more pipelines come together. 3. A set of connections in an electrical circuit. 4. The plane of contact between P-type and N-type semiconductors in a diode or transistor. See *semiconductor*.

**junction diode** A semiconductor diode made by diffusion process. The P-type material is diffused into the N-type. See *diode, junction, semiconductor*.

**junction transistor** A semiconductor transistor made by diffusion process. The P-type material is diffused into the N-type. See *junction, semiconductor, transistor*.

**Jupiter** The fifth planet from the sun in the Solar System. It is the largest planet, one of the so-called "gas giants." It is largely made up of hydrogen, ammonia, and methane at low temperature. The planet is thought to have a rocky core about the size of the earth, around which the other substances combine in solid, liquid, and gaseous form. Some astronomers think that if Jupiter had been somewhat larger, it would have become a small star. See the SOLAR SYSTEM DATA appendix.

**Jurassic period** A geologic time that started about 180 million years ago, ending with the start of the Cretaceous period about 137 million years ago. The North Atlantic Ocean widened as North America drifted westward. Various other changes occurred in the positions of the continents. The oldest rocks in the Pacific Basin date from this period. See the GEOLOGIC TIME appendix.

**juvenile hormone** In insects, a hormone that slows down growth, prolonging the larva stage. See *larva*.

**juvenile water** Seawater recycled through magma. This water is ejected with the magma, in the form of steam, when volcanoes erupt. It might remain trapped for thousands of years beneath the earth's surface. See *magma, volcano*.

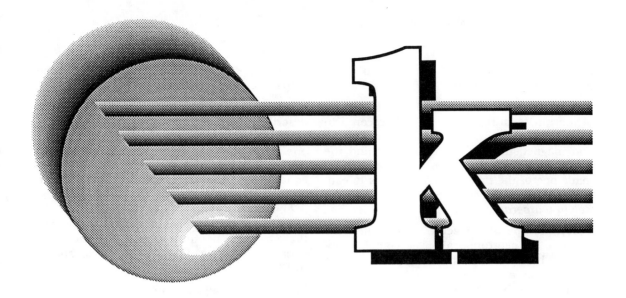

**K** Abbreviation for Kelvin (See *Kelvin temperature scale*), chemical symbol for potassium.

**k** Abbreviation for constant, kilo- (see the PREFIX MULTIPLIERS appendix).

**Kaisel, Stanley** An engineer who is known for his work in space communications. He founded a corporation specializing in the development and manufacture of microwave communications equipment.

**Kant, Immanuel** A philosopher who took a position midway between Rationalism and Empiricism, pointing out the flaws in both. See *Empiricism, Rationalism*. Also developed a theory for the creation of the Solar System, similar to the theory accepted today.

**kaolin** A clay-like substance used in industry for various purposes. Also sometimes used in medications to soothe the intestinal lining when it is irritated.

**Karnaugh map** A truth table that depicts a logical expression. There are often several, or many, ways that a complex logical expression can be written. Used by digital engineers to find the simplest combination of logic gates to get a desired function. See *logic diagram, logic function, truth table*.

**karyotype** A means of classifying the chromosomes in a cell.

**Kater pendulum** A device developed to find the intensity of the earth's gravity. This is equivalent to about 9.8 meters, or 32 feet, per second per second. But it is hard to measure by dropping weights from high places. The physicist Henry Kater got around this about 1800 A.D. with his device, which used swinging weights.

**kcal** Abbreviation for kilocalorie.

**keeper** A piece of steel or iron that is placed over a permanent magnet when the magnet is stored. This helps to prevent the slow demagnetization that otherwise takes place.

**kelp** A seaweed noted for its iodine content. It is dried and sold in powdered or tablet form as a dietary supplement of iodine.

**Kelvin** See *Kelvin temperature scale*.

**Kelvin balance** A device for measuring very small electric currents. It can sense either alternating current or direct current. It works using a mechanical balance. Coils cause attraction and repulsion when a current flows, and this upsets the balance, even for a tiny current. The extent of the deflection is measured to determine the value of current.

**Kelvin, Lord** A physicist and mathematician of the nineteenth century. He did research in geometry, especially as it pertains to electricity and magnetism. The Kelvin balance and the Kelvin temperature scale bear his name. He made one of the first serious attempts to estimate the age of the earth. See *Kelvin balance, Kelvin temperature scale*.

**Kelvin temperature scale** Also called the absolute temperature scale. The complete absence of heat is zero on this scale. All molecular motion ceases at zero Kelvin. The divisions (degrees) are the same size as they are in the centigrade scale. Water freezes at +273 degrees Kelvin. It boils at +373 degrees Kelvin (see drawing). See *centigrade temperature scale*.

KELVIN TEMPERATURE SCALE

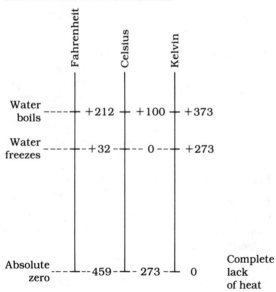

**Kelvin wave** Subsurface waves in the Pacific Ocean that occur during El Nino events. Water in the western Pacific "sloshes" back eastward, setting up currents in the entire South Pacific. See *El Nino*.

**Kepler, Johannes** A German astronomer who first put forth the idea, in 1609, that the planets orbit in ellipses rather than in circles around the sun. The sun is at one focus of the ellipse. See *Kepler's Laws*.

**Kepler's Laws** A set of mathematical rules describing the movements of the planets. 1. The planets orbit the sun in ellipses, with the sun at one focus. 2. If we draw a line connecting a planet with the sun, the line will sweep out equal areas in equal lengths of time. 3. The orbital period of a planet is in proportion to the 3/2 power (the square root of the cube) of its distance from the sun. Nowadays we know that these rules are not quite exact. There are slight variations because of relativistic effects, shown by Einstein.

**keratin** A strong substance formed from long, coiled molecules of protein. Hair is the most common example. Fingernails are another. Certain dead skin lesions are yet another. See *keratosis*.

**keratosis** A tough, horny, wart-like skin lesion that might occur in certain disease states. One form is a precancerous lesion. Some occur for unknown reasons.

**kerosene** A hydrocarbon, used as a high-performance fuel. There are 11 or 12 carbon atoms in the chain. It is a liquid, with molecules somewhat longer than those of octane but shorter than those of paraffin. See *alkane*.

**Kerr Effect** The double refraction of light in some materials when electric fields are applied. The field changes the index of refraction. The index also changes with the polarization of the light. The result might be two different rays emerging from the substance, when only one ray enters. See *index of refraction, refraction*.

**ketoacidosis** A condition in which the blood contains many ketone bodies because of incomplete metabolism of fat, and the body pH decreases. Might occur in an untreated case of diabetes mellitus. See *diabetes mellitus, ketone body*.

**ketone** Any of various different organic compounds containing carbon, oxygen, and hydrogen. They are somewhat similar to the alcohols.

**ketone body** A by-product of the metabolism of fat. When fat (triglyceride) is used as fuel in large amounts, such as when the diet is deficient in carbohydrate, or during starvation, these molecules are produced along with free fatty acids. They also accumulate in the bloodstream in an untreated diabetic. See *diabetes mellitus*.

**ketosis** A condition resulting from rapid or incomplete metabolism of fat. Can be recognized by high concentration of ketone bodies in the blood or in the urine. A person in this state might be restless and irritable. See *ketone body*.

**Kettering, Charles** An American engineer who played a major role in the development of the automobile. He is best remembered for having developed a starter that did not require a strong arm to turn a crank.

**key** 1. One pushbutton switch on a keyboard or keypad. See *keyboard, keypad*. 2. A device for manually sending Morse code. 3. An important clue for a scientist involved in a specific research program.

**keyboard** A set of pushbutton switches arranged in a standard way, and having the letters of the alphabet, digits from 0 to 9, and various function keys. The most common example is the set of keys on a typewriter or personal computer.

**keypad** A set of pushbutton switches on a calculator or telephone dialer, usually having digits from 0 to 9, and also various arithmetic operations, and perhaps mathematical functions.

**Keys, Ancel** A medical scientist at the University of Minnesota, who has done research demonstrating the relationship between dietary fat and cholesterol levels in the blood.

**kg** Abbreviation for kilogram.

**kHz** Abbreviation for kilohertz.

**kidney** Either of two small organs in the lower back (see illustration). They are responsible for filtering the blood, sending waste and excess water on to the bladder to be excreted as urine, and retaining the usable and needed fluid.

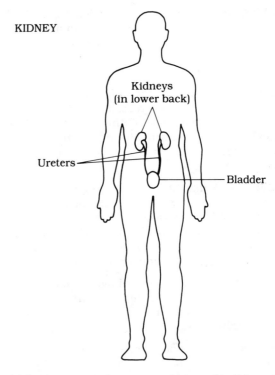

KIDNEY

**kidney stone** An accumulation of solid matter in the kidneys. There are different types. They usually contain calcium or crystallized uric acid. A few contain other substances. There may be no symptoms, or there may be pain in the lower back and painful urination.

**killer satellite** A satellite equipped with weapons designed to disable other satellites or to hit ground targets. Some satellites can carry nuclear missiles that can be delivered with almost no warning at all.

**kilo-** A prefix meaning 1000 or thousands. See the PREFIX MULTIPLIERS appendix.

**kilobyte** A unit of data, equal to 1024 bytes. See *byte*.

**kilocalorie** Abbreviation, kcal. A unit of energy equal to 1000 calories or 4187 joules. The kilocalorie is known as a "diet calorie." This is the common measure of the energy content of foods. One gram of protein or carbohydrate yields about 4 kcal when burned; one gram of fat yields about 9 kcal. See *calorie*.

**kilogram** Abbreviation, kg. The Standard International unit of mass. Equal to about 2.2 pounds in the earth's gravitational field. Mass is often given in grams (0.001 kg) or milligrams (0.000001 kg).

**kilohertz** Abbreviation, kHz. A unit of frequency equal to 1000 cycles per second or 1000 hertz. An audio tone of this frequency falls nearly in the most sensitive part of the human hearing range. It sounds moderately high in frequency. See *hertz*.

**kiloton** A unit of explosive power, representing 1000 tons (two million pounds, approximately) of TNT. It was once used to denote the force in atomic bombs. Modern atomic weapons can deliver an explosion having the power of several thousand kilotons. See *megaton*.

**kilovolt** A unit of high voltage, equal to 1000 volts. See *volt*. Used mainly when referring to high-tension power lines, or in high-power radio and television transmitting tubes.

**kilowatt** A unit of power, equal to 1000 watts. See *power, watt*.

**kilowatt hour** A unit of energy, equal to 1000 watt hours or 3.6 million joules. See *joule, watt hour* and the ENERGY UNITS appendix.

**kimberlite** A volcanic rock that comes from the earth's mantle, originating at depths as great as 150 miles. This rock is where most diamonds are found.

**kinematics** The physics of motion, but not involving forces that might be responsible for the motion.

**kinescope** 1. A television picture tube. 2. A cathode-ray tube. See *cathode-ray tube*. 3. A video tape that has been recorded directly from a television set.

**kinesis** 1. Movement or motion. 2. Motion of an organism, affected by certain environmental factors. For example, bugs move more slowly at cooler temperatures than at warmer temperatures.

**kinetic energy** Energy in the form of motion of particles or objects. Thermal energy is a good example; the warmer the temperature for a given material, the faster the molecules move. A moving car also has kinetic energy. See *energy*.

**kinetic theory** A rather complex theory in physics that describes the behavior and characteristics of matter on the basis of the movements of atoms and molecules.

**kinetophone** The first device that combined voice recordings with motion pictures, invented by Thomas Edison in 1889. Talking movies did not become practical until 1927.

**kinetoscope** One of the many inventions of Thomas Edison. It was the first motion-picture viewing machine. It flipped pictures one after the other in rapid succession. The viewer had to look through a peephole. You still might see these in amusement galleries.

**kingdom** 1. Any of three main divisions of living things: plants, certain microorganisms without nuclei, and animals. 2. Possible major categories of living things besides the plants and animals. Certain cells have characteristics of both, as do some fungi and molds.

**kin selection** A genetic resemblance that causes many species of animals to favor their immediate relatives. This is vividly apparent with human families, and especially with identical twins.

**Kipp Process** The laboratory preparation of a gas from a liquid and a solid. An example is sodium (solid) reacting with water (liquid) to get hydrogen (gas). The gas is collected by means of an arrangement called Kipp apparatus. This scheme allows control of the amount of solid that comes into contact with the liquid.

**Kirchhoff's Laws** Two laws concerning current in electric circuits. Kirchhoff's First Law: The total current flowing into any point in a direct-current circuit is the same as the total current flowing out of that point. Kirchhoff's Second Law: The sum of all the voltage drops around a circuit is equal to zero. These are shown by the examples in the drawing.

**Kitty Hawk** A beach site in North Carolina where the Wright Brothers made their first

KIRCHHOFF'S LAWS

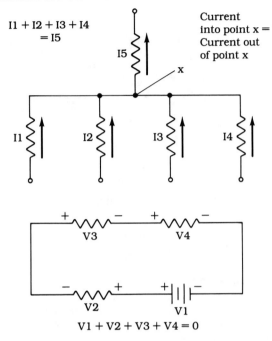

$I1 + I2 + I3 + I4 = I5$

Current into point x = Current out of point x

$V1 + V2 + V3 + V4 = 0$

flight tests at the very end of the nineteenth century and the beginning of the twentieth. The site was chosen because of its steady winds and its wide-open space without obstructions like trees, hills, or buildings.

**Klein bottle**  A topological object, the four-dimensional equivalent of a Mobius band. It has no inside or outside, as such, but just one "side." See *Mobius strip*.

**klystron**  A vacuum tube that produces microwave signals. It also can be used as an amplifier at microwave frequencies. Nowadays it has been largely replaced by solid-state devices that use much less power. See *microwave, microwave communications*.

**knock**  A problem that sometimes occurs with internal combustion engines, especially in cars and trucks, when the engine is not adjusted properly or when the fuel is of poor grade. It produces a loss of efficiency in the engine and a characteristic rattling or knocking noise.

**knot**  A unit of speed, used by sailors, and also by meteorologists and aviators to define wind speed. One knot is equal to one nautical mile per hour, or about 1.15 statute miles per hour. When wind speed is given in knots, the speed in statute miles per hour (mph) is obtained by multiplying by 0.869.

**Koch curve**  A theoretical object generated by trisecting the sides of a triangle and adding new triangles, as shown in the drawing, infinitely many times. The resulting form has infinite perimeter but finite area.

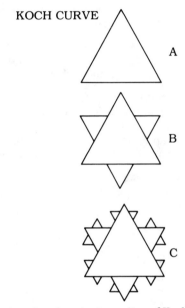

Progressive steps in generation of Koch curve.

From Gibilisco, MORE PUZZLES, PARADOXES AND BRAIN TEASERS (TAB Books, 1990).

**Koch, Robert**  A biologist who, along with Louis Pasteur, first put forth the idea that diseases are caused by microorganisms. We know today that there are many kinds of disease-producing microorganisms. And of course, some diseases are caused by other factors. See *Pasteur, Louis*.

**Kovar**  An alloy used in the manufacture of products in which glass and metal are joined. The alloy expands and contracts just about the same amount as glass as the temperature changes. Made of iron, nickel and cobalt.

**Kr**  Chemical symbol for krypton.

**Krakatoa** The most violent volcano in recent history. On August 27, 1883, this island, near Java and Sumatra, literally blew its top, with an explosive force equivalent to a modern hydrogen bomb. The sound was heard 3000 miles away; tsunamis circled the globe. Tremendous quantities of dust were blown into the atmosphere.

**Krebs cycle** An important part of metabolism in living things. It is a complex process that provides energy from protein, carbohydrate and fat that is eaten. Sometimes also called the citric acid cycle. See *metabolism*.

**krill** A small, shrimplike crustacean (see drawing) found especially in southern oceans. Whales, fishes, squid, and sea birds feed on it.

**krypton** Chemical symbol, Kr. An element with atomic number 36. The most common isotope has atomic weight 84. It is a gas at room temperature and normal atmospheric pressure. It is nonreactive. Some isotopes are radioactive.

**Kundt tube** A sophisticated device that allows measurement of the speed of sound in a liquid. It works by means of resonant effects.

KRILL

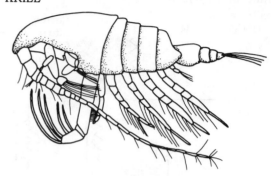

From Erickson, THE MYSTERIOUS OCEANS (TAB Books, 1988).

Knowing the wavelength and the frequency, the speed can be determined.

**kV** Abbreviation for kilovolt.

**kW** Abbreviation for kilowatt.

**kwashiorkor** A disease caused by protein deficiency. It is common in countries where starvation is rampant, such as in western Africa. The most prominent signs are extreme thinness and a bloated belly.

**kWh** Abbreviation for kilowatt hour.

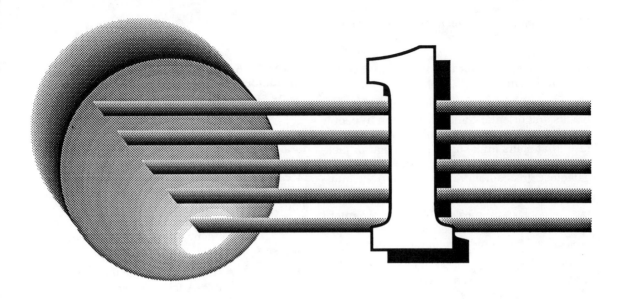

**L** Abbreviation for inductance, mole.

**l** Abbreviation for liter.

**La** Chemical symbol for lanthanum.

**labeling** A method of tracing substances in the body, or identifying them in organic tissues or cells. A radioactive isotope of the substance is injected. It can be followed by detecting the radiation.

**labile** 1. Easily changed or broken down into its constituent chemicals. 2. Pertaining to a substance in which certain atomic groups can be easily replaced by others.

**labor** 1. The process of giving birth in mammals. In humans this may last several hours and is exhausting. 2. Time spent by personnel in manufacture, testing, and repair of apparatus.

**laboratory** 1. A facility for doing scientific research and for testing prototype devices. 2. A medical facility for evaluating specimens, testing for disease conditions, and/or levels of various substances.

**labyrinth** 1. A network of underground caves and tunnels, often occurring in layers, having been carved out by water erosion over millions of years. 2. A maze. 3. A complex network or circuit.

**lactate** A compound containing lactic acid. See *lactic acid*.

**lactation** In female mammals, the production of milk after giving birth. In humans this results in so-called breast milk. It is a biological mechanism for feeding newborn infants.

**lactic acid** The acid that forms when milk sugar (lactose) is fermented. It gives yogurt and buttermilk their tangy flavor. It accumulates in the muscles during prolonged exercise, causing fatigue.

**lactobacillus acidophilus** A beneficial strain of bacteria found in the intestines. Can also be obtained in tablet or capsule form. Helps in the digestion of milk sugar (lactose) and some other carbohydrates. Also sometimes prescribed along with antibiotics, to maintain beneficial bacteria in the intestinal tract. See *lactose*.

**lactose** The sugar found in milk. It is made up of glucose and galactose. These sugars are normally split during digestion, but in some people, the enzyme for this is lacking. See *lactose intolerance*.

**lactose intolerance** A condition common in blacks and orientals, and uncommon in people of northern European descent. The enzyme lactase, that splits milk sugar, is deficient or lacking, so lactose cannot be split into glucose and galactose. Instead, it draws water into the intestine and causes diarrhea. See *lactose*.

**lag** A consistent, measurable delay. We might speak of phase lag in electronics, for example, or seasonal lag in meteorology. See *lagging*.

**lagging** Taking place later than a certain specified event or set of events. This delay is generally constant, such as a certain fraction of a second or a certain part of a wave cycle. See *leading*.

**lahar** A hot mudflow or ashflow that occurs on the slopes of an active volcano. See *volcano*.

**lake effect** A tendency for precipitation to occur near the leeward shore (downwind) of a large lake. This takes place over the Great Lakes in the winter. Cold winds pick up moisture and drop it as snow on the leeward shores (see drawing). This precipitation can be sudden and heavy.

LAKE EFFECT

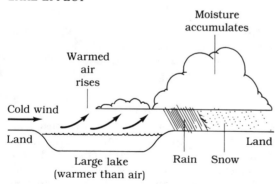

**Lamarck, Jean-Baptiste** A French scientist who developed a theory similar to Darwin's theory of evolution, but before Darwin. His famous words are "Form follows function." This means that things develop to meet needs. See *Evolutionism*.

**lambert** A unit of brightness for visible light. One lambert is the luminance of a perfectly diffusing surface that emits or reflects light with the intensity of one lumen per square centimeter. See *lumen, luminance*.

**Lambert's Law** A modification of the inverse-square law for visible light. If light falls on a surface at an angle, instead of straight down, the energy received per square meter is reduced. Thus, received light depends not only on the distance from the source, but on the cosine of the angle of incidence. See *angle of incidence, inverse-square law*.

**lamella** 1. A scale or scale-like formation in a plant or animal. 2. A fold underneath a mushroom containing spores.

**laminar flow** A type of fluid flow, in which there are no irregularities. Each molecule or atom moves at the same speed as all the others at a given location.

**laminated core** In an alternating-current transformer, a core made from several sheets of iron or steel glued together. This reduces losses in the core and makes the transformer more efficient than it would be with a plain solid core.

**lamination** 1. A process in which a material is layered to give added strength or certain characteristics. 2. One of the layers as defined in (1). 3. Any of the layers in a laminated transformer core. See *laminated core*.

**lampblack** Very fine powdered carbon. It is almost perfectly black; that is, it reflects essentially none of the visible light that strikes it, and it is not shiny. It is used as a black pigmenting agent.

**land bridge** A narrow strip of land connecting continents when the ocean level falls. Such a strip is thought to have existed during the last ice age, and humans from Asia crossed it into North America. See the illustration. See *ice age*.

**landline** Any wire or fiberoptic communications circuit, such as a telephone system.

**land mapping** The processes involved in making graphical representations of the earth's surface. Topographical mapping shows eleva-

LAND BRIDGE

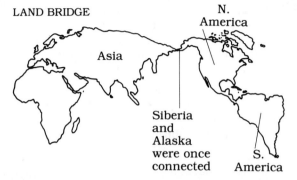

tion above sea level. Geographic mapping shows shorelines and natural barriers such as rivers. Satellites and aircraft have proven invaluable in making maps precise.

**Landsat** A series of satellites, the first of which was launched in 1972, designed for mapping the earth. See *land mapping*.

**landslide** A phenomenon in which some soil breaks loose, carrying more soil with it, and the effect builds on itself, creating a massive flow of earth. The damage is similar to that caused by an avalanche. See *avalanche*.

**lanolin** An emulsion of fat in water, used as a skin moisturizer and as the base for various skin makeups.

**lanthanide** Any of the elements having atomic numbers 57-71. These are listed in this book under: *cerium, erbium, dysprosium, europium, gadolinium, holmium, lanthanum, lutetium, neodymium, praseodymium, promethium, samarium, terbium, thulium, ytterbium.*

**lanthanum** Chemical symbol, La. An element with atomic number 57. The most common isotope has atomic weight 139. It is used in oil refining. In pure form it is a metal.

**lapilli** Also called pyroclastic fragments or cinders, ejected from an active volcano. Produces gravel-sized deposits at the base of the volcano.

**lapis lazuli** A feldspar-like rock (see *feldspar*) sometimes used for jewelry and regarded as semi-precious.

**Laplace, Pierre Simon de** A scientist of the late eighteenth century. His main contributions were in astronomy, and in the theory of probability. He developed a theory of the origin of the Solar System similar to the one we accept today.

**large intestine** See *bowel*.

**large-scale integration** A method of manufacturing integrated circuits with more than 100 logic elements per chip. Microprocessors and electronic watches use such devices. See *integrated circuit, very-large-scale integration*.

**larva** The equivalent of childhood in invertebrates. The larva hatches from the egg and must undergo maturation (metamorphosis) in order to become an adult. See *metamorphosis*.

**larynx** The voice box, in which the vocal cords are contained. Voice sounds are produced when air passes through the vocal cords in the larynx. See *vocal cords*.

**laser** Acronym for light amplification by stimulated emission of radiation. A device that produces visible light in coherent, usually parallel rays. The drawing shows a helium-neon laser, that emits red light. See *coherent light*. See also the definitions that follow.

LASER

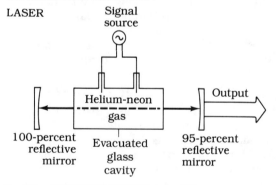

From Gibilisco, UNDERSTANDING LASERS (TAB Books, Inc., 1989).

**laser acupuncture** A means of performing Chinese acupuncture by using lasers rather than needles. This method apparently works just as well as the older way, and there are no problems with risk of infection from needles. See *acupuncture*.

**laser beam** The rays of light emitted by a laser. The beam is narrow and stays narrow for great distances from the laser. This makes the laser useful when light must be transmitted a long way with little attenuation. See *coherent light, laser.*

**laser communications** The use of modulated laser beams for transmitting information. This requires clear air; it might be more useful in space, in the future, than it is on the surface of the earth. Sometimes lasers are transmitted through glass or plastic fibers for communications purposes. See *fiberoptics, laser, laser beam, modulation.*

**laser diode** A semiconductor diode that emits visible, coherent light when a current passes through it. The output is not in a narrow beam, but is wide-angled and not very intense. However, it can be focused or collimated with lenses. Useful for fiberoptic laser communications. See *diode, fiberoptics, laser communications.*

**laser fusion** 1. The welding together of substances using lasers. 2. A hypothetical, proposed method of obtaining atomic fusion, using lasers to provide the heat. See *fusion.*

**laser surgery** The use of a laser rather than conventional surgical tools. There is no risk of infection, and lasers can often make more precise cuts with less bleeding than scalpels and other instruments. Lasers are useful in surgery in parts of the body that cannot be easily reached. There are some types of surgery in which lasers do not work well. See *endoscope.*

**latent heat of evaporation** Also called latent heat of vaporization. For a given substance, the amount of energy needed to convert one gram from liquid to vapor form, without any increase in the temperature. For water this is 540 calories per gram.

**latent heat of melting** For a given substance, the amount of energy needed to convert one gram from solid form to liquid form, without any increase in the temperature. For water this is 80 calories per gram.

**lateral inversion** Right-to-left reversal of a visual image, without turning it upside-down. This is the type of image inversion that occurs when you see your reflection.

**laterite** A soil residue made up mainly of iron hydroxide and aluminum hydroxide. There might also be magnesium hydroxide and other metal hydroxides in smaller amounts. The material is used in construction.

**latex** A viscous, usually white liquid, found in certain plants. It is commercially used as a lubricant and paint base. It can be solidified to make synthetic rubber.

**latitude** 1. The north-south displacement from the equator on the earth or some other planet. Measured in degrees from 0 to +90 (North Geographic Pole) and from 0 to −90 (South Geographic Pole). See *longitude.* 2. See *degree of freedom.*

**lattice** Any network with a regular, repeating pattern. In nature, these patterns are found in crystalline materials (see *crystal lattice*). Electronic components may be arranged in repeating networks and called lattices.

**laughing gas** See *nitrous oxide.*

**launch vehicle** The large, often multistage rocket that places a satellite or spacecraft into orbit.

**Laurasia** One of two major supercontinents that existed before the drifting began that

LAURASIA

From Erickson, VIOLENT STORMS (TAB Books, Inc., 1988).

resulted in present geography. Consisted of what is now North America, Europe and much of Asia (see drawing). See *Gondwanaland*.

**lava** Hot, molten rock that comes out of an active volcano. It flows down the slope of the volcano and usually destroys anything in its path. Before it comes out of the volcano, it is called magma. See *magma*.

**Lavoisier, Antoine** An eighteenth-century chemist. He did research into the nature of combustion, and found out exactly what it was that made things burn. He did much to make chemistry an exact science. Before his time, it had been much less scientific.

**law** A fundamental scientific rule or principle. Usually, a principle becomes a law when it is discovered by someone. Then that scientist's name is attached to it. An example is Ohm's Law; there are many others.

**Lawrence, Ernest** A physicist who, with Stanley Livingston, developed the first cyclotron at the University of California, Berkeley in the 1930s. He built a laboratory in the mid-1930s at Berkeley, and won a Nobel prize in 1939. See *cyclotron* and *Livingston, Stanley*.

**Lawson, J. D.** A twentieth-century atomic scientist who researched nuclear fission for use in atomic reactors. He developed principles for their efficient operation.

**lawrencium** Chemical symbol, Lr or Lw. A chemical element with atomic number 103. The most common isotope has atomic weight 257. It was first seen in 1961. It decays almost instantly, and is not found in nature; it occurs only in reactions.

**lb** Abbreviation for pound.

**LCD** Abbreviation for liquid-crystal display.

**LD** Abbreviation for lethal dose.

**LDL** Abbreviation for low-density lipoprotein.

**leaching** 1. The undesired removal of water-soluble nutrients from food when the food is boiled. 2. A process in the earth where excess water carries minerals and/or nutrients away from the soil.

**lead** 1. Chemical symbol, Pb. An element with atomic number 82. The most common isotope has atomic weight 208. It is a dull, grayish metal in pure form, known for its high density. It is used in many applications in science and industry. It is especially good at stopping gamma rays and X rays. 2. In electronics, a short wire that is connected to various different devices or circuits, according to need. 3. A condition of preceding. See *leading*.

**lead-acid cell** A rechargeable electrochemical cell. The positive electrode is lead peroxide, and the negative electrode is spongy lead. The electrolyte is a solution of sulfuric acid (see drawing). Several cells can be combined to make a battery, of the type usually found in cars.

LEAD-ACID CELL

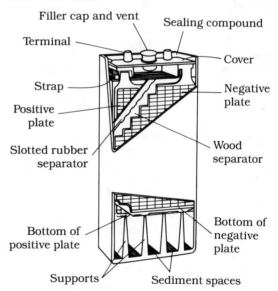

From Turner & Gibilisco, THE ILLUSTRATED DICTIONARY OF ELECTRONICS—5TH EDITION (TAB Professional and Reference Books, 1991).

**leader** 1. The initial surge of electrons prior to the large current discharge that occurs in a lightning stroke. See *lightning*. 2. A short section of material, usually plastic, attached to a length of recording tape to get the tape started in the recorder or player.

**leading** Preceding; taking place earlier than some other event to a regular, constant, measurable extent. The term is used mainly in electronics with reference to wave phase. See *lagging*.

**lead poisoning** Sickness caused by ingestion of lead. Even small amounts can produce serious effects. Lead is not known as a nutrient, but only as a poison. Symptoms include irritability and mental retardation. This form of poisoning can be a problem in buildings with old plumbing and/or old paint.

**leaf** The part of a plant responsible for photosynthesis, containing the chlorophyll and a structure of veins and capillaries. The chlorophyll produces carbohydrate from visible light, and this carbohydrate is carried by the vessels to feed the plant. See *chlorophyll, chloroplast, photosynthesis*.

**leakage** 1. A passage or loss of fluid where it is not wanted. 2. A small electrical discharge that occurs at the ends of an antenna. 3. A small flow of current when, ideally, the current should be zero.

**learning** 1. A form of memory in which mistakes are made but usually not repeated. 2. The acquiring of a certain skill or knowledge. 3. The process of improving one's skill, or increasing one's knowledge, systematically over a period of time.

**lecithin** A fat-like substance, also called a phospholipid. It is found in living tissue. It is employed in the food industry as an emulsifier.

**LED** Abbreviation for light-emitting diode.

**Leeuwenhoek, Anton van** A scientist in England who, during the late seventeenth and early eighteenth centuries, made observations with the microscope and logged his findings. He is known mainly for this research, and for his studies of living cells.

**leeward** On the side away from the wind or downwind. If the wind blows from the west, the leeward side is the eastern side.

**left-handed screw** A screw that is "backwards," so that it is tightened by turning counterclockwise. Such screws have various uses. One example is in bombs: Enemy personnel might think that they are disarming a warhead, and then it blows up.

**legume** A fruit that tends to form in pods. Peas are a good example, as are lentils. They are excellent sources of protein, fiber, complex carbohydrate, vitamins, and minerals.

**Leibniz, Gottfried Wilhelm von** A German mathematician who discovered the calculus independently of Isaac Newton, but about 10 years later than Newton. He lived during the late seventeenth and early eighteenth centuries. See *calculus* and *Newton, Isaac*.

**LEM** Abbreviation for lunar excursion module.

**lemma** In mathematics, a theorem that is proved mainly in order to simplify the proof of some other, more significant theorem. See *theorem*.

**lens** 1. A device used for focusing or bending light rays in a controlled manner. See *concave lens, convex lens*. 2. A device that refracts electromagnetic fields to focus or collimate the energy. 3. Any oblate (flattened) sphere-shaped object or cavity.

**lenticel** In a plant, a pore-like formation that lets air in and also lets various waste gases out.

**lenticular clouds** Lens-shaped, smooth-contoured clouds that may form downwind from hills or mountains (see drawing). Such clouds have sometimes been mistaken for huge flying saucers.

**Lenz's Law** When a current is generated by a changing magnetic field, or by the motion of a conductor in a magnetic field, the current itself produces a secondary magnetic field. This secondary field acts with the original magnetic field to oppose the motion. This is why, for example, a powerful engine or turbine is needed to operate a high-power electric generator.

**leprosy** Also called Hansen's disease. An infectious disease caused by a bacterium that

LENTICULAR CLOUDS

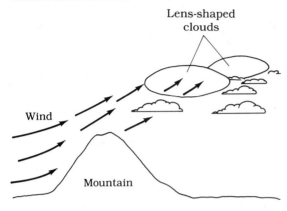

especially invades skin and mucous membranes. The main symptom is skin lesions. The nerves might also become involved. Antibiotics are used for long periods to treat the disease.

**lepton** An elementary particle with spin $1/2$, and that is subject only to weak and electromagnetic interactions. Electrons, muons and neutrinos are leptons. See *electron, muon, neutrino, spin*.

**lesion** A growth or inflamed region on or in the body, especially on the skin. A pimple is an example.

**lethal dose** 1. The single dose of a toxic substance that causes death in a certain percentage of cases, such as 90 percent. 2. The single exposure of ionizing radiation (from radioactive materials) that causes death in a certain percentage of cases, such as 90 percent.

**leucine** An essential amino acid. It must be obtained from protein in the diet, since it cannot be manufactured from other amino acids in the human body. See *essential amino acids*.

**Leucippus** One of the first philosopher/scientists to suggest that matter is made of many tiny particles, instead of being continuous. He thought of this more than 2000 years ago, in ancient times, long before there were magnifying lenses or microscopes.

**leucocyte** See *white blood cell*.

**leucoplast** In a plant cell, a colorless organelle. These organelles store nutrients. See *organelle*.

**leukemia** A disease in which cancer affects the manufacture of white blood cells. This cancer occurs in the bone marrow where the cells are made. See also *white blood cell*.

**levee** A ridge or wall of earth used for flood control. It keeps the water confined within the barrier unless, of course, the water rises too high and spills over, or the barrier is broken by something carried in the flood water.

**level** 1. Relative intensity or signal strength. 2. Loudness of sound. 3. Brightness of light. 4. An instrument used to determine the orientation of the horizon. See *horizon*.

**lever** 1. One of the six simple machines that converts a small force over a large distance into a large force over a small distance (see drawing). 2. Any device with a rod or stick that is operated by pushing to one side, such as a toggle switch or conventional wall switch.

LEVER

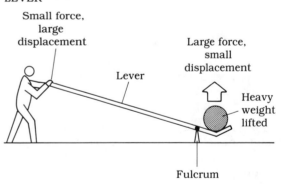

**leverage** The action of, or the force obtained by the use of, a lever. See *lever*.

**lever law** Two weights on a lever will balance when their masses are inversely proportional to their distances from the fulcrum. See *lever*.

**levitation** A magnetic phenomenon that occurs in superconductivity. An object can be suspended above another object by magnetic repulsion (see drawing). This is how a proposed

LEVITATION

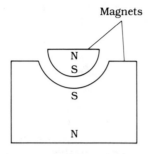

high-speed train would be suspended on its tracks. see *superconductivity*.

**LF**  Abbreviation for low frequency.

**Li**  Chemical symbol for lithium.

**library science**  The discipline of organizing information and categorizing it, so that it is easy to find again. This involves certain mathematical principles, so that the least number of steps are required to locate desired data.

**libration**  The apparent wobbling of the moon, as viewed from the earth. It keeps the same side towards us all the time—almost. But there is some back-and-forth shifting, so we can see just a little more than half of the moon's surface over time.

**lice**  Certain insect parasites that live on mammals. They are small, have no wings, and might sometimes be found on the scalp. They can be passed from person to person.

**lichen**  A hardy plant found even in the Arctic and Antarctic, consisting of certain algae and fungi growing together in a combination called symbiosis. It looks somewhat like moss. See *algae, fungus, symbiosis*.

**life cycle**  The complete series of events characteristic of the life of a certain organism. Might be expressed simply or in detail. For a human we might say, "Birth, infancy, childhood, adolescence, adulthood, old age, death."

**life expectancy**  1. The average length of time that a newborn animal will live. 2. The average length of time, in years, that a newborn human will live. 3. The average length of time, in years, that a person will live after a certain point in life. 4. The average length of time that a device will last before ceasing to function.

**life science**  All those sciences having to do with living things, their structure, evolution, and behavior. These sciences are listed in the table of sciences and engineering at the front of this book.

**life support**  1. In medicine, a machine or machines without which a hospitalized patient would probably die. Might include respirators, intravenous devices, etc. 2. The set of machines that provides a habitable environment on board a submarine or spacecraft.

**lift**  1. The effect produced when air flows past an airfoil, or water past a hydrofoil. This is a net upward force, caused by a difference in fluid pressure. See *airfoil, hydrofoil*. 2. The ability of a balloon, kite, or other device to loft an object. Usually given in pounds or kilograms.

**ligament**  An elastic, durable organic substance in the body that holds joints together and allows flexing. Consists of fibers attached to bones, or to an organ to keep the organ in its proper position.

**light**  1. Electromagnetic radiation having a wavelength of about 3900 to 7500 Angstrom units, so that it can be detected by human eyes. See *Angstrom unit* and the ELECTROMAGNETIC SPECTRUM appendix. 2. A source of visible electromagnetic radiation. 3. Having a fairly high albedo. See *albedo*.

**light-emitting diode**  A semiconductor diode that produces visible light when a current flows through it. The devices are available in various sizes, shapes and colors. See *diode, semiconductor*.

**light meter**  A device for measuring the intensity of visible light. The simplest type has a photovoltaic cell (solar cell) connected to a microammeter or milliammeter that measures the current generated by the cell. The brighter the light, the larger the current. Used mostly in photography.

**lightning**  Electrostatic discharges in the atmosphere, especially in rain clouds and sometimes in snow showers, sandstorms or dust storms. A potential difference of millions of volts builds up (see drawing). A surge of electrons, called the leader (see *leader*), ionizes the air. Then a massive flow of electrons, the return stroke, equalizes the charge difference. Voltages build up again rapidly, and the whole process might be repeated every few seconds. The high current, upwards of 100,000 amperes, can start fires and cause structural damage. It causes radio and television interference because of its broad-spectrum electromagnetic pulses. The extreme current can injure or kill people and animals. Lightning can occur as cloud-to-cloud, cloud-to-ground or ground-to-cloud. See *lightning protection*.

**lightning protection**  1. A means of reducing the probability that lightning will strike in a given area. A grounded rod may "attract" lightning away from lower, nearby objects. 2. The grounding of electrical and electronic equipment to reduce damage if lightning happens to strike.

**light pen**  A pen-shaped device that allows a computer user to "draw" on a screen. Useful for graphics or for picking out certain items from a list on a screen.

**lightplane**  A possible future aircraft capable of takeoff and landing in limited space, and designed for mass production. This has sometimes been proposed as a "flying car." There is some question of safety.

**light quantum**  See *photon*.

**light year**  The distance that light travels in one year. This is about 6,000,000,000,000 (six trillion) miles. This is the most common unit for measuring distances outside the Solar System. See *parsec*.

**lignin**  A substance that gives certain plants their woody texture. It is found also in the bran of grains, and is nondigestible.

**lignite**  A low grade of coal, containing impurities, and having a rather soft consistency compared with higher grades. See *anthracite*, *bituminous*.

**lime**  1. See *dolomite*. 2. See *limestone*.

**limestone**  Rock, mainly made up of calcium carbonate. Also can contain magnesium carbonate, and trace amounts of calcium sulfate, magnesium sulfate, and sodium chloride. This rock occurs in abundance in the earth's crust. It has a characteristic layered appearance when land is blasted away for construction.

**limit**  1. For a mathematical function, a value that the function approaches as the indepen-

LIGHTNING

From Erickson, VIOLENT STORMS (TAB Books, Inc., 1988).

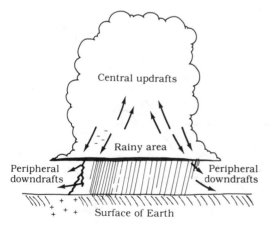

From Giblisco, VIOLENT WEATHER: HURRICANES, TORNADOES AND STORMS (TAB Books, Inc., 1984).

dent variable gets closer and closer to a certain value. In the drawing, the function approaches $y = 2$ as the value $x$ gets larger without limit ("approaches infinity"). 2. For an infinite series that converges, the sum of the whole series. Partial sums approach this value. 3. Greatest extent; maximum.

LIMIT

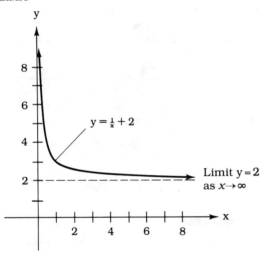

**limiter** In electronics, a device that keeps a signal from becoming stronger than a certain level.

**limnology** The science that is concerned with the study of bodies of fresh water.

**line** One of the three basic concepts in Euclidean geometry that is not strictly defined. The others are plane and point (see *plane, point*). A true line is an infinite, continuous set of points, stretching forever in two directions exactly opposite each other. Two points determine a unique line in space along the shortest path between the points.

**linear** 1. Straight; resembling a line. 2. For two variables, the property of being in proportion, so that the graph is a straight line. 3. Having an output level that is in direct proportion to the input level.

**linear accelerator** An atom smasher that uses electric and/or magnetic fields to move charge particles in a straight line at high speed. These particles then split the nuclei of target atoms.

**linear algebra** A branch of algebra concerned with the solutions of sets of linear equations. See *linear equation, linear function*.

**linear amplifier** An amplifier with power output in direct proportion to power input. If the input is doubled, so is the output, for every instant of time. Therefore, the output signal has a wave shape (envelope) that is the same as that of the input, except stronger.

**linear equation** 1. An equation whose graph is a straight line on the Cartesian plane or in Cartesian n-space. Sets of linear equations have a common solution if and only if the lines all intersect at a point. 2. An equation whose graph in n-space is a Euclidean subset of that space. This usually means that no variable is raised to a power larger than 1.

**linear function** 1. A function of the form $f(x) = y = mx + b$ in the Cartesian plane. The independent variable is $x$; the slope is $m$; the y-intercept point is $b$. The function appears as a straight line (see drawing). 2. A function whose graph in n-space is a Euclidean subset of that space. This usually means that no variable is raised to a power larger than 1.

LINEAR FUNCTION

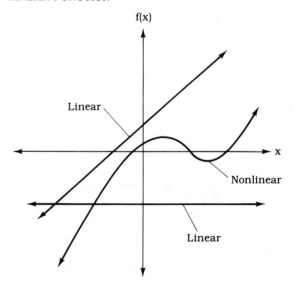

**linear motor** A type of motor that works along a straight line instead of going around and around like a conventional motor. This device is mainly experimental.

**linear programming** A method of finding an optimum solution to a problem with many variables. If all the variables can be represented in terms of linear functions, optimization can be done using linear algebra. See *linear algebra, linear function*.

**lines of flux** See *flux lines*.

**line of position** An imaginary line on the surface of the earth determined by sighting the sun with a sextant. Used in older methods of ship navigation. See *sextant*.

**line of sight** 1. Any path that a photon of visible light can follow. 2. A mode of communications, in which radio waves travel a path directly rather than by ionospheric refraction.

**line segment** 1. Two points and all the points between, along a Euclidean line connecting them. 2. Two points and all the points between, that lie along the shortest path connecting the points.

**linkage** The tendency for certain inherited traits to occur together. This happens when various groups of genes exist in a single chromosome.

**Linnaeus** Actual name, Carl von Linne. An eighteenth-century scientist remembered as the first to thoroughly classify living things.

**linoleic acid** A compound known to be essential in human nutrition. It is found in vegetable fats. Only a tiny amount is needed each day; whole grains and vegetables usually provide enough for a healthy person. See *fatty acid*.

**linolenic acid** A compound known to be essential in human nutrition, similar to linoleic acid. See *fatty acid, linoleic acid*.

**lipase** A digestive enzyme that breaks down fat in the small intestine. The resulting fatty acids and glycerol are easily absorbed. See *fatty acid, glycerol*.

**lipid** Fats and fat-like substances. See *fat*.

**lipoprotein** A substance that helps in the transport of fats and cholesterol throughout the body. Blood is water based; the lipoprotein, consisting of a protein-fat combination, serves as a sort of emulsifier for fat and cholesterol. See *high-density lipoprotein, low-density lipoprotein*.

**liposome** An artificial, tiny, cell-like particle, that can be used to treat certain diseases, particularly cancerous tumors. Poisons are carried by these particles; the particles invade diseased cells, killing them, while leaving healthy cells alone.

**liquation** A method of separating the solids out of a mixture. When there are several solids mixed into a liquid, and the solids have different melting points, the whole mixture can be heated gradually, and the solids will melt to liquids one by one. This makes it easy to take the desired substances out.

**liquefaction** 1. A phenomenon that occurs in soils when seismic waves pass through them. The waves cause the soils to lose their strength for a short time, so they flow as liquids. 2. A process in which a solid or a gas is made liquid. This might involve changes in temperature, or pressure, or both.

**liquefied petroleum** Short-chain hydrocarbons, normally gaseous, pressurized and stored as liquid. Propane is one example, used in powering portable generators. Butane is another, commonly used in cigarette lighters.

**liquid** One of the three states of matter, the other two being gas and solid (see *gas, solid*). Liquids fill containers in which they are placed, but settle because of gravity. Some liquids can be compressed, and some cannot. Some substances, at some pressures, do not have a liquid phase at all; an example is carbon dioxide at normal atmospheric pressure.

**liquid crystal** A liquid that has some characteristics of a crystal. See *crystal, liquid-crystal display*.

**liquid-crystal display** Abbreviation, LCD. A digital display, in which strips of liquid crystal material are used. When currents flow through

LIQUID-CRYSTAL DISPLAY

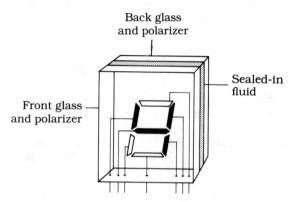

From Turner & Gibilisco, THE ILLUSTRATED DICTIONARY OF ELECTRONICS—4TH EDITION (TAB Professional and Reference Books, 1985).

the material, it darkens. Segments can be alternately darkened to form letters and numbers (see drawing).

**liquid gas** 1. A gas that is cooled so much that it turns to liquid. Examples are liquid oxygen and liquid nitrogen. 2. See *liquefied petroleum*.

**Lissajous figure** A pattern that is displayed on an oscilloscope when waves having certain frequencies are applied to the vertical and horizontal inputs. When one wave has a frequency that is a small, whole-number multiple of the other, the patterns show the exact frequency and phase relationship between the two waves.

**liter** The Standard International unit of volume. It is a little larger than a fluid quart. A cube measuring 10 centimeters (0.1 meter) on each edge has a volume of 1 liter.

**lithium** Chemical symbol, Li. An element with atomic number 3. The most common isotope has atomic weight 7. In pure form it is a reactive metal. Lithium compounds are used to relieve psychological depression.

**lithosphere** A solid outer layer of the earth's mantle, on which the continental plates ride. This layer is about 60 miles thick. See *mantle*.

**litmus paper** A strip of paper treated with a chemical that turns pink or red when immersed in an acid solution (pH less than 7), and blue or violet in an alkaline solution (pH more than 7). It does not change color at all in a neutral solution (pH = 7). Some types can be compared with standard color samples to determine pH. See *pH*.

**Little Ice Age** A period from about A.D. 1400-1850, during which weather was cooler than normal and the Arctic glaciers expanded. Sunspots were rare during this time (see *Maunder Minimum*), leading some scientists to speculate on a possible connection between weather and sunspot numbers.

**littoral** 1. Existing in shallow water. 2. The part of the ocean bottom between extreme high and low tide lines; that is, the part that is sometimes submerged but not always.

**liver** An organ in the body of many animals and of humans, responsible for filtering the blood. Also stores nutrients and makes bile. In humans the organ is located on the right-hand side of the abdomen, just below the right lung. It is about the size of a grapefruit. The cells are clustered in groups called lobules. See *lobule*.

**liver cirrhosis** See *cirrhosis*.

**liverwort** A primitive plant found in wet places. It reproduces by means of spores. It is closely related to mosses. For its place on the plant evolutionary scale, see the PLANT CLASSIFICATION appendix.

**living fossil** An organism that was thought to be extinct, and then is found alive somewhere. This occasionally happens with marine animals. The coelecanth is a recent example.

**Livingston, Stanley** A professor who helped to build the first cyclotron for smashing atoms. See *cyclotron* and *Lawrence, Ernest*.

**lixivation** A method of separating the components of a mixture by using water to dissolve those that are water soluble. This leaves the other substances behind. Those dissolved in the water can be recovered by letting the water evaporate or by boiling it off.

**lm** Abbreviation for lumen.

**load** 1. The circuit or device into which power is fed for a certain purpose in an electronic system. 2. The machinery driven by a motor.

**lobe** 1. A large portion of a body organ; for example, we might speak of the frontal lobe of the brain. 2. A part of an antenna radiation pattern, showing a direction in which signal strength or response is high.

**lobule** A group or cluster of cells in the liver, about a millimeter in diameter. Each such group gets nutrient-rich blood from the portal vein. The blood is then sent to each individual cell in the group. See *liver, portal vein*.

**local anesthesia** The "numbing" of a part of the body for a certain medical purpose. A wounded finger might be injected with novocaine or some similar local anesthetic before stitches are used to close the cut. See *general anesthesia*.

**local group of galaxies** The Milky Way (our galaxy) and several other galaxies that exist in a cluster. There are other clusters of galaxies, congregated in a so-called supercluster. These are in turn arranged in huge strings throughout the known universe. Of course "local" is a relative term, in this case meaning "within about 3,000,000 light years." See *light year*.

**local maximum** For a mathematical function, a point for which all the values on either side are smaller, within an interval containing the point (see drawing).

**local minimum** For a mathematical function, a point for which all the values on either side are larger, within an interval containing the point (see drawing).

**Locke, John** A philosopher and scientist of the seventeenth century, noted for having said that ideas are formed from what we sense, and from our reflections (thoughts).

**lockjaw** See *tetanus*.

**locus** 1. A region defined by a set of points. 2. A general location at which something occurs; for example, the intestinal locus for the absorption of vitamin B-12.

LOCAL MAXIMUM
LOCAL MINIMUM

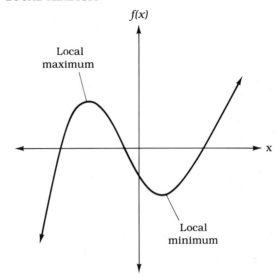

**lodestone** Iron ore (magnetite) that has become permanently magnetized. Such a stone, when suspended from a thin cord, will orient itself according to the earth's magnetic field.

**loess** A type of soil sometimes found at temperate latitudes. It is usually light to dark tan in color, loose, and accumulates after being blown by the wind.

**logarithm** 1. For a given number, the power of 10 that yields that number. This is called the base-10 logarithm. 2. For a given number, the power of the irrational number e, where e is about 2.718, that yields the number. This is called the natural logarithm.

**logarithmic curve** 1. The graph of the logarithm function. See *logarithm*. 2. Any curve that involves the logarithm function.

**logarithmic decrement** A decay curve that can be expressed according to the logarithm function. Also sometimes called exponential decay. See *logarithm*.

**logarithmic scale** A number line used either alone or combined with others to make a coordinate system (see drawing), graduated logarithmically rather than linearly. The numbers are

LOGARITHMIC SCALE
LOG-LOG GRAPH

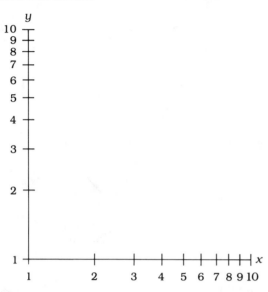

spaced according to their logarithm. Usually, the base-10 logarithm is used. See *logarithm*.

**logarithmic spiral** A special spiral in polar coordinates. It winds outward counterclockwise, in such a way that its total length, measured outward from the origin, is always equal to the length of a tangent line that intersects the radial axis. The formula is an exponential function.

**logic** 1. A branch of mathematics dealing with rigorous and consistent thought. With set theory, it forms the basis for modern scientific analysis. See *set theory*. 2. Digital electronics. See *logic gate*. 3. Rational thinking, in which conclusions are drawn from hypotheses, using agreed-on rules.

**logical analysis** See *logic, 3*.

**logical deduction** 1. A proof of a theorem using logic. 2. A process of elimination, in which a hypothesis is proved by disproving all the other possibilities.

**logic circuit** An electronic circuit, consisting of a combination of logic gates, designed to perform a specific logic operation or function. See *logic function, logic gate*.

**logic diagram** 1. A pictorial breakdown of a logic function, similar to a flowchart. 2. A truth table. See *truth table*.

**logic function** A series or set of logical operations NOT, AND, inclusive OR and exclusive OR, in a combination such that there is a certain output (true or false) for each combination of inputs. There might be several, or many, possible input combinations.

**logic gate** An electronic device that performs one of the logical functions NOT, AND, inclusive OR or exclusive OR. Additionally, there are inclusive or exclusive NOR (NOT-OR) and NAND (NOT-AND) gates.

**Logicism** A school of thought holding the belief that the rules of mathematics proceed according to a certain pattern (logic), and are not completely game-like or up-to-chance. Bertrand Russell was a proponent of this philosophy. See *Formalism, Intuitionism*.

**logistic** 1. Pertaining to computational arithmetic, as opposed to number theory or abstract algebra. 2. In some fields of science, engineering, and in the military an organizational or procedural detail. 3. Pertaining to logic. See *logic*.

**log-log graph** A graph on a plane that is a combination of two logarithmic scales (see drawing). See *logarithmic scale*. The base-10 logarithm is generally employed for both scales.

**longevity** See *life expectancy*.

**longitude** The east-west displacement from the prime meridian on the earth. Measured in degrees from 0 to +180 (Eastern Hemisphere) or 0 to −180 (Western Hemisphere). The line at +180 is the same as that at −180 and corresponds approximately to the International Date Line. See *International Date Line, latitude, prime meridian*.

**longitudinal wave** Also called a compression wave. A wave disturbance in which particles oscillate back and forth along the same line as the direction of propagation (see drawing). See *transverse wave*.

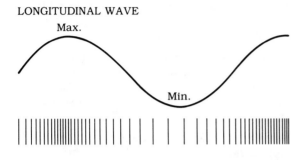

LONGITUDINAL WAVE

**LORAN** Acronym for LOng-RANge navigation. A system used by mariners to determine ships' positions. Also can be used by aviators to determine aircraft positions. Operates by means of low-frequency or medium-frequency radio pulses.

**Lorentz contraction** Also called the Fitzgerald contraction; first shown by Hendrick Lorentz and George Fitzgerald in the late nineteenth century. At high speeds, the shrinking of an object along the direction of its motion. Einstein later derived it from his relativity equations, and demonstrated the reason why it occurs. See *relativity*.

**Lorentz, Hendrick** A theoretical scientist who lived at the same time as Einstein during the early development of relativity theory. Einstein derived much of his thinking from the ideas of Lorentz. See *Einstein, Albert* and *relativity*.

**Lorentz transformation** A set of equations that converts motion and displacement from one point of view into motion and displacement from some other point of view, according to the theory of relativity. See *relativity*.

**Lorenz attractor** A geometric pattern that arises from apparently random or chaotic variables. The path of a point spirals around and around in an approximate figure-eight. The exact position cannot be predicted very far in advance, but the pattern is maintained indefinitely.

**Lorenz, Konrad** Also called the "father of ethology." Ethology is the science of animal behavior. An Austrian zoologist who showed that ducklings could be taught from birth to accept himself as their mother. This is called imprinting. See *imprinting*.

**loss** 1. See *attenuation*. 2. Power or energy that is wasted, usually as heat, rather than being delivered as intended. 3. The mathematical opposite of gain, measured in decibels (dB). See *decibel, gain*.

**loudness** The relative intensity of sound. Can be specified in general terms, or quantitatively. Sound is perceived to increase as to the logarithm of its intensity in watts per square meter. The most common unit of expressing sound level is the decibel. See *decibel*.

**loudspeaker** See *speaker*.

**Lovell, A. C. B.** The radio astronomer who oversaw the construction of the 250-foot steerable radio telescope dish at Jodrell Bank, in Cheshire, England. The project was completed in the 1950s. Lovell saw this project through despite great financial hardship.

**Love wave** A long-distance, slow-moving seismic wave, that propagates in the transverse mode (back and forth). These waves cause the greatest ground movements. Sometimes they have been detected after as long as four days following the quake itself.

**low** 1. A weather system in which atmospheric pressure is lower than the surrounding pressure. Often contains weather fronts (see *weather front*). Air circulates around it (see drawing). Generally brings precipitation and sometimes stormy weather. The influence of a single low may extend over a diameter of more than 2000 miles. 2. A condition of being decreased in level, intensity or physical position. 3. In digital electronics, the more negative of the two voltages. See *high*.

**low-calorie diet** A diet that is restricted in quantity, so that a loss of weight occurs. One takes in less food fuel than one burns, so body fat makes up the fuel difference. Any special diet requires a doctor's supervision, especially if used by children.

**low-carbohydrate diet** A diet that is restricted in the amount of carbohydrate-contain-

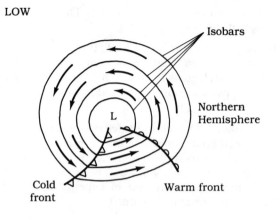

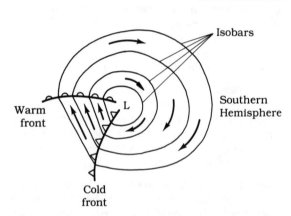

ing foods one eats. As with any special diet, a doctor's supervision is required. See *carbohydrate*.

**low-fat diet** An eating pattern in which fat is restricted severely. This is normally done under a doctor's supervision, because small amounts of certain fats are necessary for health. See *essential fatty acids*.

**low-density lipoprotein** Abbreviation, LDL. The medium of transport in the body for so-called "bad cholesterol." It is one of two cholesterol-carrying components in the blood, the other being high-density lipoprotein (HDL). See *cholesterol, high-density lipoprotein*.

**low frequency** Abbreviation, LF. The range of radio frequencies from 30 kilohertz (kHz) to 300 kHz. These are sometimes called "long waves." The wavelengths range from 1 kilometer (km) to 10 km, or about 0.6 to 6 miles.

**low-level radioactive waste** Radioactive by-products from industrial processes that are not high-level and not transuranic. The half life may still be very long. See *half life, high-level radioactive waste, transuranic element*.

**low-protein diet** An eating pattern in which protein intake is restricted. This should be done only with a doctor's supervision. Certain disease states are treated with this type of diet. A good amino-acid balance is important. See *essential amino acids*.

**LSD** Abbreviation for lysergic acid diethylamide.

**LSI** Abbreviation for large-scale integration.

**Lu** Chemical symbol for lutetium.

**lubricant** Any substance, especially petroleum-based, used to reduce friction in machinery with moving parts.

**lubrication** The application of a substance to reduce friction in machinery with moving parts. See *lubricant*.

**lumbar nerves** 1. A set of five branches of nerves that come off the lower spinal cord in humans. 2. Nerves branching off the lower spinal cord in vertebrates.

**lumbosacral plexus** The network of nerves that come off the lower spinal cord, running to the buttocks and legs. Includes the lower spinal cord itself, the lumbar nerves, the sacral nerves, the femoral nerves, and the sciatic nerve.

**lumen** Abbreviations, lm, l, lum. The Standard International unit of luminous flux. One lumen is the light emitted in one steradian of solid angle, by a source with an intensity of one candela. See *candela, steradian*.

**luminiferous ether** A hypothetical substance through which light waves were once thought to travel, the same way sound travels through air. Einstein's theory of relativity did away with the so-called ether theory, simplifying physics. See *Michelson-Morley Experiment, relativity*.

**luminance** The amount of light emitted or scattered by a surface. Usually this is expressed in candela per square meter. See *candela*.

**luminescence** Light without heat. This is the case, for example, in a phosphor that glows when X rays strike it. This phenomenon can be used to identify certain minerals with ultraviolet light, because some substances glow with characteristic colors when exposed to this energy.

**luminosity** 1. The brightness of a star. This is expressed in units called magnitude. See *absolute magnitude, apparent magnitude.* 2. The brightness of any source of light, given in candela or lumens. See *candela, lumen.* 3. An expression of the efficiency of a light source, given as the ratio of luminous flux at a certain wavelength to the total radiated flux.

**luminous flux** 1. Electromagnetic field density, for waves in the visible range. 2. The flow of visible light energy.

**luminous sterance** See *luminance*.

**Luna probes** A series of space probes launched by the Soviet Union to explore the moon. The first, Luna 1, flew within 5000 miles of the moon in 1959. Luna 2 crash-landed there, also in 1959. Luna 3 photographed the far side of the moon in November, 1959. Several more probes followed.

**lunar** Pertaining to the moon.

**lunar eclipse** See *eclipse*.

**lunar excursion module** Abbreviation, LEM. In the Apollo missions, the spacecraft that detached from the command module and descended to land on the moon. It was equipped with its own launch system for returning to the command module. See *Apollo, command module*.

**lunar orbiters** A series of space probes launched by the United States in the 1960s, that photographed the moon and took other data measurements from orbit around the moon.

**lunar phase** 1. Any of several parts of the moon's 28-day cycle around the earth. See the

LUNAR PHASE

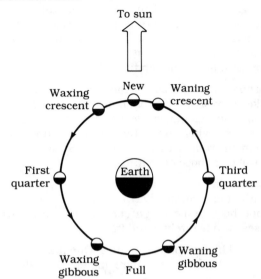

drawing for details. 2. The shape of the sunlit part of the moon as seen from the earth.

**lung** Either of two large organs in the chest cavity. These organs supply oxygen to the blood, and also help eliminate carbon dioxide and other wastes. See the drawing.

LUNG

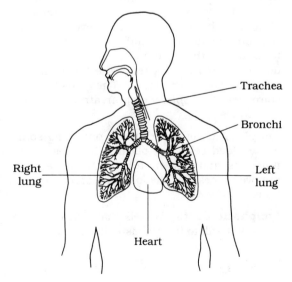

**lung cancer** Growth of malignant cells in the lungs. Smoking and air pollution are thought to be major contributors to the increase in lung cancer deaths in recent years. Also, to some extent, all cancers might be part of the aging process, given enough time. See *cancer*.

**lupus** 1. A disease of the skin, with pimple-like lesions. Sunlight often makes it worse. Most common in females age 30 and older. 2. A disease involving connective tissues, more common in women than in men, with symptoms similar to those of arthritis.

**lutetium** Chemical symbol, Lu. An element with atomic number 71. The most common isotope has atomic weight 175. It has no known uses, and is rarely found in nature.

**lux** The unit of illuminance. One lux is equal to one lumen per square meter. See *illuminance, lumen*.

**Lw** Chemical symbol for lawrencium.

**lx** Abbreviation for lux.

**lye** 1. Sodium hydroxide. 2. Potassium hydroxide. 3. A mixture of hydroxides, strongly alkaline, with high pH. See *pH*.

**Lyme disease** A complex illness transmitted by a small tick and prevalent especially in the Great Lakes and New England regions, although it has been seen almost everywhere in the continental United States. The early symptoms are like the flu, but later, nervous troubles develop. Almost any part of the body may be affected. The microorganism responsible is a spirochete (see *spirochete*). Treatment is with antibiotics.

**lymph** A clear, watery fluid containing some white blood cells and plasma. It helps to get nutrients and oxygen to body cells, and to remove wastes. See the following several definitions.

**lymphatic ducts** Vessels that run from the intestinal wall to the bloodstream, carrying fats to the blood where they can be delivered to the cells of the body.

**lymphatic system** A complex and extensive body network, consisting of lymph nodes, ducts and vessels. See *lymph, lymphatic ducts, lymph node*.

**lymph node** A place where lymphatic vessels converge, and where lymphocytes are made. See *lymphocyte*. These "glands" often become enlarged and painful during infectious diseases.

**lymphocyte** A type of white blood cell that produces antibodies to fight infections. These are manufactured by lymph nodes in great numbers when disease threatens. This is what makes your "glands" sore when you are coming down with an illness. See *antibody, lymph node, white blood cell*.

**lysergic acid diethylamide** Abbreviation, LSD. A substance that, when taken internally, causes hallucinations, delusions, and, in some people, long-term mental illness. It was developed as an experimental drug in the mid-twentieth century. It has been responsible for some deaths, because, for example, a person thought he could fly and jumped from a cliff.

**lysine** An amino acid that must be obtained from dietary protein, because it cannot be manufactured in the body from other amino acids. See *amino acid*.

**lysis** The killing of a cell. This is normal when cells get too old to function any more in the body. Antibodies kill undesirable bacteria. Certain medical treatments kill various cells, such as cancer cells.

**lysoenzyme** See *lysozyme*.

**lysosome** In cells, a structure with enzymes that help the cell utilize its food. It also helps defend the cell against invasion from bacteria or viruses.

**lysozyme** An enzyme in the body that weakens bacteria. It is especially concentrated in fluids such as saliva.

**M** Abbreviation for mass, mega- (see the PREFIX MULTIPLIERS appendix).

**m** Abbreviation for mass, meter, milli- (see the PREFIX MULTIPLIERS appendix), minute.

**mach** An expression of speed. Written *mach* followed by a number, indicating the speed relative to sound. The drawing shows an aircraft moving at mach 0.5 (half the speed of sound), mach 1.0 (the speed of sound) and mach 1.5 (one and a half times the speed of sound). At mach 1.0, about 700 miles per hour in the air, the aircraft just keeps up with sound waves in front of it. This compression causes the sonic boom. See *sonic boom*, *speed of sound*.

**Mach, Ernst** A physicist who, around the end of the nineteenth century, was one of the first to question Isaac Newton's concepts of absolute space and time. He concluded that Newton's principles are not self-evident. His work influenced Albert Einstein in the development of relativity theory. See *Einstein, Albert* and *relativity*.

**machine** 1. Any device that makes work easier. 2. Any of six simple devices, the **inclined plane, lever, pulley, screw, wedge,** or **wheel**

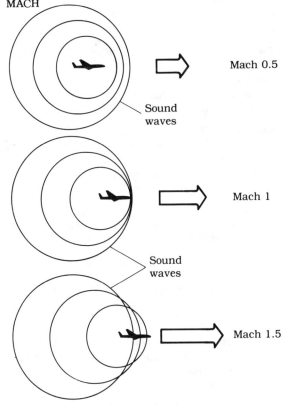

MACH

**and axle.** 3. A device or set of devices, usually mechanical, that helps in manufacturing, industry, food processing, medical care, or laboratory work. See *hardware*.

**machine language** The simple language used directly by a computer. It is tedious to program a computer in this language. This is why we have higher-order languages such as BASIC and FORTRAN. An assembler program translates between the higher-order language and the machine language. See *assembly language*.

**Maclaurin series** Any of various infinite series, that add up to certain mathematical function values. Many functions can be expressed as infinite series. See *infinite series*.

**macro-** 1. A prefix meaning "large" or "large-scale." 2. A prefix referring to whole systems or sets of systems.

**macroeconomics** The study of money-flow patterns, production, and employment on a large scale (a whole society or the world).

**macroinstruction** In a computer, an instruction in the source program or user program, that translates into several machine-language instructions. A high-level instruction. See *machine language*.

**macromolecular** 1. Consisting of very large molecules, such as long-chain hydrocarbons or deoxyribonucleic acid (DNA). 2. Having a crystal structure with comparatively large molecules.

**macromolecule** A comparatively large, or long-chained, molecule, such as certain hydrocarbons, polyethylene and nucleic acids.

**macronutrient** A substance needed by the body in comparatively large amounts. The best examples are carbohydrate, fat and protein. Certain minerals may also be considered, such as phosphorus and calcium. See *micronutrient*.

**macrophage** A large cell in the body that eats smaller cells. This helps to rid the body of unwanted microorganisms, such as disease-causing bacteria.

**macroprogram** Any computer program in a higher-order language, such as BASIC, COBOL or FORTRAN.

**macrotide** A tide whose range is more than 12 feet between highest and lowest. The nature of the coastline, and the slope of the ocean floor near the shore, determine the extent of difference between high and low tide. Extreme tides are most common in bays and gulfs. See *tide*.

**macula** A region of tissue in some part of the body, that is different from the tissue around it. These regions are a normal part of the anatomy (not caused by disease).

**Magellanic clouds** Two small irregular galaxies next to our galaxy, the Milky Way (see *Milky Way*). These appear as dim, fuzzy spots in the sky as seen in the Southern Hemisphere. They are named after the explorer Magellan who captained a ship that sailed around the world.

**magic number** In an atom, a number of protons or neutrons that result in stable elements. These numbers are 2, 8, 20, 28, 50, 82 and 126. In the case of protons, the elements are, respectively, helium, oxygen, calcium, nickel, tin and lead (there is no element with 126 protons). In the case of neutrons, various isotopes of different elements may have magic numbers in their nuclei. See the ATOMIC NUMBER AND ATOMIC WEIGHT appendix.

**magic square** A square matrix of integers, such that the sum for each row, the sum for each column, and the sum of either diagonal are all identical.

**magma** Hot, molten rock under the surface of the earth. It comes to the surface in active volcanoes (see drawing). See *mantle, volcano*.

**magnesia** See *magnesium hydroxide*.

**magnesium** Chemical symbol, Mg. An element with atomic number 12. The most common isotope has atomic weight 24. In pure form it is a metal that looks much like aluminum. It reacts readily with elements such as oxygen, chlorine and sulfur. Industrially it is combined with aluminum to form strong, light alloys.

MAGMA

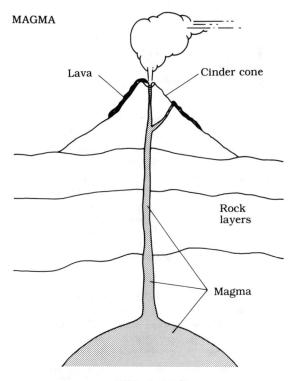

From Erickson, VOLCANOES AND EARTHQUAKES (TAB Books, Inc., 1988).

**magnesium hydroxide** An alkaline compound containing magnesium, hydrogen and oxygen. It is a white powder in pure form. It can be used as a laxative. It draws water into the intestine, causing the bowel to expel its contents.

**magnet** Any device that produces a magnetic field that is used for some purpose. The field may be stable, or it may be alternating in polarity. See *electromagnet, permanent magnet*.

**magnet coil** A coil of wire through which a current is passed in order to produce a magnetic field. There is usually a core of magnetic metal to increase the strength of the field. Speakers and microphones often contain such coils. The electromagnet is a special form of this device. See *electromagnet*.

**magnetic** 1. Pertaining to magnets and magnetism. 2. Surrounded by a magnetic field (see *magnetic field*). 3. Producing magnetic force (see *magnetic force*).

**magnetic circuit** The closed path followed by a line, or by a group of lines, of magnetic flux. See *magnetic flux*.

**magnetic compass** See *compass*.

**magnetic core** A piece of ferromagnetic material, placed within a coil or transformer to increase the inductance and/or the degree of coupling between the windings.

**magnetic coupling** The effect in which a changing current in a conductor produces a changing current of the same nature in a nearby conductor. The original current causes a fluctuating magnetic field, and this in turn induces the second current (see drawing). This is the principle by which a transformer works. See *transformer*.

MAGNETIC COUPLING
MAGNETIC FIELD

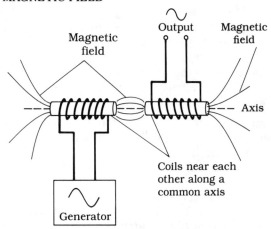

**magnetic deflection** Also called electromagnetic deflection. In a cathode-ray tube, the change in direction of the electron beam, caused by magnetic fields from deflector coils. The direction of change depends on the field polarity; the extent of change depends on the field intensity. See *cathode-ray tube*.

**magnetic dipole** 1. A pair of magnetic poles, one north (N) and the other south (S), connected by lines of flux. 2. A molecule with a pair of opposite magnetic poles, and surrounded by a magnetic field.

**magnetic disk** Either a floppy disk or a hard disk, in diameters ranging from 3.5 inches to several inches, on which magnetic particles are affixed, and used to store computer data. A typical floppy disk can store about 1.0 to 1.5 megabytes of information. See *floppy disk, hard disk*.

**magnetic equator** The line around the earth, representing the set of points on the surface that are all equidistant from the North Magnetic Pole and the South Magnetic Pole.

**magnetic field** A force field that surrounds any permanent or temporary magnet. This can be the result of orientation of magnetic molecules, such as in a piece of iron or steel; it may be caused by an electric current. See the drawing and the definition *magnetic flux*.

**magnetic field reversal** See *geomagnetic field reversal*.

**magnetic flux** The imaginary "lines of force" around any magnetic object, or around a current-carrying conductor. The density of these lines is the intensity of the magnetic field. Usually this intensity is given in gauss, teslas or webers. See *gauss, tesla, weber*.

**magnetic force** 1. The force caused by the interaction of a magnetic field and a magnetic metal such as iron or steel. This is always an attractive force. 2. The force caused by interaction of two magnetic fields. This is attractive if the poles are opposite, and repulsive if the poles are alike. 3. The force caused by a magnetic field on a moving, charged object in that field. This causes a change in the direction of the object's motion.

**magnetic inclination** The angle that geomagnetic flux lines make with respect to the earth's surface. Also, the downward pointing of a magnetic compass. At the magnetic equator, this angle is zero; at the magnetic poles it is 90 degrees.

**magnetic levitation** The use of like magnetic poles, repelling each other, to suspend an object above another. This principle is contemplated for use in high-speed trains, where the cars could be suspended above the rails, minimizing friction.

**magnetic media** Any data storage medium, such as tape or disk, in which magnetized particles are used to preserve information bits.

**magnetic moment** For a magnet, or a moving charge in an external magnetic field, the ratio of the torque to the strength of the external field. This ratio depends on the strength of the magnet (or the field produced by the moving charge), the strength of the external field, and the angle of the magnet (or the moving charge) with respect to the external field.

**magnetic north** The direction in which a compass needle points. This is along a great circle, or geodesic, toward the North Geomagnetic Pole.

**magnetic polarization** 1. The orientation of the flux lines of a magnet, or of a magnetic field produced by a moving charge. See *magnetic flux*. 2. The nature (north or south) of a magnetic pole. See *magnetic pole*. 3. The alignment of magnetic dipoles in a ferromagnetic material, causing an external magnetic field. This may be a temporary condition (see *electromagnet*) or a permanent condition (see *permanent magnet*).

**magnetic pole** 1. Any point to which magnetic lines of flux converge. Can be either north (N) or south (S). 2. Either the North Geomagnetic Pole or the South Geomagnetic Pole of the earth.

**magnetic resonance imaging** See *nuclear magnetic resonance imaging*.

**magnetic separation** A means of refining low-grade iron ore, such as taconite. The ore is crushed into fine particles, and magnets are used to attract the magnetite (see *magnetite*), leaving the other material behind.

**magnetic tape** A long strip of plastic, on which magnetic particles are affixed. It is usually gray or red to brown in color, and is available in various widths for audio and video recording and playback. It was once used to store computer data, but disks are now preferred for this. See *magnetic disk*.

**magnetism** The property of having or causing a magnetic field. This occurs when magnetic dipoles are aligned, and also when an electric charge moves. See *magnetic field, magnetic flux*.

**magnetite** A strongly magnetic, shiny black material, the richest and most important iron ore. It may be attracted away from more crude powder during refining, by the use of magnets. See *magnetic separation*.

**magneto** An electric generator that uses permanent magnets. The current is produced when wire coils are rotated with respect to the magnetic field. This may be done by rotating the coils or the magnet.

**magnetochemistry** A subscience of chemistry and physics, involved with the study of the magnetic behavior of substances.

**magnetohydrodynamics** A branch of physics and engineering, dealing with the behavior of electrically conductive fluids in magnetic fields.

**magnetometer** A device for measuring the strength and orientation of a magnetic field.

**magnetomotive force** The magnetic equivalent of electric potential; the agent responsible for causing a magnetic field. May be expressed in ampere-turns (for a coil) or in gilberts. See *ampere-turn, gilbert*.

**magneton** A unit for measuring the intensity of a magnetic moment. The classical unit, the Bohr magneton, has a value of $9.27 \times 10^{-24}$ amperes per meter squared. This is the magnetic moment of an individual electron. See *magnetic moment*.

**magnetosphere** The magnetic lines of flux that surround the earth. The field's shape is distorted by the constant flow of subatomic particles from the sun (see drawing). These particles are deflected by the flux, and this protects the earth from being bombarded by them. See *magnetic flux, solar wind*.

**magnetostriction** A force that occurs on a bar of ferromagnetic material when a magnetic field

MAGNETOSPHERE

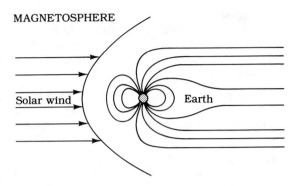

From Erickson, VIOLENT STORMS (TAB Books, Inc., 1988).

is applied to it. This is the equivalent of electrostriction in a dielectric. See *electrostriction*.

**magnetron** A type of electron tube that produces ultra-high-frequency and microwave oscillations. Used in some radio and radar transmitters at these frequencies. It works by the action of a powerful magnetic field on a stream of electrons in a near vacuum.

**magnification** The enlargement of a visual image. This is usually done with sets of lenses. See *magnification factor, microscope, telescope*.

**magnification factor** The extent to which an image is magnified, measured along any single axis. Thus a factor of two means that each dimension is doubled in apparent size; a factor of n means that each dimension is increased n times (see drawing). The apparent area therefore increases by the square of this factor.

MAGNIFICATION FACTOR

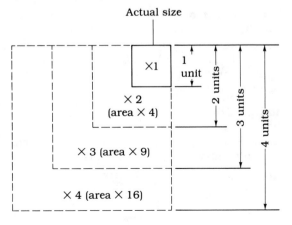

**magnitude** 1. Intensity, especially for a source of visible light or electromagnetic radiation. 2. Size or scope. 3. An expression of brilliance for a star. See *absolute magnitude, apparent magnitude*.

**mainframe** The core of a computer; the main part, containing the central processing unit, the arithmetic logic unit, and the memory and logic circuits.

**Main Sequence** The continuous, S-shaped line along, or near, which most stars are found in the Hertzsprung-Russell graph of brightness versus spectral type (color). See *Hertzsprung-Russell diagram*.

**Majaj, Amin** An Arab scientist who, in 1954, discovered a type of anemia that had been resistant to conventional treatment methods. He did this without the benefit of a lab.

**majority carrier** In a semiconductor, the predominant current carrier. In N-type material these are electrons; in P-type material they are holes. See *electron, hole, minority carrier, semiconductor*.

**make** The instant at which a signal pulse begins; the transition from "off" to "on" in a digital signal. 2. The closing of the contacts of a switch or telegraph key. See *break*.

**mal** Illness. For example, we may speak of *mal de mer* (seasickness) or *grand mal* (a severe epileptic seizure).

**mal-** A prefix meaning lack of, improper or inadequate.

**malabsorption** An undesirable condition of the digestive system, in which some nutrients are not absorbed, or are absorbed in insufficient amounts. This may be caused by allergy, poisoning, inherited genetic defects, or ingestion of some medications.

**maladaption** A failure of an organism to adapt to its environment. This generally causes a decrease in the population in a given area. Migration may occur. Extinction is possible.

**malformation** A physical defect or deformity, especially of a part of the body.

**malfunction** In machinery, the failure of a certain component. This can be a partial failure, such as reduced efficiency; it can be a total, or catastrophic, failure.

**malic acid** A mild acid found in raw fruits, especially when they are not fully ripened. It has a characteristic tart flavor and acts as a mild astringent.

**malignant** 1. Cancerous. The term is used in reference to tumors or other unnatural cell growth. 2. Highly poisonous.

**malignant melanoma** A form of skin cancer, in which a pigmented area, such as a freckle or mole, becomes cancerous. This is the most dangerous type of skin cancer because it can spread throughout the body.

**malignant tumor** A cancerous growth in the body. Especially, a localized growth, ranging in size from that of a pea to that of a grapefruit or even larger.

**malleus** One of the three bones in the middle ear that conducts sound vibrations; also called the "hammer." It is connected directly to the eardrum, and transfers its vibrations to the incus, or "anvil," and then to the stapes, or "stirrup."

**malnutrition** A lack of adequate amounts of one or more essential nutrients. The essential substances include protein, certain fatty acids, a certain minimum of carbohydrate, and various minerals and vitamins. There are different types of malnutrition for different deficiency states.

**malt** Processed grain, used in the making of beer and liquors. The starch is split partially by soaking in water.

**Malthusian model** A theory of population growth and decline, first put forth by Thomas Malthus. Population increases geometrically, but food supply only goes up arithmetically, until some crisis point is reached (see drawing). There are those today who say that such a crisis point for the human population is not far off. See *Malthus, Thomas*.

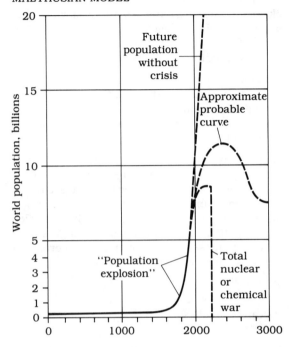

MALTHUSIAN MODEL

**Malthus, Thomas** A nineteenth-century English economist, known for his model of human population versus food supply. Eventually, with unchecked population growth, food shortages and starvation would occur. Malthus saw that it would be necessary to keep the population growth limited to only an arithmetic increase. See *Malthusian model*.

**maltose** A sugar, whose molecules consist of two glucose molecules bound together. Digestive enzymes break starch down into maltose, and then split this bond, yeilding glucose. See *glucose*.

**mammal** Any vertebrate, the female of which is capable of feeding its young directly via mammary glands. They give birth to live young. These animals are warm-blooded and usually have hair. Humans are in this class.

**mammary gland** A milk-secreting gland in female mammals. In humans these glands are in the breasts. Milk tends to be produced after giving birth, intended by nature for feeding the baby.

**mammatocumulus clouds** Literally, mother clouds. The main cloud or clouds in a severe thunderstorm, especially one that is accompanied by hail, damaging winds, or a tornado.

**mammogram** A photographic analysis of the breasts of the human female, made using X rays, ultrasound or similar techniques. It is a valuable aid in the early detection of breast cancer.

**Man Amplifier Project** A research-and-development program, concerned with the perfection of robots that mimic human movements, but with much greater strength. A person might then lift a lever, and the machine would lift a 10-ton piece of military hardware.

**Mandelbrot, Benoit** One of the pioneers, and probably the best known, of the science of chaos theory. He was born in Poland in 1924 and fled in the 1930s to France because of the Nazis. He had an independent spirit, and made his own way, eventually finding a niche at International Business Machines (IBM) and using computers to probe into the behavior of systems. His methods led to significant discoveries such as the set that bears his name. See *chaos theory, Mandelbrot set*.

**Mandelbrot set** An infinitely complicated graph on the complex-number plane, resulting from a very simple mathematical equation. Named for Benoit Mandelbrot, a researcher in the field of chaos theory. The whole set looks something like a plant bud. When examined in detail, it has seahorse-shaped curls, spirals, and infinitely many small similes of itself. No matter how much magnification is used, there is always great complexity. This has been regarded by some scientists as an indicator of the true nature of the physical universe. See *Mandelbrot, Benoit*.

**mandible** 1. True, movable jaws, as those of an insect. 2. The lower, movable jaw in humans and many other mammals (see drawing).

**maneuverability** The ease and speed with which a motor vehicle, boat, ship or aircraft can change direction.

**MANDIBLE**
**MAXILLA**

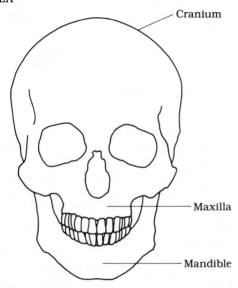

**manganese** Chemical symbol, Mn. An element with atomic number 25. The most common isotope has atomic weight 55. It is a metal in pure form. It has various industrial uses.

**manganin** An alloy made with copper, manganese and nickel. It is used to make wire for heating elements and electrical resistors, since it is only a fair conductor.

**Manhattan Project** The scientific program, conducted during the Second World War, that led to the development of the atomic bomb. The first bombs were crude and very small compared with the hydrogen weapons we have today.

**mania** A behavioral disorder, in which the main symptoms are restlessness, irritability, excitability and insomnia. In the extreme case, a patient may have to be restrained to prevent injury to himself/herself or others.

**manic-depressive psychosis** A severe mental illness in which mood states change from unreasonably happy to unreasonably sad. The cycle may occur every few weeks or months. It is an extreme "up-and-down" mood swing. It is a psychosis because the patient often has delusions, such as that the world is coming to an end tomorrow.

**mannitol** A carbohydrate, with a sweetness like sugar but metabolized more slowly. This makes it useful for use in candies and chewing gums, for example, that do not promote tooth decay as readily as sugar does.

**mannose** A simple sugar with the same chemical formula as (but a different molecular arrangement than) glucose. It is found in nature bound together in chains, similar to starches, in various plants.

**man-o-war** A large jellyfish with tentacles that may be several feet long, and an air-filled, usually bluish sac that keeps it afloat (see drawing). The man-o-war is actually several organisms living in a colony together. The tentacles kill small fish for food by means of stingers. The poison in these stingers can be dangerous to humans who come into contact with them.

MAN-O-WAR

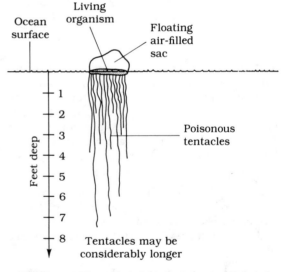

**mantissa** The part of a logarithm to the right of the decimal point. In base-10 logarithms, for numbers greater than 1 and less than 10, the mantissa is the same as the logarithm. For other numbers, the mantissa is different from the logarithm by some whole-number amount. See *logarithm*.

**mantle** The part of the earth that lies underneath the crust and above the core. It is made up of rocks containing much iron and magnesium. It is believed to be in a continuous state of convective motion, causing continental drift. See *continental drift*.

**Mantoux test**  A simple test to determine whether or not a person is infected with tuberculosis. A chemical is injected into the skin; 48 hours later the site is examined. A lump indicates probable infection, but not necessarily active tuberculosis.

**manufacturing engineering**  A branch of engineering concerned with the efficiency of factories. This involves not only the quality of the machinery, but the workability of the environment for those employed in the factory.

**map**  A diagram of some physical object, but especially of a region of the earth's surface. It may highlight certain features, such as elevation above sea level, the shape of a coastline, or the type of vegetation. Satellites are used in mapping of the earth and other planets.

**mapping**  1. A function between point sets, in which values from one set are assigned to values in the other set. 2. The diagramming of features of the earth's surface, or of some other planet's or moon's surface. 3. The diagramming of the features of any physical object, especially in space, such as the galaxy at radio wavelengths.

**marble**  A hard rock, usually white or light gray but also sometimes with shades of yellow or pink. It is metamorphosed limestone or dolomite, and consists mainly of calcium and magnesium compounds. It is used in elegant buildings.

**Marconi, Gugliolmo**  An engineer who sent the first wireless message across the Atlantic in 1901. There is some debate as to whether he should be called the "Father of Radio," since Nikola Tesla, a Russian, actually demonstrated its principles before Marconi. See *Tesla, Nikola*.

**maria**  Flatlands on the moon. The word means "seas." Before telescopes revealed their true nature, these regions were thought to be oceans, because they look darker than the surrounding terrain, as we would expect water to look from such a great distance away.

**Marianas Trench**  The deepest trench known in the oceans of the world, located near the Mariana Islands in the North Pacific Ocean (see drawing). The ocean is nearly seven miles deep there.

MARIANAS TRENCH

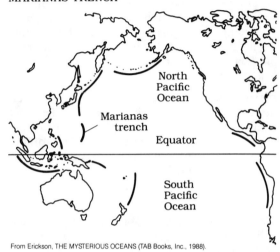

From Erickson, THE MYSTERIOUS OCEANS (TAB Books, Inc., 1988).

**mariculture**  The commercial raising of marine species, particularly for food, and to feed other fish that are in turn used for food. It is a fancy term for ocean-fish farming.

**marijuana**  See *cannabis*.

**marine biology**  A branch of biology, concerned with the physiology of ocean-dwelling animals. This includes not only fish, but plankton, shrimp, and all the strange creatures, many still unknown, living at great depths.

**marine engineering**  The field of engineering concerned with the design of boats, ships, and other water-borne vehicles.

**Mariner probes**  A set of space probes, launched by the United States in the 1960s and 1970s to explore Mercury, Venus and Mars. See the SPACE PROBES appendix.

**marrow**  See *bone marrow*.

**Mars**  The fourth planet from the sun in our Solar System. It has a thin atmosphere and is a rusty color. For a long time, it was thought that intelligent life dwelt there. Now, most scientists think that even microorganisms may not exist there. Its surface is constantly exposed to high levels of ultraviolet, and most of the water is locked up as ice. See the SOLAR SYSTEM DATA appendix for comparison with other planets.

**marsh gas** Flammable gases that form from decomposing organic materials, especially in swampy areas. Methane is probably the most abundant. These gases occasionally burn, causing flashes of light at night, and sometimes people mistake this for unidentified flying objects (UFOs).

**marsupial** A mammal with a pouch, in which newborn young live. The kangaroo is the most commonly recognized example. They probably originated in North America about 100 million years ago, and migrated to South America, then to Antarctica, and finally to what is now Australia before it broke away from Antarctica. Opossums are a North American example.

**Marx, Karl** One of the best known social scientists. He lived during the nineteenth century, and believed that capitalism could not survive. For a time it seemed as if perhaps he was right. Nowadays it seems like he was wrong. The main result of his thinking was communism, still followed in some parts of the world. Economically, however, communism is inefficient; and psychologically, it provides little or no motivation for the people to work.

**mascon** A distortion in the gravitational field of a planet, caused by uneven density of matter beneath the surface. This has been noticed on the moon.

**maser** A device for producing coherent electromagnetic energy at microwave range. The word is an acronym for microwave amplification by stimulated emission of radiation. See the illustration. See *coherent radiation, laser*.

**masking** 1. The blotting-out of one sound or noise by another. Usually this is because the other noise has more energy. It can also happen because of distraction to the listener. 2. A condition in which a medication or nutrient interferes with the diagnosis of a disease.

**mass** 1. The amount of matter in an object, measured according to the amount of inertia

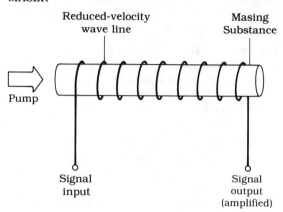

Traveling-wave type maser functional diagram

that it has. The standard unit is the kilogram. In the earth's gravitational field, a kilogram (kg) of mass weighs about 2.2 pounds. 2. A material object. 3. In industry, a term meaning "large-scale" or "in large quantities." Thus we speak of mass production, for example. 4. A region of tissue in the body that is more concentrated or dense than the tissue around it.

**mass-action law** The speed of a chemical reaction depends on the molar concentrations, or active masses, of the chemicals.

**mass defect** For an atomic nucleus, the difference between the mass of the nucleus, as compared with the sum of the masses of the protons and neutrons if they were each all by themselves.

**mass distortion** The relativistic effect that causes mass to be increased at very high speeds. As the speed approaches the speed of light, the mass increases without limit (see graph). It therefore takes great energy to accelerate any material object to extreme speeds. To get an object to attain the speed of light, an infinite amount of propulsion energy would be required. This is why no material object can be accelerated to the speed of light. See *relativity*.

**mass extinction** Extinctions of many species in a geologically short time, such as might be caused by a chemical war or by an asteroid hit-

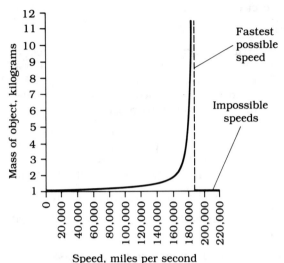

MASS DISTORTION

Speed, miles per second

ting the earth. There have been several of these events in the earth's geologic history.

**massless particle** 1. Any particle that can be accelerated to the speed of light because relativistic mass distortion does not limit its speed (see *mass distortion*). 2. A particle with spin but with essentially zero mass, and therefore with extreme penetrating power. See *spin*.

**mass production** An industrial process in which large numbers of identical items are made. Recently, computers and robots have been put to work for this, so that people are free to do more interesting work.

**mass spectrometer** A device that uses an electric or magnetic field to deflect moving, charged particles. The particles are deflected according to their speed, their charge and their mass. If the speed and charge are known, the mass can be calculated, and the particles identified.

**mastectomy** Surgical removal of the female breast. This is sometimes necessary in cases of breast cancer.

**mastitis** Inflammation of the female breast. This may be caused by any of various different conditions, including cancer.

**mastoid process** A porous part of the skull connected with the middle ear. It is partly responsible for conductive hearing, affecting the way that you hear your own voice.

**mathematical logic** A branch of mathematics, concerned with the logical foundations on which all rational thought is based. This is the field in which Kurt Godel specialized, proving his famous Incompleteness Theorem. See *Incompleteness Theorem*.

**mathematical model** A representation, in mathematical terms, of a material object, phenomenon or process. This representation is never truly exact, since the physical universe cannot be precisely measured; but a good model predicts things reliably.

**mathematics** The science concerned with numbers, equations, shapes, and logic. There are many specialties in this field, each with its own set of definitions, axioms and theorems, and many of which are essential for scientific study of the universe. See *applied mathematics, pure mathematics*. See also the table of sciences in the front of this book, for a list of all the branches of mathematics defined here.

**matrix** 1. A set of numbers arranged in a rectangle, and used to solve linear equations. See *linear algebra*. 2. Any array of objects in a generally rectangular pattern.

**matrix algebra** See *linear algebra*.

**Maudslay, Henry** An engineer of the late eighteenth century who is best known for having developed precision machine tools in England.

**Maunder minimum** The long-lasting sunspot minimum that occurred from about A.D. 1645-1715, in conjunction with the coldest part of the Little Ice Age. During this period, there were few sunspots; the 11-year peaks did not occur. See *Maunder, Walter*.

**Maunder, Walter** The English astronomer of the nineteenth century who, upon examining historical records, noticed that there was a period from about A.D. 1645-1715 during which there were not many sunspots. He noticed that this coincided with a period of unusually cool weather. See *Maunder minimum*.

**maxilla** The upper jaw, which in humans and many other mammals is not movable. It is where the upper teeth are set (see drawing for mandible).

**Maxim, Hiram Percy** The founder of the American Radio Relay League (ARRL), an organization of amateur radio operators, in the early twentieth century. Amateurs have made significant discoveries in radio, but they have had to fight at times to keep their privileges.

**Maxim, Sir Hiram** An engineer and inventor of the nineteenth and early twentieth centuries. He contributed to the development of powered flight. He also invented a machine gun.

**maximum** 1. The peak, or greatest, value or level in a variable quantity. 2. See *local maximum*. 3. Optimum.

**maxwell** Abbreviation, Mx. A unit of magnetic flux equivalent to ten billionths (0.00000001) of a weber. It is sometimes considered to be one line of flux. See *magnetic flux, weber*.

**Maxwell, James Clerk** A nineteenth century physicist who developed theories and equations concerning electromagnetic waves. He is best known for his mathematical analysis of these waves. See *Maxwell's Equations*.

**Maxwell's Equations** A set of equations developed by James Clerk Maxwell concerning electromagnetic energy. They can be summarized as follows. 1. The work needed to move a unit magnetic pole completely around a closed path is proportional to the total current linking the path. 2. The voltage induced in any non-moving, closed loop is proportional to the rate of change of the magnetic flux in the loop. 3. The total electric flux surrounding a charged object is equal to the charge quantity. 4. Magnetic flux lines have no beginning or ending points; they are always complete, closed loops.

**Mayer, Jean** A Harvard researcher in the field of nutrition, who concluded that the main cause of obesity is not overeating, but lack of exercise, especially in young people.

**mcg** Abbreviation for **microgram.**

**McCormick, Cyrus** An American engineer of the nineteenth century, who perfected a machine for harvesting grain. Although it needed horses to pull it, the design was used in later machines that were fully mechanized.

**McLellan, William** An American engineer, who, in 1960, built a prototype microminiature electric motor. It was about the size of a grain of sand. He was at least 20 years ahead of his time.

**Md** chemical symbol for mendelevium.

**MDT** Abbreviation for Mountain Daylight Time.

$m_e$ Abbreviation for electron rest mass.

**Mead, Margaret** An anthropologist who conducted a study in 1957, concluding that young people of the time thought very highly of scientists. This was just after the end of World War II, and was probably because of the development of the atomic bomb that brought a quick end to the war.

**mean** Average. Usually used with respect to a large, or even an infinite, set of values, or to a mathematical function over some interval.

**mean free path** For molecules, atoms or subatomic particles in constant motion, the average distance, for a specific type of particle, between collisions.

**mean free time** For molecules, atoms or subatomic particles in constant motion, the average time, for a specific type of particle, between collisions.

**mean temperature** The average temperature over a given interval of time, such as a period of several days (see drawing).

**measles** 1. Rubeola, a fairly serious childhood disease with fever and skin rash. Also called red measles. 2. Rubella, also called German measles, less severe in childhood but more dangerous in adulthood, especially in pregnant women.

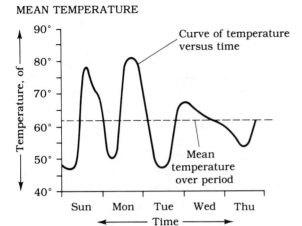

MEAN TEMPERATURE

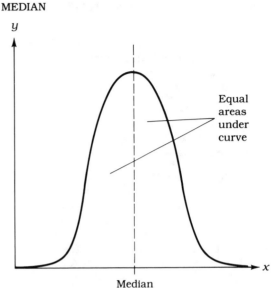

MEDIAN

**measurement** 1. The process of determining the size of a physical quantity, in universally defined units. This is always an approximation, since there are always some errors. 2. An expression of size or quantity in known units.

**mechanical engineering** A branch of engineering concerned with the design and efficient operation of machines having moving parts.

**mechanics** 1. The science of motions of, and among, physical objects, and the ways in which the motions depend on forces such as gravitation and electrical charge. 2. The principles according to which a certain type of machine works. 3. Hardware, especially of machines with moving parts.

**mechanization** 1. The use of machines, rather than human labor, to perform a task or set of tasks. 2. The general progress that humankind has made in substituting machines for people. This applies especially to boring, tiring or dangerous work.

**median** 1. For a set of values, that value such that there are just as many smaller values as larger ones. 2. In a statistical distribution, the value of the independent variable such that the area under the curve to the left is equal to the area under the curve to the right (see drawing).

**median eye** In some animals, especially certain reptiles, an organ resembling an eye, found on top of the head. Although it seems to be made like an eye, it apparently cannot see. It may be a vestigial organ. See *vestigial organ*.

**median nerve** One of the nerves in each arm of the human body, responsible for movement and sensation in part of the hand and fingers.

**medical chemistry** A branch of chemistry, and also of medical science, concerned with development of chemicals to serve as drugs for treating disease conditions.

**medical engineering** A large branch of engineering, concerned with the development of machines for life support. These include respirators, artificial kidneys (dialysis machines), and heart pacemakers. Under development are new types of heart valves, artificial eyes, hearing aids and microminiature motors. At the forefront of this technology is the idea that perhaps tiny robots can be made to act as antibodies against certain viruses and bacteria.

**medical physics** A branch of biology interrelated with physics. Scientists in this field study such things as the effect of high-speed atomic particles on living tissue.

**medical science** The science of finding new ways of diagnosing and curing disease; especially developing medications and treatment procedures. It encompasses all of the above several fields.

**medicine** 1. A branch of science concerned with diagnosing and curing diseases, especially in humans. 2. Medication; a substance taken to relieve disease states or symptoms.

**medium** 1. Average or mean. 2. Not especially high or low in value. 3. A material or environment through which some wave or particle disturbance travels.

**medium frequency** The range of radio frequencies from 300 kilohertz (kHz) to 3000 kHz or 3 megahertz (MHz). The wavelengths are from 1000 meters down to 100 meters. In this range, signals can travel for up to about 200 miles in the daytime, and thousands of miles at night. See the ELECTROMAGNETIC SPECTRUM and RADIO SPECTRUM appendices.

**medulla** The center of a body organ in animals, or the stem of plants.

**medulla oblongata** The region of the brain where the spinal cord connects. Autonomic body functions and reflexes are controlled by this part of the brain. See *brain*.

**mega-** 1. A prefix meaning "million" (1,000,000) or "millions." See the PREFIX MULTIPLIERS appendix. 2. A prefix meaning "very large." 3. In computer practice, a prefix meaning $2^{20}$, or 1,048,576.

**megabyte** A unit of information equal to $2^{20}$ bytes, or 1,048,576 bytes. See *byte*. A typical floppy disk, 5.25 inches or 3.5 inches in diameter, has room for approximately this much information.

**megahertz** Abbreviation, MHz. A unit of frequency equal to 1,000,000 hertz. A signal at 1 MHz is in the middle of the standard amplitude-modulation (AM) broadcast band. See the RADIO SPECTRUM appendix for comparison of frequencies.

**megaton** A unit of explosive power, equivalent to the detonation of one million (1,000,000) tons of TNT all at once. There are hydrogen bombs capable of delivering several megatons, enough to destroy a city the size of Denver or San Francisco.

**megavolt** Abbreviation, MV. A unit of electric potential, equal to one million (1,000,000) volts. Some high-tension power lines carry several megavolts. Lightning strokes have upwards of 100 MV.

**megawatt** Abbreviation, MW. A unit of power equal to one million (1,000,000) watts. This is about the average electric power consumed by a town having 300 homes. Power generators have outputs measured in megawatts.

**megger** A device with a high-voltage source, used for measuring large values of electrical resistance.

**meiosis** A process in sexual reproduction, in which the number of chromosomes per cell is halved. The cells divide without reproduction of chromosomes, preparing them for sexual union. If not for this process, the fertilized egg cell would end up with twice the number of chromosomes that it should have.

**Meissner Effect** The disappearance of internal magnetic flux when a conducting substance is cooled to the temperature needed for superconductivity. See *superconductivity, superconductor*.

**melanin** Also the skin pigment that causes freckles and suntanning. This is the body's response to ultraviolet radiation; it reduces the amount reaching deeper skin layers. It is thought to be a protection against sunburn and also from getting too much vitamin D.

**melanism** In certain animals, the production of a large amount of melanin, causing the entire animal to be dark or even black. This may occur as a defense against high levels of ultraviolet radiation.

**melanoma** A dark, pigmented area in the skin, similar to a common freckle but usually larger and darker.

**meltdown** A type of nuclear-reactor accident. The heat from the reactor causes the housing and containment unit to leak. This may release radioactive materials into the earth, or into water and air near the reactor.

**melting point**  For certain solids, the lowest temperature at which they become liquid. This is often a specification for metals. See the MELTING POINTS appendix.

**membrane**  1. A thin layer of tissue found in the body of an animal or human, that keeps substances from flowing uncontrollably from place to place. Also may protect against invasion by germs. 2. Any thin layer of material that controls the flow of fluids or gases between different places. 3. A barrier in space or time.

**membrane bone**  Bone that develops within a soft part of the body, and that serves a specific purpose. The cranium is probably the best example. At birth it has not yet completely hardened.

**memory**  1. The capacity to recall past events, and to behave accordingly. Most animals display this ability. 2. In electronics circuits, the capacity to store information for later use.

**mendelevium**  Chemical symbol, Md. An element with atomic number 101; the most common isotope has atomic weight 256. It is not found in nature; it is the result of nuclear experimentation by humans.

**Mendel, Gregor**  An Austrian monk who, in the mid-nineteenth century, first theorized that certain traits can be inherited, both for animals (including humans) and for plants. See *Mendelian Laws*.

**Mendelian Laws**  The rules concerning genes and the way in which traits are passed along by them. Based on the theories of Gregor Mendel (see *Mendel, Gregor*). There are two rules: (1) Each trait is passed along by two factors in different cells; (2) Pairs of these factors split without being affected by other pairs.

**Mendelism**  The theories of Mendel. See *Mendel, Gregor* and *Mendelian Laws*.

**meninges**  Membranes that serve to protect the brain and the spinal cord.

**meningitis**  Inflammation of the meninges (see *meninges*). This takes the form of an infectious disease. Symptoms include headache, fever and a stiff neck. The disease may be acute or recurrent (chronic). Treatment is usually with antibiotics.

**menopause**  In the human female, the time in life at which menstruation ceases. The time varies but it is usually around age 40. There are various hormone changes associated with it. See *menstruation*.

**menstruation**  The time during which the female egg is replaced. In humans this is about every 30 days. If the egg has not been fertilized, the uterus ejects it. This monthly cycle usually starts at age 11-13 and ends around age 40. This age span is the time in life during which a female is theoretically capable of bearing children. See *menopause*.

**menthol**  A substance used widely as a flavoring agent, and also as a home remedy for sore throats and nasal congestion. It is used in some skin preparations and shave creams.

**Mercator map**  A projection of the earth's surface features onto a flat plane. The map is made as if a light were shone through a globe onto a surrounding cylinder. The result is a good representation of regions near the equator, but exaggerated size for features near the poles (see drawing). The poles themselves do not appear on the map at all; they are infinitely far away upwards and downwards.

MERCATOR MAP

**Mercury** The closest planet to the sun in our Solar System. Its surface temperature rises to more than 800 degrees Fahrenheit. There is little or no atmosphere, and the surface is covered with craters, so that the planet looks much like our moon. See the SOLAR SYSTEM DATA appendix for comparison with other planets.

**mercury** Chemical symbol, Hg. A metallic element with atomic number 80. The most common isotope has atomic weight 202. It is the only metal that is liquid at room temperature. This makes it useful in a variety of electrical applications. It is also used in thermometers. It is quite toxic, even in small amounts.

**mercury cell** An electrochemical cell or "battery," with a steel case, and electrodes of mercuric oxide and zinc. It produces about 1.4 volts, and this stays constant for the life of the cell.

**mercury-vapor lamp** A bright blue-white illuminating lamp that makes use of the emission characteristics of mercury. The small amount of mercury is heated by a filament, until it boils into vapor. This vapor glows in the visible and ultraviolet ranges when an electric current passes through it. The drawing shows the construction of this type of lamp.

**meridian** Any of the lines of longitude on the earth, connecting the North and South Geographic Poles. See *longitude*.

**meristem** A part of a plant that grows rapidly. This especially applies to the ends of branches and roots, but also to certain other parts, depending on the particular plant.

**Merton Rule** Named after Merton College at Oxford University in Cambridge, England. Suppose an object has constant acceleration. Then the distance that it goes, in a certain time, is the same as the distance that another object would go, if it were moving at a constant speed equal to the average speed of the first object.

**mesa** A land formation that remains after erosion over millions of years by water. The land around it is carved away, leaving a flat-topped hill. There may be several or many of them scattered over a region.

**mesh** 1. A closed current path in an electrical circuit. 2. A set of windings in certain types of electrical transformers.

**mesh analysis** A mathematical process, in which the currents and voltages in a circuit are determined using Kirchhoff's Laws and various calculations. This usually involves a set of linear equations. See *Kirchhoff's Laws*.

**meso-** 1. A prefix meaning "intermediate" or "typical" or "average." 2. A prefix meaning "in the center of."

**mesoderm** The part of a mammal's embryo that develops into the skeleton, muscles, and circulatory system.

**mesometeorology** The study of medium-scale weather phenomena such as squall lines and thundershowers.

**mesomorph** A type of person who is regarded as average or typical, or with the so-called classical physique. This is muscular but not muscle-bound, and not overweight or underweight. See *ectomorph, endomorph*.

MERCURY-VAPOR LAMP

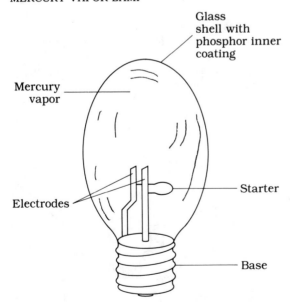

**meson** Any subatomic particle that is made up of pairs of quarks and their antimatter counterparts. They are thought to be the "glue" that keeps atomic nuclei from flying apart, as they otherwise would because of repulsion among the positively charged protons.

**mesophile** An organism that lives in moderate temperatures, neither very hot nor very cold. They probably evolved from thermophiles. See *thermophile*.

**mesophyll** The living cells that comprise a plant leaf.

**mesophyte** Any plant that requires good soil, and moderate amounts of water, in which to grow. Literally, the word means "average plant."

**mesosphere** 1. The part of the earth's atmosphere from about 24 miles to 48 miles up. This is above the stratosphere and below the thermosphere and ionosphere. See *atmosphere*. 2. The rigid part of the earth's mantle, above the outer core. See *mantle*.

**Mesozoic era** The time of middle life in the geologic history of the earth. It began about 230 million years ago and extended until about 65 million years ago. The dinosaurs flourished near the beginning of this time. Later, the mammals and birds evolved. See the GEOLOGIC TIME appendix.

**mesquite** A plant that can send its roots very deep, to get to the water table even in a dry environment. Found in such places as Texas and New Mexico.

**meta-** 1. A prefix meaning "after" or "following." 2. A prefix meaning "more involved than" or "extended in scope." 3. A prefix referring to something that has been dehydrated.

**metabolic** Pertaining to the use of food and nutrients by the body, especially as a source of energy for body functions and for physical activity. See *metabolism*.

**metabolism** 1. The conversion of food calories (protein, fat and carbohydrate) to energy for use by the body. 2. The rate at which food calories are consumed or used by the body in converting to energy. 3. The conversion of one type of food calories to another, such as excess carbohydrate into fat.

**metabolite** A material by-product of metabolism. This might be carbon dioxide and water in the case of the complete, efficient metabolism of carbohydrate. In the case of partially metabolized fat or protein, there are waste products that must be eliminated.

**metacarpus** Any of the small bones in the hand connecting the fingers to the wrist.

**metacenter** In a boat or ship, the point where the midline meets the upward force caused by flotation. In a stable vessel, this point is above, or nearly above, the center of gravity.

**metal** Any of various elements that have fair to good conductivity both for heat and for electrical current, and varying degrees of ductility and malleability. Some of the more common metals and their properties are given in the METALS appendix.

**metal fatigue** The weakening of a metal after it has been subjected to stress for a long time. This may cause the metal to break apart. Some alloys are resistant to fatigue and are used in important applications such as aircraft fuselages and boat hulls.

**metallic crystal** A solid with properties of both a crystal and a metal (see *crystal, metal*).

**metallography** The study of metals and their structures, especially on a small scale. This is important in determining how the substances, especially newly developed alloys, will behave on a large scale.

**metalloid** A substance that has some of the properties of metals, but not all, so that it cannot be regarded as a true metal. The most common are the semiconductors. See *semiconductor*.

**metallurgical engineering** A field of engineering involved with refining of metals, and the formulation and use of metal alloys, especially for industry.

**metallurgy** The study of metals and their behavior. This may include **metallography** and **metallurgical engineering.**

**metal-oxide semiconductor** Abbreviation, MOS. 1. A material formed from the oxide of a metal, and that acts as a semiconductor (see *semiconductor*). 2. An electronic device, such as a transistor or integrated circuit, made using metal oxides instead of silicon or other conventional semiconductors.

**metamathematics** A recently developed field of mathematical logic, involved with the interpretation of mathematical symbols. We know that some statements cannot be proven either true or false, but we cannot identify such statements. Some mathematicians hope that this new branch of their science might help them to tell whether or not a statement is of this type.

**metamorphic rock** Rock that has been changed by nature over a long time, so that its properties are altered, usually by high temperatures and pressures. An example is marble, which is metamorphosed limestone or dolomite.

**metamorphosis** 1. The process by which an insect matures; for example, a caterpillar into a moth or butterfly. 2. A change of state that occurs naturally.

**metaphysics** 1. The unknown or partly known frontiers of physics. 2. Unproven and/or controversial theories and ideas, especially pertaining to supernatural phenomena, such as attempts to communicate with aliens by means of mental telepathy, or travel "outside of the body."

**metatarsus** Any of the small bones in the foot, connecting the toes with the ankle.

**metazoan** A multicellular animal species that evolved about 750 million years ago. At this time there was about 7 percent as much oxygen in the earth's atmosphere as there is today. The earliest of these species were groups of cells that lived together in a colony for common purposes such as feeding and protection from the environment.

**meteor** A rock or pebble from space that burns up in the atmosphere before it reaches the surface of the earth. At night this may produce the effect of a "shooting star" or "falling star." See *meteorite, meteoroid.*

**meteoric dust** Fine particles in space, ranging from the size of sand grains down to microscopic particles.

**meteorite** A meteor that does not completely vaporize in the earth's atmosphere, thereby reaching the ground partly intact. See *meteor.* These objects provide clues as to the origin of our own planet and of other planets.

**meteorite crater** The result of the impact of a large meteorite on a landmass of the earth or some other planet. The crater is usually several times the diameter of the meteorite itself. See *meteorite.*

**meteoroid** A rock or pebble in interplanetary space. Most such objects follow orbits around the sun. If they get close enough to the earth, they are pulled down by the earth's gravity, and then they become meteors or meteorites. See *meteor, meteorite.*

**meteorology** The science of the behavior of weather systems, and of forecasting the weather for a given place hours, days or weeks in advance. Also includes the measurement of temperature, barometric pressure, wind direction, wind speed, relative humidity, cloud cover and other properties of the atmosphere.

**meter** 1. The Standard International unit of length, equal to about 39.37 inches. 2. A device for measuring physical quantities, especially electric currents, voltages and power.

**methanal** See *formaldehyde.*

**methane** A simple hydrocarbon, found as a gas within the earth as a component of so-called "natural gas." It is flammable, burning efficiently to produce carbon dioxide, water and energy. The illustration shows the basic molecular structure.

**methanol** Also called wood alcohol. It is the product of oxidized methane. It is used as a solvent; it is poisonous if taken internally.

METHANE

$CH_4$

○ Carbon atom

● Hydrogen atoms

—— Bonds (CH)

**method** 1. A way of doing scientific work. 2. A means of manufacturing, especially on a large scale. 3. A set way of solving a mathematical or scientific problem or equation.

**methyl** 1. Pertaining to chemicals that contain a molecule of two, three or four hydrogen atoms bonded to one carbon atom. 2. A chemical made from methane by any of various processes. See *methane*.

**methyl alcohol** See *methanol*.

**methylbenzene** Also known as toluene. It is produced from coal, and is an important ingredient in explosives. It can also be used as an industrial solvent.

**methylcellulose** A carbohydrate substance that cannot be digested, but absorbs water and swells in the stomach and intestines. It is used in some weight-reducing programs because it promotes a feeling of fullness without providing calories. However, it can also cause blockages in the digestive system if not taken with sufficient water.

**Metraux, Rhoda** A researcher who, along with Margaret Mead, conducted a survey and found that scientists were highly regarded in the 1950s. See *Mead, Margaret*.

**metric system** A system of units based on the meter (see *meter*), kilogram and second, rather than the older English system of foot, pound and second. Nowadays the common system is called the Standard International system. See the STANDARD INTERNATIONAL SYSTEM OF UNITS appendix.

**metric ton** See *tonne*.

**metrology** 1. The branch of experimental science concerned with measurement of quantities. 2. The development of specialized systems for measuring certain quantities.

**MF** Abbreviation for medium frequency.

**Mg** Chemical symbol for magnesium.

**mg** Abbreviation for milligram.

**mho** See *siemens*.

**MHz** Abbreviation for megahertz.

**mica** Minerals composed of silicon and oxygen, and characterized by a shiny, sheet-like appearance. When you break these sheets apart in the dark, you can see sparks. The individual sheets are transparent or translucent, usually clear or yellowish in color. This material has good dielectric properties, and has a variety of industrial applications.

**Michell, John** A mathematician and astronomer of the eighteenth century, who thought that the escape velocity from an extremely dense object might exceed the speed of light. Then the object would look black, and anything that fell into it would never be able to get back out of its gravitational influence. It would be a *black hole*. Nowadays we know that such objects may really exist, in the form of stars that have undergone gravitational collapse. See *black hole, gravitational collapse*.

**Michelson-Morley Experiment** An experiment conducted by the physicists Albert Michelson and Edward Morley, late in the nineteenth century. It was designed to determine the motion of the earth with respect to the "ether" (see *luminiferous ether*). One ray of light was sent in a direction thought to be sideways relative to the flow; one ray was sent along with the flow. It turned out that the rays traveled at identical speeds. This came as a surprise, suggesting there wasn't any "ether" at all. Albert Einstein discarded the idea of "ether" when he began work on his relativity theory. See *relativity*.

**micro-** 1. A prefix meaning "one millionth" (0.000001). 2. A prefix meaning "very small" or "miniaturized."

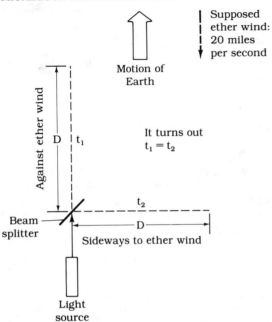

MICHELSON-MORLEY EXPERIMENT

**microbiology** A branch of biology concerned with very small, and microscopic-sized, living cells and organisms.

**microburst** See *downburst*.

**microcircuit** Any electronic circuit that has been greatly reduced in physical size as compared with the same circuit a few years ago. A good example is an integrated circuit. See *integrated circuit*.

**microcontroller** A microcomputer that is mostly, or completely, contained in a single integrated circuit, and used for controlling the operations of devices. See *integrated circuit, microcomputer*.

**microdissection** The dissection of individual cells. This requires a microscope and special microminiature tools.

**microearthquake** A small or localized earth tremor. May occur all by itself, or as an aftershock or foreshock. See *aftershock, foreshock*.

**microeconomics** Economics on a small scale. The best example is personal finances. For example, in the case of a single person the result of overborrowing may be bankruptcy. On a larger scale, different symptoms occur, such as inflation and recession. See *macroeconomics*.

**microelectronics** A branch of electronics, concerned with circuits that are greatly miniaturized. See *integrated circuit, microcircuit*.

**microfilament** A form of microfossil; a tiny, threadlike structure that is thought to be of bacterial origin. See *microfossil*.

**microfossil** A tiny fossil, produced by an individual cell or by a primitive or simple organism. Often these fossils are so small that they need a microscope to be seen in detail.

**microgram** Abbreviation, mcg. A tiny unit of mass, equal to one millionth (0.000001) gram, or one-thousandth (0.001) milligram. See *gram*.

**microlithography** A method of manufacturing integrated circuits, using ultraviolet light and etching techniques. Complex patterns are produced at tiny sizes on photographic film. The ultraviolet light shines through the film, producing the circuit itself. See *integrated circuit*.

**micrometeorology** A branch of meteorology, concerned with small-scale weather phenomena, such as dust devils and windshear. See *meteorology*.

**micrometer caliper** An instrument used to accurately measure very small distances, down to millionths of a meter (microns). The most

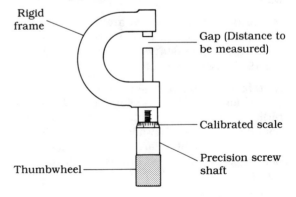

MICROMETER CALIPER

common device employs a screw with a thread pitch that has been accurately reproduced (see drawing). A calibrated scale then allows measurement of the distance.

**microminiaturization** Miniaturization to an extreme degree. This is a recent, and growing, field of engineering. A good example of the result of this research is a motor so small that a microscope is needed in order to see its parts. It has been suggested that this technology might someday make it possible to build "robot antibodies" to hunt down and destroy viruses such as the one that causes acquired immune deficiency syndrome (AIDS).

**micronutrient** A vitamin or mineral needed only in very small amounts, such as micrograms (mcg). Examples are folic acid and vitamin B-12. See the RECOMMENDED DAILY ALLOWANCES appendix.

**microorganism** A living thing that is too small to be resolved without the aid of a microscope or magnifying glass. Primarily, a cell or small group of cells. See *cell*.

**micron** A unit of length equal to one millionth (0.000001) meter, or one-thousandth (0.001) of a millimeter. Used for expressing the sizes of some microorganisms, the diameters of thin wires and fibers, and other microscopic distances.

**microphone** A transducer that converts sound waves into electrical impulses. There are various different types, used for different applications. The most common type uses a coil and magnet. The magnet is attached to a diaphragm that vibrates when sound waves strike it. This results in small currents in the coil, having the same waveshapes as the sound. These currents are then amplified. See *transducer*.

**microphonics** In a high-fidelity amplifier or radio receiver, a tendency for physical vibration to cause unwanted sounds in the output. In the worst cases this can produce feedback with "howling," making the circuit useless.

**microphyll** A tiny leaf, with a simple vascular system, seen in certain plants.

**microprocessor** An integrated circuit that functions as the central processing unit for a computer. See *central processing unit, integrated circuit*.

**microprogram** A small computer program that is used for a certain purpose, not necessarily having anything to do with the main program. Microprograms are often put in the computer as firmware, or permanent programs.

**microscope** A device for providing great magnification of visual images. It consists of a set of lenses. In its simplest form there is an objective, placed near the object to be observed, and an eyepiece, that focuses the image from the objective so that it is suitable for viewing with the eye. The object is usually lit from beneath (see drawing). Typical microscopes may make an object appear up to about 1500 times actual size. A special type of microscope, called the electron microscope, can magnify more than a visual microscope because of the properties of high-speed electrons. See *electron microscope*.

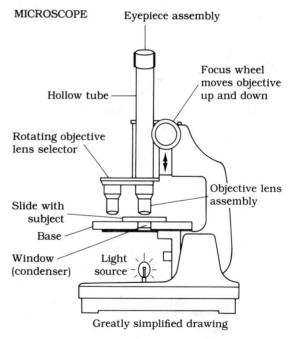

MICROSCOPE. Greatly simplified drawing.

**microscopic** Pertaining to objects or organisms too small to be resolved without the aid of a microscope.

**microsecond** A unit of time equal to one millionth (0.000001) second. It is used in conjunction with high-speed electronic circuits, and also with radio waves and light waves traveling over short distances. A ray of light travels about 300 meters, or 990 feet, in one microsecond.

**microtome** A device used to prepare biological samples for viewing under a microscope. It cuts the samples into slices as thin as, and sometimes thinner than, a single cell. The samples are therefore translucent or transparent, so that the backlighting provides a good image.

**microvilli** The fine, hairlike filaments in the small intestine, though which nutrients pass into the bloodstream. These cover the larger villi. See *villi*.

**microvision** A technology in which it is possible to land an aircraft when there is zero visibility, such as a dense fog. Electronic devices, called *beacons*, are placed along the edges of the runway. The signals from these devices are received and assembled, by means of a computer, into a "picture" of the runway for the pilots of the aircraft.

**microwave** Pertaining to electromagnetic radiation having very short wavelengths, but longer than infrared. Generally, the lower frequency limit is about 3 gigahertz (GHz) or a wavelength of about 10 centimeters (cm). The upper frequency limit is at the longest infrared wavelengths. See *infrared* and the ELECTROMAGNETIC SPECTRUM and RADIO SPECTRUM appendices.

**microwave background** The constant, faint radiation from all directions in space, observed in the microwave part of the electromagnetic spectrum. First noticed by Arno Penzias and Robert Wilson of the Bell Laboratories in the mid-1960s, this radiation is thought to be redshifted remnants of the Big Bang. See *Big Bang*.

**microwave communications** Radio communications at microwave frequencies, or those frequencies above about 3 gigahertz (GHz) but below the infrared region. See *microwave*.

**microwave oven** A device that uses high-energy radio emissions in the microwave frequency range to cook food. The radio energy is dissipated in the cells of the food to be cooked. This results in just as much heating, just as fast, inside the food as outside. It also allows foods to be cooked in less time than with conventional ovens.

**mid-Atlantic ridge** The longest mountain range in the world, extending from north to south along the midline of the Atlantic Ocean floor (see drawing). It is a midocean ridge, responsible for spreading of the sea floor and the generation of new oceanic crust.

MID-ATLANTIC RIDGE

From Erickson, THE MYSTERIOUS OCEANS (TAB Books, Inc., 1988).

**midbrain** The part of the brain in the mammal embryo that later evolves into the brainstem. See *brain*.

**middle ear** The part of the ear inside the eardrum, but outside the inner ear. In this region are found the bones commonly called the *ossicles*, or the hammer, anvil and stirrup. This part of the ear is susceptible to infection, especially in children. It connects to the throat via the *eustachian tube*. See *ear* for illustration.

**midocean ridge** A range of mountains under the ocean, that tends to occur near the center line of the ocean. Responsible for spreading of the sea floor, and the generation of new oceanic crust. The best example is the mid-Atlantic ridge. See *mid-Atlantic ridge*.

**midrange** The audio frequencies near the middle of the human hearing range, above bass and below treble. These are approximately the human-voice or baseband frequencies of 300 hertz (Hz) to 3000 Hz.

**midrange speaker** A speaker designed for reproducing sound in the midrange. See *midrange*.

**migraine** A severe form of headache that some people get periodically. It is often on just one side of the head, and may last for hours or days. Sometimes stomach upset happens along with it.

**migration** 1. Instinctive behavior of certain animals, especially birds that fly to different places during different seasons. Many species of fish also move to warmer or cooler water as the seasons change. 2. The movement of animal species in response to human activities. For example, pollution may drive fish or birds to cleaner places, sometimes nearby, sometimes far away.

**mil** A small unit of length or distance, equal to one-thousandth (0.001) inch or 0.0254 millimeter.

**Milankovich model** A theory about the cause of ice ages. Various periodic changes in the earth's orbit cause more or less solar radiation to be received by our planet. Cool summers in the Northern Hemisphere, more than severe winters, would be most likely to cause ice ages. There are three cycles superimposed on each other, with periods of 100,000 years, 41,000 years and 22,000 years.

**Milankovich, Milutin** A Yugoslavian astronomer who made calculations concerning the variations in the earth's orbit around the sun. In the early twentieth century he proposed a theory for the cause of ice ages, based on his determinations. See *Milankovich model*.

**milk** The liquid nutrient produced by a female mammal, intended by nature to feed her offspring. It contains protein, fat and carbohydrate in ideal amounts for the offspring of that particular species. The nutritive values of different animals' milk varies somewhat, according to what is best for the species.

**milk of magnesia** See *magnesium hydroxide*.

**milk sugar** See *lactose*.

**Milky Way** The spiral galaxy in which our Solar System exists. The galaxy has a spinning motion, and lies mostly in one disk-shaped region about 100,000 light years in diameter (see drawing). Our sun is located in one of the spiral "arms," about $3/4$ of the way out from the center to the edge. On a clear, moonless night, away from city lights, our part of the galaxy can be seen as a milky haze across the sky. See *galaxy*.

MILKY WAY

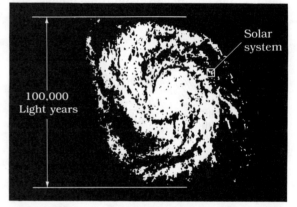

**Miller-Urey Experiment** An experiment conducted by chemists Stanley Miller and Harold Urey in the 1950s, to simulate conditions on the primordial Earth. It was theorized that a spark, such as lightning, created amino acids. The experiment showed that this indeed can occur.

**milli-** Abbreviation, m. A prefix meaning one-thousandth (0.001).

**millibar** A unit of atmospheric pressure, equal to 1000 dynes per square centimeter. Normal atmospheric pressure is approximately 1000 millibars or 1 bar. Weather maps often show the pressure in millibars rather than in inches of mercury. See *barometric pressure*.

**milligram** Abbreviation, mg. A unit of mass, equivalent to one-thousandth of a gram (0.001 g). This unit is used by chemists and physicists in laboratory experiments. Some vitamins and minerals are needed in amounts measured in milligrams.

**milliliter** Abbreviation, ml. A unit of fluid measure or volume, equivalent to one cubic centimeter, or 0.001 liter. One fluid ounce is about 30 ml. See *liter*.

**millimeter** Abbreviation, mm. A unit of length, equal to one-thousandth (0.001) meter. One inch is equal to about 25.4 mm. See *meter*.

**millipede** An arthropod with a many-segmented body, each segment having two pairs of legs. They are often seen in old buildings, preferring darkness and dampness. They range in length from one inch to about three inches. See the drawing.

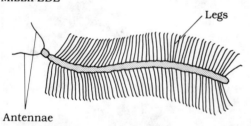

MILLIPEDE

**millivolt** Abbreviation, mV. A unit of electrical potential, equal to one-thousandth (0.001) volt. This is a small voltage, not anywhere near enough to produce electrical shock, but is a fairly large voltage compared to the microvolt levels normally found in radio antennas. See *volt*.

**milliwatt** Abbreviation, mW. A unit of power, generally used to express low levels of audio-frequency or radio-frequency power. Equal to one-thousandth (0.001) watt. See *power, watt*.

**mimic** An animal that has evolved a certain feature or set of features very similar to those of some other animal, even though they are different species. This tends to protect the mimic from the same environmental or predatory hazards as the other animal.

**min** Abbreviation for minimum, minute.

**mineral** 1. Any inorganic substance. 2. Pertaining to anything that is inorganic; that is, nonliving. 3. Any of various elements or compounds found in the earth's interior.

**mineral acid** An inorganic acid, such as sulfuric, hydrochloric, or nitric; comprised of a simple combination of mineral elements.

**mineral mapping** The determination of the locations of various types of rocks in the earth, using aircraft and satellites. Radar allows penetration beneath the surface for identifying structures and classifying rock units. Satellite radar can be used to map the ocean floor. Satellite photographs can locate faults and various mineral deposits.

**mineralogy** The study of minerals, including their identification and behavior. Closely connected with geology in the study of how various minerals are deposited and formed in the earth.

**mineral oil** A heavy, indigestible oil, used as a mild laxative. It works by lubricating the contents of the bowel. It should not be taken with alcohol, and is best used only with a doctor's recommendation.

**mineral spring** A place where groundwater, laden with minerals such as iron, calcium and magnesium comes to the surface. Such springs have often been regarded as having medicinal properties, even when there is little scientific reason to believe it. Sometimes people travel long distances just to bathe in certain springs of this kind.

**minim** 1. A tiny unit of volume, equal to 0.0616 milliliter. Used in pharmaceutical applications and sometimes also in chemistry. See *milliliter*. 2. A very small amount.

**minimal** 1. Very small. Pertaining to a phenomenon that does not have much effect because of its small value. 2. Reduced to the smallest possible value, or the lowest practical level.

**minimum** 1. The smallest value that a parameter normally attains. 2. See *local minimum*.

**mining** The removal of minerals from within the earth for various human needs, particularly industrial applications. There are various different methods. In strip mining, open pits are dug and the minerals simply removed. In shaft mining, holes and tunnels are dug, creating underground caverns much deeper than strip mines. Other methods of mining, involving less damage to the environment and less risk to human life, are being developed.

**minority carrier** In a semiconductor material, a current carrier that is present only in very small numbers, and does not contribute much to the current flow. In P-type material, electrons are the minority carriers; in N-type material (and in most electrical conductors), holes are the minority carriers. See *electron, hole, majority carrier, semiconductor*.

**minute** 1. A period of time equal to 60 seconds or 1/60 hour. 2. A unit of angular measure equal to 1/60 degree or 60 seconds of arc. 3. Extremely small.

**Miocene epoch** A portion of the Cenozoic era that began about 24 million years ago, and extended till about 5 million years ago. During this time the apes appeared and the Alps were formed. See the GEOLOGIC TIME appendix for comparison with other epochs.

**mirage** A type of optical illusion, created by the reflection of light at small angles between masses of air having much different temperatures. The drawing shows a common instance of sky reflected from the hot air boundary near a hot surface; the observer may think this reflection is water, because water also reflects the sky.

**Mira type star** A type of variable star, named after a particular star called Mira in the constellation Cetus. They are red giants that grow brighter and dimmer in a regular, but long, cycle.

**mirror** A device that reflects all of the energy striking it, especially in the visible range, and in such a way that images are reproduced. See *concave mirror, convex mirror, first-surface mirror*.

**mirror galvanometer** An instrument that measures very small electric currents, or very tiny changes in currents. In order to detect microscopic changes in the position of the meter needle, a beam of light is bounced off a mirror attached to the meter movement. This makes the meter needle in effect very long, but with zero mass (see drawing). See *galvanometer*.

MIRAGE

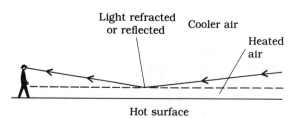

MIRROR GALVANOMETER

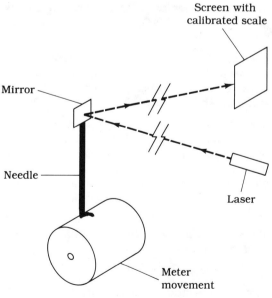

Functional diagram

**mis-** A prefix meaning "in error," similar to **mal-**.

**miscarriage** Discharge of an embryo from the womb of a female mammal, long before the proper time for birth. This may occur before the embryo is even recognizable as a "baby." Sometimes this condition is deliberately induced; then it is technically an abortion. See *abortion*.

**mite** A tiny, spider-like creature (arachnid) that is a pest and often a carrier of diseases.

**mitochondrion** A small part of a cell that is important in respiration and metabolism. They appear as fragments in the cytoplasm. Generally, the higher the metabolic rate in a cell, the more mitochondria it has. See *cell, metabolism, respiration*.

**mitosis** Cell reproduction that occurs as more or less "simple" splitting or duplication. The chromosomes divide first; then the cell itself splits. The result is two cells, each with the same number of chromosomes as the first cell. See *cell, meiosis*.

**mixer** In electronics, a device in which signals at two different frequencies are combined, producing signals at the sum and difference frequencies. Either of these two output signals may then be amplified and used for some other purpose.

**mixing** 1. The combining of two electronic signals in a mixer (see *mixer*). 2. The combining of two or more chemical substances, without a reaction taking place.

**mixture** A combination of two or more chemical substances, in which there is no reaction between/among them, but where they remain more or less evenly distributed. Air, for example, is a mixture of gases, mainly nitrogen and oxygen, with traces of carbon dioxide, argon and other elements.

**MKS system** See *Standard International System of Units*.

**ml** Abbreviation for milliliter.

**mm** Abbreviation for millimeter.

**Mn** Chemical symbol for manganese.

$m_n$ Abbreviation for neutron rest mass.

**Mo** Chemical symbol for molybdenum.

**mobile telephone** A radio telephone unit, used in a car, truck, boat or plane. It has a dialer and can access the same lines as an ordinary telephone unit. See *cellular mobile radio telephone*.

**mobility** 1. The ability of something to move around, or to be moved around, in its environment. 2. The ease with which a charge carrier moves through a substance. See *majority carrier, minority carrier*. 3. For a living organism, the capacity to move under one's own power.

**Mobius strip** A strip that has only one side and one edge. It is a three-dimensional figure, not just a two-dimensional one. The drawing shows how you can make one.

MOBIUS STRIP

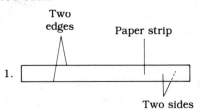

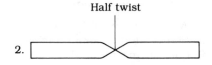

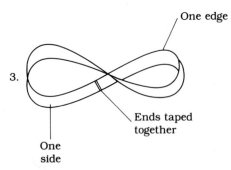

**model** 1. A mathematical representation of some physical process, object or phenomenon. 2. A small, exactly proportional object that represents a larger object or device, and may be used to determine how the larger device will behave. An example is a small plastic aircraft in a wind tunnel. 3. See *prototype*.

**modem** Acronym for *modulator/dem*odulator. A device that converts one form of signal to another, such as sound to electrical impulses, and also does the reverse, often simultaneously. An example is the telephone unit that you set the receiver into, in order to use a home computer on the telephone lines. Another example is the wired-in unit used with facsimile (FAX) machines connected to the telephone lines.

**moderator** 1. A material that helps regulate the activity of a nuclear reactor. It works by slowing the neutrons down and keeping the reaction controlled. 2. The leader of a group in a group therapy session. See *group therapy*.

**modified Mercalli scale** A scale for rating the intensity of earthquakes, based on their effects.

**modular construction** A method of building, where individual "units" are prefabricated and then assembled together to make the final machine or electronic device. Sometimes a set of modules can make more than one device, depending on how they are put together.

**modulation** 1. A means of getting a carrier wave (see *carrier*) to convey information. This is done by systematically varying some property of the wave, such as its level (amplitude), frequency or phase, or by chopping it into defined pulses. See *amplitude modulation, frequency modulation, phase modulation, pulse modulation*. 2. The regular, back-and-forth variation of a quantity.

**modulator** In electronics, a device that controls some quality of a carrier wave, in order to convey data. There are two inputs, one for the carrier and one for the data itself. The output contains the modulated carrier. See *modulation*.

**modulator/demodulator** See *modem*.

**modulo** Pertaining to the base in a number system. The usual decimal system (base 10) is modulo 10. Octal numbers are modulo 8. Binary numbers are modulo 2. The modulo is equal to the number of different digits that exist in a system.

**modulus** 1. A numerical expression of some physical property, such as hardness. 2. The absolute value of an electrical impedance. See *impedance*. 3. A constant, used to convert logarithms of one base into logarithms of some other base. See *logarithm*. 4. See *modulo*.

**Moho** See *Mohorovicic discontinuity*.

**Mohole Project** The drilling of a hole all the way to the bottom of the earth's crust, and into the mantle. First conceived by the oceanographer Walter Munk in 1957. The idea was to drill the hole in the ocean bottom. This is where the crust is thinnest. It would also be easier to assemble the apparatus with a floating ship and underwater equipment, than it would be on dry land. See *Mohorovicic discontinuity*.

**Mohorovicic discontinuity** Discovered by Andrija Mohorovicic, an abrupt transition inside the earth, where the crust gives way to the mantle (see drawing). This occurs at a depth ranging from a few miles to as much as 45 miles. The crust "floats" on this mantle, in a manner somewhat similar to the way ice packs float on the Arctic Ocean. The continents drift around on the mantle over periods of millions of years. See *continental drift, mantle*.

MOHOROVICIC DISCONTINUITY

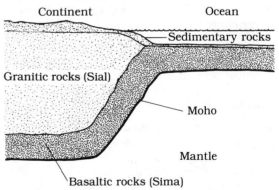

From Erickson, VOLCANOES AND EARTHQUAKES (TAB Books, Inc., 1988).

**Mohs' hardness scale** A quantitative scheme for evaluating the hardness of various solids. It is based on what scratches what;

harder minerals scratch softer ones. Any given substance can be assigned a hardness from 1 to 10 on this scale. The standard materials are listed in the table, along with their hardness numbers. Sometimes this scale is expanded with diamond given a hardness of 15; this provides somewhat more precision.

MOHS' HARDNESS SCALE

| Hardness | Standard mineral |
|---|---|
| 1 | talc |
| 2 | gypsum |
| 3 | calcite |
| 4 | fluorite |
| 5 | apatite |
| 6 | orthoclase |
| 7 | quartz |
| 8 | topaz |
| 9 | corundum |
| 10 | diamond |

**moire pattern** A wavy-lined pattern formed when two grids or regular matrices are superimposed. You can generate patterns of this sort by placing two pieces of window screen one over the other and then looking through them. On a smaller scale these patterns sometimes occur as a result of wave interference with visible light.

**mol** Abbreviation for **mole.**

**molar** 1. An expression of a property per unit quantity (generally 1 mole) of a substance. See *mole.* 2. A tooth in the rear of the mouth, used for grinding and pulverizing food. Found in mammals.

**molarity** An expression of the amount of a chemical per unit volume or unit quantity. Usually given in moles per liter. May also be called concentration. See *mole.*

**mole** A fixed number, about $6.02 \times 10^{23}$. In this sense it is the same as saying that 12 is "a dozen" or 144 is "a gross." Used in chemistry. Sometimes abbreviated mol. Also defined as the number of atoms in 0.012 kilograms of carbon-12.

**molecular** 1. Pertaining to molecules. 2. Made of molecules. See *molecule.*

**molecular biology** The study of reactions among atoms that affect biological processes.

**molecular bond** The bond between atoms of a molecule, that keeps the atoms from separating. It is the result of the sharing of electrons between or among the atoms. This bond is broken in the event of a chemical reaction.

**molecular distillation** A method of distillation at low temperatures and low pressures. This allows better control of the process than is the case with distillation at high temperatures.

**molecular flow** A type of gas flow in which the individual molecules travel in irregular paths. Therefore, they go much farther, for a given length of the guiding tube or pipe, than they would if they went in geometrically straight lines. This is most common when the pressure is low, and the molecules move farther between collisions with other molecules.

**molecular medicine** A branch of chemistry at the molecular level, involved with the diagnosing and curing of diseases. We might observe, for example, the effects of certain molecules on viruses and bacteria, hoping to find some chemical that will hinder the viruses or bacteria without harming normal body cells.

**molecular motions** The essentially random movements of molecules in a substance. The movement becomes more energetic as the temperature is raised. Even in solids, the molecules are always moving. But they move farther in liquids and farther still in gases. Collisions occur among the molecules; these are more frequent as the temperature and/or pressure increase. See *Brownian motion, molecule.*

**molecular physics** 1. A branch of physics concerned with the behavior of molecules. 2. The physics of the motions of molecules and how this motion affects the overall characteristics of a substance.

**molecular volume** The volume of one mole of a substance; that is, of $6.02 \times 10^{23}$ atoms or molecules of that substance.

**molecular weight** For a given substance, the sum of the atomic weights of the atoms in one molecule. This depends on the isotopes of the atoms that make up the molecule, so there may be several different molecular weights for a single chemical.

**molecule** A combination of atoms in an element or compound, in natural form. Sometimes a single atom can make a molecule. Two atoms of the same element may join to form a molecule; this is the situation with most of the oxygen in the atmosphere. Two or more different elements may combine, sometimes in extremely complicated structures, to make large molecules. Paraffin is an example of this type of molecule.

**mole fraction** The proportion of a particular chemical in a mixture, defined according to number of molecules. For example, if we have one mole of mixture ($6.02 \times 10^{23}$ molecules), and we have $6.02 \times 10^{22}$ molecules of a certain substance in this sample, then the mole fraction is $1/10$ or $0.1$ or 10 percent.

**mollusk** Also spelled mollusc. An invertebrate, often having a shell. They may be found in the water or on land. They include such animals as snails (on land) and squids (in the water). For their place in the evolutionary tree, see the ANIMAL CLASSIFICATION appendix.

**molybdenum** Chemical symbol, Mo. An element with atomic number 42. The most common isotope has atomic weight 98. In pure form it is a hard, silver-colored metal. It is used to harden steel.

**moment** For a rotational motion, the product of the applied force and the radial distance (see drawing).

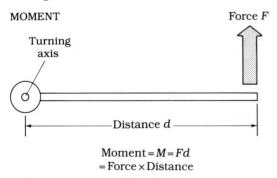

MOMENT

Turning axis

Force F

Distance d

Moment = M = Fd
= Force × Distance

**moment of inertia** An expression of the effective mass of an object, in a rotating system rather than in a straight-line-motion system. While mass has inertia (see *inertia, mass*) that causes it to resist linear motion, it also has resistance to rotational motion. A relationship can be stated: torque equals moment of inertia times angular acceleration (T = Ia). See *acceleration, angular momentum, angular motion, torque*.

**momentum** The product of mass and speed (scalar), or mass and velocity (vector), for an object moving at constant speed in a straight line. See *angular momentum*.

**monitor** 1. A television type cathode-ray-tube unit, used for viewing computer data. 2. A video picture-tube unit in a closed-circuit security system. 3. Any instrument that graphically displays some aspect of the operation of a system. 4. To periodically check the operation of a system by observing instruments. 5. To electronically eavesdrop on communications.

**mono-** A prefix meaning single or one. Compare **multi-**.

**monochromatic light** Electromagnetic energy in the visible range, consisting of just one wavelength or a narrow band of wavelengths. Such light is typically of a bright color, matching some part of the conventional color spectrum.

**monoclonal** Descended or derived from a single cell.

**monoculture** In agriculture, the practice of cultivating just one crop and no others.

**monocyte** A white blood cell that performs the task of eating bacteria and waste products in the system of an animal or human. It is an important part of the defense mechanism against disease. See *white blood cell*.

**monomer** A single molecule of a substance that often generates chains of several or many molecules. When the molecules link together, the properties change; this is called polymerization. See *polymer, polymerization*.

**mononucleosis** An infectious disease, common in teenagers and young adults. Its symptoms include fatigue, sore throat, fever and weight loss. It is not usually serious but it can require up to several weeks of bedrest.

**monoplane** An aircraft with just one set of main wings. It was chosen as the preferred type of aircraft just prior to World War II. Before that, biplanes were common. Monoplanes have greater speed capability and are more rugged, as well as being simpler, than biplanes. See *biplane*.

**monopole** 1. A single, isolated electrical pole, either positively charged or negatively charged. 2. A single, isolated magnetic pole. According to conventional theory, a true magnetic monopole never occurs. 3. A type of communications antenna, characterized by a single radiating element operating against electrical ground.

**monosaccharide** A simple sugar, having just one molecule. The most common of these are glucose and fructose. Glucose is the carbohydrate that provides much of the energy for the body in mammals. See *glucose*.

**monosodium glutamate** Abbreviation, MSG. A chemical added to food to enhance the flavor. In some people, it causes mild unpleasant symptoms such as irritability, headache and fatigue. In most people there are no apparent side effects.

**monostable circuit** 1. An electronic circuit having one stable state. 2. A multivibrator that delivers one output pulse each time it gets an input pulse. See *multivibrator*.

**monotreme** An egg-laying mammal found in Australia. Spiny anteaters and platypuses are two examples.

**monounsaturated fat** A fat whose molecules contain two adjacent carbon atoms, each missing one hydrogen atom. See *polyunsaturated fat, saturated fat*.

**monsoon** A pattern of winds that occurs on a seasonal basis, bringing heavy rains to a certain region. India is well known for its summer monsoon season. Similar effects are observed at low latitudes throughout the Northern Hemisphere in the summer months (see drawing).

**Moog synthesizer** An electronic device, used to create musical notes and to imitate the sounds of various musical instruments. It can also produce unique sounds, unlike any conventional musical instrument. Used by some rock bands.

MONSOON

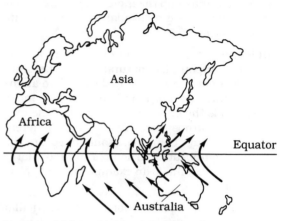

Summer pattern of winds shown (July, August in Northern Hemisphere)
From Erickson, VIOLENT STORMS (TAB Books, Inc., 1988).

**moon** 1. The large natural satellite orbiting the earth at a distance of approximately 240,000 miles. 2. Any natural satellite around a planet, generally regarded as too big to be an asteroid.

**moonbounce** A means of radio communication in which the moon is used to reflect signals. This requires sensitive receivers, large antennas

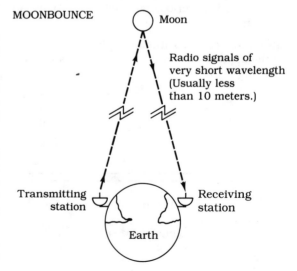

MOONBOUNCE

and high-power transmitters. It is done by some radio amateurs, especially at wavelengths of 2 meters and 3/4 meter, corresponding to frequencies of 144 and 432 megahertz.

**moraine**  A ridge of debris that is deposited at the melting margin of a glacier as it retreats. The debris results from erosion caused by the glacier.

**morphine**  A derivative of opium, used to relieve severe pain. It is highly addictive and is generally used only in hospitals. Side effects include nausea and disorientation. See *opium*.

**morphogenesis**  The evolution of the physical characteristics of an individual living thing, or of a group or species.

**morphology**  The study of the physical characteristics of individual living things, or of groups or species.

**Morse code**  Either of two telegraph codes, using long and short pulses ("dashes and dots") in combinations to represent letters of the alphabet, numerals and punctuation marks. The American Morse code is not used very often nowadays, except by some wire telegraph operators. The International Morse code or Continental Code is used by radio amateurs. See the MORSE CODE appendix.

**Morse, Samuel F. B.**  The American experimenter who is credited with the development of the wire telegraph during the nineteenth century.

**MOS**  Abbreviation for **metal-oxide semiconductor.**

**mosaic**  1. A combination of identical shaped objects, making up a larger whole. 2. A grid or matrix of devices, such as sensors in a television camera tube. 3. A large photograph assembled from numerous smaller ones. 4. A multicellular organism having cells of two or more genotypes. 5. A plant disease in which the leaves turn yellow or brown in a grid-like or matrix-like pattern.

**Moscow Papyrus**  An important papyrus, or scroll document, concerning ancient Egyptian mathematics. It dates from about 1890 B.C. It shows that volumes were calculated for complex figures such as truncated pyramids, and that the ancient Egyptians had well-developed three-dimensional geometry.

**moss**  A fairly primitive green plant, appearing something like carpet from a distance but having small leafy stems when seen up close. For their place on the plant evolutionary tree, see the PLANT CLASSIFICATION appendix. These plants are hardy, and can grow in places where many others cannot. They are often found on rocks in moist, shady places.

**motion picture**  A common "movie," made by flashing still pictures at a rapid rate through a projector. The human eye and brain smooths out the jerky motion and sees a realistic moving picture as a result. Developed by Thomas Edison around the end of the nineteenth century.

**motion sickness**  Symptoms that some people get when they ride in cars or trains, and especially in aircraft or boats. Headache, dizziness and sometimes nausea occur. The problem is often outgrown; there is medication available to lessen the symptoms.

**motion study**  Analysis of the movements of the human body, for purposes such as designing a more efficient running shoe or for improving the stroke of a competitive swimmer. Nowadays, computer graphics assist in this. Video tape is processed to display aspects of the motion, such as thrust vectors.

**motivation**  A term used in psychology, in reference to certain aspects of behavior. Physical factors often affect how often an animal will do something. Strangely, desire is frequently reduced by depriving an animal of something, such as sugar. The converse can also hold: Supplying alcohol may drive an animal to consume the substance till death takes place.

**motor**  1. A device for converting electrical energy into mechanical, rotational motion. It works by the action of electric currents in magnetic fields, with resulting forces. 2. Pertaining to movements of the body and to the actions of internal organs and glands.

**motor/generator** 1. An electromechanical device that can convert electrical energy into mechanical, rotational energy, and can also do the reverse. Many motors will work as small generators without modification. 2. An electric motor that is connected directly to a generator. The input and output of such a device are both electric, but the output is different. May be used, for example, to convert one level of direct-current voltage to another.

**motor neuron** A neuron responsible for transferring an impulse in the body to an organ or muscle, so that organ or muscle will do something in particular. A neuron that facilitates motion. See *neuron*.

**moulting** The shedding of the outer layer of skin, or hard covering, in a reptile or arthropod. A snake is a good example of a reptile that periodically does this.

**Mountain Daylight Time** Mountain Standard Time plus one hour, or Coordinated Universal Time minus six hours. Generally used from early April till the last weekend in October in the United States. See *Coordinated Universal Time, Mountain Standard Time*.

**Mountain Standard Time** Coordinated Universal Time minus seven hours. Used along and somewhat either side of the longitude line 105 degrees west. This runs through the mountainous territory of the western United States. Used from the end of October till the beginning of April in most applicable places. See *Coordinated Universal Time, Mountain Daylight Time*.

**Mount Palomar Observatory** An observatory in California, where a 200-inch-diameter reflecting telescope is located. For some time, this was the largest telescope in the world. It has been used to take many astronomical photographs, especially of other galaxies.

**Mount St. Helens** A volcano in southwestern Washington state in the United States that erupted in 1980. It was the first volcano to come alive in the continental United States in more than 60 years. It was a massive eruption, creating a vast dust cloud in the atmosphere. The event also triggered earth tremors, mudflows, and avalanches. It caused almost 3 billion dollars ($3,000,000,000) in damage to property and killed at least 62 people.

**Mount Wilson Observatory** An important observatory in California. It has a 100-inch-diameter reflecting telescope and other devices.

**mouse** 1. A small rodent generally regarded as a pest, and found throughout tropical and temperate latitudes. 2. A device used with a computer to move the cursor on the screen. The operator simply pushes it around on a flat surface and the cursor moves accordingly. See the drawing.

MOUSE

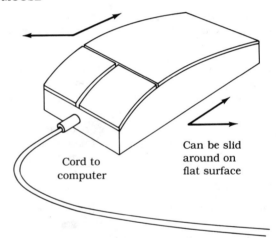

Cord to computer

Can be slid around on flat surface

**moving-coil device** 1. A dynamic speaker or microphone, in which the coil is attached to the diaphragm. See *microphone, speaker*. 2. A permanent-magnet generator or motor in which the magnet is fixed and the coil rotates.

**moving-magnet device** 1. A dynamic speaker or microphone, in which the magnet is attached to the diaphragm. See *microphone, speaker*. 2. A permanent-magnet generator or motor in which the coil is fixed and the magnet rotates.

**$m_p$** Abbreviation for proton rest mass.

**MSG** Abbreviation for monosodium glutamate.

**MST** Abbreviation for Mountain Standard Time.

**mucilage** A plant-derived gum that is often used as an adhesive for paper and cardboard. It is not strong and is not suitable for industrial applications.

**mucosa** The lining of the esophagus and intestines; also of the bronchi and windpipe. It is lubricated by mucus. It has millions of tiny cilia and villi that serve the purpose of absorbing nutrients (in the intestines) and keeping the surface clean and free of germs (in the lungs). See *cilium, mucus, villi.*

**mucous membrane** The linings of the nose and throat, lubricated by mucus and kept moist by body fluids. See *mucosa, mucus.*

**mucus** A secretion that is usually thick and viscous. It helps to carry away harmful microorganisms and serves to keep the membranes or mucosa clean. See *mucosa, mucous membrane.*

**mudflow** A form of landslide that occurs when earth, laden with water and therefore made heavier, breaks away and slides down a slope. Large mudflows can devastate the landscape in their paths, knocking down trees and even buildings.

**Muller, Erwin** The inventor of a device called the field ion microscope, that produced the first photographs showing atoms. The device can magnify several million times. The high energy and short wavelengths allow greater resolution than is possible with other types of microscopes.

**multi-** A prefix meaning "several" or "many." Compare mono-.

**multimeter** An electronic meter that measures current, voltage and resistance. May also have provisions for measuring other electronic variables such as capacitance or inductance.

**Multinomial Theorem** An expansion of the Binomial Theorem. It deals with the expansion of expressions like $(x+y+z)^n$, where n is some integer, perhaps very large. This is a complex mathematical theorem; a text on advanced algebra should be consulted for details. This theorem was originally proven and developed by Leibniz, around the time of Isaac Newton.

**multiple** 1. A number that, when divided by some whole number, yields some specified number. Thus, for example, 20 is a multiple of 2 ($2 \times 10 = 20$). 2. Several or many. We may say that an antenna radiation pattern has multiple lobes, for example.

**multiple sclerosis** A disease of the nervous system that strikes some people in young adulthood. It causes general deterioration of coordination and sensation. This disease is not very well understood, and a cure, or preventive vaccine, has yet to be found.

**multiplex** A process in which two or more signals are sent simultaneously over a single circuit. This may be done by sending the signals on different frequencies, or by splitting them up into time slots.

**multiplication** 1. A mathematical operation in which the product of two or more numbers is calculated or expressed. 2. In electronics, amplification of a harmonic frequency rather than the fundamental frequency. See *harmonic.* 3. In electronics, a digital process in which the frequency of a train of pulses is increased by some whole-number factor.

**multiplication table** A table of arithmetic products. Usually this is given for the first few natural (counting) numbers. See the MULTIPLICATION TABLE appendix.

**multiplier** A factor by which a number or variable is multiplied.

**multivariable** Pertaining to an equation having two or more variables. Sometimes these equations occur in sets; this is especially true of linear equations.

**multivibrator** An electronic amplifier whose output is deliberately fed back to the input, making it into an oscillator. Usually has two transistors that alternately switch on and off at a frequency determined by a resistor and capacitor, or a resistor and inductor. The output is a rectangular or square wave. See *amplifier, oscillator.*

**mumetal** An alloy used for magnetic shielding and in some transformers. It magnetizes easily and has high permeability, similar to ferrite. See *ferrite, magnetic core, permeability, transformer.*

**Munk, Walter** The oceanographer who conceived the Mohole Project — to drill all the way through the earth's crust. He thought of it one morning with friends at breakfast. See *Mohole Project, Mohorovicic discontinuity.*

**muon** A negatively charged subatomic particle, with the same charge as an electron. It is about 200 times as massive as an electron.

**muscle** In an animal, tissue that can contract, making movement possible. Some muscles are *voluntary*; your biceps are an example. Some are *involuntary*, such as those that cause peristalsis in your stomach and intestines. Muscle cells are made up of proteins *actin* and *myosin*, interwoven to form *myofibrils*. These cells are in turn arranged in *fibers*. See *actin, myofibril, myosin.*

**muscle fiber** An individual filament or strand of muscle cells. See *muscle.*

**muscular dystrophy** An inherited disease in which muscle wasting occurs. The victim eventually becomes so weak that he or she cannot walk; death ultimately results from respiratory failure or heart failure.

**mushroom** A fungus with a characteristic "toadstool" shape, that reproduces by means of spores. The spores are formed in folds underneath the cap. The fungus prefers dark, warm, moist environments. Some varieties are poisonous. See the PLANT CLASSIFICATION appendix for the place of the mushrooms on the evolutionary tree.

**mushroom cloud** The characteristic cloud produced by a massive explosion. It may grow to the top of the troposphere, where it is sheared off by upper-level winds. Updrafts occur near ground zero. A blast wave travels outward along the surface. See the illustration. This is the type of cloud produced by nuclear detonations.

**music** Simple or complex combinations of audio tones, sometimes with accompanying

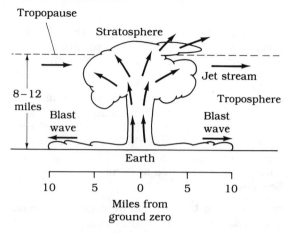

MUSHROOM CLOUD

voices, arranged to have a certain psychological effect.

**mussel** A shelled mollusk, found in rivers and in the ocean. It somewhat resembles a clam. It is considered a food delicacy by some people.

**mustard gas** A gaseous compound containing carbon, chlorine, hydrogen and sulfur. It causes skin blisters and is a general irritant. It is made by some governments for use chiefly as a chemical weapon.

**mutagen** Any chemical or environmental agent that increases the probability of mutations in offspring, especially of animals or humans. Radioactive materials are well known for this. Certain chemicals, such as pesticides, are also sometimes suspected. See *mutation.*

**mutant** An animal or plant having a mutation, especially an abnormality that is noticeable and that affects its way of life. See *mutation.*

**mutation** An abnormal characteristic in an animal, plant or human, caused by an error in chromosome reproduction. Mutations are thought to play a major role in the process of natural selection, since some mutations, by coincidence, make an organism better suited to deal with its environment.

**mutual capacitance** 1. The tendency for two electrically conducting objects to act as a capacitor when they are near each other. 2. The amount of capacitance between two nearby electrical conductors. Usually measured in picofarads; sometimes in microfarads. See *capacitance.*

**mutual inductance** 1. The tendency for a current in an electrical conductor to induce currents in nearby electrical conductors. See *inductance.* 2. The amount of inductance between two nearby electrical conductors. This is greatest in the case of coils having a common axis, and in particular, in transformers. 3. The extent of coupling between the windings of a transformer. See *transformer.*

**mutualism** Behavior between or among different species, resulting in an improvement in the living conditions for all species concerned. The species are often at greatly different evolutionary levels. For example, certain bacteria, called lactobacilli, help humans to digest milk sugar; the environment in the intestines is ideal for these organisms. Thus, both the lactobacilli and the host human benefit.

**Mv** chemical symbol for mendelevium.

**MV** Abbreviation for megavolt.

**mV** Abbreviation for millivolt.

**MW** Abbreviation for megawatt.

**mW** Abbreviation for milliwatt.

**mycosis** A fungal infection, causing disease. "Athlete's foot" and "swimmer's ear" are examples. A more serious condition can be caused by fungal infection of the gastrointestinal system.

**myelin** A fat-like substance that protects nerves. It also helps the nerves to send their signals. Excessive consumption of alcohol injures this layer and causes irritability ("hangover").

**mylar** A synthetic, clear or silvery material with exceptional strength and durability. It is used to tint windows and helps conserve energy by preventing radiation heat loss in buildings. It has a wide variety of industrial uses.

**myocardial** Pertaining to the heart muscle. See *heart.*

**myocardial infarction** A death of part of the heart muscle. This is sometimes called a "heart attack," although heart failure may occur for various reasons. Symptoms typically include intense, smothering pain in the chest and left shoulder, and possible fainting. In extreme cases death may occur rapidly. See *heart.*

**myofibril** Any of numerous filaments that make up a muscle cell. These filaments are formed by interweaving of two proteins. See *actin, myosin.*

**myoglobin** A waste product, resulting from muscle death. Some of this is normal and occurs all the time. A serious injury may result in elevated levels of this substance in the bloodstream. It is eliminated through the kidneys and urinary system.

**myopia** See *nearsightedness.*

**myosin** One of two proteins that make up the fibers in muscle cells. The two proteins fit together in a manner similar to the way a zipper works. See *actin, muscle.*

**myxedema** A disease in which the thyroid gland is underactive to a severe degree. The result is inactivity, sluggishness, sensitivity to cold, dry skin, and sometimes mental retardation. See *thyroid.*

**myxovirus** A virus that contains ribonucleic acid (RNA) and is comparatively large in size. The common viruses that cause influenza are of this type.

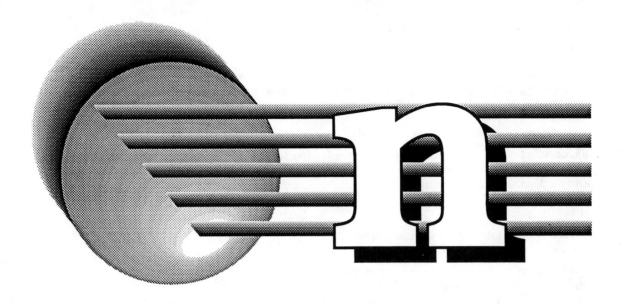

**N** Abbreviation for newton; chemical symbol for nitrogen.

**n** Abbreviation for nano- (see the PREFIX MULTIPLIERS appendix), natural number.

**Na** Chemical symbol for sodium.

**$N_a$** Abbreviation for mole.

**NAD** Abbreviation for nicotinamide adenine dinucleotide.

**nadir** The direction straight downwards, toward the center of the earth (or of whatever planet you might be on). This can be determined by means of a weight suspended by a string (see drawing). See *zenith*.

**nano-** 1. A prefix meaning billionth or billionths ($10^{-9}$ or 0.000000001). See the PREFIX MULTIPLIERS appendix. 2. A prefix meaning "extremely small." See *nanotechnology*.

**nanometer** A tiny unit of distance equal to one billionth ($10^{-9}$) of a meter. Sometimes, wavelengths of visible light and infrared energy are expressed in these units.

**nanotechnology** A new field of science involved with the design and construction of microminiature devices. In particular, microscopic-sized motors and machines are being developed. This is a rapidly advancing field. The potential applications are widespread.

**Napierian logarithm** See *logarithm*.

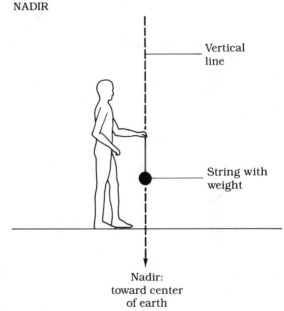

NADIR

**narcolepsy** A disorder in which a person becomes overwhelmingly sleepy during the day. In severe cases the person might fall asleep while sitting or standing, possibly while operating machinery. Medical treatment is available.

**narcotic** 1. Any strong drug that generally requires a doctor's prescription for use. Opium and its derivatives are the most commonly prescribed narcotic drugs. 2. A depressant type street drug, such as heroin.

**NASA** Abbreviation for National Aeronautics and Space Administration.

**nasal cavity** A region in an animal's or human's head, through which air passes when breathing is done through the nose. This region is where the sense of smell originates, as air passes over nerve endings that can detect minute traces of various chemicals.

**National Academy of Sciences** A group first founded in America in 1863 by Congress. Its purpose is to do experiments and make reports to the government on scientific subjects. Members are chosen according to their accomplishments in scientific research.

**National Aeronautics and Space Administration** The part of the United States government that makes decisions and carries out programs involved with the exploration of outer space. The Mercury, Gemini, and Apollo projects were overseen by this agency, as has been the Space Shuttle program and various space probes. See the SPACE PROBES appendix.

**National Electric Code** Guidelines used for safety in electrical wiring. The main purposes of these rules are to minimize hazards to people (such as electrical shock risk) and to protect property (especially against fire).

**National Hurricane Center** A United States government agency with headquarters based in Coral Gables, Florida, a suburb of Miami. This agency prepares forecasts and warnings concerning tropical disturbances throughout the world. It operates a system of satellites and aircraft, and also has numerous observing outposts. See *hurricane*.

**National Institutes of Health** A government-supported agency headquartered in Bethesda, Maryland. It conducts programs of research in the health sciences and in medicine.

**National Reclamation Act** A government action facilitated by President Theodore Roosevelt in 1902. This act concerned the management of water resources in the United States. It provided funds to build dams to supply towns such as Phoenix, Arizona with drinking water.

**National Science Foundation** A government-sponsored organization that supports long-term scientific research. It was begun in 1950. It administers grants to individual scientists and institutions.

**National Weather Service** Originally called the United States Weather Bureau. A nationwide (United States) network of observation stations, connected by wire and radio links, that keeps track of weather systems as they move across the North American continent. Provides updated weather forecasts for specific places every few hours.

**natural food** A term once used to describe unprocessed or lightly processed food. Actually it can apply to any food, because all foods come from nature. This has resulted in its use in advertising, to such an extent that its meaning has been almost completely lost.

**natural gas** 1. Hydrocarbon gases that exist under pressure beneath the earth's surface, and that are useful as fossil fuels. The principal gas is methane, with some ethane and butane. See *butane, ethane, methane*. 2. Methane gas used as a fuel and provided via pipelines for heating and other energy needs. It is given a characteristic scent, artificially, so that it can be easily smelled in case of a leak.

**natural group** A set of species all descended from a common ancestor. This definition depends on how far back in time we go. The more distant the ancestors, the more species can be considered a natural group.

**natural logarithm** See *logarithm*.

**natural number** Denotation, N. 1. Any number of the set of nonnegative integers {0, 1, 2, 3, ...}. 2. Any number of the set of positive integers {1, 2, 3, 4, ...}.

**natural resource** Any material or energy that is derived from nature. This includes fossil fuels, sunlight, and hydroelectric energy; also included are metal ore deposits, minerals, wood, water, and air. These are just a few examples.

**natural selection** An evolutionary process in which the fittest survive. Mutations occur; some of these result in improved adaptation to the environment. These animals or plants tend to gain dominance over the others. The drawing is a flowchart simplification of this process. Over millions of years, entirely new species might evolve by this process. According to the theory of Evolutionism, this is how humans came to populate this planet. See *Creationism, Evolutionism*.

NATURAL SELECTION

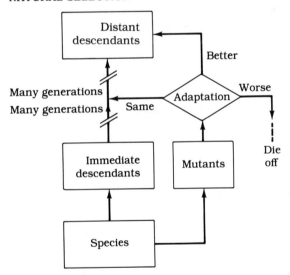

**nausea** The sensation of an upset stomach that generally precedes vomiting. It is a symptom of various infectious diseases and also might occur with certain types of poisoning. Vomiting may occur without nausea.

**nautical** Pertaining to the ocean and to ocean-going vessels.

**nautical almanac** A data book that gives information of use to navigators of ocean-going vessels. It tells what celestial objects are in what positions, as seen from various parts of the world.

**nautical chart** A special map used by navigators of ocean-going vessels. Courses are plotted on these charts, designed ideally so that shortest routes (great circles) can be drawn as straight lines. The chart also shows water depth, shipping hazards, buoys and markers, and currents. See *great circle*.

**nautical mile** 1. A distance of 2000 yards or 6000 feet (for approximate purposes). 2. A distance of 6080 feet (for precise purposes). 3. The distance over the earth's surface that is subtended by one minute of arc ($1/60$ degree) of latitude change going either directly north or directly south. Used by navigators of ocean-going vessels.

**Nautilus submarine** 1. The United States nuclear-powered submarine that went underneath the Arctic Ocean ice, passing over the North Geographic Pole in 1958. 2. A submarine built for Napoleon Bonaparte in 1801.

**nautilus** A marine animal known for the manner in which it builds its shell. It lives in ever-larger chambers, building them in a spiral with a characteristic shape. Also called a chambered nautilus.

**Naval Research Laboratory** A group of researchers and an institution associated with the United States Navy. Their job is to design and build efficient ships and submarines. Their work has resulted in benefits for shipping in peacetime as well as in wartime.

**navigation** The process of finding the best course for a ship or aircraft and then following that course. See *nautical almanac, nautical chart, navigational satellite, navigational table, radiolocation, radionavigation, sextant*.

**navigational satellite** A satellite that allows the captain of a ship or aircraft to precisely determine the position and heading of the craft. This is done with the aid of computers and radio signals, so that the information can be directly read by the navigator without the need for manual observations or calculations.

**navigational table** A table that the navigator of an ocean-going vessel uses to make computations for location and course. The table is compiled by the United States Navy, and allows precise navigation by means of the sun. See *sextant*.

**navigation network** The complete system of satellites, radio transmitters and computers that make modern navigation of ships and aircraft possible. See *navigational satellite, radiolocation, radionavigation*.

**Nb** Chemical symbol for niobium.

**Nd** Chemical symbol for neodymium.

**Ne** Chemical symbol for neon.

**Neanderthal man** The first of the species Homo sapiens, or modern man. The first fossils of this species were found in Germany in 1856. They were more heavily built than we are, and they were evidently much stronger also. They lived in western Europe and central Asia from about 130,000 years ago to 35,000 years ago. They were stone-age people. See *Homo sapiens*.

**neap tide** The tide that takes place when the moon is at its first quarter and third quarter. This is when its gravitational effects act at right angles to those of the sun. Therefore, the lunar and solar tides are opposing, and the range of the tides is smallest. See *spring tide, tide*.

**near infrared** See *infrared*.

**nearsightedness** A condition of abnormal sight in which nearby objects appear in good focus, but faraway objects are blurred. This can be caused by an inflexible lens or by a misshapen eyeball, or both (see drawing). See *farsightedness*. This problem is usually easy to correct, using glasses or contact lenses.

**near ultraviolet** See *ultraviolet*.

**nebula** A vast cloud of gas and dust in interstellar space. Some nebulae block the light from behind them, so they look dark; others glow because of constant bombardment with ultraviolet and X rays from nearby stars. The distant galaxies were thought to be gas nebulae before their true nature was discovered early in the twentieth century. Galaxies are still sometimes referred to as nebulae. See *galaxy*.

**necromancy** A pseudo-science concerned with the communication with spirits of dead people. In ancient times it was thought that the future could be predicted in this way. Today, some occult practices are similar to this. Scientists in general do not believe it.

**necrosis** Death of body tissue. This can occur because of malnutrition and also because of lack of oxygen. It can also be the result of injury, such as frostbite.

**negative charge** See *charge*.

**negative feedback** See *feedback*.

**negative number** Any real number with a value less than zero. Usually this is denoted by putting a minus sign in front of the digital value; for example, $-3.445$. All negative real numbers can be expressed as $-1$ times some positive real number. See *real number*.

NEARSIGHTEDNESS

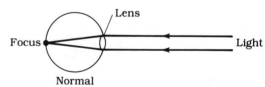
Normal

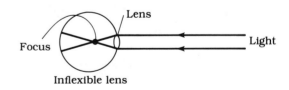
Inflexible lens

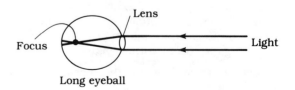
Long eyeball

**nematode** The scientific term for roundworm. There are numerous varieties. Some of them are responsible for diseases in humans, especially in tropical climates. They are primitive life forms.

**Nemesis** A hypothetical dim star that might pass near the Solar System every 26 million years. According to one theory, this could cause comets to be thrown out of the Oort cloud and careen toward the inner Solar System. Some of them might strike the earth and cause catastrophes such as mass extinctions. See *Oort cloud*.

**neo-** 1. A prefix meaning new or recent. 2. A prefix in chemistry meaning related to.

**neodymium** Chemical symbol, Nd. An element with atomic number 60. The most common isotope has atomic weight 142. It is a soft metal in pure form. It is used to tint glass and to make a laser for certain medical applications. See *neodymium-YAG laser*.

**neodymium-YAG laser** A device used in laser surgery. The abbreviation YAG stands for yttrium-aluminum-garnet, which is a crystal in which the neodymium solution is contained. The laser produces infrared radiation. It is driven by energy from arc lamps. The efficiency is quite low. See *laser*.

**Neogene** A geologic age that began 26 million years ago and extends to the present. This is the recent part of the Cenozoic era. It is sometimes called the "age of mammals," because this is when the mammals flourished. See the GEOLOGIC TIME appendix.

**Neolithic** Pertaining to the more recent part of the stone age when humans began farming as a means of obtaining food. The first settlements were near rivers: the Nile, the Tigris, and the Euphrates in the Middle East, and the Indus, and Hwang Ho in Asia.

**neon** Chemical symbol, Ne. An element with atomic number 10. The most common isotope has atomic weight 20. This element is a gas at room temperature and atmospheric pressure. It does not react easily. It is present in small amounts in the atmosphere, from which it is obtained.

**neon lamp** An illuminating device that operates by exciting neon gas. When a current flows through the gas, the lamp glows with an orange hue. The glow is not especially bright. The construction is simple (see drawing).

NEON LAMP

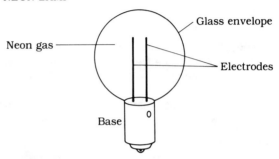

**neoprene** A synthetic rubber, known for its durability and its resistance to petroleum solvents.

**nephritis** See *nephrosis*.

**nephron** A small apparatus in the kidney that performs the task of getting rid of wastes and allowing water, as needed, to be retained by the body. It is a kind of filtering device. A single kidney has millions of them.

**nephrosis** Inflammation and/or infection of the kidneys. Can be caused by strain resulting from toxic substances in the bloodstream. Certain drugs can be responsible. It also can occur in conduction with other diseases or injuries. When severe, the condition is called nephritis. Kidney failure can take place; the condition requires medical treatment and can be extremely serious.

**Neptune** The eighth known planet from the sun in our Solar System. It is a large, cold, bluish planet considerably larger than the earth. The American space probe Voyager II flew by this planet in 1989, sending back numerous photographs. See the SOLAR SYSTEM DATA appendix.

**neptunium** Chemical symbol, Np. An element with atomic number 93. The most common isotope has atomic weight 237. This element does not occur naturally. It occurs as a result of the manufacture of plutonium for making nuclear bombs.

**Nernst Effect** A phenomenon that takes place when a strip of metal is heated and placed in a magnetic field. If the heating is not uniform, and if the magnetic flux is at right angles to the strip, a voltage develops between the edges of the strip.

**Nernst-Ettinghausen Effect** The opposite of Nernst Effect (see *Nernst Effect*). For certain materials, when a current flows in them and they are in a magnetic field, they will become warm. The current must be in a direction nearly at right angles to the magnetic field.

**nerve** A cell or fiber that converts visible light, chemical stimuli, pressure, and temperature changes into impulses and sends these impulses to the brain. These are the sensory nerves. Also, a cell or fiber responsible for the voluntary or involuntary movements of muscles. These are the motor nerves. See *neuron, nerve ending, nerve fiber.*

**nerve cell** See *neuron.*

**nerve ending** The termination of a sensory nerve that detects the effects we perceive as heat/cold, pain, itching, visible light, sound, taste, and smell.

**nerve fiber** The long strand-like components of nerve cells that, in combination, transmit the sensory and motor impulses, something like a cable sends electricity, but more slowly. The fibers of individual nerve cells are called axons. See *axon, dendrite, nerve, neuron.*

**nerve gas** Any of various gases that disable people by affecting the nervous system. They are manufactured with the intent of use in the event of warfare, and are thus classified as chemical weapons.

**nerve impulse** The electrochemical wave that travels along a nerve fiber. It moves at anywhere from about one foot to 350 feet per second, depending on the intensity and type of sensation or motor command.

**nervous system** The entire apparatus of the brain, spinal cord, and all of the nerve cells in

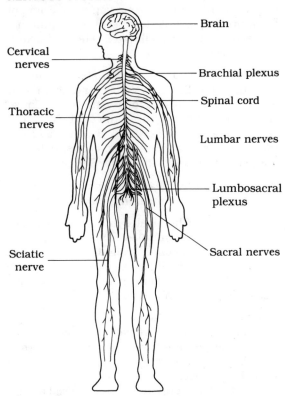

NERVOUS SYSTEM

the body, and the impulses that travel along the nerve fibers.

**net** 1. Denoting the contents of a shipment, not including packing material. 2. A group of communications stations organized for the purpose of circulating messages. 3. Pertaining to the practical consequences of an action. We might speak of the net effect of adding ethanol to gasoline, for example.

**net primary production** 1. The energy trapped by all of the photosynthetic plants in a certain ecological system. 2. The energy trapped by all the photosynthetic plants in the entire world. See *photosynthesis.*

**net tonnage** A measure of the useful, or revenue-producing, part of a ship. This is actually a volume measure, and includes passenger and cargo space.

**network** 1. Any system that can be precisely or approximately mapped, and exists for a specific purpose. 2. An electronic device or system constructed for some specific application; a circuit or set of circuits. 3. In communications, a set of radio or wire stations or subscribers, and the radio links or wire cables that interconnect them.

**neuritis** A condition in which the nerves are inflamed and irritated. This might occur in various disease states. The main symptom is a tingling sensation in the fingertips and toes. In moderate cases, pain might also be felt. When severe, numbness and paralysis might occur.

**neurogenic** 1. Pertaining to nerve function, especially when it is affected in some way that causes abnormality. 2. Caused by abnormal nerve function.

**neurology** The branch of medicine that is concerned with the workings of the nervous system. In particular, the specialty concerned with diagnosing and treating nervous-system disorders.

**neuron** An individual nerve cell. There are several different kinds, depending on the location in the body and on the function. Basically they can be categorized as *sensory*, responsible for sensation; *motor*, responsible for muscle movement; and *connecting*, that send impulses back and forth between the spinal cord, the brain, and various body parts. Some nerve cells have long filaments called fibers or axons that transmit the impulses. See *axon, dendrite*.

**neurophysiology** The study of the physical aspects of the nervous system, such as the exact chemical and electrical processes that occur in the transmission of sensations and motor commands.

**neurosis** A behavioral disorder that interferes with efficient daily living, but that does not involve a loss of contact with reality. Almost any person has one or more mild neuroses or "hangups." They often do not require any treatment, although it is helpful for a person to know what his/her neuroses are.

**neuter** 1. Having characteristics of neither sex. 2. To sterilize, especially in the case of an animal, thereby making it impossible for the animal to produce offspring.

**neutering** The process of sterilizing an animal, especially a pet, so that it cannot produce offspring. This is usually a simple and routine operation, done by a veterinarian. In addition to sterilizing the animal, the operation often moderates its behavior, so that the animal is more tame.

**neutral** 1. Having no net electrical charge; having equal numbers of positive and negative charge carriers. 2. Having a pH of 7; neither acidic nor alkaline (see *pH*). 3. Having no magnetic polarization.

**neutralization** 1. The rendering of a chemical to have a pH of 7 by adding acid (in the case of a base) or a base (in the case of an acid). See *pH*. 2. The rendering of an object so that it has zero electrical charge. 3. An electronic process in which an amplifier is adjusted so that it will not oscillate.

**neutrino** A subatomic particle, emitted in large numbers by stars and by the sun. It is believed to have a theoretically zero mass. This gives it great penetrating power, making it hard to detect. Nonetheless neutrinos were found in the 1950s by means of apparatus located in mine shafts.

**neutron** The neutral subatomic particle that, along with positively charged protons, comprises the nucleus of an atom. It has just about the same mass as a proton. See *electron, proton*.

**neutron bomb** A type of bomb that causes little or no destruction to buildings and other inanimate objects, but that emits huge numbers of high-speed neutrons, causing death to living organisms (see drawing). This type of bomb is controversial but has been developed for possible wartime use.

**neutron number** The number of neutrons in the nucleus of an atom. This is approximately the atomic weight minus the atomic number.

**neutron radiation** Bombardment by high-speed neutrons. This occurs during the blast of a neutron bomb (see *neutron bomb*) and also in certain particle accelerators.

## NEUTRON BOMB

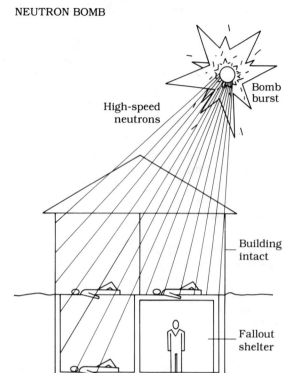

High-speed neutrons / Bomb burst / Building intact / Fallout shelter

**neutron rest mass** The mass of a single neutron when it is not in motion. This value is $1.675 \times 10^{-24}$ gram, or $1.675 \times 10^{-27}$ kilogram. When a neutron is accelerated to high speed, its mass is increased because of relativistic effects. See *mass distortion*.

**neutron star** A star that has collapsed under the influence of its gravitation, after having exhausted its supply of nuclear fuel for energy. The electrons are driven into the nuclei of the atoms, so that only neutrons remain. The whole star might have several times the mass of the sun, yet be only a few miles across. If it gets dense enough, the star becomes a black hole. Rapidly spinning neutron stars may emit pulses of intense radio energy. These are called pulsars. See *black hole, pulsar*.

**newton** Abbreviation, N. The Standard International unit of force. It gets its name from Isaac Newton, the physicist who defined much of classical physics. It is equal to one kilogram meter per second per second. That is, if a mass of one kilogram is accelerated in a straight line at a rate of one meter per second per second, the force is one newton.

**Newtonian mechanics** Also called classical mechanics. The mechanics devised by Isaac Newton. It was modified by the development of Einstein's relativity theory. See *Newton's Laws of Motion*.

**Newtonian physics** The laws of physics when relativity is not taken into account. Developed by Isaac Newton. For comparatively "slow" speeds, less than about 10 percent of the speed of light, these rules are adequate for most applications. See *Newton's Laws of Gravitation, Newton's Laws of Motion*.

**Newton, Isaac** One of the greatest physicists of all time. In the seventeenth century he developed the calculus, and a theory of gravitation that is still used today except when extreme accuracy is needed. He earned the Lucasian chair of mathematics at the University of Cambridge, England when he was just 27 years old. This chair has recently been held by such notables as Stephen Hawking, the author of the bestselling book *A Brief History of Time*.

**Newton's Laws of Gravitation** 1. Every mass in the universe exerts an attractive influence on every other mass. 2. For two objects, the force of gravitation acting between them is directly proportional to the product of their masses, and inversely proportional to the square of the distance between their centers of mass.

**Newton's Laws of Motion** Three rules of physics that are accepted for speeds less than about 10 percent of the speed of light. At greater speeds there is considerable error, and relativity must be considered. The three rules are: (a) Unless acted on by an outside force, an object continues at the same speed in a straight line; (b) When a force acts on an object, the change in momentum occurs in the same direction as the force and in direct proportion to the intensity of the force; (c) For every action, there is an equal and opposite reaction.

**Ni** Chemical symbol for nickel.

**niacin** One of the B complex of vitamins, sometimes called vitamin B-3. Deficiency causes pellagra. Might occur in a form called *nicotinic acid* or in a form called *niacinamide*. See *niacinamide, nicotinic acid, vitamins* and the VITAMINS AND MINERALS appendix.

**niacinamide** A form of niacin. It is the most common formulation for large-dosage tablets of the substance. See *niacin, nicotinic acid* and the VITAMINS AND MINERALS appendix.

**Nichrome** An alloy that has considerable electrical resistance, and therefore is only a fair conductor. It consists of nickel and chrome, hence the trade name. This alloy is made into wire, and used for heating elements. It might also be used to make resistors for electrical equipment.

**nickel** Chemical symbol, Ni. An element with atomic number 28. The most common isotope has atomic weight 58. It is a metal in pure form, occurring along with iron in ore deposits. It is used in various industrial alloys.

**nicotinamide adenine dinucleotide** Abbreviation, NAD. A substance that comes from niacin in its nicotinic acid form. It is important in food metabolism. It attracts hydrogen atoms, and therefore acts to remove hydrogen from various substances in the body.

**nicotine** A drug with stimulant properties, found in tobacco smoke. In minute amounts it causes elevation of the blood pressure and, in some people, a feeling of well-being; it makes others sick. In pure form it is a deadly poison, one drop being sufficient to kill a human being almost instantly.

**nicotinic acid** A form of niacin. If taken in large doses it causes most people to develop a temporary "flush" or "hot flash." Therefore, niacinamide is more often used when it is not necessary to take more than a few milligrams at a time. See *niacin, niacinamide*, and the VITAMINS AND MINERALS appendix.

**nimbostratus clouds** Low clouds that produce drizzle, snow, or rain, but are usually not associated with violent weather. See the illustration. These are the clouds you see on a dreary day in autumn.

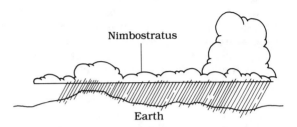

NIMBOSTRATUS CLOUDS

**niobium** Chemical symbol, Nb. An element with atomic number 41; the most common isotope has atomic weight 93. It is a soft metal, used in industry especially in welding.

**nit** 1. The egg of a common louse. This is often the medium for the spread of body lice. 2. A unit of visible-light intensity, equal to one candela per square meter.

**nitrate** 1. A radical or compound containing nitrogen. 2. A radical with three atoms of oxygen and one of nitrogen, found in various compounds, especially organic substances. It is commonly found in sewage pollutants. See *nitrate pollution, radical*.

**nitrate pollution** A common form of water pollution resulting from the dumping of untreated or inadequately treated sewage into rivers and the ocean. Some nitrates stay in sewage even after it is treated. Too many nitrates in the water can cause eutrophication. See *eutrophication*.

**nitric acid** A compound containing hydrogen, nitrogen, and oxygen that occurs as a powerful acid. It is a liquid in its natural form. It dissolves many substance that other acids cannot. In fact one of its main uses is as a constituent of a powerful solvent. See *aqua regia*.

**nitride** A certain type of compound containing nitrogen. The nitrogen acts as the negatively charged part of the compound.

**nitrite** A radical with two atoms of oxygen and one of nitrogen. Compounds containing this radical are used to preserve food, especially processed meats. There has been some concern that these substances can be converted to carcinogens in the intestines. See *carcinogen*.

**nitrogen** Chemical symbol, N. An element with atomic number 7; the most common isotope has atomic weight 14. It is a gas at room temperature and pressure, and comprises about 78 percent of the earth's air at the surface. In liquid form it is used in medicine to freeze off skin lesions such as warts. Industrially it is used in various applications such as welding. It is not an especially reactive gas, and in some situations will suffice as an inert gas. Nitrogen is an important constituent of organic compounds, including protein.

**nitrogen cycle** An important natural cycle in which the environment of the earth is stabilized organically. See the illustration. Nitrogen enters the earth from the air and also from the deaths of living things; it is returned to the air by a process called enitrification. This is a greatly simplified rendition of this cycle.

NITROGEN CYCLE

N   Free nitrogen
O   Nitrogen oxides
●   Nitrates
    Denitrification
    Death and decomposition

**nitroglycerin** A compound made by the reaction of nitric acid, sulfuric acid, and glycerol. It is well known as an explosive.

**nitrous oxide** A gas consisting of a compound of two nitrogen atoms and one oxygen atom. It is sometimes used as an anesthetic, mixed with oxygen. In pure form it can be dangerous because it deprives the brain of oxygen.

**NMRI** See *nuclear magnetic resonance imaging*.

**No** Chemical symbol for nobelium.

**Nobel, Alfred** A wealthy Swedish inventor of the nineteenth century who created the famous Nobel Prize. He made most of his money on inventions related to explosives. This was important especially in warfare. He also experimented with synthetic materials, testing substitutes for rubber, leather and various other products. See *Nobel Prize*.

**Nobel, Immanuel** The father of Alfred Nobel. See *Nobel, Alfred*.

**nobelium** Chemical symbol, No. An element with atomic number 102. The most common isotope has atomic weight 254. This element is not found in nature. It is a by-product of nuclear reactions. All isotopes are radioactive and decay very quickly (half life less than one minute).

**Nobel Prize** An award given to professionals who make outstanding contributions to humankind. The award was founded by Alfred Nobel (see *Nobel, Alfred*). Awards are given in various fields of sciences and arts. They are not based on nationality or race in any way; only merit is considered.

**noble gas** Any of the elemental gases helium, argon, krypton, neon, and xenon. These gases are inert; that is, they do not generally react with other elements.

**noctilucent clouds** High-altitude stratospheric clouds that can be seen glowing long after sunset, especially during the summer months at high latitudes. They glow because the sun shines on them, because they are so high, even though the earth's surface is dark.

**node** 1. A point toward which pathways converge, or from which pathways branch out. 2. In electronics, a local minimum, especially pertaining to current or voltage on an antenna or transmission line. See *local minimum*. 3. A lymph gland, especially when it is inflamed because of infection in the body. 4. In plants, a point where a leaf or leaves join the stem. 5. In astronomy, a point where the orbit of an object crosses the plane of the ecliptic. See *ecliptic*.

**noise** 1. Sound disturbances that tend to be annoying to listeners. 2. Sound energy that occurs over a wide band of frequencies. 3. Unwanted electrical currents in any circuit. These can be caused by natural disturbances, like lightning, or by human-made systems, such as electric motors. They also can result from spontaneous fluctuations such as thermal effects.

**nomograph** A straight line calibrated in two scales for comparison. One scale is on the left or the top side of the line; the other is on the right or the lower side. By this means it is easy to convert from one quantity (such as inches) to the other (such as centimeters).

**non-** 1. A prefix meaning not. 2. A prefix meaning nine or times nine.

**non-Euclidean** Pertaining to geometry in which the axioms of Euclid do not hold. See *Euclidean, non-Euclidean geometry, non-Euclidean space*.

**non-Euclidean geometry** 1. Geometry on surfaces that are not flat. 2. Geometry in spaces that do not obey the normal Euclidean rules; that is, spaces that are non-Euclidean subsets of some space in more dimensions. See *Euclidean geometry, non-Euclidean space, Parallel Postulate*.

**non-Euclidean space** Any space in which the rules of Euclidean geometry do not generally apply. The drawing shows Euclidean space and two examples of non-Euclidean spaces. The measures of the interior angles of a triangle do not always add up to 180 degrees if the space is non-Euclidean. See *Euclidean space*.

NONEUCLIDEAN SPACE

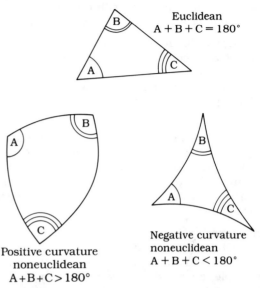

Euclidean
$A + B + C = 180°$

Positive curvature noneuclidean
$A + B + C > 180°$

Negative curvature noneuclidean
$A + B + C < 180°$

**nonillion** The number obtained by writing 1 followed by 30 zeros, or 1000 followed by nine sets of three zeros.

**nonionizing radiation** Any form of radiant energy that does not cause atoms to ionize. Generally this includes electromagnetic energy at radio, infrared, and visible wavelengths. At higher wavelengths ionization begins to take place. See *ionizing radiation*.

**nonlinear** 1. Pertaining to an equation with a graph that is not a Euclidean subset of the coordinate space. In a plane, this would mean that the graph is not a straight line. In three-space, it would mean that the graph is "warped" (not "flat"). 2. In electronics, pertaining to an amplifier in which the amplification factor, or gain, changes when the input level changes. See *gain, linear*.

**nonlinear circuit** 1. An electronic circuit in which the output waveform has a shape that is different from the input waveform. 2. An electronic circuit in which the gain changes when the level of the input signal changes. See *gain*.

**nonlinear equation** In mathematics, an equation that is not linear. See *linear equation*. Examples are equations in which some of the variables are raised to a power, and equations containing logarithms, exponential expressions, and trigonometric expressions.

**nonlinear function** A function whose graph is not a Euclidean subset of the coordinate space in which it is graphed. Such functions contain variables raised to powers, logarithms and exponential expressions, or trigonometric expressions.

**nonrenewable resource** 1. A natural resource that is limited in supply, so that care must be used if depletion is to be avoided. 2. A natural resource that is almost gone, and that cannot be replenished by any known, efficient means. See *renewable resource*.

**nonterminating decimal** A decimal that cannot be fully written out. An example is the decimal rendition of $1/3 = >0.333...$ . It may be possible to write the whole decimal expansion by implying what the numbers are, as in the case of $1/3$; with some numbers, however, this cannot be done. See *nonterminating, nonrepeating decimal*.

**nonterminating, nonrepeating decimal** A decimal that cannot be fully written out, even by implying what the sequence is. The digits do not follow any pattern, but instead, are "random" in the sense that the only way to determine each digit is to calculate it. All irrational numbers are of this type. See *irrational number*.

**norepinephrine** A hormone that works along with epinephrine to increase the heart rate and raise the blood pressure. This occurs when there is sudden stress, such as fright or anger. See *epinephrine*.

**normal** 1. Typical or average; representing the usual state of affairs. 2. Intersecting a surface at a right angle. See *normal line*. 3. In chemistry, a condition of having one gram equivalent of dissolved substance per liter of solution. 4. In chemistry, having neither acid hydrogen nor alkaline hydroxide ions. 5. Pertaining to a standard statistical distribution. See *normal distribution*.

**normal distribution** A characteristic curve of statistics. It is sometimes called the bell-shaped curve because of its appearance (see drawing). The vertical center line is called the median. In the normal distribution this is also the average or mean. The curve is symmetrical relative to this line, although the steepness may vary. See *average, median*.

NORMAL DISTRIBUTION

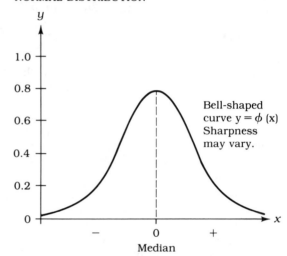

**normal fault** The boundary between two sections of the earth's crust that are pulled apart. One section slides downward along a plane that is often slanted. Actually this is a misnomer, because most faults result from compression. Thus, so-called normal faults are not the most common kind. See *reverse fault*.

**normal line** A line that is perpendicular to a surface at a given point. If the surface is not flat, then the line is perpendicular to the tangent plane at the point where the line passes through the surface.

**northern lights** See *Aurora Borealis*.

**North Star** See *Polaris*.

**notation** 1. A standard, accepted way in which information is written. 2. A method of writing mathematical expressions, equations and functions, with certain letters generally standing for unknowns, certain letters for constants, some for integers, some for real numbers, and so on.

**note** 1. A reference statement in a technical document. 2. An audio tone, usually consisting mainly of one frequency within the human hearing range. 3. The sound made by a Morse code pulse in a radio receiver.

**nova** The flare-up of a star, causing a seemingly new, bright star to appear in the sky for a while. The star might become hundreds or even thousands of times brighter than normal for a few hours or days. See *supernova*.

**Np** Abbreviation for neper; chemical symbol for neptunium.

**nuclear** 1. Pertaining to the nuclei of atoms. 2. Pertaining to atomic power or reactions. 3. Radioactive.

**nuclear chemistry** A branch of inorganic chemistry concerned with the reactions that occur within individual atomic nuclei. This may include fission and fusion, but more often refers to the changing of isotopes by adding or subtracting neutrons. See *fission, fusion, isotope*.

**nuclear energy** 1. The energy produced by the decay of unstable atomic nuclei. This is in the form of radioactivity. See *alpha particle, alpha rays, beta particle, beta rays, gamma rays, neutron radiation*. 2. Useful energy derived from controlled atomic reactions. See *nuclear power plant*.

**nuclear fallout** See *radioactive fallout*.

**nuclear fission** See *atomic fission*.

**nuclear force** The "glue" that keeps the protons in an atomic nucleus from flying apart because of their mutual electrical repulsion. They are all positively charged, and like charges repel, but the nuclear force is a much stronger attracting factor.

**nuclear fusion** See *fusion*.

**nuclear magnetic resonance imaging** A medical technique used to look inside the body without surgery. In recent years this has been used in place of X rays. It works by causing atoms to vibrate in a certain way, emitting radio waves that are detected, and whose distribution is displayed on a monitor screen. This provides detailed images of internal body organs and structures.

**nuclear physics** A branch of physics concerned with the behavior of particles within atomic nuclei. This especially includes the effects of bombarding nuclei with high-speed particles.

**nuclear power plant** An installation for generating electricity from atomic energy. The heat from nuclear reactions is used to boil water, making steam. The steam, under pressure, drives turbines (see drawing). Fission reactions are used for this, and waste products are produced. Fusion would be cleaner, but it has not yet been controlled. See *fission, fusion*.

NUCLEAR POWER PLANT

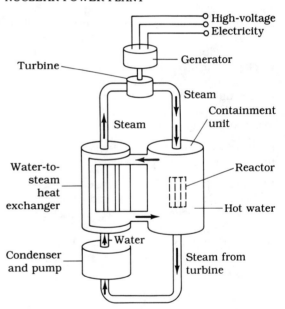

**nuclear propulsion** A means of getting thrust for a spacecraft by using nuclear reactions. This is still in the conceptual stages. One method would use nuclear bombs; another would use continuous fusion.

**nuclear radiation** Any of the types of ionizing radiation resulting from nuclear reactions. See *alpha particle, alpha rays, beta particle, beta rays, gamma rays, ionizing radiation, neutron radiation*.

**nuclear reactor** A device that uses a controlled atomic reaction to obtain energy for human needs. Generally, this energy occurs as heat, which is converted into electricity by steam turbines. See *nuclear power plant*. A by-product from reactors is plutonium, used in making nuclear weapons. All reactors thus far developed have used fission. Research is being done into the possibility of using fusion, which does not produce harmful waste products. See *fission, fusion*.

**nuclear submarine** A submarine that gets its propulsion and other power from a nuclear reactor. Most of the important military submarines are of this kind. See, for example, *Nautilus submarine*.

**nuclear warfare** Any warfare in which nuclear weapons are used. Tactical nuclear war involves limited use of small atomic weapons to wipe out enemy troops. Strategic nuclear war would involve the use of large bombs (upwards of 5 megatons apiece), and long-range missiles with multiple warheads. This could cause the destruction of whole cities, and perhaps of whole societies.

**nuclear waste** See *high-level radioactive waste, low-level radioactive waste*.

**nuclear winter** The change in weather throughout the world that would probably take place after a large-scale strategic nuclear war. The drawing gives an indication of the temperature distribution we might expect. There would be so much debris in the atmosphere that the

NUCLEAR WINTER

From Erickson, VIOLENT STORMS (TAB Books, Inc., 1988).

sun's energy would be blocked, and it would be as cold and dark as winter, even in the middle of the summer.

**nucleonics** A branch of engineering concerned with the handling of nuclear materials, including radioactive wastes.

**nucleotide** Any of the small molecules that link together to form deoxyribonucleic acid (DNA) and ribonucleic acid (RNA). See *deoxyribonucleic acid, ribonucleic acid*.

**nucleus** 1. The dense, central part of an atom made up of protons and neutrons. See *atom, Bohr atom, neutron, proton, Rutherford atom*. 2. The central part of a cell that contains the deoxyribonucleic acid (DNA). See *cell*. 3. The core of a galaxy. See *galaxy*.

**null** 1. The cancellation of two signals having equal amplitude and frequency, but opposite phase. See *phase*. 2. A local minimum for a positive quantity, or a local maximum for a negative quantity, at which the value is zero or very near zero. See *local maximum, local minimum*. 3. A subscript numeral zero ($_0$). 4. Having no elements. See *null set*.

**null set** In mathematics, the set containing no elements. It is sometimes written as a numeral zero with a slash through it; also can be denoted { }.

**number** 1. A set, built up from other sets according to a prescribed mathematical scheme. Consult a text on advanced number theory or set theory for details. 2. An expression of quantity or magnitude. A number should not be confused with a numeral. See *complex number, imaginary number, integer, irrational number, natural number, numeral, rational number, real number*.

**number field** First defined by J. Dedekind about 1879. A set of numbers that forms a group that obeys the properties of commutativity for addition and multiplication, and for which multiplication is distributive with respect to addition. See *commutative law, distributive law, group*.

**number line** 1. A line drawn and marked off with numerals intended to show magnitude or time. The scale might be linear, with equal differences represented by equal lengths, or it might be nonlinear, with the lengths computed according to some nonlinear function of the number values. 2. The hypothetical geometric line, on which each point corresponds to exactly one real number, and each real number corresponds to exactly one point on the line. It is not known whether this is true of all geometric lines. It can be assumed true and called the *continuum hypothesis*. See the drawing.

NUMBER LINE

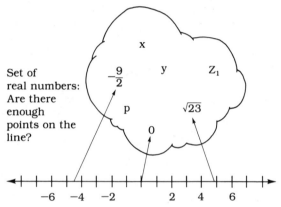

Each point on line corresponds
to a unique
real number.

**number system** Any scheme of numbers devised for some mathematical or scientific purpose. See *complex number, imaginary number, integer, irrational number, natural number, rational number, real number*.

**number theory** A branch of mathematics, concerned with the nature and behavior of numbers and number systems. All numbers are ultimately built up from simple sets, with zero being the empty set and the foundation for all other numbers. See *number*.

**numeral** The written symbol for a number. See *number*.

**numeration** The scheme by which numbers are represented as numerals. We use a base-10 (decimal) system, known as the Arabic, or Hindu-Arabic, system. There are other systems. See, for example, *binary number, octal number*.

**numerical analysis** The mathematical study of numerical algorithms for solving problems.

**nutation** A shift in the extent to which the earth's axis is slanted. Presently it is at an angle of 23.5 degrees relative to the ecliptic. But sometimes it is slanted as much as 24.5 degrees, and at other times only 22.5 degrees. The cycle completes itself every 41,000 years. For the past 10,000 years the tilt has been getting less. This affects the climate. See *ecliptic*.

**nutrient** 1. Any organic substance that is needed or utilized by a living organism. 2. Protein, essential fatty acids, carbohydrates, vitamins, or minerals in the human diet.

**nutrient value** See *nutritive value*.

**nutrition** 1. The study of the metabolism of various substances by humans or animals, and the amounts and ideal proportions of these substances that are needed in the diet. 2. The adequacy of a diet, particularly for humans. We speak of "good nutrition" or "bad nutrition."

**nutritional deficiency** See *malnutrition*.

**nutritional quackery** 1. The selling of useless or even phony substances with claims that they have nutritive value. 2. The promoting of a nutritional supplement with the claim that it will cure serious diseases such as cancer, without any scientific evidence that this is true. 3. The advertising of vitamin, mineral or other nutritional supplements, in such a way as to mislead the public about their functions and effectiveness.

**nutritional science** The branch of science concerned with nutrition, and especially for human beings. See *nutrition*.

**nutritional supplement** Concentrated nutrients, particularly vitamins, minerals, amino acids, or essential fatty acids. These are sold in tablet or liquid form, and sometimes also as natural products such as desiccated liver or brewer's yeast.

**nutritive value** 1. A general expression concerning the vitamin and mineral content of a certain food. 2. A qualitative expression of the ratio of essential nutrients to calories for a given food.

3. A table showing the nutrient content of a standard portion of a food, indicating how much of the Recommended Daily Allowance it supplies for each nutrient. See the RECOMMENDED DAILY ALLOWANCES appendix.

**nylon** A synthetic material, developed by polymerization of protein-like molecules. It has exceptional durability and can withstand wear that many other fabrics cannot. See *polymer, polymerization.*

**nymph** An intermediate stage in the development of some insects. The insect can move around, but it does not yet have wings (in the case of a flying insect) and cannot produce offspring.

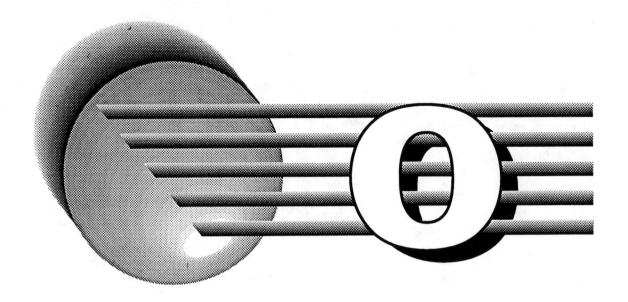

**O** Chemical symbol for oxygen.

**oasis** A place in the desert where water is provided and where plants grow and wildlife can take refuge. Might occur when a natural spring comes to the surface; can also be created by pumping water to the surface.

**obesity** A condition of extreme overweight. For an adult, this generally refers to having a weight more than 25 percent higher than one's ideal weight.

**objective lens** 1. In a microscope, the convex lens that is nearest the sample to be examined. It produces an image that is clarified by the eyepiece. 2. In a refracting telescope, the convex lens that gathers the light and focuses an image to be magnified by the eyepiece. See *eyepiece, microscope, telescope.*

**objective mirror** In a reflecting telescope, the concave mirror that gathers the light and focuses an image to be magnified by the eyepiece. See *eyepiece, telescope.*

**oblique** 1. Existing at an angle; slanted with respect to some reference line. 2. Receding.

**oblique drawing** A type of illustration in which the front of an object is drawn as if it were exactly facing the viewer, but with the top and sides also visible. Perspective is ignored. All dimensions are to scale; this exaggerates the apparent depth of the object. See the illustration. See *isometric drawing.*

OBLIQUE DRAWING

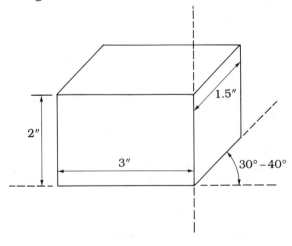

**oblong number** Any of the numbers obtained by multiplying a natural number by its successor. The general expression is $n(n+1)$ for $n = 1, 2,$

3, ... . The first few such numbers are 2, 6, 12, 20, 30.

**observatory** A facility housing one or more telescopes. Some observatories have radio telescopes while others have optical, infrared and/or ultraviolet telescopes. The location of an observatory is important. Radio telescopes must not be hindered by manmade interference, so they are often located in remote places. Optical telescopes require altitude and are therefore usually on mountain summits.

**obsession** A type of behavior where a person puts an abnormal amount of thought or effort into one thing, to such an extent that it interferes with life's routine. See *obsessive-compulsive neurosis*.

**obsessive-compulsive neurosis** A behavioral disorder in which a person does unnatural or unproductive things. For example, someone might feel compelled to count every speed-limit sign or every red light encountered, on every trip, no matter where. This condition often can be helped by a psychiatrist.

**obstruction** 1. A condition of blockage, especially in a pipe or tube through which fluid or gas flows. This might be partial, restricting the flow but not cutting it off entirely. Or it might be total, so that the flow is stopped until the object causing the blockage is removed. 2. The object that causes a blockage as described above.

**Occam's Razor** A general rule that states that when there are several theories to explain the behavior of a system, the simplest theory is the best. This principle was used by Copernicus when he declared that the sun must be at the center of the Solar System.

**occipital** Pertaining to the back of the head or brain. We might speak of the occipital bone of the cranium, or the occipital lobe of the brain, for example.

**occluded front** See *weather front*.

**occlusion** An obstruction. Used especially in reference to the arteries that supply blood to the heart muscle. See *obstruction*.

**occult** 1. In astronomy, to pass in front of, thereby eclipsing, some object. See *occultation*. 2. The unscientific pursuit of supernatural phenomena such as extrasensory perception.

**occultation** The passing of some astronomical object in front of another, so that the one behind is eclipsed. This can be partial or total. An example is a planet passing in front of a star. By means of time measurement, astronomical positions can be precisely determined by this means.

**ocean current** Large-scale circulations in the oceans. These follow, to some extent, similar currents in the atmosphere. The drawing shows the general patterns that these currents usually follow in the world's oceans.

OCEAN CURRENT

From Erickson, VIOLENT STORMS (TAB Books, Inc., 1988).

**oceanic crust** The crust underneath the oceans. Normally the earth's crust is thinnest beneath the oceans of the world. But there are irregularities. The illustration shows the major undersea mountain ranges. See *crust*.

OCEAN CRUST

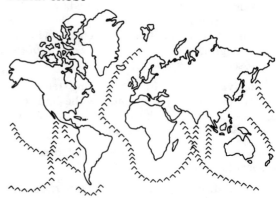

From Erickson, THE LIVING EARTH (TAB Books, Inc., 1989).

**ocean mining** A comparatively new endeavor in which minerals are extracted from seawater. There are also probably rich deposits beneath the ocean floor, in the oceanic crust. At this time, mining these reserves has not been done on a large scale because of the high cost.

**oceanography** The study of the evolution, characteristics, and life forms that inhabit the oceans of the world. This can range from mapping the ocean floor to finding new organisms and communicating with dolphins. This is an important scientific field for the future, because the oceans have a major influence on environment throughout the world, and human activities could adversely affect the oceans.

**oct-** A prefix meaning eight or in groups of eight.

**octal** Having eight elements, sides, or faces usually in a symmetrical arrangement.

**octal number** A number in the base-8 system. In this system one counts 1, 2, 3, 4, 5, 6, 7, 10, ..., 17, 20, 21, and so on. There are no digits 8 or 9. The octal expression for a large quantity is therefore a "bigger" number than the decimal expression. This system is sometimes used by computers. See *modulo*.

**octane** A hydrocarbon and a member of the alkane series, with eight carbon atoms and 18 hydrogen atoms, arranged as shown in the

OCTANE

$C_8H_{18}$

○ Carbon atoms

∘ Hydrogen atoms

— Bonds (CH)

drawing. It is a flammable liquid at room temperature and at atmospheric pressure. The octane level is sometimes used as a general indicator of the quality or grade of gasoline fuel.

**octave** 1. A two-to-one audio frequency range, such as the interval between a tone of 440 hertz (Hz) and one of 880 Hz. 2. Any interval covering some electromagnetic frequency f and the frequency 2f; or, alternatively, some wavelength w and half this wavelength, w/2.

**octillion** The number obtained by writing a 1 followed by 27 zeros, or 1000 followed by eight sets of three zeros.

**ocular** Pertaining to the eyes or the function of the eyes.

**OD** Abbreviation for overdose.

**o.d.** Abbreviation for outside diameter. See *diameter*.

**odd-even** Pertaining to an atomic nucleus having an odd number of protons and an even number of neutrons. The first designator refers to the proton count and the second to the neutron count. See *even-even, even-odd, odd-odd*.

**odd number** Any integer that is not divisible by 2; or, any even number plus 1.

**odd-odd** Pertaining to an atomic nucleus having an odd number of protons and an odd number of neutrons. The first designator refers to the proton count and the second to the neutron count. See *even-even, even-odd, odd-odd*.

**Oe** Abbreviation for oersted.

**oersted** Abbreviation, Oe. A unit of magnetic field intensity, not often used. It is equal to about 79.6 amperes per meter.

**Oersted, Hans** A physicist who discovered the relationship between electric currents and magnetic fields, in 1820. See *electromagnetism*.

**offshore breeze** A local phenomenon that occurs when the land is cooler than a large body of water. The air over the land is cooled and sinks, while the air over the water is warmed and rises. This creates a convection pattern in

OFFSHORE BREEZE
ONSHORE BREEZE

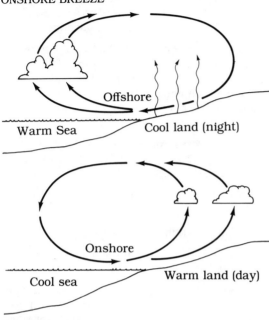

From Erickson, VIOLENT STORMS (TAB Books, Inc., 1988).

which surface winds blow offshore, that is, from land to water (see drawing). See *onshore breeze*.

**ohm** The standard unit of electrical resistance and also of reactance and impedance. Generally, ohms can be considered as volts per ampere. For alternating current the definition is more complex. See *impedance, Ohm's Law, reactance, resistance*.

**ohmmeter** A device for measuring direct-current resistance, using a battery and an ammeter. The battery drives a current through the resistance to be measured; the current depends on this resistance. The ammeter is calibrated in ohms instead of in amperes. See *ohm, Ohm's Law*.

**Ohm's Law** A general rule stating the relationship among voltage, current, and resistance in a direct-current circuit. When the voltage is E volts, the current is I amperes, and the resistance is R ohms, the relationship is $E = IR$. It can also be stated as $I = E/R$ or $R = E/I$.

**oil drilling** 1. Exploration for oil deposits by drilling wells into the earth and hoping to strike oil. There are ways to determine, for a given location, whether or not oil is likely to be struck once drilling is begun. 2. Tapping of underground oil reserves, by means of drilled wells. Only one well out of 10 actually strikes oil.

**oil refining** The process of converting crude oil into usable forms, such as motor oil, gasoline, heating oil and kerosene. Non-fuel products can also be obtained by this means; petroleum jelly, a cosmetic, is an example.

**oil shale** A soft rock, also called kerogen, that contains considerable amounts of crude oil. This source of fossil fuel has not been tapped very much yet, but might become important as other sources of crude oil are depleted.

**oil slick** A layer of oil that floats on water after an oil spill. The oil spreads out to cover a large area. It interferes with oxygenation of the water and also poisons fish and other organisms that live in the water. When it washes ashore it fouls the beach. See *oil spill*.

**oil spill** The result of a ruptured oil-carrying ship. Millions of gallons might be discharged into the ocean, a river, or a large lake causing environmental damage as well as lost oil revenue. See *oil slick*.

**Olbers' Paradox** An apparent contradiction obtained if we assume that the universe is infinite. We should, according to this theory, see a star at some finite distance for every point in the heavens, making the whole sky as bright as the sun. This paradox neglects absorption of energy by interstellar nebulae (see *nebula*). Also, there is some doubt as to its mathematical validity. Besides this, most scientists no longer think the universe is infinitely large.

**oleic acid** A fatty acid that is common in practically all foods. It is a major source of fuel in the human diet. See *fatty acid*.

**olfactory** 1. Pertaining to the sense of smell. 2. Pertaining to the nerves and membranes in the nasal cavity that are responsible for detecting odors in the air.

**Oligocene epoch** A time that began about 37 million years ago and ended about 24 million years ago. It was the earliest epoch in the Neogene period, the time of recent life. See the GEOLOGIC TIME appendix.

**omega minus particle** See *Eightfold Way*.

**Omega navigation system** A system that was originally put together by the United States Navy for communication with submarines. It uses very low radio frequencies that can penetrate underwater. It has long range and works regardless of geomagnetic storms or weather conditions.

**omniscope** A device invented by Simon Lake in the late nineteenth century, to allow submarine personnel to see over wide angles on the surface. It was the predecessor of the modern periscope. See *periscope*.

**omnivore** An animal that eats vegetable and animal foods; that is, a plant-eater and meat-eater. Humans are in this category (except for strict vegetarians). See *carnivore, herbivore*.

**one-celled organism** Any life-form, such as a bacterium, that carries out all of its functions as a single cell. See *cell*.

**onshore breeze** A local phenomenon that takes place when the land is warmer than a large body of water. Air over the water is cooled, and sinks; air over the land is warmed, and rises. This creates a convection pattern, with surface winds blowing onshore, that is, from water to land (see drawing). See *offshore breeze*.

**Oort cloud** The swarm of comets that lies at the periphery of the Solar System. It extends outward as far as one light year, or about six trillion (6,000,000,000,000) miles. This is about $1/4$ of the way to the nearest star.

**opaque** 1. Not allowing visible light to pass through, reflecting and/or absorbing it all. 2. For a specific wavelength or form of energy, not allowing passage, but instead, reflecting and/or absorbing it all.

**open-hearth process** A method of steel manufacturing that produced a better quality of steel than previous, older methods. Developed by Siemens-Martin, and put in general use by John Fritz in the late nineteenth century.

**open-heart surgery** A critical medical procedure, used as a last resort in cases of advanced heart disease or severe injury. The operation is performed on the heart as it beats in plain sight of the surgeons. The patient is under anesthesia.

**open star cluster** A star cluster in which the concentration of stars is not great enough to make the center appear as a solid glowing ball (see drawing). See *globular star cluster*.

**open system** A set of machines, in which materials and energy are brought in from the outside environment, and other materials and energy are discharged into the environment. Most power plants are such systems.

**operational** 1. For a machine or system, the condition of being workable and functioning. 2. Pertaining to the operation of a machine.

**operational amplifier** An integrated circuit containing several transistors and associated components, and capable of being wired in various ways to amplify signals. The amplification factor (gain) is adjustable.

**operational research** The process of perfecting the functioning of a device or system after it has been placed in service. The first set, or lot, might need design improvements, the nature of which can be learned only after some hours of real-time use.

**operator** 1. The person (or sometimes a computer) who oversees the functioning of a device or system. 2. In mathematics, a notation that tells us that we are to do something to a number or variable. For example, the *j operator* indicates that we must multiply the given quantity by the square root of $-1$.

**ophiolite** A piece of the oceanic crust, thrust up on land. They are mined for their metal ore content. Some of them are 3.6 billion (3,600,000,000) years old.

**opiate** Any drug that is derived from opium. This includes morphine and heroin. It also includes the better-known codeine and paregoric. Most opiates are used to relieve pain. They also might slow down the functioning of the intestines. In larger doses they can be dangerous. Therefore, they always require a doctor's prescription.

**opium** The juice obtained from a certain poppy, called the opium poppy. It is a bitter liquid and ingestion causes drowsiness and delirium. It has addictive properties. This substance is used to make various prescription narcotic drugs.

**opposition** 1. For a planet, lying in a position such that its phase (through a telescope) appears exactly full. This means that the earth is directly between that planet and the sun. 2. For two waves having the same frequency, the condition of being exactly out of phase. See *out of phase*. 3. In contradiction to; acting against. We might say that friction acts in opposition to the motion of a car on a road.

**optical fiber** A glass or plastic filament that transmits light, either to convey a visual image or to transmit data. Optical fibers are not susceptible to interference, as are wires. They are also corrosion-resistant. Thousands or even millions of signals can be sent over a single fiber. See *fiberoptics*.

**optical disk drive** A disk drive in which the information is encoded or retrieved by means of

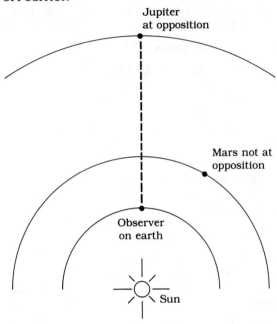

OPPOSITION

light and light-sensitive devices. The other common means is magnetic recording. See *disk drive*.

**optical maser** See *laser*.

**optical telescope** See *telescope*.

**optic nerve** The nerve that leads from the retina of the eyeball to the brain. It carries the impulses that the brain processes into a perceived image.

**optics** 1. The branch of physics involved with visible light and its behavior, especially with lenses and mirrors. Geometry is the major form of mathematics used. 2. The lenses and/or mirrors used in a microscope, telescope or other precision optical instrument.

**optimization** 1. A mathematical process in which the greatest possible value, or yield, is obtained by manipulating certain variables. 2. The process of making optimum. See *optimum*.

**optimum** Most efficient. We might speak of the optimum magnification for looking at a certain object in the sky. Or we might determine the optimum frequency on which to set up a radio communications link between two fixed points.

**optimum working frequency** In radio, the frequency on which the least number of received-signal errors occurs. There is always one and only one such frequency, determined by experimentation. This frequency might change with time.

**optocoupler** A device that allows connection of two electronic circuits, so that the circuits do not interfere with each other's operation. The coupler consists of a light-emitting device and a photocell (see drawing).

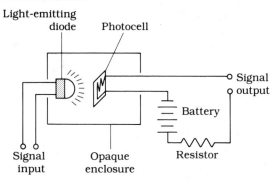

OPTOCOUPLER

**optoelectronics** The field having to do with the uses of visible light in electronics. Especially, this concerns modulated light and the use of light beams to control circuits and systems. See, for example, *fiberoptics, laser*.

**oral** 1. Pertaining to the mouth and its anatomy. 2. Conveyed by means of spoken words.

**oral medicine** The field of medicine concerned with the anatomy of the mouth, and with the diagnosis and treatment of diseases that occur in the mouth. This includes dentistry.

**orbit** 1. The path that an object takes in space when it is held completely under the influence of gravitation of some moon, planet, or star. This assumes it is not on a path to fall. The orbit is approximately an ellipse, with the center of the moon, planet or star at one focus of the ellipse. See *orbital period, orbital velocity*. 2. The theoretical path taken by an electron in an atom, when that electron is bound to the atom. 3. The socket in which the eyeball rests.

**orbital** 1. Pertaining to the orbit of a satellite around some large, massive object such as a moon, planet, or star. 2. An electron shell in an atom. This is a sphere representing the median position of the electron relative to the nucleus.

**orbital cycle** Any of the variations in the earth's orbit that may affect climate. See *Milankovich model*.

**orbital period** The time it takes for a satellite to complete one revolution in orbit. See *orbital velocity*.

**orbital speed** The speed that a satellite moves as it orbits some moon, planet, or star. The more massive the moon, planet, or star, the greater the speed. Also, the greater the altitude of the satellite, the slower the speed. The speed does not depend on the mass of the satellite.

**orbiting astronomical observatory** Any of the satellites placed in earth orbit by the United States and the Soviet Union, and that have been used to examine the distant stars and galaxies at various wavelengths.

**order** 1. A recognizable pattern in nature. 2. A category in the classification of living things. A group of orders is a class. 3. A mathematical expression for the number of times an operation is repeated. 4. An expression for the type of a chemical reaction. Consult a chemistry text for details.

**order/chaos** Two primary theories of the nature of the universe. Is it orderly, or is it disorderly? There are arguments either way. Disorder on a small scale often seems to create order on a large scale. This is an interesting dichotomy and there is much philosophical discussion about it among scientists.

**ordered** 1. Arranged in a preset way. 2. For a set of numbers, having the characteristic that their order or sequence is important. The set {2, 3, 4} is not ordered because it doesn't matter which number comes first, second or third. But the triple $(x,y,z)=(2,3,4)$, in Cartesian three-space, is ordered.

**order of magnitude** 1. The power of 10 to which an expression is raised in scientific notation. See *scientific notation*. 2. A difference factor of 10. Thus, 25 is one order of magnitude larger than 2.5, which is one order of magnitude larger than 0.25, and so on.

**ordinal number** A number in a sequence. Any natural number is its own ordinal, if we call the set N = {1, 2, 3, ...}. These are *finite ordinals*. The first infinite ordinal is the number we would obtain if we could count all the way to the end of the set N. See *cardinal number*.

**ordinate** In a Cartesian coordinate system, the axis where the dependent variable is shown. This is usually the vertical axis. See the accompanying drawing. See also *abscissa, Cartesian coordinates, Cartesian plane, dependent variable, function, independent variable*.

ORDINATE

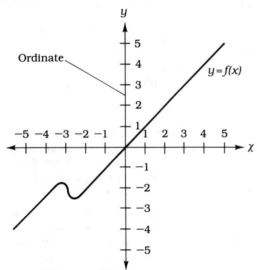

**Ordovician period** A time period during the Paleozoic era. It began about 500 million years ago and lasted till about 435 million years ago. During this time, fish first appeared in the seas of the world. See the GEOLOGIC TIME appendix.

**ore** The crude form of a mineral, found in crustal rocks of the earth. Ores for specific minerals have names, such as bauxite for aluminum and hematite or taconite for iron.

**Oresme, Nicole** A mathematician of the fourteenth century. One of the first to analyze acceleration, applying mathematics to physical phenomena.

**organ** 1. A part of the body with a specific function or set of functions. In humans, examples are the liver, pancreas, and kidneys. 2. A part of a plant with a specific function. Examples are leaves, stem, and roots.

**organ donor** A person who agrees to give up an organ to someone who needs it. A living person normally has two such organs and can do all right without one; kidneys are a good example. Sometimes people agree that when they die, their organs are to be made available to those who might need them. This allows donation of hearts, livers and other essential single organs.

**organelle** Any small particle in a cell that performs a definite task. In this sense, they are to the cell as organs are to the body of an animal. See *organ*.

**organic** Pertaining to, or containing, complex carbon-based chemicals involved in life processes. See *inorganic, organic chemistry, organic compound*.

**organic chemistry** The branch of chemistry concerned with compounds and reactions that involve molecules containing carbon, such as the hydrocarbons, fats, and proteins. See *biochemistry*.

**organic compound** A carbon-based compound, such as a hydrocarbon or carbohydrate. Examples are fossil fuels, fats, and amino acids.

**organic farming** A method of agriculture in which no chemical fertilizers are used, but only fertilizers made from animal manure, composted leaves, or other nonindustrial sources. Chemical pesticides and weed-killers are also avoided.

**organic molecule** See *organic compound*.

**organism** A living thing, complete and self-contained. The simplest of these are cells and viruses. The most complex include animals capable of reasoning. We humans are one such animal; dolphins might be another. There is a theory in which larger organisms can be thought to exist, such as the entire planet earth (see *Gaia Hypothesis*).

**organ of corti**  A part of the anatomy of the ear. It contains many microscopic hairs that move along with sound disturbances. These hairs stimulate the auditory nerve to send signals to the brain, to be interpreted as sound. See *ear*.

**organ transplant**  The removal of an organ from a healthy person, and the surgical implantation of that same organ in a person who needs it. For this to work, the people must be compatible, so that the recipient's body does not reject the new organ.

**origin**  1. The roots of a species. 2. Beginning; the first step, such that it is not possible to inquire further into the past regarding a phenomenon. We might speak of the origin of the universe, for example. 3. In a coordinate system, the point where all the axes intersect. This is usually, but not always, the point where all the variables are equal to zero (see drawing). See *abscissa, ordinate*.

ORIGIN

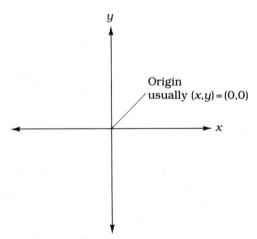

**Orion**  A constellation that lies roughly on the celestial equator, and is shaped as shown in the drawing. It is visible in the winter in the Northern Hemisphere, and is historically a symbol of the winter months in the night sky.

ORION

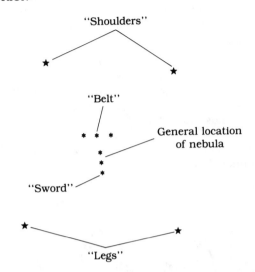

**Orion Nebula**  A nebula in the central part of the constellation Orion. It is a fuzzy spot in a small telescope. When time-exposure photographs are taken, it is resolved into a complex, glowing cloud of gas and dust. It is one of the most famous nebulae.

**orogeny**  An episode of mountain building in geologic history. Actually, such "episodes" are not abrupt, but take place over millions of years.

**ortho-**  1. A prefix meaning standing straight up. 2. A prefix meaning correct or proper. 3. A prefix meaning ideal or optimum. 4. A prefix referring to certain chemical structures; consult a chemistry text for details.

**orthogonal**  Existing at a right angle to a plane surface; perpendicular. If the surface is not flat, then an orthogonal line is perpendicular to the tangent plane at the point where the line passes through the surface.

**orthomolecular**  Pertaining to a branch of medicine in which nutritional supplements are used in large quantities, like drugs, to treat diseases. An example is the use of niacinamide (vitamin B-3) to treat schizophrenia. This is still a largely experimental field of medicine.

**Os**  Chemical symbol for osmium.

**oscillate** 1. Move back and forth in a regular, uniform motion. 2. Produce a wave disturbance as a result of positive feedback. See *oscillation, oscillator*.

**Oscillating Universe Theory** A theory that the universe exploded in a fireball about 15 million years ago, and will eventually implode or collapse, and then will explode again. The theory holds that this has happened countless times before and will repeat itself countless times more (see drawing). See *Big Bang, Big Bounce, Big Crunch*.

OSCILLATING UNIVERSE THEORY

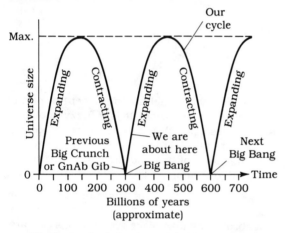

**oscillation** 1. A back-and-forth, regular motion. One complete cycle is completed in the period, usually denoted T. The number of cycles per second, or hertz, is the frequency, usually denoted f. 2. An alternating-current disturbance resulting from positive feedback in an electronic amplifier. The frequency typically ranges from a few hertz to many millions or billions of hertz. See *oscillator*.

**oscillator** An electronic amplifier in which positive feedback is used, so that an alternating-current wave disturbance occurs. Some oscillators have a fixed frequency; others have variable frequency. There are several different methods of determining the frequency, which may range from a few hertz to many millions or billions of hertz.

**oscilloscope** A device for observing the wave shape for an alternating current. It plots instantaneous amplitude versus time on a cathode-ray-tube screen.

**osmium** Chemical symbol, Os. An element with atomic number 76; the most common isotope has atomic weight 192. In its pure form it is a bluish metal. It occurs naturally along with platinum.

**osmosis** A process that takes place when there are two solutions, separated by a membrane, and the solutions have different concentrations of some chemical. The concentration of each chemical tends to even out, so that the concentrations are the same on either side of the membrane.

OSMOSIS

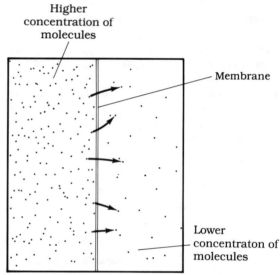

**osmotic pressure** The pressure required to prevent osmosis. See *osmosis*.

**ossicle** Any of the three bones in the middle ear that transfer sound energy from the eardrum to the organs that convert the vibrations into nerve impulses. These bones are called the hammer, anvil, and stirrup because of their shapes. See *ear*.

**ossification** 1. The formation of bone, by the building-up of minerals, particularly calcium and phosphorus. 2. The deposition of calcium and other minerals in soft tissue. This is not always good. For example, calcium might be laid down in arterial plaques, worsening a case of arterial disease ("hardening of the arteries").

**osteo-** A prefix meaning bony or concerning the bones.

**osteopath** Chiropractor. A medical person who works with disorders involving the skeleton of the body.

**osteoporosis** Literal meaning, "porous bones." A disease that affects some people as they age, especially women. There is a mineral loss from the bones, causing increased risk of fracture. The treatment methods have been improved in recent years.

**Otis Brothers** The engineers who installed the first electric elevator in an office building in the United States around the end of the nineteenth century.

**otology** The branch of medicine concerned with the diagnosis and treatment of problems with the ear.

**Otto, Nikolaus** A German engineer of the nineteenth century. He is credited with having built the first working gasoline engine. This led to the development of efficient automobiles.

**ounce** 1. A unit of mass, equal to $1/16$ pound or about 28.4 grams. See *gram, kilogram*. 2. A unit of fluid measure, equal to $1/16$ pint or about 1.8 cubic inches or 29.6 milliliters. See *liter, milliliter*.

**outer ear** The part of the ear that can be seen. This includes the cartilage structure and the canal that leads to the eardrum.

**outgassing** The escape of gas into the atmosphere from within the earth. It is thought that this is how much of the earth's original atmosphere was formed.

**out of phase** 1. For alternating-current waves, the property of having the same frequency but a phase difference of 180 degrees (see drawing). If the amplitudes of the waves are equal, the sum of their strengths is zero. 2. For two events, one having a maximum level when the other has a minimum level, repeatedly. 3.

OUT OF PHASE

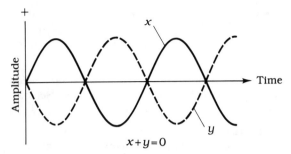

For two effects, a tendency to act against each other.

**output** 1. The signal, voltage or current that is produced by an electronic circuit. 2. The terminals in a circuit from which a signal, voltage, or current emerges. 3. The information that comes from a computer.

**outside diameter** See *diameter*.

**ovary** The organ in the female that produces eggs. Various hormones are also secreted by this organ. The female mammal has two ovaries. For illustration of their location, see *endocrine glands*.

**overdose** 1. A dose of drugs or vitamins far in excess of the ideal or recommended dosage. 2. To administer an excessively large amount of a drug or vitamin, either to oneself or to someone else.

**overflow** 1. The condition in which a river rises to a level beyond flood stage, bursting its banks. 2. In an electronic memory circuit, the condition that occurs when there is more data than the memory has room for.

**overshoot** In electronics, a "spike" that follows a pulse having opposite polarity to that of the pulse itself.

**overweight** The result of insufficient exercise in combination with excessive food intake. Common in developed countries. When severe, it is called obesity. See *obesity*.

**ovulation** In the female mammal, the movement of the egg from the ovary into a position where it is ready to be fertilized by a sperm cell. In humans this occurs about once a month.

**ovum** The egg cell in the female mammal.

**oxbow** A lake that is formed when a river changes course, leaving some water standing. Such a lake often has a curved shape, hence the term.

**oxidant** A substance that causes, or speeds up, oxidation of some other substance. See *oxidation*.

**oxidation** The combination of a substance with oxygen. This might occur slowly, as with exposed copper or iron; or it might take place rapidly, as with wood burning. Flammable gases oxidize fastest of all; then the process is usually called combustion. See *combustion*.

**oxide** A compound resulting from the combination of some element with oxygen. Examples are iron oxide (rust), copper oxide and carbon monoxide. The first two of these are solids and the third is a gas at normal atmospheric temperature and pressure.

**oxo-** Pertaining to a compound in which certain atoms are bound to oxygen. In particular, pertaining to an acid of this type in which the hydrogen ions are attached to oxygen atoms.

**oxygen** Chemical symbol, O. An element with atomic number 8; the most common isotope has atomic weight 16. This element is a gas at standard temperature and pressure. It comprises about 21 percent of the atmosphere at sea level. It is necessary for humans, animals and many other life forms. Oxygen is produced by plants from carbon dioxide.

**oxygen mask** 1. A mask worn by pilots flying high-altitude aircraft when the cabin is not pressurized. It delivers oxygen for breathing. 2. A mask worn by hospital patients when they need oxygen-enriched air, such as in cases of pneumonia or smoke inhalation. It delivers pure oxygen at an adjustable rate.

**oxytocin** A hormone that is secreted by the part of the brain called the hypothalamus. It is thought to be responsible for starting the process of labor in the female when a baby is ready to be born. See *hypothalamus, labor*.

**oz** Abbreviation for ounce, ounces.

**Ozma Project** The first serious attempt to find radio signals from possible alien civilizations. It was begun by Frank Drake at the Green Bank, West Virginia radio telescope observatory in 1959. It was named after the imaginary land of Oz.

**ozone** A form of oxygen in which three atoms are bound together to make a molecule. It is not a common form, but is important in various ways. It has a pungent odor similar to chlorine bleach. See *ozone depletion, ozone hole, ozone layer, ozone pollution*.

**ozone depletion** The destruction of the ozone layer that is thought to be taking place because of human activities, especially the use of chlorofluorocarbon gases. See *chlorofluorocarbon*.

**ozone hole** A gap in the ozone layer of the earth's atmosphere. The drawing shows an example of a localized ozone hole. The main problem is a large, circular region over Antarctica that has been getting larger in recent years. It is feared that this hole may soon exist over a substantial part of the earth. See *ozone depletion, ozone layer*.

OZONE HOLE
OZONE LAYER

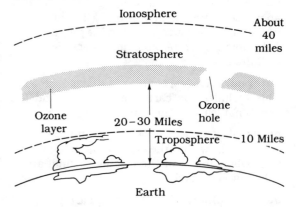

**ozone layer** A part of the stratosphere at an altitude of 20 to 30 miles (see drawing), where solar radiation causes some of the oxygen atoms to combine into groups of three, rather than the usual two. This form of oxygen is called ozone, and it absorbs much of the ultraviolet energy from the sun. Without this thin layer of ozone, the sun's ultraviolet rays would cause damage to many forms of life on the surface. See *ozone, ozone hole.*

**ozone pollution** A phenomenon in which solar energy causes ozone to form at the surface of the earth. This process is accelerated by automobile exhaust and other pollutants. It causes a foul smell like laundry bleach. Extreme levels cause eye and lung irritation. It is common in large cities, especially in the northeastern United States. This ozone cannot help replenish the ozone layer because it stays near the surface. See *ozone, ozone layer.*

**P** Abbreviation for peta- (see the PREFIX MULTIPLIERS appendix), phosphorus, point.

**p** Abbreviation for pico- (see the PREFIX MULTIPLIERS appendix).

**Pa** Abbreviation for pascal; chemical symbol for protactinium.

**PABA** Abbreviation for para-aminobenzoic acid.

**pacemaker** See heart pacemaker.

**Pacific Daylight Time** Pacific Standard Time plus one hour, or Coordinated Universal time minus seven hours. Generally used from early April till the last weekend in October in the United States. See *Coordinated Universal Time, Pacific Standard Time*.

**Pacific Standard Time** Coordinated Universal Time minus eight hours. Used along and somewhat either side of the longitude line 120 degrees west. This encompasses the extreme western continental United States. Used from the end of October till the beginning of April in most applicable places. See *Coordinated Universal Time, Pacific Daylight Time*.

**Packard, David** One of the cofounders of the Hewlett-Packard Company. He graduated from Stanford in 1934. See *Hewlett, William*.

**pad** 1. A material used in a disk type automobile brake to cause friction. 2. In electronics, an attenuator that has only pure resistance, has no reactance, and has a constant input and output impedance. 3. The foot of an animal. 4. An airstrip or platform from which certain aircrafts and rockets can take off and land.

**paddlewheel** See *water wheel*.

**pahoehoe lava** A volcanic lava that, when it cools off, forms rope-like structures.

**palate** The roof of the mouth, separating the mouth from the nasal cavity.

**paleo-** 1. A prefix meaning ancient. 2. A prefix meaning early. 3. A prefix meaning pertaining to fossils.

**Paleogene epoch** The earlier portion of the Cenozoic era, starting about 65 million years ago and ending about 24 million years ago. It encompasses the Paleocene, Eocene, and Oligocene epochs. Horses and whales are among the species that proliferated during this time. See the GEOLOGIC TIME appendix.

**paleomagnetism** The study of the geomagnetic field's history. The locations of the earth's magnetic poles have shifted in the past; the polarity also reverses. See *geomagnetic field, geomagnetic field reversal, geomagnetic pole.*

**paleontology** A branch of earth science concerned with the study of ancient life forms, mainly based on fossil remains.

**Paleozoic era** The time of ancient life. It began about 575 million years ago and ended about 230 million years ago. The drawing shows the positions of North America, South America and Africa at this time, and the mountainous regions of the continents as they were then. See the GEOLOGIC TIME appendix.

PALEOZOIC ERA

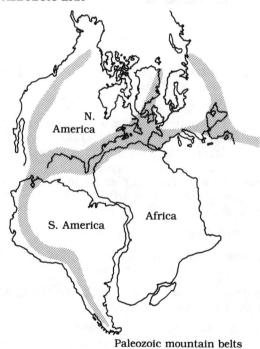

Paleozoic mountain belts

From Erickson, VOLCANOES AND EARTHQUAKES (TAB Books, Inc., 1988).

**palladium** Chemical symbol, Pd. An element with atomic number 46. The most common isotope has atomic weight 108. It is found naturally along with copper ore, and is used in manufacture of some kinds of ornamental jewelry.

**Palomar Observatory** See *Mount Palomar Observatory.*

**palpitation** A sudden irregularity in the heartbeat. Usually this is felt as a fluttering in the chest; sometimes dizziness occurs. There are many possible causes. It is not always serious, but a doctor should be consulted if these symptoms take place.

**palsy** 1. Inability to control the movements of a certain part of the body. 2. A nervous condition causing lack of control of certain parts of the body.

**pancake ice** A form of ice that sometimes occurs in the polar seas or on large lakes. The ice takes on the appearance of gigantic pancakes or lily pads several feet in diameter.

**pancreas** A small organ located just beneath the liver in the human body. It secretes the hormone insulin, responsible for metabolizing blood glucose. One form of diabetes occurs when this organ does not produce enough insulin. See *insulin.*

**pancreatitis** A disease involving inflammation of the pancreas. It is quite painful, nauseating and might require the surgical removal of the pancreas. See *pancreas.*

**Pangaea** The ancient supercontinent consisting of all the present-day continents in a single land mass (see drawing). The continents eventually split off from this mass and drifted around. See *continental drift.*

**pangamate** A chemical that is sometimes called a vitamin, but is not generally recognized as such. It is thought to help the body utilize oxygen. It has been used in some countries as a supplement for athletes.

**panoramic radio receiver** A special type of radio receiver that allows monitoring of a whole band of frequencies at the same time. A video screen shows frequency on the horizontal axis and relative signal strength on the vertical axis. Signals appear as "pips" on this screen. See *spectrum analyzer.*

**Panthalassa** The huge ocean that existed in the time of Pangaea. See *Pangaea.*

PANGAEA

From Erickson, VIOLENT STORMS (TAB Books, Inc., 1988).

**pantothenic acid**  A B-complex vitamin. See *vitamins*.

**papain**  An enzyme found in the papaya plant. It helps with digestion of protein.

**Papin, Dennis**  A physicist who first demonstrated that steam pressure can be used to displace a piston. James Watt later used this principle to invent the steam engine. See *Watt, James*.

**Pappus**  Also known as Pappus of Alexandria. An ancient Greek mathematician and philosopher. He lived around A.D. 300. He is remembered for his work in geometry, and also for noting that nature often evolves in a mathematical way for greatest efficiency.

**papule**  A red, small, raised skin lesion, such as occurs in measles or mild acne, or in some allergy reactions.

**para-**  1. A prefix meaning similar to or resembling. 2. A prefix meaning adjacent to or alongside. 3. A prefix meaning defective. 4. A prefix denoting a certain atomic structure in benzene rings; consult a chemistry text for details.

**para-aminobenzoic acid**  A member of the B-complex of vitamins. Also an ingredient in sunscreens. It tends to act as a block to damaging ultraviolet rays when it is applied to the skin. See *vitamins*.

**parabola**  The curve formed on a plane when a cone intersects the plane in a certain way. If you shine a flashlight onto a flat surface at a slant, so that the far edge of the large light cone just leaves the surface, the edge of the light cone will trace out a parabola on the surface. The Cartesian graph and equation for a parabola, with axis along the abscissa, are shown in the illustration.

PARABOLA

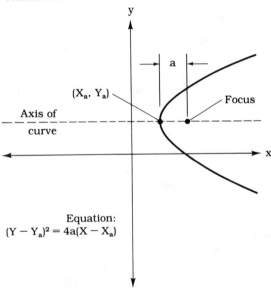

**parabolic**  1. Having the shape of a parabola. See *parabola*. 2. Having the shape of a paraboloid. See *paraboloid*. 3. Pertaining to a parabola or paraboloid, or to the shape of a parabola or paraboloid.

**parabolic mirror**  A concave mirror with a precise focus (see *concave mirror*). It is shaped like a paraboloid. See *paraboloid*.

**parabolic reflector**  1. A wire screen or piece of sheet metal, bent into the shape of a paraboloid and used as a reflecting medium for a dish antenna. 2. A parabolic mirror. See *parabolic mirror*.

**paraboloid** 1. A three-dimensional curve, resulting from the rotation of a parabola around its axis of symmetry. See *parabola*. 2. Any section of such a curve, but usually, the bowl-shaped section that lends itself to reflectors and dish antennas. See *parabolic mirror, parabolic reflector*.

**Paracelsus** A German doctor of the sixteenth century, one of the first to use chemicals for treating illnesses.

**paraffin** A long-chain alkane molecule, consisting of many carbon atoms linked together and bonded to hydrogen atoms. A typical molecule has 30 carbon atoms and 62 hydrogen atoms. See *alkane*.

**parallax** 1. The difference in images seen by the right eye, as compared with the left eye. This occurs for objects up to a few hundred feet away, because the line of sight is different for each eye. 2. The difference in images seen from two different points along a base line perpendicular to the line that runs to the distant objects. 3. The difference in the background of stars as caused by the revolution of the earth around the sun. See *parsec*.

**parallel** 1. For two straight geometric lines, existing in the same plane but not intersecting. 2. For two flat geometric planes, to be disjoint (not intersecting).

**parallel circuit** A form of electrical circuit in which two or more components are connected like rungs of a ladder, so that they all get the entire voltage from the source. The drawing shows bulbs in parallel; this is the method used in standard house wiring. See *series circuit*.

PARALLEL CIRCUIT

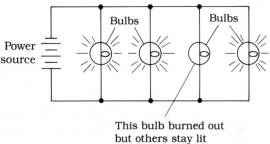

This bulb burned out but others stay lit

**parallel data transfer** The sending of data (signal information) along multiple lines. Each bit for a given word is sent along a different line. This speeds up the process as compared with sending the bits one after the other along a single line. See *serial data transfer*.

**parallelepiped** A three-dimensional figure with parallelograms as faces. See *parallelogram*.

**parallelogram** A geometric plane polygon having four sides, such that opposite sides have equal length and are parallel, and opposite angles have equal measure. The square and rectangle are special cases.

**Parallel Postulate** One of the axioms of Euclidean geometry. It says that if there is a straight line and a point not on that line, then there is one and only one straight line through the point that is parallel to the first line. While this might seem obvious, there are types of geometry that can be developed and that are consistent, even if this postulate is assumed to be false. We might allow no lines, or infinitely many lines, through the point and parallel to the original line. See *Euclidean, Euclidean geometry, Euclidean space, non-Euclidean, non-Euclidean geometry, non-Euclidean space*.

**paralysis** The inability to move a certain part of the body.

**paramagnetic material** A substance that increases the number of magnetic flux lines per square centimeter or square meter, but not as much as a ferromagnetic material. Such a material keeps the same permeability even if the magnetic field strength changes. See *ferromagnetic, permeability*.

**parameter** A factor that is variable and has an effect on one or more phenomena or other variables. Time is a commonly specified parameter.

**parametric equations** A set of equations, all of which depend on the value of a single parameter (see *parameter*). For example, we might plot an object's position in (x,y,z)-space, and find parametric equations that define the object's x, y and z coordinates individually as functions of time.

**paranoia** An irrational fear of persecution. The person with this delusion might think, for example, that secret agents are following him/her, when in reality there is no reason to believe this to be the case.

**paranoid psychosis** A severe form of paranoia, in which a person thinks he/she is being persecuted, and might react with violence against imagined enemies. Such a person has a misperception of reality and can be a danger to him/herself or others. This condition usually does not involve hallucinations, but it may progress into paranoid schizophrenia. See *paranoid schizophrenia*.

**paranoid schizophrenia** Delusions of persecution and/or grandeur, often with hallucinations, such as the voice of the devil. A person with this disorder might be a danger to him/herself or others.

**parasite** An organism that lives off of some other organism, called the host. Some parasites are external (such as the leech), and some are internal (such as the tapeworm).

**parasitic** 1. Having the characteristics of a parasite. 2. In electronics, an unwanted oscillation that sometimes takes place in an amplifier. It usually occurs on a frequency different from the desired one, and it robs the amplifier of power that would otherwise go into the desired signal.

**parasympathetic nervous system** A part of the involuntary, or autonomic, nervous system that balances and sometimes acts against the sympathetic nervous system. See *autonomic* and *sympathetic nervous system*.

**parathyroid** A group of four endocrine glands near the thyroid. For illustration of location, see *endocrine glands*. These glands produce a hormone that regulates blood calcium. This level is important in nerve and muscle function.

**parietal bone** Either of two bowl-shaped bones that make up part of the cranium. They are situated generally in the right rear and left rear sections of the skull.

**parity** 1. A condition of being standardized, or equivalent. 2. For an integer, the property of being even or odd. 3. Having or not having a counterpart. For example, an electron has a counterpart in the form of a positron. 4. For a digital signal, the property of having either an even number of bits or an odd number of bits.

**Parkinson's disease** A progressive, degenerative disorder involving the nerves and muscles, that affects mainly older people, but sometimes young people. The characteristic symptom is tremor in various parts of the body, especially the hands.

**paroxysm** A series of violent spasms. An example is a coughing fit.

**parsec** A unit of distance used by astronomers. It is defined as the distance for which the parallax caused by the earth's revolution is equal to one second of arc (*pa*rallax *sec*ond). See the drawing. This is about 3.26 light years. See *light year, parallax*.

PARSEC

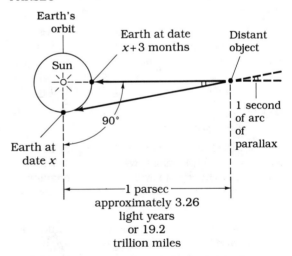

**partial derivative** In calculus, a derivative in a multivariable equation, taken with respect to just one of the variables. See *derivative*.

**partial eclipse** See *eclipse*.

**particle** 1. Any tiny fragment of matter. 2. See *molecule*. 3. See *atom*. 4. A subatomic unit such as an electron, neutron, proton, quark, photon, neutrino, tachyon or other of the increasingly complicated set of matter and energy units.

**particle accelerator** A device that uses electric and/or magnetic fields to propel subatomic particles, usually electrons, protons, or alpha particles to high speeds for the purpose of atom smashing. See *cyclotron, linear accelerator*.

**particle model** 1. The theory that visible light, and all forms of electromagnetic energy, can be treated as a barrage of particles. See *photon*. 2. The theory in physics concerning the particle nature of matter.

**particle physics** The science of the behavior and interaction of subatomic particles, both of matter and of energy.

**particle/wave dichotomy** In physics, the duality of the nature of electromagnetic energy. In some ways it behaves like a barrage of particles; in other ways it acts as a wave disturbance. Sometimes the particle model is better for explaining things, and sometimes the wave model is better.

**partition** 1. In mathematics, the breaking-up of the region under a curve into rectangles, whose areas are summed to approximate the area under the curve. 2. In chemistry, a ratio of solvent concentrations among different liquids.

**pascal** A unit of pressure expressed in newtons per square meter.

**Pascal, Blaise** A seventeenth-century mathematician and physicist who did research in probability theory. He evaluated the behavior of thrown dice. He also did research in the behavior of fluids and gases.

**Pascal's Law** A rule concerning fluids in sealed chambers. When a pressure is applied to the fluid, that pressure goes in all directions to an equal extent. The drawing shows a piston being pushed in, and other pistons moving out because of compression of the fluid. The sideways piston gets as much pressure per unit surface area as the vertical piston.

**passive** 1. In electronics, pertaining to a component that operates without the application of any power other than that of the input signal. 2. Pertaining to a substance that forms its own insulating layer when exposed to a reactive chemical. An example is aluminum, that forms an oxide layer that prevents further oxidation in the atmosphere.

**passive component** An electronic component that functions using only the power of the input signal. An example is a semiconductor diode. It can act as a rectifier, detector, or mixer, while needing no power supply.

**Pasteur, Louis** A nineteenth-century biologist, one of the first to teach that diseases are caused by microorganisms.

**pasteurization** A process in which food, particularly milk, is heated to kill harmful bacteria before being bottled or canned.

**patella** The floating bone in the knee joint, often called the kneecap.

**patent** 1. A registration of an invention with the government to prevent others from making and selling the same device. Each patent is issued a number or numbers. 2. To protect one's rights to make and sell an invention by registering it with the government.

PASCAL'S LAW

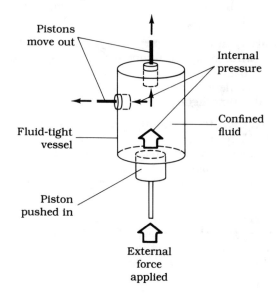

**pathology** 1. The study of diseases. 2. The set of symptoms and manifestations that go along with a given disease.

**Paul, Frank** An illustrator who, along with publisher Hugo Gernsback, made science fiction popular and entertaining during the early part of the twentieth century. Especially fascinating were his depictions of deep-sea and outer-space exploration. Science-fiction stories and novels have had wide appeal ever since.

**Pauli Exclusion Principle** See *exclusion principle*.

**Pavlov, Ivan** A Russian scientist and psychologist known for his research into stimulus/response behavior. Won a Nobel Prize in 1904 for his work. Perhaps the best known of his projects involved training a dog to salivate at the ringing of a bell. The bell was rung just before food was brought. Eventually, the dog would react as if food were coming, simply on hearing the sound of the bell.

**payload** The material, such as warheads, carried by an aircraft, rocket, or missile.

**Pb** Chemical symbol for lead.

**Pd** Chemical symbol for palladium.

**pdl** Abbreviation for poundal.

**PDT** Abbreviation for Pacific Daylight Time.

**peak** 1. A local maximum value for a phenomenon that varies with time, particularly an electronic current. We might speak of a positive peak (see *local maximum*) or a negative peak (see *local minimum*). 2. The highest, or maximum, instantaneous value that a variable attains. See *peak value*. 3. The narrow bow or stern of a boat or ship.

**peak-to-peak value** The amplitude, or strength, of an alternating-current wave, as measured between the positive peak and the negative peak (see drawing). See *peak*.

**peak value** Either the maximum positive or maximum negative instantaneous value of an alternating-current wave, measured relative to zero (see drawing). See *peak*.

**peat** Partly decomposed plant remains, that are thought eventually to form coal when subjected to long-term subsurface pressure.

**pectin** A vegetable substance that helps to hold plant cells together. It is jelly-like in pure form, and in fact it is used to make food jellies. It is sometimes used in medicinal preparations to help relieve diarrhea.

**pedalfer** A type of soil prevalent in the eastern United States. The topsoil is sandy, light-colored and acidic. The subsoil is rich in various minerals, especially iron and aluminum. It is especially abundant in coniferous forests.

**pedocal** A type of soil that is whitish because of its high level of calcium, both in the topsoil and the subsoil.

**pedal plane** An aircraft that operates entirely under human power. It is run like a bicycle. Such craft must be made of extremely lightweight, yet strong, materials. Well-conditioned athletes have flown them for miles in good weather.

**pedology** A branch of earth science concerned with the study of the soil. This is especially important in agriculture.

**Peking man** A variety of Homo erectus whose remains were found in China near Peking (now called Beijing). This ancestor of modern humans is thought to have been a cave dweller from 500,000 years ago till about 250,000 years ago. They were hunters.

PEAK-TO-PEAK VALUE
PEAK VALUE

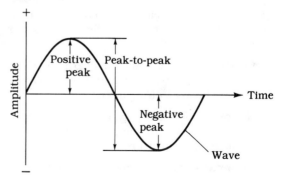

**pellagra** A deficiency disease that occurs when a person does not get enough niacin (vitamin B-3). There are mental and nervous symptoms, and often a swollen, sore tongue, along with fatigue and general debilitation. This disease is rare in civilized countries because food is enriched with niacin.

**Peltier Effect** A thermoelectric effect that takes place with certain dissimilar metals. When a current is passed through the junction, the region around the junction loses heat, and therefore cools off.

**pelvis** The pair of large, bowl-shaped bones that make up the hips, serving to support the lower abdominal organs such as the intestines and bladder, and connecting the legs to the trunk of the body.

**pendulum** A weight suspended from a cord or rigid rod, and set up so that it will swing back and forth in a single geometric plane. The period, or time per round trip, is a function only of the length of the cord or rod.

**pendulum clock** A timekeeping device making use of the regularity of a pendulum swing. The pendulum is given a small boost on each swing to keep it moving. This boost can be provided by a battery and motor, or by a tension spring. Nowadays, this type of clock is used mainly as a decorative timepiece, because quartz-crystal standards are far more accurate.

**penicillin** An antibiotic effective for many different types of microbial infections. It is found in common blue molds. It was discovered in 1928 by a Scotsman, Alexander Fleming.

**penis** The male organ that serves as a vehicle for the discharge of urine and the ejaculation of sperm.

**penstock** A large pipe beneath a dam, through which water flows and drives turbines to produce hydroelectric power.

**pentagon** 1. A plane polygon with five sides. 2. A regular, convex plane polygon having five sides, all of equal measure.

**pentagonal number** Any positive integer in the sequence defined by $n(3n-1)/2$, when n is a natural number in the set $\{1, 2, 3, ...\}$. The first few numbers in this sequence are 1, 5, 12, 22.

**pentode** See *vacuum tube*.

**penumbra** 1. The part of a shadow in which part, but not all, of the light from a source is eclipsed. 2. The lighter, outer portion of a sunspot. See *sunspot*.

**Penzias, Arno** One of the researchers at the Bell Labs in Holmdel, New Jersey who discovered the background radiation from the primordial fireball in the mid-1960s. This microwave radiation is thought be red-shifted energy from the Big Bang about 15 billion (15,000,000,000) years ago. See *Big Bang* and *Wilson, Robert*.

**pepsin** An enzyme in the stomach that is important in the digestion of protein-containing foods. Sometimes taken orally in small amounts to aid digestion.

**peptic** Pertaining to the stomach and stomach lining.

**peptide** A by-product of the digestion of protein. Chemically formed from amino acids linked by a certain type of bond. See a text on organic chemistry for details.

**per-** 1. A prefix meaning contained all through. 2. A prefix referring to a substance having the greatest possible amount or concentration of a chemical element.

**per** A word meaning for every. Often used formally in place of an, as in miles per hour.

**perception** A physical awareness of something based on detection via any of the five senses of sight, hearing, touch, taste, or smell.

**perception test** 1. A test conducted to detect the threshold level of one of the five senses. 2. A test conducted to detect the range over which a given physical sense operates; for example, we might test to find the range of audio frequencies a person can hear. 3. A test conducted by engineers to determine the best kind of instrumentation to use for operators of sophisticated equipment, such as aircraft.

**perennial plant** Any plant that maintains its life function for three years or more. Trees and grass are examples.

**perigee** 1. The minimum distance of the moon, or an earth-orbiting satellite, from the earth (see drawing). 2. The condition of the moon or an earth satellite being at its smallest distance from the earth. See *apogee*.

**perihelion** 1. The minimum distance of the earth or a planet or solar-orbiting satellite, from the sun (see drawing). 2. The condition of a planet or solar satellite being at its smallest distance from the sun. See *aphelion*.

PERIGEE
PERIHELION

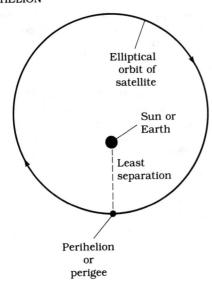

**period** 1. For a cyclic or repeating phenomenon, the length of time for one complete cycle to occur. 2. The time required for an orbiting satellite to make one complete revolution.

**periodic** Repeating on a regular basis, with a defined time for each cycle.

**periodic table of the elements** A table that groups the chemical elements in a certain way that is useful to chemists. See the PERIODIC TABLE OF THE ELEMENTS appendix.

**periscope** A device used in a submarine to observe events at the sea surface. It consists of mirrors or prisms and a long tube fitted with an eyepiece. Also can be used in an underground bunker to observe events at the surface.

**peristalsis** The contractions of the intestines that push food and waste along (see drawing). Also, the contractions of the esophagus that pass food from the mouth, after swallowing, to the stomach.

PERISTALSIS

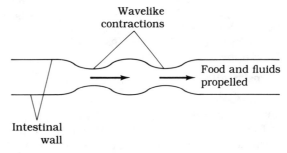

**peritoneum** The lining of smooth muscle that holds the abdominal organs in place.

**peritonitis** Infection and inflammation of the peritoneum (see *peritoneum*). This might occur if an ulcer becomes perforated, or in the event of a puncture wound or bullet wound. It is a serious and life-threatening condition.

**permafrost** Frost that stays in the ground for 12 months of the year. It is common above the Arctic Circle and in Antarctica. It makes the construction of buildings, and various other human activities, difficult.

**permalloy** An alloy made of iron and nickel, and having a high magnetic permeability. See *permeability*. These alloys are used in a variety of electronic circuit applications.

**permanent magnet** A ferromagnetic material with the magnetic dipoles aligned, so that a magnetic field is produced even in the absence of an electric current. Usually such "magnets" are made of iron or steel, or various alloys containing iron.

**permeability** 1. In electronics, the ratio of magnetic flux density in a given material to that in air. Air has permeability. 1. Paramagnetic materials have permeability somewhat more than 1; ferromagnetic substances have permeability much greater than 1. Diamagnetic materials have permeability slightly less than 1. See *flux density*. 2. In geology, a measure of the ease with which a fluid can flow through an earth material.

**permeance** The ease with which a substance allows a magnetic field to pass through.

**Permian period** A geologic time period during which marine waters retreated from the land, leaving large salt and gypsum deposits. Began about 290 million years ago and ended about 230 million years ago. See the GEOLOGIC TIME appendix.

**permittivity** See *dielectric constant*.

**permutation** The order in which the elements are arranged in a set. The more elements in the set, the greater the number of possible permutations. Used in statistics and probability theory.

**peroxide** 1. A compound having an excess of oxygen, or an extra oxygen atom for every molecule. 2. See *hydrogen peroxide*.

**perpetual motion** Motion without friction, and without any barriers to stop it. In the real universe, there is no such thing. All motion must end after some finite length of time.

**perpetual-motion machine** A theoretical machine that operates forever without any new energy being supplied. A hypothetical example is shown in the drawing. In reality there can be no such machine.

PERPETUAL-MOTION MACHINE

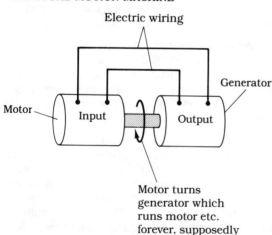

This kind of thing never works!

**personality** The set of behavioral characteristics that make a person unique. Some psychologists believe that certain people can have multiple personalities, with different ones dominating at different times.

**personality disorder** An abnormality in the behavioral characteristics that make up an individual's personality. Multiple personality disorder (MPD) is an example of a severe type of personality derangement. Milder forms include neuroses. See *personality*.

**perspective** The effect resulting from the fact that objects appear smaller as they get farther away. This causes parallel lines to seem to converge at some distant point. Important in art, and in some architectural and engineering drawings.

**perspective drawing** A drawing that takes perspective into account, so that objects appear realistic. This is done at the expense of geometric precision, because distances must be distorted to incorporate perspective. A perspective drawing of an object appears in the same proportion as a photograph (see illustration). See *isometric drawing, oblique drawing*.

PERSPECTIVE DRAWING

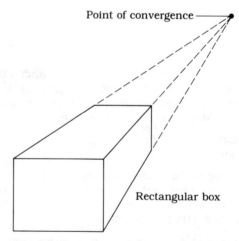

**perspiration** The liquid that comes from the pores when the body temperature rises above normal to a certain extent. The evaporation of the liquid helps to cool the body. It is secreted by the sweat glands.

**perturbation**  An irregularity in the orbit of a satellite.

**Peru current**  Part of the greater Pacific Ocean circulation. Surface water is driven offshore from the vicinity of this South American country by the trade winds. This causes cold, nutrient-rich water to rise to the surface. This current is disrupted during an El Nino event. See *El Niño*.

**pesticide**  Any of a variety of chemicals used to kill unwanted insects or worms. Especially, the term applies to such chemicals used by farmers.

**peta-**  See the PREFIX MULTIPLIERS appendix.

**petit mal**  A mild form of epilepsy, in which a seizure might pass unnoticed to the sufferer and even to observers. A loss of consciousness occurs for a few seconds or minutes; the person might stare vacantly into space, seeming only to be deep in thought. See *epilepsy, grand mal*.

**petri dish**  A small dish in which nutrient material, such as agar, is placed, and in which bacteria are cultured. This can be done, for example, to find out what (if any) bacteria are responsible for certain disease symptoms in a patient.

**petrification**  A process in which cells, especially of dead plants, become calcified and thereby turn to stone. Sometimes fossils of plants or animals are found in this state.

**petrol**  The British word for gasoline or kerosene, used as fuel for internal-combustion engines.

**petroleum**  1. A hydrocarbon, liquid fossil fuel. Also called crude oil. 2. Any of the various products obtained from the refining of crude oil.

**petrology**  A branch of earth science that is concerned with the formation, origin, history, and structure of rocks.

**pewter**  An alloy of tin and sometimes small amounts of lead, once used for dishes, pots and pans, and silverware. Not generally used today, because tin and lead are recognized as toxic even in trace amounts.

**pH**  A quantitative measure of acidity and alkalinity. The scale ranges from zero to 14, with neutral represented as 7. Values less than 7 are acid; greater than 7 indicates alkalinity. The scale is logarithmic and is based on the activity of hydrogen ions. For each unit decrease in pH, the hydrogen ion activity increases by a factor of 10.

**phagocyte**  A cell that acts as part of the body's defense against disease-causing microbes. The cell surrounds its prey and consumes it, in a manner similar to the action of an amoeba.

**phalanges**  The bones in the fingers and toes, making up the movable and discrete parts or digits.

**Phanerozoic eon**  The geologic time span from the first appearance of complex life until the present. This encompasses the Paleozoic, Mesozoic, and Cenozoic eras. It began approximately 575 million years ago with the first sea plants. See the GEOLOGIC TIME appendix.

**pharmacology**  The branch of medicine concerned with the effects of drugs on the human body.

**pharmacy**  1. The dispensing of medication according to prescription from a doctor. 2. See *pharmacology*. 3. A supply of medicines, some or all of which require a doctor's prescription to be sold or dispensed.

**pharynx**  The voice box, in which the vocal cords are contained. See *vocal cords*.

**phase**  1. The shape of the illuminated part of the moon's disk as seen from the earth. 2. The shape of the illuminated part of any celestial object as seen from a given vantage point. 3. See *phase difference*. 4. Stage, in reference to something that is changing; we might speak of the early phase of the measles, for example.

**phase angle**  See *phase difference*.

**phase difference**  For two wave disturbances having identical frequency, a displacement in the positions of the waves when plotted as amplitude versus time (see drawing). This is measured in degrees from 0 to 360, or from

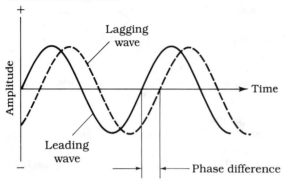

PHASE DIFFERENCE

−180 (half a cycle lagging, or behind) to +180 (half a cycle leading, or ahead).

**phase-locked loop** An electronic circuit that maintains a constant, stable frequency, by detecting slight changes and correcting them before they become large. It works because small frequency changes also cause phase changes, which are more easily detected by simple circuits.

**phase modulation** Variation of the phase of a radio signal in order to convey information. The effect is similar to that of frequency modulation (see *frequency modulation*). Phase modulation is sometimes used instead of direct frequency modulation, because the electronic circuits are simpler.

**phase opposition** See *out of phase*.

**phasor** 1. See *complex number*. 2. A highly sophisticated mathematical expression, similar to a vector but involving more elements and giving more information. See a text on advanced vector analysis. See *vector*.

**phenol** A type of organic compound, similar to alcohol but having an acid pH.

**phenolphthalein** A substance that turns red in the presence of alkaline solutions. It is also used in some laxative pills and candies.

**phenology** The study of recurring natural phenomena, and their relationship to changes in the climate. The Milankovich model of ice ages is a good example (see *Milankovich model*). The change in sunspot numbers is thought to have an effect on climate; the sun seems to get slightly more intense when the sunspot numbers are the highest.

**phenylalanine** An amino acid. It is not manufactured in the human body from other amino acids, so it is required in the diet. This substance is used to make artificial sweeteners having very few calories.

**pheromone** A scent given off by an animal or person, so subtle that it cannot be consciously detected, but that nonetheless affects the way in which other people react to the person. This appears especially true between the sexes.

**phlegm** Mucus discharged by the lungs. In allergic reactions and respiratory infections, large amounts of this substance are produced, causing the person to cough often and clear the lungs of foreign matter.

**phloem** In plants, the artery-like structures that carry food from the leaves to the growing parts, the stem, branches, and roots. See *xylem*.

**phobia** An unreasonable, often intense fear of something. Common phobias are fears of heights (acrophobia), enclosed places (claustrophobia) and water (hydrophobia). This last is not to be confused with the disease of rabies occasionally called hydrophobia.

**phon** A unit of sound intensity. At a frequency of 1 kilohertz, or 1000 cycles per second, the phon is the same as the decibel (see *decibel*). At other frequencies the phon is different from the decibel, because phons are determined by taking into account the fact that people's ears have different sensitivity at different frequencies.

**phone** 1. See *telephone*. 2. Communications lingo for voice transmission via radio.

**phonograph** An archaic term for an audio disk player, of the type that uses a cartridge with a needle that moves along a groove in the disk. The standard speed is $33 1/3$ revolutions per minute (RPM).

**phonon** A theoretical sound quantum, or unit sound particle. It is equal to the product of Planck's constant and the frequency of the sound wave in hertz.

**phosphate** A compound containing phosphorus and oxygen, along with some other element. Generally there are four oxygen atoms for every atom of phosphorus. There are many substances of this kind used in industry.

**phosphate pollution** Pollution of water as a result of the dumping of chemicals containing phosphates. This is a major problem in some regions. It can cause eutrophication as well as general fouling of the water supply. See *eutrophication, phosphate*.

**phosphide** A compound containing phosphorus and some other element that attracts one or more electrons from the phosphorus.

**phosphine** A gas formed from phosphorus and hydrogen. There are three hydrogen atoms for every atom of phosphorus. It has uses in the manufacture of semiconductor materials to make diodes and transistors. In pure form it is poisonous.

**phosphor** A material that glows when it is struck by high-speed subatomic particles, or when it is subjected to ionizing radiation such as X rays or gamma rays. These substances are commonly used to coat the inside surfaces of the screens of cathode-ray tubes and television picture tubes.

**phosphorescence** Light produced by certain substances when they are bombarded by high-speed subatomic particles or by ionizing radiation, such as X rays or gamma rays. The electrons are raised briefly to higher energy levels in the substance. When they fall back, light is emitted.

**phosphorus** Chemical symbol, P. An element with atomic number 15. The most common isotope has atomic weight 31. This element is abundant in living tissue, and forms an important part of bones in vertebrates. It has a wide variety of industrial uses.

**phot** A unit of illuminance, equal to one lumen per square centimeter. Equal to 10,000 lux. The lux is more often used. See *lux*.

**photo-** A prefix used for terms referring to light, its behavior, or its characteristics.

**photocathode** A light-sensitive electrode in a phototube. When light strikes the electrode, it emits electrons. See *phototube*.

**photocell** 1. A device whose electrical resistance varies depending on the amount of light that strikes it. See *photoconductive cell, photodiode, phototransistor, phototube*. 2. See *photovoltaic cell*.

**photochemical haze** A form of smog with the appearance of a natural haze. Various human-made pollutants react in the presence of sunlight, especially the ultraviolet wavelengths. Ozone is a common component of this haze; it results from the grouping of oxygen atoms into triples. See *ozone pollution*.

**photochemical reaction** Any chemical reaction that is caused by, or accelerated by, the presence of visible light or ultraviolet. Perhaps the commonest example is the reaction that darkens photographic film.

**photoconductive cell** A semiconductor device whose electrical resistance decreases when exposed to light. Cadmium compounds are generally used to make the devices.

**photoconductivity** The property of some materials to become more conductive to electrical currents when exposed to visible light, infrared and/or ultraviolet. Many semiconductors have this property, some much more than others.

**photocopy** A photograph of a document or drawing, made using a special machine and printed on ordinary paper. Machines that make such copies have become very common.

**photodetector** Any electronic device that can convert visible light, infrared or ultraviolet energy into electricity or into an electrical signal. Such devices include the following: photoconductive cell, photodiode, photoelectric cell, phototransistor, phototube, photovoltaic cell.

**photodiode** A semiconductor diode with a transparent window through which visible light, ultraviolet, or infrared energy can shine onto the junction. This causes a change in the conductivity of the diode, depending on the intensity of the radiant energy. See *diode*.

**photoelectric cell** Any device that converts visible light, infrared or ultraviolet, and sometimes X rays, into electrical current, or into changes in electrical resistance. See *photodetector*.

**photoelectric effect** 1. The property of some materials, especially semiconductors, to change their electrical resistance depending on the intensity of visible light, infrared or ultraviolet striking the substance. 2. The property of certain semiconductor junctions to produce an electrical current when struck by visible light, infrared or ultraviolet, and in some cases, X rays and gamma rays.

**photogrammetry** The science concerned with making precise measurements of objects on the earth's surface, using photographs taken straight down from aircraft.

**photograph** A permanent, fixed rendition of a real image on a light-sensitive film. Such a rendition might be in shades of gray only, in which case it is called black-and-white; it might render the colors in either true color or false color. Some renditions show wavelengths that cannot be seen by the unaided eye, such as infrared or ultraviolet. Using special instruments it is even possible to make images at X-ray and gamma-ray wavelengths. See *real image*.

**photography** The art and science of making photographs at visible, infrared, ultraviolet, X-ray, or gamma-ray wavelengths. Usually this term refers to visible-light photographs. See *photograph*.

**photometer** A device that measures visible-light intensity. The most common method is to use a photovoltaic cell and a milliammeter. Another method uses a battery, a photocell and an ammeter or milliammeter.

**photometry** The science involved with the measurement of visible-light intensity at various wavelengths. Generally, an attempt is made to measure brightness according to the way it is perceived by the human eye.

**photomicrograph** A photograph taken through a microscope, so that the image is greatly enlarged.

**photomosaic** A matrix of tiny, round dots on the photocathode of a television camera tube. The visible image falls on this matrix, and an electron beam scans it to produce the electrical video signal. See *image orthicon, photocathode, vidicon*.

**photomultiplier** A vacuum-tube device that greatly amplifies changes in the intensity of visible light. It uses a photocathode and several electrodes in order to change the visible light into a high-energy stream of electrons.

**photon** A single packet of electromagnetic energy. In some ways it can be thought of as a particle traveling at the speed of light (about 186,000 miles, or 300,000 kilometers, per second). Each photon has an energy that depends on the frequency of the wave disturbance. This energy is given by the product of Planck's constant and the frequency of the wave. The higher the frequency, and the shorter the wavelength, the greater the energy each photon has. See the ELECTROMAGNETIC SPECTRUM appendix.

**photonics** The science involved with the transmission of visible-light signals. Modulated light and fiberoptics are the most important practical applications of this technology.

**photoreceptor** Living cells that are sensitive to light, and whose chemical activity varies in the presence of light. Sometimes infrared or ultraviolet have effects, also. The cells in the retina of your eye and the chloroplasts in plant leaves are two common examples.

**photoresist** A light-sensitive substance that is used to make printed circuits and integrated circuits. When ultraviolet light is shone on it, this substance changes the effects of certain chemical solvents. A film mask, with the desired circuit image printed on it, can be used to make highly sophisticated etchings on a semiconductor wafer using this technology.

**photosensitivity** 1. The tendency of some substances to react chemically in the presence of visible light, ultraviolet or infrared, and in some cases, X rays or gamma rays. 2. A condition in which a person is abnormally susceptible to sunburn. Might occur as a side effect with some drugs, and also in some disease conditions. 3. Abnormal sensitivity of the eyes to light. Might occur as a side effect with some drugs, and also in some disease conditions.

**photosphere** The brilliant, visible layer of the sun that we see in projected images or photographs of the sun. The brilliance is so great that looking directly at it, without any eye protection, will cause permanent eye damage. This sphere is approximately 864,000 miles in diameter, or a little more than 100 times the diameter of the earth (see drawing).

PHOTOSPHERE

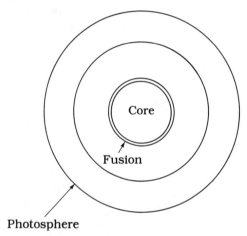

From Erickson, VOLCANOES AND EARTHQUAKES (TAB Books, Inc., 1988).

**photostat** See *photocopy*.

**photosynthesis** A process that occurs in certain plant cells, called chloroplasts. Visible light, along with atmospheric carbon dioxide and soil nutrients, is converted into carbohydrate that is used and stored. See *chlorophyll*, *chloroplast*.

**phototransistor** A semiconductor transistor whose electrical resistance varies with changes in the intensity of visible-light energy striking the junctions. Ultraviolet and infrared energy can also cause these resistance changes.

**phototube** A vacuum tube whose resistance varies depending on the intensity of visible light that falls on its cathode. The cathode emits more electrons as the light intensity increases. These tubes are also sensitive to other wavelengths from infrared to gamma rays.

**photovoltaic cell** Also called a solar cell. A semiconductor device that produces an electric current, directly from visible light striking it. Large arrays of these cells can produce hundreds of watts of useful electric power from direct sunlight. In recent years this technology has been improving and getting cheaper, so it is becoming feasible for alternative energy.

**phreatophyte** Also called a well plant, because it sends roots deep into the soil, all the way down to the water table. The existence of such plants in an arid region indicates water beneath the surface, and thus they can be used by people hoping to obtain water by well-drilling.

**phrenology** A quasi-science concerned with the evaluation of the shape of the human skull, and how it relates to various traits. Bumps and dents in various locations are supposed to indicate, for example, an easygoing character or a predilection for sweet foods.

**phylum** A group of animals that shares the same general body plan. Contains one or more classes.

**physical anthropology** See *anthropology*.

**physical chemistry** The branch of chemistry that involves observable effects and relationships of chemicals with other phenomena, such as electricity and magnetism.

**physical geography** The branch of geography concerned with the features of the earth's crust and surface in different locations. An example is the classification of terrain as grassland, forest, desert, tundra, mountains, and ice. Also, about three-fourths of the earth's surface is water.

**physical science** Any science concerned with the behavior of things as opposed to the behavior of complex organisms. Physics is one branch of this large field; chemistry, astronomy, and electronics are others.

**physical state** For matter, the condition of being solid, liquid or gas. Sometimes plasma is also considered a state of matter. See *gas, liquid, plasma, solid*.

**physicist** A scientist who teaches, does research in, or does experiments in physics. See *physics*.

**physics** The branch of science concerned with the nature and behavior of matter, energy, and forces in the universe.

**physiology** 1. The study of the functioning of a living organism, or of some particular organ in a living organism. 2. The branch of medicine that deals with the functions of, and relationships between, the different organs of the human body. 3. The study of the functioning of a particular organ of an animal or human body.

**phyto-** A prefix meaning plant or pertaining to plants.

**phytoplankton** Tiny plant organisms that are an important source of food for various types of ocean-dwelling fish. This makes them crucial in the food chain for the whole planet.

**pi** For a circle in a plane, the ratio of its circumference to its diameter (see drawing). This value is generally remembered as 3.14 or 3.14159. It is, however, an irrational number. Some people have used computers to calculate the digits of this number to thousands or even millions of decimal places.

PI

$\pi = \text{PI} \approx 3.1416$

**Picard, Jean** A seventeenth-century French astronomer who, using a telescope, was the first to accurately determine degrees of longitude on the earth. His data was used by Isaac Newton and others.

**pico-** A prefix meaning one trillionth. See the PREFIX MULTIPLIERS appendix.

**pictorial diagram** A method of showing the interconnection of the components in an electrical or electronic circuit, in which the components are drawn as they actually look. Not often used. The schematic diagram is preferred by engineers because it is far easier to draw and to read. See *schematic diagram*.

**pie graph** An illustration showing the relative proportions of various effects or physical quantities. The drawing is an example showing a hypothetical budget for combating pollution.

PIE GRAPH

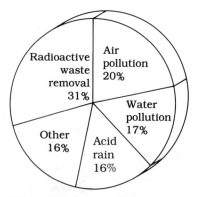

HYPOTHETICAL ANTI-POLLUTION BUDGET ALLOCATION FOR A MAKE-BELIEVE PLACE.

**piezoelectric crystal** A piece of dielectric material, usually quartz, that converts mechanical vibration into electric currents, and vice-versa. Such crystals have numerous applications, including frequency standards for clocks and oscillators, and sensitive transducers. See *piezoelectric effect*.

**piezoelectric effect** The property of certain crystal materials to convert alternating electrical currents into mechanical vibrations, and vice-versa.

**piezometric surface** The level to which water should, in theory, rise in a well under its own pressure. Sometimes this level is above the surface, in which case no pump is needed. See *artesian well*.

**pig iron** Iron that is cooled from the molten state during metal manufacture, and shaped into bricks. This iron contains impurities, but these are later removed in making the desired alloy or refined pure iron.

**Pilkington, Alastair** A British engineer who perfected a new way to make plate glass. He floated molten glass on top of molten metal. Then the glass was allowed to cool and solidify.

**pillow lava** Volcanic lava that has cooled and formed large rocks that look something like pillows. This is most common on the ocean floor.

**pinna** Also called the auricle. The part of the ear that is outside the skull and gathers sound.

**pint** A unit of fluid measure equal to 16 fluid ounces or half a quart. See *ounce*.

**pipeline** Any long, hollow tube used to transport liquid or gas from one place to another. The substance carried might be oil, water, natural gas, or even sludge containing various minerals for refining.

**Pirie, N. W.** A British biological scientist who developed a method of concentrating the protein from plant leaves. The leaves are crushed, and the juice is taken out. Then the protein is separated by coagulation. The method was developed during World War II as a means of feeding soldiers threatened with possible starvation at the hands of the Germans.

**pistil** The part of a plant flower that acts as the female organ in reproduction. See *stamen*.

**piston** A movable device, contained inside a cylinder, that is pushed back by pressure from heated gas. Used primarily to convert pressure into mechanical motion, such as in an internal combustion engine.

**piston engine** A device using one or more pistons (see *piston*) in one or more cylinders, to convert pressure into mechanical motion. The internal combustion engine is the most common example. See *internal combustion engine*.

**pitch** 1. For sound, the perceived frequency. 2. For a helix, the number of revolutions per unit length. 3. A black, tar-like substance obtained from plants. 4. Degree or extent of slant. 5. In aircraft, the extent to which the nose points up or down. 6. For a propeller, the slant of the blades.

**pith** 1. The central tissue in a plant stem. Food and water can be stored by the plant in this tissue. This part of some plants has medicinal and industrial uses. 2. A surgical procedure in which an animal's spinal cord is cut. Used in laboratory research.

**pituitary gland** An endocrine gland situated at the base of the skull (in humans). It serves as a sort of control center for the whole endocrine system. Especially, it secretes hormones that regulate growth and maturation. See *endocrine glands*.

**pituitary hormone** Any of various hormones secreted by the pituitary gland. Especially refers to the hormone that regulates growth. Too little of this results in dwarfism; too much causes giantism. The secretions of this hormone are controlled by information contained in the genes as soon as conception has taken place.

**pixel** Acronym for picture (pix) element. A single element of a television type video picture or an image transmitted via satellite.

**placenta** The tissue that connects an embryo to the wall of the uterus in a pregnant mammal. See *embryo, uterus*.

**placer** 1. A rock deposit that is left behind by a glacier as it melts and retreats. 2. A deposit of ore that is enriched by the action of a stream.

**plague** 1. Any serious epidemic of infectious disease that causes widespread death. 2. An infectious disease known as bubonic plague that ravaged Europe during the Middle Ages. It can still occur today, but epidemics have been prevented in recent times.

**Planck, Max** A physicist in Germany in the late nineteenth century. He is known for his work in developing the quantum theory of light and other electromagnetic energy (see *quantum mechanics, quantum physics*).

**Planck's constant**  A physical constant that defines the amount of energy contained in a photon (see *photon*). This energy depends on the frequency, and therefore also on the wavelength, of the disturbance. If f is the frequency in hertz, e is the energy in joules per photon, and h is Planck's constant, then e=hf. The accepted value for h is $6.626 \times 10^{-34}$ joule-second.

**plane**  1. A primitive concept in geometry, along with the point and line. A flat surface that extends infinitely far in all directions. 2. Any flat surface. 3. An aircraft.

**planet**  1. Any object more than a few hundred miles in diameter that orbits a star. 2. One of nine known large, spherical masses orbiting the sun. See the drawing and also the SOLAR SYSTEM DATA appendix.

PLANET

Sun

• Mercury

o Venus

o Earth

• Mars

Jupiter

Saturn

O Uranus

O Neptune

• Pluto

From Erickson, THE MYSTERIOUS OCEANS (TAB Books, Inc., 1988).

**planetoid**  See *asteroid*.

**plankton**  A tiny, water-dwelling organism, that is a food for many fishes. Some are plants (phytoplankton) and some are animals (zooplankton). See *phytoplankton, zooplankton*.

**plant classification**  See the PLANT CLASSIFICATION appendix.

**plant genetics**  The science of traits and characteristics of plants, and how plants can be specially cultivated to have certain desired characteristics, based on information contained in the ribonucleic acid (RNA) in the nuclei of the cells. See *genetics*.

**plant protein**  See *vegetable protein*.

**plaque**  1. A hard film that forms on the teeth when a person neglects to get proper dental care. Excessive accumulations of this material can lead to tooth and gum disease. 2. A growth in an artery, composed of various substances found in the blood. There are different types of plaques, some of which are dangerous because they can increase the chances of heart attacks.

**plasma**  1. A gas in which the electrons are not bound to individual nuclei, but wander among the atomic nuclei. This causes the gas to be an electrical conductor. 2. The part of the blood that is a liquid; that is, not containing the red cells, white cells or other microorganisms.

**plasma physics**  A branch of physics involved with the behavior of plasmas.

**plastic**  A synthetic material usually made from polymerized chemicals. They are generally categorized as either thermoplastic, that melt when heated, and thermosetting, that remain hard even under intense heat.

**plastics**  The technology involving the formulation of new plastics, and the manufacture and use of plastic materials.

**plastic surgery**  An operation or surgical technique in which certain visible body features are changed, especially in the face. A facelift is a common example.

**plate**  1. A large section of the earth's crust. 2. The anode of a vacuum tube. See *vacuum tube*.

**plateau** An elevated, usually inland, flat part of the earth's surface. Often surrounded by large hills or mountains. Much of Tibet is a good example of this type of terrain.

**platelet** A small cell in the blood that assists in the function of clotting. When these are deficient, internal bleeding is more likely. If there are too many or if their activity is too high, unwanted clots may form in blood vessels.

**plate tectonics** A theory concerning the behavior of the earth's crust. It explains the major surface features in terms of the interaction of crustal plates. The illustration shows an example. The black blobs are volcanic magma.

PLATE TECTONICS

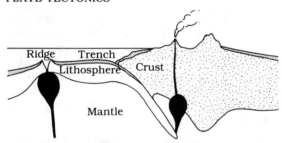

From Erickson, VIOLENT STORMS (TAB Books, Inc., 1988).

**platinum** Chemical symbol, Pt. An element with atomic number 78. The most common isotope has atomic weight 195. In pure form it is a precious metal. It is used in jewelry and also in various industrial applications.

**Plato** An ancient philosopher who existed at about the same time as Aristotle (2000 years ago). He was concerned with the nature of reality. Does reality consist of anything and everything that we can think of? Or is it only what we can detect with the senses? The debate still goes on today.

**Pleiades** Also called the "Seven Sisters." A constellation in the Northern Celestial Hemisphere recognizable as a fuzzy group of stars. People with good vision can see seven of them with the unaided eye. Through binoculars or a small telescope, the group resolves into dozens of stars. Observatory photos reveal that they are young stars developing from gas and dust.

**Pleistocene epoch** The time of the most recent major ice age. Began about two million years ago and ended only about 10,000 years ago. The drawing shows the extent of glacial coverage at the height of this time. See also the GEOLOGIC TIME appendix.

PLEISTOCENE EPOCH

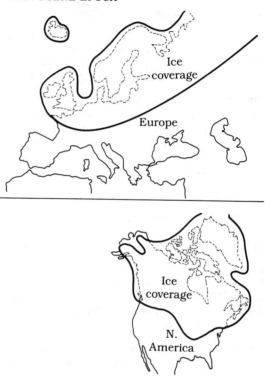

**pleura** A membrane that encloses the lungs. It has two layers with a small space in between.

**plexus** Any region in the body where nerves converge into a network. For example, the solar plexus lies in the upper central abdomen; the brachial plexus is in the chest and shoulder.

**Pliny** An ancient Roman scholar. He believed various things that we would find silly today; for example, that eating hippopotamus snouts would enhance a man's virility.

**Pliocene epoch** A geologic time that began about five million years ago and ended about two million years ago. The mastodons appeared during this time. At the close of this epoch, the most recent major ice age began. See the GEOLOGIC TIME appendix.

**Pluto** The outermost known planet in the Solar System. It is a perpetually frozen world. From it, the sun would appear hardly more significant than a very bright star, a point of light with roughly the brilliance of our full moon. See the SOLAR SYSTEM DATA appendix.

**pluton** An underground mass of igneous rock that is not as old as the rocks that surround it. This type of formation occurs when molten rock oozes into a gap between existing solid rocks.

**plutonium** Chemical symbol, Pu. An element with atomic number 94. The most common isotope has atomic weight 242. This element does not occur in nature, but only as a by-product of human-devised nuclear reactions. It is used in the manufacture of atom bombs.

**Pm** Chemical symbol for promethium.

**pneumatic** Pertaining to compressed air, and the use of compressed air in various types of machinery.

**pneumatic tube** A long cylinder that is used to transfer memos and letters over short distances. Compressed air literally blows the paper through the tube. You might have seen these at drive-in bank tellers.

**pneumococcus** A spherical bacterium that proliferates in a certain form of pneumonia. See *pneumonia*.

**pneumonia** An infection of the lungs, caused either by bacteria or by viruses. Also might occur when fluid accumulates after surgery, or when certain foreign substances enter the lungs. It is a dangerous disease with high fevers and coughing, and it often requires hospitalization. It can be fatal if not treated. May develop as a complication of influenza, especially in weak and elderly patients.

**P-N junction** The plane boundary between P-type and N-type semiconductor materials. A diode has one such junction; a transistor has two. The junction conducts when the N-type material is more negative than the P-type (forward bias); it does not conduct when the polarity is the other way (reverse bias). See *diode, semiconductor, transistor*.

**Po** Chemical symbol for pollonium.

**podzol** A soil that occurs in forest terrain, especially when the climate is cool and wet. The top layer is grayish and the lower layer has minerals that have filtered down from the top.

**Poincaré, Henri** A noted French mathematician of the late nineteenth and early twentieth centuries. He worked in all fields of mathematics. He was fond of exploring the frontiers of knowledge in all of his research and theoretical work. He had a flexible mind, causing some of his critics to say that he lacked thoroughness and rigor.

**point** 1. A geometric concept, considered as an undefined or primitive entity. Has zero dimension but exact position. 2. A moment in time. 3. The position of a number on a number line. 4. The position corresponding to an ordered n-tuple in an n-dimensional coordinate system. 5. A unique set of conditions in a system, representable on a graph.

**poison** Any substance that is lethal if taken or applied in a large enough dose.

**polar air mass** Cold air over the north or south pole. See *polar front*.

**polar coordinates** A two-dimensional system for uniquely determining the position of a point

POLAR COORDINATES

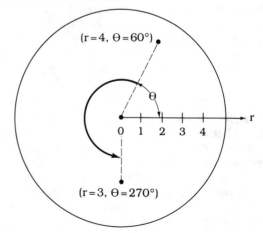

on a plane. The two values are the radius and the angle (see drawing). To avoid ambiguity, the radius is always nonnegative, and the angle is always nonnegative but less than 360 degrees. Sometimes different rules are used. See *Cartesian coordinates*.

**polar easterlies** Prevailing winds in the high latitudes of the earth, both in the Arctic and in the Antarctic. They generally blow from east to west. See *polar air mass*.

**polar front** The boundary between a polar air mass and a temperate air mass. The jet stream is usually found at or very close to this boundary. See *polar air mass*.

**polar ice cap** 1. Arctic or Antarctic ice on the earth. The size of the Antarctic ice cap is limited because any expansion would melt into the sea. In the Arctic, ice has advanced and retreated with the ice ages. See *ice age*. 2. Polar ice on Mars that advances and retreats with the seasons.

**Polaris** More often called the North Star. The star in the Northern Hemisphere around which all of the other stars seem to revolve as the earth turns on its axis.

**Polaris missile** A guided nuclear missile capable of being fired from submarines. These missiles can be used even if a preemptive strike destroys all of the land-based missile silos in the United States.

**polarity** See *electrical polarity, magnetic polarity*.

**polarization** The orientation of electromagnetic waves in space. In radio, we might have horizontal, vertical, elliptical, or circular orientation. With shorter wavelengths we usually specify the orientation of the plane in terms depending on the surroundings.

**polarized light** Visible light whose waves are all aligned in the same direction. All of the wave planes are parallel. The light from a common bulb is unpolarized, but it can become partly or completely polarized when it reflects from surfaces or is passed through special filters.

**polar-orbiting satellite** A satellite whose orbit takes it over, or nearly over, the geographic poles of the earth. Such a satellite is capable of observing every point on the earth's surface, given enough time.

**polar wandering** The changes that have taken place in the positions of the geographic poles of the earth, with respect to the continents, over long periods of time (see drawing). This might be the result of continental drift or of a cosmic catastrophe such as an asteroid impact. Most likely, both have played some role.

POLAR WANDERING

Possible varieties of polar wandering.

From Erickson, VOLCANOES AND EARTHQUAKES (TAB Books, Inc., 1988).

**pole** 1. A point towards which lines of flux converge, or from which lines of flux emanate. 2. A concentration of negative or positive electrical charge. 3. A point on a rotating object at which the axis intersects the surface. 4. On the celestial sphere, a point around which all of the stars seem to revolve once a day.

**polio** A disease, now largely eradicated by vaccinations, that caused paralysis of various parts of the body. It used to occur in children and was responsible for many people's becoming permanently crippled. Vaccinations are still important, because without them, the disease could spread again.

**pollen** Tiny particles in plants containing the male reproductive cells. These are released into the air and are also carried by insects.

**pollen count** A quantitative indicator of the concentration of various types of pollen in the air, given as an approximate number of grains per unit volume. It is important to some people with allergies to specific pollens. Ragweed and grass pollens are two common examples.

**pollination** The plant equivalent of egg fertilization in animals. The transfer of pollen (see *pollen*) to the female reproductive part of a flower. This might occur through the air, or it might be carried out with the help of insects.

**pollutant** Any human-made substance that fouls the environment, with possible danger to one or more species of living things. See *pollution*.

**pollution** Human activity that releases chemicals into the environment that cause contamination, sometimes harmful. This can occur in the atmosphere, the water supply, and the soil. Sometimes noise is considered pollution. Radiation can be still another type.

**polonium** Chemical symbol, Po. An element with atomic number 84. The most common isotope has atomic weight 209. In nature, it is found with uranium in very small amounts. It releases considerable heat as it degenerates radioactively.

**poly-** A prefix meaning several, in groups, or multiple.

**polyester** A plastic with various industrial applications, such as dielectrics for capacitors and energy-conserving window films. It is commonly known by the trade name Mylar.

**polyethylene** Polymerized ethylene. It is a translucent, flexible plastic. It is used in the manufacture of cables, especially of the coaxial type, because it is resistant to moisture and it has low dielectric loss.

**polygon** A two-dimensional geometric figure with three or more straight sides. In a regular polygon, all of the sides are the same length, and all of the vertex angles have the same measure. See the drawing.

**polygonal number** Also called a figurate number. An ancient concept involving geometric points arranged in polygons. There are unique number sequences called triangular, quadrangular, pentagonal, hexagonal, etc. Refer to a text on mathematics history for details.

POLYGON

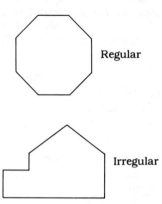

**polygraph** A lie detector. It monitors various changes in body functions that indicate stress. A person is questioned, and the indicators show the general stress level. It is assumed that lying causes a person to show a higher stress level. There is some controversy over how accurate this really is.

**polyhedron** A three-dimensional geometric figure with four or more flat faces. In a regular polyhedron, all of the faces are identical, and the object has symmetry and is convex. The drawing shows two examples of well-recognized regular polyhedra.

POLYHEDRON

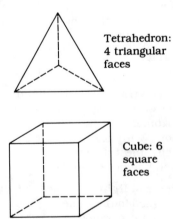

**polymer** Any material made up of long chains of molecules. These substances have various industrial uses. Sometimes they form when they are not desired (an example is the result of trying to use pure protein powder to make biscuits). Some polymers are found in nature, and many are synthetically made for specific applications.

**polymer chemistry** A branch of chemistry dealing with the structure and synthesis of polymers. See *polymer*.

**polymerization** The attaching of molecules together in long chains. This causes major changes in the properties of substances. See *polymer*.

**polymorphism** The existence of multiple forms of something. In biology there can be different forms of a species; in chemistry, different forms of an element or compound.

**polynomial** A mathematical expression that is a sum of terms, often with exponents. An example is $3y^3 - 8y^2 + 4y - 2$.

**polyp** 1. A swollen tumor in the body that protrudes and has a definite form and shape. 2. A coelenterate with a mouth, and having tentacles surrounding the mouth that serve to gather food.

**polyphosphate** A food additive often used in processed meats and various frozen foods. Also used in some cheeses and packaged bakery goods.

**polysaccharide** Food starches, consisting of three or more sugar molecules, usually glucose, attached together. These molecules are split in digestion and absorbed into the bloodstream, mainly as glucose. See *glucose*.

**polyunsaturated fat** A form of fat that has several bonds without any hydrogen atoms attached. These are usually liquids at room temperature. Most vegetable fats are of this type. Processed oils, such as corn oil and safflower oil, are mostly polyunsaturated.

**polyurethane** A polymer that is most often thought of as "foam rubber," used for such applications as the stuffing of furniture. Some forms, however, are a solid plastic. Made from polymerized protein by-products.

**polyvinyl chloride** Abbreviation, PVC. A plastic often used for electrical conduit and various other construction purposes. It is white (although it can be dyed various colors), durable, and easily cut to fit specific construction needs.

**population** 1. The number of a given species in a certain defined geographical region. For example, the number of deer in Wisconsin. 2. The number of atoms or subatomic particles having a defined energy state.

**population control** Any effort to keep the population of a given country, or of the world or a region of the world, from increasing exponentially. See *population explosion*.

**population explosion** The exponential increase in the earth's human population that has taken place over the past several centuries. Recently it has become apparent that this cannot continue without catastrophes, such as mass starvation, pandemics of disease, and/or devastating wars. See *Malthusian model*.

**porcelain** A hard, brittle, white ceramic made from various hard minerals. It appears solid but is actually porous on a microscopic scale. It is extensively used to make eating utensils, and in sinks, bathtubs, and tiles. It has excellent dielectric properties, and is used to make capacitors and insulators.

**porosity** The extent to which a filtering substance allows fine particles to pass through. Most filters will allow the passage of particles smaller than a certain diameter. The porosity can be expressed in terms of this diametric measure. Larger particles are trapped by the filter.

**port** 1. In electronics, a terminal or set of terminals, into which data can be fed, or from which data can be obtained. 2. On a ship, the left side or the direction to the left as seen while facing forward. 3. In shipping, a place where ships are loaded, unloaded, and stored. 4. An opening in the side of a ship; a window.

**portal** Pertaining to the underside of the liver, where various blood vessels enter. See *portal vein*.

**portal hypertension** High blood pressure in the portal vein, usually resulting from liver disease. If left untreated, this problem can have serious complications. See *portal vein*.

**portal vein** The large vein that carries blood enriched with nutrients from the intestines to the liver. See *liver*.

**Portuguese man-o-war** See *man-o-war*.

**positional astronomy** An early form of astronomy concerned with plotting the positions of the moon and planets with respect to the stars. In this way it was possible to predict tides, eclipses, and other celestial phenomena.

**positive charge** See *charge*.

**positive feedback** See *feedback*.

**positron** The antimatter counterpart of an electron. It has the same mass as an electron, but opposite electric charge. See *antimatter, electron*.

**post-** A prefix meaning after or later than.

**postulate** 1. A statement that is taken as true on faith, usually because it appears intuitively obvious. These statements are sometimes called axioms in rigorous mathematics. 2. To formulate an axiom. See *axiom*.

**potash** A compound containing potassium, especially the alkaline potassium hydroxide.

**potassium** Chemical symbol, K. An element with atomic number 19. The most common isotope has atomic weight 39. In pure form it is a metal. It is an important mineral in the body, being in balance with sodium. Salts of this element are common in nature.

**potassium bromate** A food additive used to bleach flour and also to help make it easier to bake. Some people are concerned that this chemical might reduce the nutritional value of the flour by destroying substances such as vitamin E.

**potential** 1. Electrostatic charge buildup, capable of producing an electric current. 2. Pertaining to stored energy, capable of being used to do work. 3. The capacity for something to perform a defined task or to have a certain effect.

**potential difference** An electrical voltage; measured in volts. Between two points, a difference in the electrostatic charge. If a conductive or semiconductive path is provided between the two points, a current will flow.

**potential energy** The capacity to do work. A mass that is elevated has this capacity; an example is water at the top of the dam in a hydroelectric power plant. See *energy*.

**potentiometer** A variable, adjustable electrical resistor. It has three contacts: two at either end of a resistive element, and one that slides along the length of the element. These devices are commonly used for volume controls, tone controls, level adjustment controls, and for various other applications in electronic devices.

**pound** Abbreviation, lb. A unit of weight, equivalent to about 0.454 kilograms in the gravitational field of the earth at the surface. Divided into 16 ounces for smaller weight measures. Not generally used by scientists, who prefer the kilogram, the Standard International unit of mass.

**poundal** A unit of force. When a mass of 0.454 kilograms (one pound in the earth's gravitational field at the surface) is accelerated at a rate of one foot per second per second, the force is one poundal. This unit is almost never used by scientists; the newton is standard. See *newton*.

**power** 1. The rate at which energy is expended, used, or dissipated. Measured in watts, or joules per second. 2. An exponent.

**power line** A transmission line for delivering electrical energy from one place to another. Often these lines are charged to several thousand volts. In some cases they can carry more than a million volts. There are usually three wires, with alternating currents at 60 hertz, each differing by $1/3$ cycle of phase (120 degrees). This minimizes the electromagnetic radiation from the line, and high voltage optimizes its efficiency. See *transmission line*.

**power-of-10 notation** See *scientific notation*.

**power plant** A facility where various forms of energy are converted into electricity. This energy may come from the burning of fossil fuels, or from nuclear fission, or from falling water. In the future, it might also come from geothermal heat, solar radiation and nuclear fusion.

**power supply** 1. Any place where energy is stored for possible use. Especially, this pertains to electrical energy. 2. A device that provides a specific voltage or voltages, at certain maximum current ratings, for electronic apparatus.

**power transistor** A transistor that can handle large currents, so that it can be used in high-power electronic devices. See *transistor*.

**Pr** Chemical symbol for praseodymium.

**pragmatism** The trial-and-error or raw experimental approach to making scientific discoveries. This is the oldest method of inquiry, and has been used since the beginning of civilization.

**praseodymium** Chemical symbol, Pr. An element with atomic number 59; the most common isotope has atomic weight 141. It is used in making flint, and also in oil refining. In pure form it is a metal.

**pre-** A prefix meaning before or prior to, or early.

**preamplifier** A sensitive electronic circuit that amplifies weak signals for use by other amplifiers. We might put such an amplifier between a radio receiver and its antenna, to increase the sensitivity of the radio set. Or we might use one between a microphone and an audio amplifier to allow us to detect very faint sounds.

**Precambrian era** The first four billion (4,000,000,000) years of the existence of the earth. The earth is about 4.6 billion years old, so this era accounts for most of the life span of our planet. Divided into the Archean eon, from 4.6 billion to 2.5 billion years ago, and the Proterozoic eon, from 2.5 billion to 575 million years ago. See the GEOLOGIC TIME appendix.

**precession** The "wobbling" of the earth on its axis, in the same way that a top wobbles as it spins. The axis of the earth describes a double cone for each complete cycle of this wobbling, and the position of the North Celestial Pole therefore describes a circle on the celestial sphere. One revolution is completed every 26,000 years.

PRECESSION

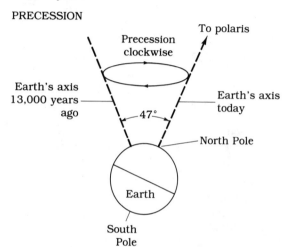

**precipitate** The material that settles out of a suspension when the suspension is allowed to stand undisturbed for a length of time, or when it is centrifuged.

**precipitation** 1. Rain, snow, sleet, hail, or any other form of water that falls from the sky under the appropriate weather conditions. 2. The settling-out of solid material from a suspension that is allowed to stand undisturbed for a length of time, or is centrifuged.

**precipitation-evaporation balance** The extent to which precipitation in a region exceeds, or falls short of, evaporation. This depends on many factors, not just on the annual rainfall (see drawing). Positive balance means that precipitation is greater than evaporation; negative balance means the opposite is true.

**precipitation probability** For a given period of time in a given location, the percentage of land area that can be expected to receive measurable precipitation. This factor is usually given in daily weather forecasts. It indicates the

PRECIPITATION-EVAPORATION BALANCE

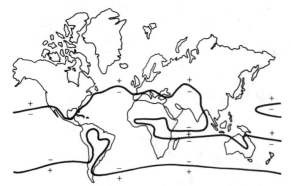

+: Precipitation is greater than evaporation
−: Precipitation is less than evaporation
From Erickson, VIOLENT STORMS (TAB Books, Inc., 1988).

likelihood, based on past occurrences of identical weather conditions, that any given place will get rain that day or night.

**precision** 1. The degree of exactness, or accuracy, of a measurement. 2. The ability for a measuring instrument to provide accurate indications. 3. Having high accuracy.

**predictive science** Any application of scientific principles that is used to forecast, in advance, what will take place. It is often said that if a theory *predicts* events with accuracy, then it is a good theory or model.

**prefabrication** A method of construction, in which modules are first put together, and then later assembled into whole units or systems. This makes it easier to manufacture devices in large numbers, and it also makes maintenance and repair procedures simple and efficient.

**prefix multiplier** A word prefix that precedes a unit of quantity, to denote a fraction of that unit. For example, *milli-* means one-thousandth or 0.001; therefore one *milliwatt* is equal to 0.001 watt. Sometimes scientific notation is used instead; we would then say $1 \times 10^{-3}$ watt. See *scientific notation* and the PREFIX MULTIPLIERS appendix.

**prefix multipliers** See the appendix PREFIX MULTIPLIERS appendix.

**prefrontal squall line** The showers and thunderstorms that form in front of a fast-moving cold front. These storms often merge into a

PREFRONTAL SQUALL LINE

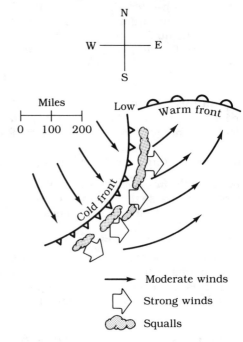

continuous band, or line (see drawing). High winds, and occasionally tornadoes, are a possibility in such weather systems.

**prescription** 1. A written document prepared by a medical doctor authorizing the use of a certain medication in assigned amounts. 2. The administration of medication for a certain symptom or disease. 3. The recommended treatment for a specific illness or malady.

**presence** 1. In communications, the clarity of a signal; the quality of its sound. This has a psychological effect, especially for voice communications. It can influence the accuracy of reception by making it easy for the receiving operator to listen to the signal. When a signal has good presence, it sounds as if the other person is right there in the same room. 2. In video, the sharpness and contrast of an image.

**preservation** 1. Environmental conservation. 2. The use of, or application of, preservatives. See *preservative*.

**preservative** Any chemical substance that slows or prevents the progress of decay or spoilage, especially in foods and wood products.

Some of these chemicals are produced naturally. Many are synthetic, and some are a cause for controversy because of possible harmful effects when consumed by people. Even so, the incidence of illness from eating spoiled foods has been practically eliminated since these chemicals have become common.

**pressure** Force per unit area. Usually expressed in newtons per square meter. Sometimes also given in pounds per square inch, or millimeters of mercury. See *barometric pressure*.

**pressure gradient** The change in pressure that exists per unit of displacement. Often used in meteorology. The steeper the gradient, the more quickly the barometric pressure changes over a given number of miles. A steep gradient usually means high winds. On a weather map this can be seen as isobars that are close together.

**pressure suit** A flight suit, space suit, or other airtight garment that provides normal, or nearly normal, atmospheric pressure, temperature, and oxygen content inside.

**pressure wave** A disturbance that precedes an aircraft moving at less than the speed of sound. The wave travels out ahead of the plane at the speed of sound for a short distance, moving the air out of the way. If the aircraft moves at the speed of sound or more, it keeps up with, or exceeds the speed of, the pressure wave, and a sonic boom results. See *sonic boom, supersonic*.

**pressurization** 1. The compression of a liquid or gas for some purpose, such as fuel storage. 2. The maintenance of a certain pressure level in an aircraft or space vessel.

**prevailing wind** A wind that blows from the same direction most of the time. There are various belts of these winds at various latitudes on the earth (see drawing). In the polar and equatorial regions, these winds are generally from east to west; at temperate latitudes they are most often from west to east. But there are frequent local exceptions.

PREVAILING WIND

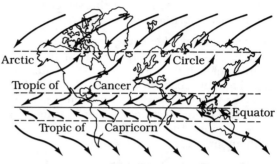

This is a simplification. Local variations are frequent.

⟶ General wind direction

**primary colors** For illumination, the colors red, blue, and green. Equal amounts of all three of these colors, shone on a surface that does not absorb any visible wavelengths, will produce gray to white. Any shade at any saturation can be produced by combining these three colors in various intensities.

**primary pigments** For reflecting materials, such as chalk or paint, the colors red, yellow and blue. Also applies to color filters. Equal amounts of these colors, when combined, produce gray to black. Any shade, at any saturation, can be obtained by combining these three pigments in various amounts.

**primate** Humans, and also the apes, monkeys, and baboons. They are all mammals and all share similar characteristics of manual dexterity and organ placement. According to one theory, human beings are the descendants of apes. According to another theory, humans were created independently of the other primates. See *Creationism, Evolutionism*.

**prime meridian** The line corresponding to zero degrees longitude, running from the North Geographic Pole southwards through Greenwich, England and on to the South Geographic pole, following a great circle halfway around the earth.

**prime number** Any positive integer that is divisible only by itself and 1 to produce another whole number. The first few of these numbers are 1, 2, 3, 5, 7, 11, 13. There are infinitely many of these numbers in the set of natural numbers. See *prime number theorem*.

**prime number theorem** For a given whole number n, as n gets larger without limit, the number of prime numbers (see *prime number*) less than n approaches the value n/(ln(n)), where ln is the natural logarithm function. This indicates that there are infinitely many prime numbers, because the value of n/(ln(n)) increases without limit as n increases without limit.

**printed circuit** A circuit board plated on one or both sides with metal foil that has been etched away in certain regions, so that electronic wiring is obtained. Components, such as resistors, capacitors, inductors, transistors, diodes, and integrated circuits are mounted on the circuit board, and the metal foil serves to interconnect the components to make a complete module.

**printer** 1. An accessory for a computer or word processor, that puts graphics or text on paper as hardcopy. 2. The mechanical device that transfers ink to paper in the aforementioned accessory device.

**pro-** 1. A prefix meaning before or ahead of. 2. A prefix meaning preceding or earlier than. 3. A prefix meaning to act with.

**proalgae** The first green-plant photosynthesizers that appeared in the seas of the earth billions of years ago. They were probably intermediate between bacteria and algae. They lived only in shallow water where sunlight could reach. According to the theory of evolution, all life on the earth today must have evolved from these organisms.

**probability** 1. The chances that something will take place, based on a large number of samples. Can be expressed as a fraction between 0 and 1, or as a percentage between 0 and 100. For example, the probability that a six-faced die will come up 3 on a given toss is $1/6$, or 0.167, or 16.7 percent. 2. An estimate of the likelihood of an event. This is not a mathematically precise use of the term, because in the real world, things either do happen, or else they don't. See, for example, *precipitation probability*.

**probability theory** A branch of statistics concerned with the chances that various events will happen. Can be used to predict certain physical phenomena when there exist a large number of identical objects, or sets of conditions. An example is the behavior of a gas in a confined space.

**probe** 1. In electronics, a device used to pick up a small voltage or electromagnetic field, to obtain a signal that can be used for various purposes. 2. An unmanned spacecraft that is used for exploration. 3. An unmanned undersea vessel, used for exploration. 4. To explore. 5. To look for an irregularity or abnormality, such as a cyst in the body.

**productivity** 1. The performance of a worker in industry based on various measurable variables, such as the number of units produced daily. 2. The performance of a factory, based on the number of units produced daily or monthly. 3. In economics, a measure of the total goods that are made by a society or a part of the society, over a certain period of time.

**progesterone** A female sex hormone. It gets the uterus ready to contain a zygote. It prevents more eggs from being produced once a female mammal is pregnant. See *uterus, zygote*.

**prognosis** The outlook for a patient with a specific disease based on experience gained from seeing large numbers of patients with the disease in the past. This outlook can change depending on various factors, such as quality and thoroughness of the treatment, and the general health of the patient prior to onset of the disease.

**program** A sequence of instructions for a computer. There are various different programs needed for the operation of any computer. Some are permanently in memory (see *firmware*) and some are changeable by the operator (see *software*).

**programmable** Pertaining to any device whose operation can be changed by means of a program furnished by the operator. See *program*.

**programmable calculator** An electronic calculator that can do small sequences of operations automatically, in much the same way a computer works, but with far less memory. Different sequences of operations are chosen by the operator. See *program, programmable.*

**programmable controller** A type of computer used to control machines. Some of these machines can do boring, monotonous tasks that were once done by people. A manufacturing plant may have many of these computers, all linked together into a network.

**programming** 1. The set of instructions used by a computer. See *program.* 2. The process of providing a computer or programmable calculator with the sequence of instructions that it needs to do a certain kind of work.

**progression** 1. A sequence of numbers with a defined function that determines each succeeding number. For example, if we double a number over and over, we get the sequence 1, 2, 4, 8, 16, 32, 64, ... . 2. Advancement, according to a well-known and fairly predictable process. We might speak, for example, of the progression of liver disease.

**projection** 1. Any object or device that protrudes from a generally uniform surface. 2. The visual production of a large image on a screen, from a small film or sample, or via a telescope. 3. A particular scheme for making a map. 4. In mathematics, the outline of an object as it would appear "in shadow" on a flat surface. Thus a sphere might be projected onto a plane to appear as a circle (see drawing).

**Project Moho** See *Mohole Project.*

**proline** An amino acid. It is not needed in the diet, because it can be synthesized in the body from other amino acids. See *amino acid.*

**promethium** Chemical symbol, Pm. An element with atomic number 61. The most common isotope has atomic weight 145. It is radioactive and decays quite rapidly. In pure form it appears as a metal.

**prominence** See *solar prominence.*

**proof** 1. A demonstration of the validity of a mathematical theorem. 2. A demonstration of the correctness of a scientific theory. 3. A cheap photograph, made to see if the negative is as good as desired. 4. The amount of alcohol in a solution. One "proof" is equivalent to about $1/2$ percent by volume of pure ethyl alcohol.

**propagation** 1. The traveling of an effect from one place to another. 2. The spreading of a species over a certain geographic area. 3. The spreading of disease-causing organisms. 4. The travel of a radio wave through the atmosphere or through space.

**propane** A hydrocarbon and a member of the alkane series, with three carbon atoms and eight hydrogen atoms, arranged as in the drawing. Highly flammable, it is a gas at room temperature and atmospheric pressure. It is a liquid

PROJECTION

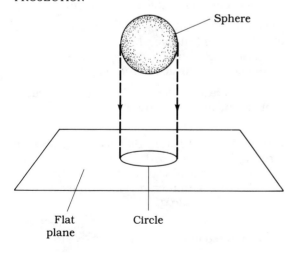

when pressurized (liquid propane or LP gas). It is commonly used for portable generators and other applications.

**propeller** Also called an airscrew (in aircraft) or a water screw (in boats and ships). A fan-like set of blades that pushes air or water in one direction, thereby causing a reaction that results in propulsion. Also might produce lift by blowing air over the wing of an aircraft. See *airfoil*.

**proper divisor** For some integer m, an integer n greater than 1 and less than m, such that m/n is an integer. That is, a proper divisor "goes into" a number and yields a whole number different from 1 and different from the original number.

**proper motion** 1. For a planet, its movement compared with the background of stars. 2. The movement of any celestial object relative to the background of stars.

**proportion** 1. Fractional part of. 2. Ratio or quotient.

**propulsion** 1. The method of obtaining motion for any type of vehicle, such as a car, truck, boat, or aircraft. This can be done by means of engines and wheels, or by propellers, jets or rockets. There might be other methods as yet undiscovered. 2. The use of the action/reaction principle to move an object for some purpose.

**prostate gland** The gland that produces semen, the vehicle for sperm cells in the male mammal. It is a common cause of trouble later in life if it becomes enlarged.

**protactinium** Chemical symbol, Pa. An element with atomic number 91. The most common isotope has atomic weight 231. It is radioactive and occurs naturally along with uranium.

**protein** Any of various compounds containing carbon, hydrogen, oxygen, nitrogen, and sometimes sulfur. Most of the tissue in animal bodies consists partially of protein; there is also a variable amount of fat. Proteins are formed by various combinations of amino acids. They can be obtained from almost any whole food, animal or vegetable. Protein is needed in the diet but not in great amounts. See *amino acid, animal protein, protein balance, vegetable protein*.

**protein balance** The ratio of protein taken in the diet, compared with protein excreted in the urine and feces. If the balance is positive, the body is getting enough protein; if it is negative, the body is losing protein. Generally about 40 to 60 grams of protein are needed by a healthy human each day, in order to maintain a positive balance.

**protein synthesis** The manufacture of protein from amino acids obtained in the diet. Certain amino acids are essential to humans; the others can be made from them. See *amino acid, essential amino acids, protein*.

**Proterozoic era** The time of earliest life, from about 2.5 billion years ago till 575 million years ago. During this time, the violent volcanic activity of the Archean died down, and the land mass stabilized into the giant supercontinent Pangaea. See *Pangaea* and the GEOLOGIC TIME appendix.

**proto-** A prefix meaning preceding or an early stage of.

**protocol** 1. The pre-planned scheme for an experiment or set of experiments. 2. A piece of scientific information, based on sensory evidence. 3. The way in which data is formatted in an electronic communications system.

**protocol fact** A scientific truth, directly observable and measurable. An example is the magnitude of a star; another example is the diameter of a virus. See *protocol plane*.

**protocol plane** The set of all scientific truths that can be observed and measured. See *protocol fact*.

**protogalaxy** A galaxy that is in the process of forming. Consists mostly of hydrogen gas. The nucleus, or central region, congeals into stars first. Gas in the outer regions condenses into stars later. See *galaxy*.

**proton** A positively charged subatomic particle, contained in the nucleus of every atom. The number of protons in a nucleus is the atomic number for a chemical element. This number does not change regardless of changes in isotope. A hydrogen nucleus, and therefore a positive hydrogen ion, consists of a single proton. See *electron, neutron, proton rest mass*, and the ATOMIC NUMBER AND ATOMIC WEIGHT appendix.

**proton rest mass** The mass of an individual proton. Equal to about $1.673 \times 10^{-24}$ gram or $1.673 \times 10^{-27}$ kilogram. This mass increases when the particle is accelerated to extreme speeds, such as is done in cyclotrons and in linear accelerators for atom smashing. See *proton*.

**protoplanet** A planet in the process of formation. According to the most commonly accepted theory, the planets in our Solar System condensed, because of gravitational effects, from rings of dust and rock around the young sun.

**protoplasm** The interior matter in a cell, including the nucleus and cytoplasm. See *cell*.

**protostar** A star in the early stages of its formation, before nuclear fusion begins at the core. A cloud of gas, mostly hydrogen, and interstellar debris condenses because of gravitation among the atoms and particles. As the pressure rises, the temperature increases. Finally the temperature is great enough to start fusion, and the protostar becomes a young star.

**prototype** A device that is built carefully by engineers and technicians, and perfected by testing and by trial-and-error, until it works just as desired. Also called a test unit or a pre-production model. Then plans are drawn up for mass production.

**protozoan** Primitive living organisms, some of which are cells. Others are even lower on the evolutionary scale than cells.

**protractor** A tool for measuring angles on drawings. Usually consists of a half-circular piece of rigid, clear plastic, graduated in degrees of arc (see drawing).

**Proxima Centauri** The nearest star to the earth, other than the sun. It is part of the star system Alpha Centauri and is about 4.3 light years away from the Solar System. It is in the Southern Celestial Hemisphere.

PROTRACTOR

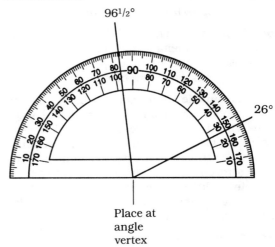

**proximal** Pertaining to the nearer portion of a part of the body. We might speak of the proximal part of the arm, for example, in reference to the upper arm. Or we might speak of the proximal part of the small intestine, meaning the portion through which foods pass first.

**pseudo-** A prefix meaning imitation or similar to.

**PST** Abbreviation for Pacific Standard Time.

**psychiatry** A branch of medicine, dealing with abnormal mental behavior, its diagnosis and treatment. This may or may not involve the use of drugs that affect the central nervous system and brain. See *psychoanalysis*.

**psychoanalysis** A method of treating emotional and mental illness, largely developed by Sigmund Freud in the early twentieth century. Basically consists of letting the patient "talk problems out." Often, by identifying the reasons for a certain emotional problem, the problem can be better controlled. This method is sometimes more effective than drugs for treatment of mental problems.

**psychology** 1. The science of behavior, in animals and in humans. 2. A branch of medicine dealing with stimulus/response activity, thought processes and emotions, and the modification of human behavior.

**psychophysics** The study of the effects of physical phenomena on the brains and nervous systems of animals and people. Overlaps between physics and psychology.

**psychosomatic** Pertaining to illnesses with their roots in emotional and/or mental problems. For example, the death of a loved one can cause depression, and this in turn can cause chronic diarrhea in some people. Sometimes illnesses are entirely imagined, not actually existing physically. However. in an attempt to treat this imagined disease, a patient may incur side effects that produce real illness.

**Pt** Chemical symbol for platinum.

**pt** Abbreviation for pint.

**Ptolemaic model** An ancient model of the Solar System, in which the earth was at the center. The moon, sun, planets, and stars all revolved around the earth (see the drawing). In order to explain the retrograde motions of some planets, complex orbits-within-orbits were invented. See *deferent, epicycle*.

PTOLEMAIC MODEL

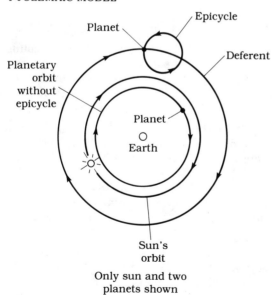

Only sun and two planets shown

**Ptolemy** A philosopher and astronomer of ancient times, best remembered for his complex, sophisticated geocentric model of the universe. See *Ptolemaic model*.

**ptyalin** An enzyme in saliva that breaks starches down into simple sugars. This is the first process in the digestion of complex carbohydrates.

**Pu** Chemical symbol for plutonium.

**puddling** A process of combining clay with water and sand, so that it is watertight. This process was devised by James Brindley in England in the middle eighteenth century, for the purpose of building canals. It was the precursor of modern concrete.

**pulley** One of the simple machines. It consists of a wheel and axle, attached to a fixed bearing with a rope or cable running through. Pulleys can be arranged in sets to provide increased lifting power.

**pulmonary** Pertaining to the lungs, and to the systems of the body involving the lungs.

**pulmonary artery** The artery leading from the right ventricle of the heart to the lungs. Oxygen-poor blood flows through this artery, to receive oxygen from the lungs.

**pulmonary vein** The vein leading from the lungs to the left auricle of the heart. Oxygen-rich blood flows through this vein, to be pumped though the heart and out to the tissues of the body.

**pulp** 1. The membrane separating the sections of a fruit. 2. Finely pulverized wood, used for making paper. 3. The innermost part of a tooth, where the blood vessels and nerve endings are.

**pulsar** A rapidly-rotating, dense, burned-out star that produces regular bursts of radio energy because of electromagnetic effects. These bursts can be heard or graphed using radio telescopes. They occur at intervals ranging from a fraction of one second to several seconds. See *neutron star*.

**pulsating-universe theory** See *oscillating-universe theory*.

**pulsation**  1. Regular changes in intensity, for any source of energy.  2. A single occurrence, or pulse, of energy in a series of pulses. See *pulse*.

**pulse**  1. A burst of energy from a source of sound, radio waves, or other electromagnetic field.  2. The rate at which the heart beats, determined by placing a finger over an artery and counting the number of beats felt in one minute. Normal resting pulse for a healthy adult is between about 50 and 75 beats per minute.

**pulse modulation**  Any of several methods of conveying information as a stream of radio pulses. The intensity, duration, or spacing of the pulses may all be modulated. The drawing shows an example of digital pulse-intensity modulation. See *amplitude modulation, frequency modulation, phase modulation*.

PULSE MODULATION

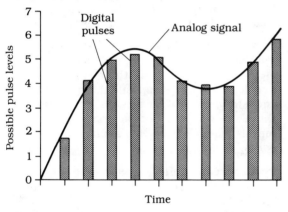

**pumice**  1. Material ejected from volcanoes, that is full of gas-containing cavities and is therefore very light in weight.  2. Finely ground volcanic material, used for abrasive applications in cleaning teeth and in grinding lenses.

**pump**  1. A device that brings liquid up from underground or removes it from a tank, sending it through pipelines. Uses an electric or gas-powered motor.  2. A machine that removes gas from a chamber, either dispersing it into the atmosphere or moving it to some other enclosure.

**pupa**  In insect metamorphosis, the stage during which the larva changes into an adult. The insect does not move nor eat during this time. An example is the cocoon phase of the development of a moth.

**pupil**  The opening in the eye through which light enters. Its size changes as the iris adjusts to the level of light. In dim light the pupil is large, or dilated; in bright light it contracts. Some drugs affect the sizes of the pupils. See *eye*.

**pure mathematics**  A branch of mathematics in which no concern is given to possible practical applications. Only the concepts, axioms, theorems and techniques are dealt with. Some mathematicians devote their whole careers to this aspect of their craft.

**pure tone**  A sound wave at a single frequency, represented by a sine wave. There are no harmonics or sounds at any other frequency. The result of listening for a long time to such a note is ear fatigue.

**purification**  The process of rendering a substance free of contaminants, so that all, or almost all, of the molecules in a sample are those of the desired substance.

**purification plant**  Any facility, such as an oil refinery or seawater desalination apparatus, that produces an uncontaminated substance.

**purine**  A protein-like compound, found in various organic substances, that can be converted into certain amino acids. These substances promote excretion of excess water in edema (water retention) by the body. They are also important in the production of deoxyribonucleic acid (DNA) and ribonucleic acid (RNA). See *amino acid, deoxyribonucleic acid, ribonucleic acid*.

**pushdown stack**  Also called a first-in/last-out memory. Data is fed into the memory, and comes out again in reverse order. See the drawing. See *first-in/first-out*.

**push-pull**  A type of amplifier circuit in which two transistors are used. One transistor works on the positive half of the input cycle, and the other transistor works on the negative half. The result is an output wave with the same shape as the input wave, but stronger.

PUSHDOWN STACK

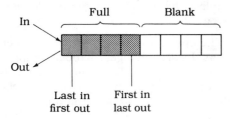

From Gibilisco, ENCYCLOPEDIA OF ELECTRONICS (TAB Books, Inc., 1985).

**push-push** A type of amplifier or frequency-multiplier circuit, using two transistors. One transistor works on the positive half of the input cycle, and the other transistor works on the negative half. One half of the wave is inverted because of the way in which the transistor outputs are connected. Therefore, even multiples of the input frequency are amplified, and odd multiples are suppressed.

**putrefaction** Spoilage, or the process of decay. So called because of the bad smell it often produces when organic materials rot.

**PVC** Abbreviation for polyvinyl chloride.

**P-wave** In an earthquake, the primary or pressure wave. It travels via compression of the crustal rock.

**pyloric sphincter** The valve that connects the stomach to the small intestine. It regulates the flow of food into the intestine, where digestion is completed and where absorption eventually takes place.

**pylorus** The muscle that forms the pyloric sphincter. See *pyloric sphincter*.

**pyramid** A polyhedron with five faces. The base is square, and the other four faces are triangles that converge to a vertex lying directly above the center of the base. Some people think that pyramids have unique energy-concentrating properties. Most scientists dismiss this idea since it has not been generally verified by experiments.

**pyridoxine** Also called vitamin B-6. An essential, water-soluble nutrient, found in a wide variety of foods. See the VITAMINS AND MINERALS appendix.

**pyro-** A prefix meaning heat and/or fire. 2. A prefix denoting a certain kind of acid that results from removal of hydrogen and oxygen from some other acid.

**pyroclastic** Pertaining to fragments of material that are explosively released from a volcanic vent.

**pyrometer** 1. A device used to measure extremely high temperatures. 2. A thermometer used to measure the temperature of molten rock, such as volcanic lava.

**Pythagoras** A philosopher, astronomer, mystic and prophet of ancient Greece (around 500 B.C.). He lived at about the same time as Buddha, Confucius and Lao-Tse. He founded a cult-like group based on philosophy and mathematics. He is best remembered for his famous theorem concerning right triangles, but he made many other contributions in geometry as well. See *Pythagorean Theorem*.

**Pythagorean Theorem** A well-known geometrical principle that applies to right triangles. The square of the length of the hypotenuse, or side opposite the right angle, is equal to the sum of the squares of the lengths of the other two sides. See the drawing. The theorem is named after Pythagoras. See *Pythagoras*.

**Pythagorean triple** A set of real numbers a, b, and c, such that $a^2 + b^2 = c^2$. Thus the numbers can be assigned to the lengths of the sides of some right triangle. See *Pythagorean Theorem*.

PYTHAGOREAN THEOREM

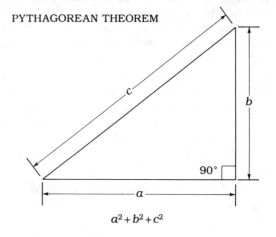

$$a^2 + b^2 + c^2$$

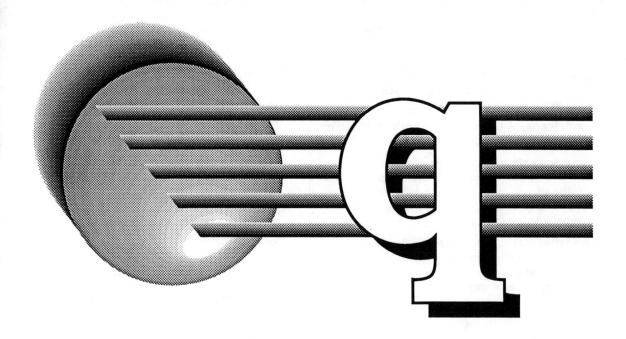

**Q** Abbreviation for the set of integers. See *integer*.

**Q factor** 1. A quantitative measure of efficiency in a coil, capacitor, or tuned circuit. It is proportional to the ratio of energy stored to energy lost. 2. A measure of the degree of selectivity in a tuned circuit. See *selectivity*.

**qt** Abbreviation for quart.

**quack** 1. A person who claims to be a medical doctor, but in fact is not. 2. A medical doctor who acts in an unprofessional or illegal manner.

**quack diet** A diet advertised as promoting weight loss, but that is either medically unsound, or else simply does not work. Many popular fad diets are of this type.

**quack nutrition** The advertising and selling of unproven and/or unnecessary substances, usually with claims that they will get rid of various illnesses or result in "super health."

**quad-** A prefix meaning four or in groups of four.

**quadrangle** See *quadrilateral*.

**quadrant** 1. One quarter of the Cartesian plane. The first quadrant has x and y both positive. In the second quadrant, $x < 0$ and $y > 0$; in the third quadrant, both x and y are negative; in the fourth quadrant, $x > 0$ and $y < 0$. 2. A section of a plane, a region, or of space marked off in some way via a coordinate system.

**quadrat** A small geographical area used for environmental sampling, when it is impractical to take samples from a whole region.

**quadratic equation** An algebraic equation in one variable, in which there exists an exponent of 2 (the variable is squared). Examples are $x^2 + 2x + 1 = 0$ and $-4x^2 = -4$.

**quadratic formula** A general formula for solving quadratic equations. For the equation $ax^2 + bx + c = 0$, the general solution is given by either of the following: $x = (-b + (b^2 - 4ac)^{1/2})/(2a)$, or $x = (-b - (b^2 - 4ac)^{1/2})/(2a)$.

**quadratrix** A geometric curve used by Hippias around 400 B.C. in an attempt to square the circle. This is one of the geometric constructions generally regarded as impossible according to the rules of compass and straightedge. Hippias' method did not precisely follow the standard construction rules.

**quadrature** For two waves of the same frequency, the property of being 1/4 cycle (90 electrical degrees) out of phase.

**quadriceps** A set of muscles in the thigh that helps move the legs and maintain balance.

**quadric surface** The three-dimensional analog of a conic section. A three-space surface resulting from an equation in two variables of second degree. Ellipsoids, hyperboloids and paraboloids are examples.

**quadrilateral** Any polygon with four sides lying in a single geometric plane. The drawing shows some examples. The square, rectangle, parallelogram and trapezoid are special cases.

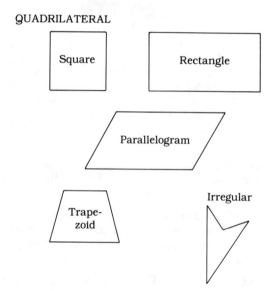

**quadrillion** The large number represented by a 1 followed by 15 zeros, or 1000 followed by four sets of three zeros.

**quadrivium** The four branches of mathematics, as defined by the ancient mathematician Archytas: arithmetic, geometry, music, and astronomy. Nowadays we depict this field much differently.

**quadrophonics** Four-channel stereo. The channels are: Right front, right rear, left front and left rear (see drawing). Thus, it is possible not only to portray linear motion, such as a

QUADROPHONICS

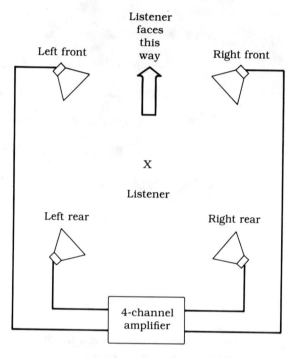

train moving from left to right, but revolution, such as a circling aircraft.

**qualitative** Pertaining to results that indicate the general nature of a phenomenon, but do not depict it precisely in terms of measured values. See *quantitative*.

**quality** 1. A general expression for good manufacturing; reliability. 2. A specific property of an object or phenomenon.

**quality control** Also called quality assurance or quality engineering. A procedure in manufacturing, in which units are tested to be sure that they work before they are sold or distributed. Usually this consists of testing a certain percentage of the units, such as one in every ten, chosen at random or systematically. If the proportion of nonfunctional units (new failures) is too high, the manufacturing process is changed accordingly.

**quantitative** Pertaining to numerical representations of phenomena, such as the results of measurements. Examples are a temperature of

45.5 degrees centigrade, or an acceleration of 9.8 meters per second per second. See *qualitative*.

**quantity** 1. The number of items in a lot. 2. A large amount; a substantial volume of something. 3. A measure of the amount of electrical charge, given in coulombs. See *coulomb*.

**quantization** 1. The existence of energy in discrete units, called quanta or *photons*. 2. The existence of matter in particle units. The smallest known particle is the quark. 3. The division of space or time into interlocking units. Space and time might actually be continuous, or they might exist in smallest possible intervals, so that a very small object could be resolved only approximately, as shown in the drawing. See *photon, quark*.

QUANTIZATION

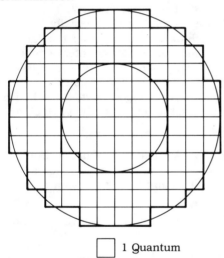

☐ 1 Quantum

From Gibilisco, REACHING FOR INFINITY (TAB Books, Inc., 1990).

**quantum** 1. A single unit of electromagnetic energy, usually represented as a packet called a photon. The energy in a single photon depends on the wavelength. See *photon, Planck's constant*. 2. A single unit of time, mass or space, considered according to some theories as the smallest possible such unit.

**quantum chemistry** A branch of chemistry concerned with interactions of fundamental particles as they pertain to chemical reactions.

**quantum jump** An increase or decrease in the energy level of an electron in an atom, such that a photon is either absorbed or emitted at a certain wavelength that depends on the extent to which the electron gains or loses energy.

**quantum mechanics** The study of phenomena that occur as a result of the quantization of matter and energy. This especially involves the changes in the energy levels of electrons within atoms. See *quantization, quantum, quantum jump*.

**quantum number** 1. The energy level of an electron, as given by its orbital level and various other factors including the existence of external electric or magnetic fields. See *electron, quantum jump*. 2. The spin of a particle. See *spin*.

**quantum physics** The branch of physics that is concerned with the behavior of fundamental particles. The uncertainty principle is an example of one of the results of this branch of physics. See *uncertainty principle*.

**quantum state** 1. The set of quantum numbers for an electron or other elementary particle. 2. The general energy level of a system, as defined in terms of quanta.

**quantum statistics** A branch of physics that deals with the behavior of a system according to quantum theory. See *quantum, quantum physics*.

**quark** A subatomic particle considered to make up protons and neutrons. There are different types of quarks with names such as up, down, strange, charmed, top, and bottom. Some scientists think these might be elementary particles, the smallest particles in the universe. See *elementary particle, neutron, proton*.

**quart** A unit of fluid measure equal to 32 fluid ounces, 1/4 gallon, or 2 pints. It is a little less than one liter. See *liter, ounce*.

**quartic equation** An algebraic equation, usually in one variable, in which the largest exponent is 4. An example is $6x^4 + 3x^3 - 13x - 5 = 0$.

**quartz** A crystallized form of silicon dioxide. It has an appearance similar to diamond and can be made into inexpensive jewelry; it scratches glass just as diamond does, but an expert can easily tell the difference between quartz and diamond. It is clear in pure form, and colorless, but there often exist impurities that cloud it or discolor it. It is used in a wide variety of industrial applications, particularly in electronics, where it is made into oscillator frequency standards and is used for the piezoelectric elements of transducers. See *piezoelectric crystal, piezoelectric effect, quartz crystal, quartz timepiece.*

**quartz crystal** A piezoelectric crystal, made from quartz cut into a thin wafer. The frequency at which this crystal is most sensitive depends on the thickness of the wafer, and also on the axis in which it is cut. Such crystals are used for electronic oscillators. They can also be used to make transducers. See *piezoelectric crystal, piezoelectric effect.*

**quartz timepiece** A clock or watch that uses a quartz-crystal oscillator as a time standard. These time standards have become very easy to make, and can be found in watches costing only a few dollars. They are generally accurate to better than one second per week.

**quasar** Also called a quasi-stellar radio source. An object that looks like a star in a conventional optical telescope, but is very far away and has the energy emission of a whole galaxy. The first of these objects to be identified were strong sources of radio emissions, and this is what drew attention to them. Some of these objects are actually radio quiet, so the name does not always precisely apply. Some astronomers think these objects are protogalaxies. See *protogalaxy.*

**quasi-** A prefix meaning resembling.

**quasi-stellar** Literally, star-like. See *quasar.*

**Quaternary period** The Pleistocene and Holocene epochs, beginning about two million years ago and extending to the present time. Humans appeared at the start of this time, when the last great ice age began. See the GEOLOGIC TIME appendix.

**quaternion** A complex number according to a theory developed by William Hamilton in the nineteenth century. There were three types of complex number, called i, j, and k, such that $i^2 = j^2 = k^2 = ijk = -1$. The theory of quaternions is sophisticated and a text on advanced algebra is recommended for study.

**quenching** A process in metal manufacture, in which it is dipped into a cold liquid when still hot. This affects its final state, making it useful for the desired application. 2. The extinguishing of a fire or flame, especially by means of water or powder that deprives the flame of oxygen. 3. In a radiation counter, a process that keeps the discharge from taking place continuously, so that the device can count particles effectively. 4. In electronics, a method of preventing continuous oscillation in a circuit.

**quicksilver** See *mercury.*

**quinine** A substance obtained from the cinchona plant. It has a bitter, unpleasant taste. It has traditionally been used as a treatment for malaria.

**quint-** A prefix meaning five or in groups of five.

**quintillion** The large number represented by a 1 followed by 18 zeros, or 1000 followed by five sets of three zeros.

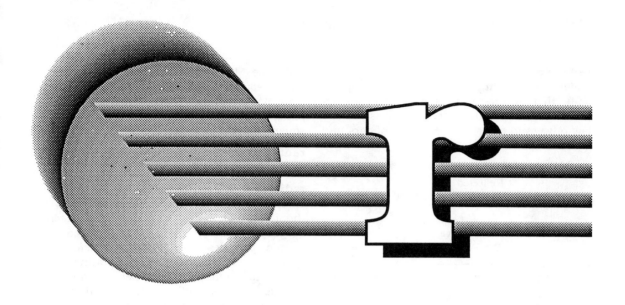

**R** Abbreviation for gas constant, resistance, roentgen.

**r** Abbreviation for radius, revolution.

**Ra** Chemical symbol for radium.

**rabies** A viral infection of mammals, particularly carnivores and omnivores. It is carried by body fluids. The main symptoms are nervous-system derangement, paralysis, and ultimately death in almost all cases. In humans the disease is sometimes called hydrophobia, because the victim cannot drink fluids. Shots are available for people who suspect they have been exposed; these shots must be received before symptoms begin.

**race** Within a single species, a group with distinct features. These differences are easy to see in humans, but they can exist in animals also.

**rad** 1. A unit of ionizing radiation dose, equivalent to 0.001 joule per kilogram. 2. Abbreviation for radian. 3. Abbreviation for radio. 4. Abbreviation for radix. 5. Abbreviation for radical.

**radar** Acronym for *radio detection and ranging*. 1. A system that usually operates in the ultrahigh-frequency or microwave part of the radio spectrum, and is employed for detecting the position and movement of flying objects, such as enemy aircraft and missiles. It also can be used to track storm systems, because precipitation reflects the waves. It also can be used to measure distances, to make precise maps, and for various other purposes. See the following several definitions. 2. A system that is operated at high-frequency radio, and that is used to track the approximate movements of aircraft and missiles at long distances from over the horizon.

**radar astronomy** 1. The use of precision, high-power radar to measure the distances to the moon, other planets, and other objects in the solar system. 2. The use of radar on board unmanned probes, to map planets whose surfaces are obscured by clouds. Venus has been extensively explored in this way.

**radar detector** A device that can be used in a car or truck to warn of the presence of police radar. See *radar speed measurement*.

**radar jamming** A method of disabling military radar to allow aircraft or missiles to pass through enemy territory without the enemy being able to tell exactly where they are. Usually, this technique operates in a wedge-shaped region called a corridor.

**radar speed measurement** A technique in which radar is employed to measure Doppler shift, thereby indicating the speed of an approaching (or retreating) object. This is used by law officers to measure the speeds of cars and trucks (see drawing).

RADAR SPEED MEASUREMENT

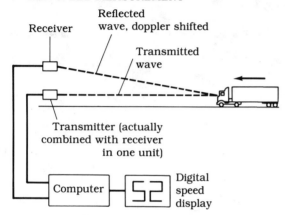

**radial** 1. Pertaining to, or existing along, a line that extends outward from some central point. 2. One of several or many wires in a transmitting antenna system, lying under or parallel to the surface of the earth, and serving as a reflecting plane for radio waves. 3. Pertaining to the thin bone in the lower arm.

**radial engine** An internal combustion engine, in which the cylinders are arranged in a circle. It was used in early aircraft, especially in fighters during World War I. It provided more horsepower for its weight than the in-line engine. But it had a tendency to act as a gyroscope and hinder flight. It also sprayed its oil onto the pilots. See *internal combustion engine*.

**radial nerve** One of the nerves that lead from the brachial plexus through the arm to the hand. The other two are the median nerve and the ulnar nerve. See *brachial plexus*.

**radian** A unit of angular measure in which the arc length is equal to the radius of the circle for which the angle is determined. This is an angle of about 57.3 degrees. Used in theoretical physics and engineering.

**radiance** 1. The amount of energy given off by a point source, within a solid angle of one steradian. See *steradian*. 2. The amount of energy passing through a surface, given in watts per square meter.

**radiant** Pertaining to energy transmitted in the form of radiation. See *radiation*.

**radiation** 1. Energy transmitted in the form of electromagnetic waves. This also can be thought of as a barrage of photons. See *electromagnetic wave*, *photon*. 2. High-speed subatomic particles. See *alpha particle*, *beta particle*, *neutron*, *proton*. 3. Penetrating, ionizing electromagnetic energy. See *gamma rays*, *X rays*. 4. A form of heat transfer that takes place by means of infrared energy. Unlike conduction or convection, this form of heat transfer can occur even in a vacuum. See *conduction*, *convection*, *infrared*.

**radiation poisoning** 1. Exposure of the body to large and damaging amounts of ionizing radiation. 2. Short-term physiological symptoms resulting from exposure to large amounts of ionizing radiation. These include burns, nausea, hair loss, coma, and death, depending on the extent of exposure. 3. Long-term effects of exposure to high levels of ionizing radiation. These can include cancer, chromosome damage, decreased immunity to disease and generally shortened life span.

**radiation resistance** The apparent electrical alternating-current resistance that an antenna shows. This depends on how large the antenna is. It is important in calculating antenna efficiency and in designing antenna systems.

**radiation sickness** See *radiation poisoning*.

**radiation therapy** A method of treating certain forms of cancer, by exposing the tumors to ionizing radiation. The idea is to kill the cancer cells without harming normal cells. This method of treatment works well for certain kinds of cancer, but not so well for other kinds.

**radical** 1. The symbol used to indicate roots of numbers or variables. 2. Root-like, or pertaining to the roots of a plant. 3. A set of atoms linked together that has certain properties.

**radio** 1. Pertaining to communications by electromagnetic means, usually voice or digital-code transmission not involving video. 2. A receiver or transmitter for communicating via electromagnetic waves.

**radio-** A prefix meaning radiation or pertaining to radiation.

**radioactive** Pertaining to a substance that emits ionizing radiation, such as alpha particles, beta particles, gamma rays, or X rays as it decays. See *alpha rays, beta rays, gamma rays, ionizing radiation, X rays.*

**radioactive dating** A means of determining how old a sample is by the concentration of radioactive isotopes it contains. The most common radioisotope that is used for this purpose is carbon-14. See *radioisotope.*

**radioactive decay** A process in which certain isotopes of various elements gradually degenerate into more stable isotopes. Sometimes the more stable isotope has the same atomic number, and sometimes it does not. Ionizing radiation is emitted by matter undergoing this process. The speed of the process is measured according to the half life, which might be just a fraction of a second, or millions of years. See *half life, ionizing radiation.*

**radioactive fallout** Debris from a nuclear explosion. The blast vaporizes the bomb housing itself, and some of the earth beneath if the

RADIOACTIVE FALLOUT

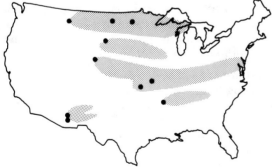

Heavy fallout patterns from large-scale nuclear attack on American missile bases.

From Erickson, VIOLENT STORMS (TAB Books, Inc., 1988).

explosion occurs close to the surface. This material becomes radioactive and then recondenses, and can be scattered over large areas of the earth, endangering many forms of life for long periods of time. After a large-scale nuclear attack on the United States, the regions shown in the drawing would be especially subject to this fallout.

**radioactive isotope** See *radioisotope.*

**radioactive waste** By-products of nuclear processing and nuclear power generation that emit ionizing radiation and are therefore dangerous. It can be very difficult to find safe ways to get rid of this waste, because in some cases it has a half-life of many human generations. See half-life, high-level radioactive waste, low-level radioactive waste.

**radioactivity** The emission of ionizing radiation. See *alpha rays, beta rays, gamma rays, ionizing radiation, X rays.*

**radio astronomy** The observation, mapping, and exploration of the heavens at radio wavelengths. This is done mostly at wavelengths shorter than about 1 meter. See *radio telescope* and the RADIO SPECTRUM appendix.

**radiobiology** A branch of biology concerned with the effects of ionizing radiation on living things.

**radiochemistry** A branch of chemistry involved with radioactive compounds. These compounds have one or more atoms that are radioisotopes. See *radioisotope.*

**radio direction finder** A radio receiver with a directional antenna that has a sharp null in a certain direction. This is usually a loop antenna (see drawing) with a calibrated azimuth scale. The loop is turned till a null occurs in the received signal. Bearings are taken from two or more locations and lines drawn on a map to find the source of the signal. See *radiolocation.*

**radio frequency** Electromagnetic radiation having a wavelength shorter than about 100 kilometers (a frequency of 3 kilohertz), but longer than infrared, or about 0.1 millimeter (a frequency of 3000 gigahertz). See the ELEC-

RADIO DIRECTION FINDER

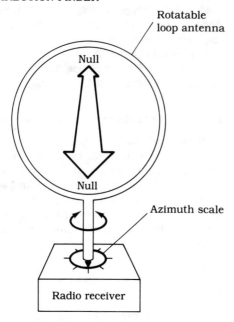

RADIOLOCATION

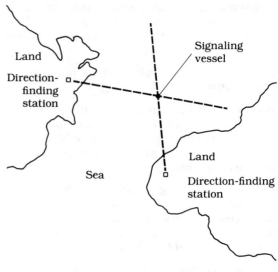

TROMAGNETIC SPECTRUM and RADIO SPECTRUM appendices.

**radio galaxy** A galaxy that is not unusually bright in the visible-light range, but that emits a great amount of energy at radio wavelengths. See *galaxy*.

**radiography** A means of producing a photograph from ionizing radiation, instead of visible light. An X-ray picture, such as a dentist makes, is an example of this technique. Usually, X rays are used, although other types of ionizing radiation also can be employed. See *ionizing radiation, X rays*.

**radioisotope** An isotope of an element that gives off ionizing radiation. Some elements have no such isotopes; some have many. In general, the higher the atomic number of an element, the more radioisotopes it has. See *isotope*.

**radiolocation** A method by which a ship captain or aircraft pilot can determine his/her location. Direction finders are used with two or more shore stations (see drawing). This method can also be used by shore personnel to locate a ship or aircraft. There are several systems in operation for this purpose that use other methods besides that shown in the illustration.

**radiology** A branch of medicine concerned with the use of ionizing radiation to treat certain disease conditions, particularly cancer. See *radiation therapy*.

**radiolysis** The occurrence of chemical reactions in the presence of high-speed subatomic particles, gamma rays, or X rays. A simple example is the way that photographic film is exposed by ionizing radiation.

**radio map** A map of the heavens made using a radio telescope instead of an optical telescope. The nature of the radio sky depends on the wavelength at which the map is made, and also on the resolving power of the radio telescope. See *radio telescope*.

**radiometer** A device that measures the ability of light to generate heat energy. The simplest such device consists of three or four vanes that are free to rotate inside a glass bulb. The vanes are black on one side and white on the other side. Light causes the assembly to rotate (see the drawing).

**radiometric dating** See *radioactive dating*.

**radionavigation** The use of radio communications and waves to assist a ship captain or an

## RADIOMETER

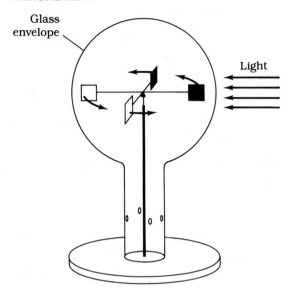

## RADIOSONDE

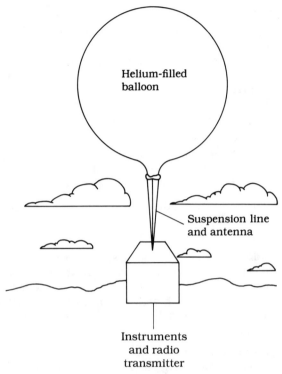

aircraft pilot to plot and follow a prescribed course. Much of this is done nowadays using satellites and computers. Ground-based signals can also be used. See *radiolocation*.

**radionuclides** Highly radioactive elements that generated heat in the primordial earth, melting it from the inside out. This caused the heavier elements to gravitate toward the center of the earth. This is the reason why the earth is still hot inside. This heat can prove to be a useful source of energy in the future. See *geothermal energy, geothermal power plant*.

**radiosonde** A set of weather instruments that is attached to a balloon and released into the atmosphere. It has a radio transmitter that sends telemetry back to receiving stations, to monitor conditions in the troposphere and lower stratosphere. See the drawing.

**radio spectrum** The part of the electromagnetic spectrum that is normally used for radio communications, radio astronomy, and radar. See the ELECTROMAGNETIC SPECTRUM and RADIO SPECTRUM appendices.

**radio telescope** A highly sensitive radio receiver connected to a large directional antenna, and used for observing the heavens at radio wavelengths. The most common wavelengths are those shorter than 1 meter and longer than about 0.1 millimeter. These wavelengths correspond to frequencies of 3 kilohertz to 3000 gigahertz. Huge dish antennas are often used. A special, sensitive amplifier called a maser provides low noise and high amplification. The larger the antenna for a given wavelength, the better the resolution of the device. See *maser*.

**radioteletype** The transmission of teletype by means of radio. There are two signals at slightly different frequencies, called mark and space. Rates of speed vary from about 60 words per minute to hundreds of words per minute.

**radio wave** An electromagnetic wave whose frequency and wavelength fall within the radio spectrum. These are wavelengths shorter than about 100 kilometers but longer than 0.1 millimeter. See the ELECTROMAGNETIC SPECTRUM and RADIO SPECTRUM appendices.

**radium** Chemical symbol, Ra. An element with atomic number 88. The most common isotope has atomic weight 226. This was the first radioactive element to be identified. It was discovered by the Curies at the end of the nineteenth century. It was once used in watch dials to allow visibility in darkness.

**radius** 1. For a circle or sphere, the distance from the center point to the periphery or surface. 2. Half of the diameter of an object. See *diameter*. 3. The thin bone in the lower arm.

**radix** 1. The modulo of a numbering scheme. See *modulo*. 2. The main point or region from which something originates or emanates.

**radome** A large, usually spherical enclosure that protects radar antennas from the weather. It is made of some material that is impervious to the elements, such as fiberglass, but that will allow microwave radio signals to pass through. See *radar*.

**radon** Chemical symbol, Rn. An element with atomic number 86. The most common isotope has atomic weight of 222. This is a gas at normal temperature and atmospheric pressure. It is a product of radium decay. This gas has recently been implicated as a potential cause for lung cancer and other health problems. It tends to accumulate in basements and lower floors of buildings, in regions where the subsurface rocks contain significant radioactive materials.

**rainband** A band of showers and thundershowers that exists in the outer circulation of a hurricane. There are usually several of these in a spiral-shaped pattern around the eyewall or wall cloud. See *hurricane*.

**rainfall** 1. Precipitation, melted so as to determine the equivalent in the form of liquid water. 2. The number of inches of liquid precipitation (or the equivalent) that falls within a given period of time, usually in a year.

**rain forest** A forest that receives abundant rainfall. In particular, a tropical forest. Such a region might get more than 100 inches of rainfall annually. In recent years, there has been concern among environmentalists and also among some scientists, that too much of the earth's rain forest land is being cleared for farming and industry. This could cause climate changes, the nature of which are hard to predict.

**ramjet** A type of jet engine with no compressor. The movement of the engine through the air acts to compress the air. Fuel is injected into this air, and ignited; this causes expansion of the compressed air. The hot gas then is ejected from the back of the engine (see drawing).

RAMJET

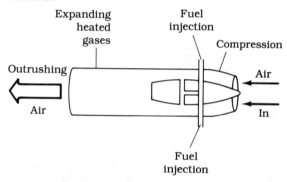

**ramp wave** A sawtooth wave, in which the level rises gradually and falls almost instantly. See *sawtooth wave*.

**Ramsden, Jesse** A craftsman of the eighteenth century, who was one of the first to manufacture items in large quantities rather than one at a time. His technique was an early version of mass production. See *mass production*.

**random numbers** Any sequence of digits 0 through 9, in which there is no pattern and such that it is impossible to predict the next one in the sequence. In reality there may be no such thing as truly random numbers, since so-called pseudorandom numbers are generated by means of algorithms that must be devised by some orderly scheme. Numbers of this type are sometimes needed for statistical analysis.

**range** 1. In mathematics, the set of values for the dependent variable, for which a function is defined. See *dependent variable, domain, function*. 2. The maximum possible distance over which a radio communications system will work. 3. The maximum distance over which a radar will function. 4. For a set of values, the difference between the maximum and minimum. 5. A place for testing various types of apparatus, such as radio antennas.

**Rankine temperature scale** An absolute temperature scale not often used. The degrees are the same size as those of the Fahrenheit scale, but zero Rankine is given as absolute zero. This is about −459 degrees Fahrenheit. Degrees Rankine can be obtained from degrees Kelvin, by multiplying the Kelvin temperature by 1.8. Degrees Fahrenheit are obtained by subtracting 459. See *Fahrenheit temperature scale, Kelvin temperature scale*.

**rapid eye movement** Constant changes in the position of the eyeballs during dream sleep. One theory holds that the eyes move to follow the objects being dreamed about, just as if the person were awake.

**rare earth** See *lanthanoid*.

**raster** The rectangle of lines that is seen on a television picture tube when there is no signal. This is the standard set of lines that are scanned by the picture tube and receiver circuitry.

**rate meter** A device that measures the intensity of ionozing radiation, for example, in rads per hour.

**Rationalism** A philosophy whose main theme is that ideas can exist independently of people's minds. Thus, for example, a topological space, or a set of numbers, has an existence all its own, whether people think of it or not. Plato and Descartes held this view. See *Empiricism*.

**rational number** A number that can be expressed as a quotient of two integers, where the denominator is not zero. See *integer, irrational number, real number*.

**raw materials** The ores, fossil fuel deposits, and other matter that are used to produce industrial goods.

**raw ore** Unprocessed metal ore. For example, hematite is raw iron ore; bauxite is raw aluminum ore.

**raw sewage** Sewage that has not been treated. This might be, for example, the contents of a portable toilet.

**ray** 1. In geometry, a half line, that may or may not include the end point. It extends infinitely in one direction and only a finite distance in the other, for any given point on the half line. 2. A narrow beam of radiant energy, especially of short electromagnetic wavelength. 3. A species of marine creature with a flattened body. May be up to several feet across. Some have stingers (stingrays).

**Rayleigh, Lord** Winner of the Nobel physics award in 1904 for discovering the element argon. Some people consider him to be the last great classical physicist.

**Rayleigh wave** A form of seismic wave, in which individual particles of the earth move in elliptical patterns. This is similar to the movement of water molecules in wind-driven waves.

**rayon** A fiber that is sometimes used to make articles of clothing. It is processed from wood. Its primary advantage is that it is made from natural sources; it weakens, however, when it gets wet.

**Rb** Chemical symbol for rubidium.

**rd** Abbreviation for rad.

**RDA** Abbreviation for recommended daily allowance.

**RDF** Abbreviation for radio direction finder.

**Re** Chemical symbol for rhenium.

**reaching** In sailing, a means of traveling more or less sideways relative to the wind direction. In a close reach, the angle of the bow relative to the wind direction is less than 90 degrees; in a beam reach it is about 90 degrees; in a broad reach it is more than 90 degrees (see drawing).

**reactance** The opposition that an inductance or capacitance offers to sine-wave alternating current. It does not dissipate power, but causes energy to be stored and then released later. Inductive reactance stores energy as a magnetic field, and capacitive reactance stores it as an electric field. Reactance is measured in ohms.

REACHING

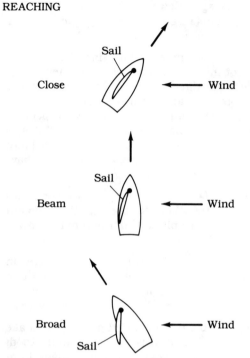

See *capacitive reactance, impedance, inductive reactance.*

**reaction** 1. In chemistry, the combining of elements or compounds to produce new compounds. Often, energy is gained or lost as well. 2. Nuclear decay, fission or fusion. See *fission, fusion.* 3. In physics, the force that exactly opposes an action. See *action-reaction.* 4. The behavior of an organism, in response to a certain stimulus.

**reaction engine** An engine that produces thrust by means of the action-reaction principle, rather than by driving wheels or propellers. Jets and rockets are examples. See *action-reaction, jet engine, rocket engine.*

**reactive** 1. In chemistry, pertaining to a substance that reacts easily with other substances. Chlorine is an example of a reactive element. 2. Containing reactance. See *reactance.*

**reactor** 1. See *nuclear reactor.* 2. An inductor or capacitor, or an equivalent circuit. See *capacitive reactance, inductive reactance, reactance.*

**real image** The image obtained by focusing from a convex lens or a concave mirror. This is the kind of image that falls on a camera film or on the retina of your eye. It is always inverted (upside down) relative to the actual object or scene. See *virtual image.*

**real number** Any rational or irrational number. According to the contiuum hypothesis, the set of real numbers can be paired off one-to-one with the points on a geometric line or line segment. See *irrational number, rational number.*

**real time** Pertaining to a system in which data is processed as it becomes available. Events are acted on as they occur. This applies especially to computer systems.

**reaper** A harvesting machine invented by Cyrus McCormick in the 1830s. It revolutionized farming, although it took a few years for the device to become widely accepted.

**Reber, Grote** An electronics engineer who was interested in radio astronomy as a hobby. He built a 31-foot parabolic dish antenna in his back yard. His work led, in the 1940s, to the construction of large radio telescopes. See *radio astronomy, radio telescope.*

**receiver** 1. In communications, a device that converts electromagnetic energy, usually in the radio spectrum, into audible and/or visual signals. 2. The handset of a telephone unit.

**reception** 1. The conversion of radio signals into audible and/or visual signals. 2. The quality of a signal at the output of a receiver. See *receiver.*

**receptor** 1. A sensor; anything that receives an impulse or a signal. 2. A sensory pick-up device in the body. Nerve endings are a good example. Taste buds in the tongue, and the rods and cones in the eye, are other examples.

**recessive trait** A reproductive trait that tends to appear less often than would be dictated by pure chance. These traits "lose out" to dominant traits. Straight hair, fair skin and blue eyes are examples of recessive traits in humans. See *dominant trait.*

**reciprocal** 1. For a given nonzero number x, the value 1/x. 2. Opposite to. 3. Existing in pairs.

**reciprocity theorem** A rule for the behavior of electrical circuits. It concerns the relationship between current, voltage and resistance. In general, if two different circuits behave in exactly the same way under all conditions in a given application, they can be thought of as being identical circuits for that application.

**recombination** 1. The return of equilibrium to a semiconductor when the voltage is removed. A charge causes an excess of either electrons or holes; this excess disappears when the charge is taken away. 2. A change that takes place in the genetic makeup when an egg is fertilized. This is why children never look exactly like their parents.

**recombination time** In a semiconductor material, the time required for equilibrium of charge carriers to be restored, after a voltage is removed. See *recombination*.

**recommended daily allowance** The optimum amount of a nutrient for a person per day, as determined by the United States Food and Drug Administration. See the RECOMMENDED DAILY ALLOWANCES appendix.

**reconnaissance** The gathering of intelligence, especially by means of sophisticated observing devices. Electronic spying is a good example. Satellites and aircraft are used extensively for this purpose. By this means, we can verify to what extent another country is honoring a weapons-control treaty, for example.

**recording disk** A disk, often magnetic and sometimes optical or mechanical, on which audio, video or digital information can be stored for future use. The advantage of the disk over a tape is that the data can be accessed more quickly. No two bits of data are separated by more than the diameter of the disk. But there is a limit to the amount of data that can be stored in a given volume. This can be a problem with complicated signals such as video. See *recording tape*.

**recording tape** A long strip of material, usually magnetic, on which audio, video or digital information can be stored for future use. Accessibility is more difficult with tape, as compared with disk, because the tape must be wound or rewound, sometimes for long distances, to get at specific data bits. See *recording disk*.

**rectification** 1. In electronics, a process of generating direct current from alternating current by cutting off or inverting one-half of the cycle. 2. Correction of a malfunction.

**rectifier** 1. A semiconductor diode designed for alternating-current rectification. 2. An electronic circuit that causes rectification of a current or signal. See *rectification*.

**rectum** The chamber at the end of the large intestine, into which feces pass just before they are expelled from the body.

**redbed** Rocks of an early age in the earth's geologic history that formed on the land. They are red because of their iron-oxide content. This oxide proves that the atmosphere of the earth contained oxygen at an early time.

**red blood cell** Also called an erythrocyte. There are about 700 red cells for every white blood cell (see *white blood cell*). The erythrocytes carry oxygen to the tissues of the body, and also carry carbon dioxide away from the tissues to be exhaled by the lungs. They are manufactured in bone marrow and live about three months.

**red dwarf** A small star with a spectrum whose greatest intensity is in the visible red range. This is the kind of star that forms when there is just barely enough hydrogen mass to start fusion at the core. They are stable and live the longest of any stars, perhaps up to 20 billion years. If the planet Jupiter was a little bigger, it would have become one of these type stars.

**red giant** A very large star, usually having evolved from a sun-like star when hydrogen fuel starts to run out. The greatest intensity of the spectrum is in the visible red or infrared range. Our sun will eventually become such a star, possibly getting so large that it envelops the orbit of the earth.

**red shift** A change in the wavelength of electromagnetic radiation, especially visible light, coming from distant celestial objects. The spectral lines move toward the longer wavelengths, or the red end of the visible spectrum (see drawing). This occurs when an object is moving away from us. It can also take place when the radiation passes through a strong gravitational field. All distant galaxies and quasars show a red shift that is proportional to their distance from us; this has been taken as proof that the universe is expanding. See *absorption line, Big Bang, blue shift, Hubble constant*.

RED SHIFT

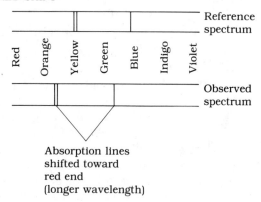

Absorption lines shifted toward red end (longer wavelength)

**reduce** 1. To make less in quantity; to minimize. 2. To lose body fat by restricting calories and/or exercising. 3. To simplify.

**reductio ad absurdum** In logic and mathematics, a method of proving theorems. Suppose we want to prove that statement X is true. We assume that X is false, that is, that −X is true. From this we derive a contradiction. Therefore, we know that −X is false, and, by double negatives, that X is true.

**reduction** 1. The process of reducing. See *reduce*. 2. An obsolete term for removal of oxygen from a chemical. 3. In chemistry, a reaction in which electrons are acquired by a substance. Thus the charge becomes more negative; it is "reduced."

**reef** A buildup of coral, found along some continental shelves, but more often around seamounts and atolls. These coral structures are home to many different kinds of life. The structure may moderate large waves such as storm surges and tsunamis, providing some protection to shore dwellers.

**reflectance** 1. For radiant energy striking a reflecting surface, the ratio of the reflected energy to the incident energy. Can be as low as zero or as great as 1. Also can be expressed as a percentage. For visible light it is called albedo (see *albedo*). 2. In an electronic transmission line, the ratio of actual current to optimum current. It is an expression of the mismatch between the line and the load.

**reflected wave** 1. The electromagnetic or sound wave that is not absorbed by, and not transmitted through, an object. 2. The electromagnetic wave not accepted by the load at the termination of a transmission line.

**reflecting telescope** See *telescope*.

**reflection** A condition in which energy is neither absorbed by, nor transmitted through, an object. This is usually partial, although with certain devices designed for the purpose, it can be near 100 percent.

**reflection law** Usually stated, "The angle of incidence is equal to the angle of reflection." This holds for a smooth reflecting surface. If the surface is not flat, the angles are determined with respect to a plane tangent to the surface at the point(s) of incidence. See the drawing.

REFLECTION LAW

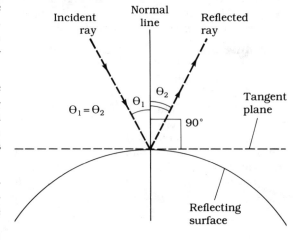

**reflective** Pertaining to a surface that reflects most or all of the energy striking it at certain wavelengths. See *reflection*.

**reflector** A surface deliberately designed to reflect all, or almost all, of the energy that strikes it. Some devices of this kind can focus energy for certain purposes. See, for example, *concave mirror, parabolic reflector.*

**refracting telescope** See *telescope.*

**refraction** A change in the direction of electromagnetic wave propagation, because of a change in the nature of the medium through which the wave travels. The drawing shows an example of the visible effect of refraction in a pool of water. The rod appears bent because the light from under water changes direction when it passes through the pool surface. See *index of refraction.*

REFRACTION

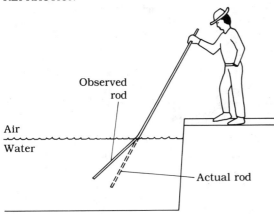

**regeneration** 1. Positive feedback in an electronic circuit. See *feedback.* 2. Replacement of body cells following an injury. An example is the healing of a cut.

**register** An electronic memory circuit designed for storage of information for short periods of time. There are various different configurations. See, for example, *first-in/first-out, pushdown stack.*

**regression** 1. The opposite of progression; a falling back, a return to more primitive conditions. 2. In a species, a tendency to become less,

rather than more, sophisticated and advanced with the passage of time. This might happen to humanity following a general nuclear war, for example. 3. Movement in a backwards direction, or opposite to the normal flow. 4. In geology, the retreat of the sea from the land.

**regulation** 1. The process of ensuring that an electronic circuit operates within certain limits. The voltage or current may be kept constant; these are voltage regulation and current regulation, respectively. 2. The adjustment of a quantity to maintain certain conditions. We might, for example, control the flow of water through a pipeline.

**relation** 1. In mathematics, a mapping of one set of values onto another set. This is not always a one-to-one correspondence; a single value x in set X might have two corresponding values, $y_1$ and $y_2$, in set Y. Or there might be three, a hundred, or even infinitely many y values for a single x value. A function is a special type of relation. See *function.* 2. A sibling, parent, child, grandparent, grandchild, uncle, aunt, or cousin of an individual in an animal species.

**relative** 1. Depending on point of view. For example, to a person from northern Alaska, 70 degrees Fahrenheit might seem very warm; to someone from the Solomon Islands it would be cool. See *absolute.* 2. An individual of a species that is a close relative of another individual of the same species. See *relation.* 3. A species that is similar to, and evolved along the same lines as, another species.

**relative humidity** For air at a given temperature, the proportion (as a percentage) of water vapor in the air, as compared with the maximum amount of water vapor the air can hold without condensation taking place. This varies from zero (no water vapor) to 100 percent (condensation occurs).

**relativity** A theory developed by Albert Einstein in the early twentieth century. The special theory evolved from the postulate that light should seem to travel at the same speed—about 186,000 miles per second—no matter what the viewpoint of the observer. But this theory did not take accelerating reference frames into account. That was done by Einstein later, when he formulated his general theory. Some of the effects predicted by relativity, such as time dilation and the bending of light by gravity, have been observed and measured.

**relay** 1. In electronics, a device that acts as a remote-control switch. It uses a magnetic coil to open and close a set of contacts. 2. To receive a signal and then send it again. This used to be done by human operators in long-distance telegraph and radiotelegraph circuits. Nowadays it is done by repeaters and satellites.

**reliability** 1. The probability that, after a given length of time in use, a device will still be working. Measured as a percentage from 0 to 100. 2. A general expression for the ability of a machine to continue working for long periods with minimal maintenance.

**reluctance** For magnetic fields, the equivalent of resistance; the opposition that a medium offers to a magnetic field.

**REM** Abbreviation for rapid eye movement.

**rem** Acronym for radiation equivalent man. A unit of ionizing radiation that takes into account the differences in effect for different kinds of radiation. One rad of alpha rays can cause several times as much damage as one rad of X rays. In general, the shorter the wavelength, the more rems per rad; also, high-speed heavy particles do more damage than X rays. See *rad*.

**remanence** 1. Magnetic flux that stays in a ferromagnetic substance after the magnetizing force is removed. 2. The flux density that remains in a ferromagnetic substance after the magnetizing force is removed.

**remanent magnetism** Permanent magnetization in certain ferromagnetic rocks in the earth. Such rocks, called lodestones, behave as small permanent magnets independently of the geomagnetic field.

**remote control** Operation of a machine from some location away from the machine itself. This can be accomplished by radio and television, or by wire. In some cases visible light, infrared, or ultraviolet can be used. Examples range from the armchair television control unit (using infrared, usually) to "smart" missiles that can be piloted via television to their targets.

**remote sensing** 1. The ability to tell something about an object without being in direct contact with it. 2. The recording and analysis of radio waves, infrared, visible light, ultraviolet, and X rays from a distance.

**Renaissance** A period of enlightenment in science and the arts, following about 1000 years during which religious belief controlled all aspects of life in European society. With respect to science, this period began in the sixteenth century with the rise of experimental science. Before this time, theories were often simply made up by aristocrats, and nonbelievers were ridiculed or even punished.

**renal** Pertaining to the kidneys or to kidney function.

**renal artery** The artery that supplies blood to the kidney. In the human being there are two of them, branching off from the aorta, one to each kidney. See *kidney, renal vein*.

**renal vein** The vein that returns blood from the kidney to the inferior vena cava. In the human being there are two of these veins, one for each kidney. See *kidney, renal artery*.

**renewable resource** 1. A resource whose supply is infinite for all practical purposes. Geothermal and solar energy are examples. 2. A resource that can be replenished through responsible management. Wood is an example. See *nonrenewable resource*.

**rennet** An enzyme used to curdle milk for making cheese. Obtained from the stomach of a cow. Also sometimes called rennin.

**repeater** A device that receives a signal via wire, fiberoptics, or radio and retransmits it, usually at the same time. See the drawing.

**repeating decimal** A number written in decimal form, such that a sequence of digits is repeated infinitely many times. Examples are $1/3 = 0.3333...$ and $125/999 = 0.125125125...$ .

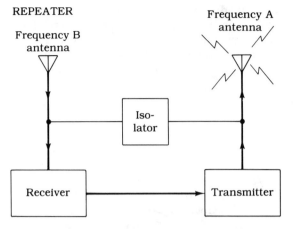

REPEATER

**reproduction** 1. The process by which animals produce offspring, or by which plants produce new plants. 2. Cell division. 3. A recording of audio, digital, or visual data. 4. The quality of recorded audio, digital, or visual data.

**reproductive system** The organs responsible for producing offspring, particularly in animals. See *ovary, ovum, sperm, testicles, uterus, zygote.*

**reptile** A cold-blooded vertebrate that moves on short legs or by crawling on its belly. Familiar examples include lizards, snakes, and alligators. For their place in the animal kingdom, see the ANIMAL CLASSIFICATION appendix.

**research** 1. The gathering of data to test, modify, or formulate a theory or to solve practical problems. 2. Theoretical exploration of a subject or phenomenon.

**reservoir** 1. A body of water, sometimes naturally formed but often artificial, that is used for generating hydroelectric power and also as a supply for public use. 2. Any quantity of liquid that is held as a supply for some purpose.

**resistance** 1. The opposition that a substance offers to the flow of electric current. Measured in ohms. 2. The opposition that a substance offers to the conduction of heat. 3. The opposition that a system offers to a force. 4. See *friction.*

**resistivity** Electrical resistance per unit length, area, or volume. Given in ohms per meter, ohms per square meter, or ohms per cubic meter. See *resistance.*

**resistor** An electronic component designed to offer a specific amount of resistance to the flow of current. These devices are manufactured in a variety of ways. Some are variable. See *potentiometer, resistance, rheostat.*

**resolution** 1. The ability of a device to distinguish between two objects sensed. Applies to telescopes, radio telescopes, radar, and microscopes in particular. Also called resolving power. 2. The satisfactory solving of a problem.

**resolver** A device that senses the position of a component in a servo system. An example is a selsyn that might be used to show the direction in which a television antenna is pointed. The device resembles a motor.

**resonance** 1. In acoustics, a condition in which sound waves are reinforced, such as in an enclosure of a certain size. This enhances the volume at a certain frequency or frequencies. See *resonant frequency.* 2. In electronics, a condition where there are equal but opposite amounts of reactance, so that a circuit tends to store and release energy at a specific frequency or frequencies. 3. A condition in which an object is of such a size that vibrations are amplified. This can cause mechanical failure in an aircraft, for example. It might also occur at an extremely low frequency in the water of a bay during a severe storm, producing a flood.

**resonant frequency** The frequency or frequencies at which resonance occurs in acoustical or electronic circuits. These frequencies are often harmonically related. See *harmonic.*

**resource** A substance or commodity available from nature for human needs. See *nonrenewable resource, renewable resource.*

**respiration** A process that occurs in living organisms, in which energy is furnished from gases in the atmosphere. In humans and animals, the essential gas is oxygen. In some living organisms it is carbon dioxide. Some organisms can thrive without respiration as we know it.

**rest energy** The rest mass of an object, multiplied by the speed of light squared. That is, the energy that would result from an object at rest if it were all converted to energy. See *rest mass*.

**rest mass** The mass (see *mass*) of an object when it has zero velocity and zero acceleration with respect to the point of view of an observer. This mass increases at great speeds because of relativistic effects.

**resuscitation** A means of reviving a person whose breathing and/or heartbeat has stopped. The most common method is called cardiopulmonary resuscitation (CPR) and is taught by the American Red Cross. It involves a combination of heart massage and mouth-to-mouth breathing.

**retardation** 1. A condition of being inhibited in development. Underdeveloped brain function (mental retardation) is probably the most commonly thought-of example. 2. A process of holding back, or preventing the progress or movement of an object or system.

**retina** The part of the eyeball that contains the receptors for vision. In the human these are rods and cones. See *eye*.

**retinol** The form of vitamin A that is directly used by the body. It exists in many fish liver oils and also in the livers of most animals. See the VITAMINS AND MINERALS appendix.

**retro-** A prefix meaning backwards or reversed or to work backwards in space or time.

**retrograde** Moving in the opposite direction from the norm. For example, most planets rotate counterclockwise as seen from above their North Poles. But Venus rotates clockwise, so its rotation is said to be retrograde. See *retrograde motion, retrograde orbit*.

**retrograde motion** The occasional backward movement of certain planets relative to the distant stars. This causes the path of the planet to describe a small loop. The Ptolemaic model explained this using epicycles. But it is easily shown with the heliocentric theory. See *epicycle, geocentric theory, helicentric theory, Ptolemaic model*.

**retrograde orbit** An orbit that takes a satellite in a path contrary to the rotation of the body around which it is going. In the case of a satellite around the earth, this would mean an orbit that goes from east to west (see drawing).

RETROGRADE ORBIT

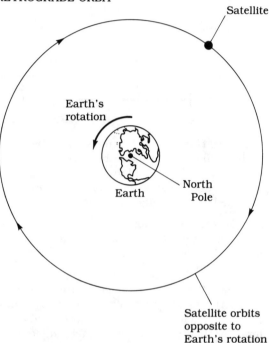

Satellite orbits opposite to Earth's rotation

**retro rocket** A small rocket engine that slows an orbiting spacecraft down for reentry into the earth's atmosphere, or for descent to the moon or another planet.

**reverberation** 1. In acoustics, the reflection of sound from surfaces in an enclosure; echoing. 2. In audio electronics, a process in which echoing is simulated. 3. Resonance in an acoustic device. See *resonance*.

**reverse** 1. A condition of moving backwards, or contrary to the usual direction. 2. In electronics, a term used to indicate reflected energy along transmission lines. 3. To cause to move in a backward direction, or contrary to the norm.

**reverse fault** A crack in the earth's crust resulting from compression of the crust. This causes rocks to slide along each other, and earthquakes are the result. See *normal fault*.

**reverse osmosis** A method of desalinating seawater. It is done under high pressure. See *active absorption, osmosis*.

**revolution** 1. Motion in a circle or ellipse around some central point. 2. One complete cycle of such movement, so that the object gets back to its starting point. 3. A radical change in ways of scientific thought or procedures.

**revolutions per minute** A measure of angular speed, given as the number of 360-degree cycles of a rotating or revolving object in one minute of time. This is equal to 9.54 times the angular speed in radians per second.

**revolutions per second** A measure of angular speed, given as the number of 360-degree cycles of a rotating or revolving object in one second of time. It is equal to 0.159 times the angular speed in radians per second.

**revolve** 1. Synonym for orbit, in the sense that an orbiting object also revolves. 2. To move in a circle or ellipse around some central point.

**Reye's syndrome** A disease condition that sometimes occurs as a complication of a bad cold or flu. The brain and nervous system are affected. Delirium may occur. There are also many other possible symptoms. The disease can be fatal. Children should not be given aspirin or aspirin-containing medications when they have a bad cold or the flu, because this appears to increase the chances of this syndrome developing.

**RF** Abbreviation for radio frequency.

**Rh** Chemical symbol for rhodium.

**rhenium** Chemical symbol, Re. An element with atomic number 75. The most common isotope has atomic weight 187. It is a metal and occurs naturally along with molybdenum. This element is being used in superconductivity experiments. See *superconductivity, superconductor*.

**rheostat** A variable resistor made with a doughnut-shaped coil of resistive wire. A sliding contact, connected to a rotatable shaft, adjusts the number of turns of wire that are placed in the circuit, and this controls the resistance. In electronics, potentiometers are more often used. See *potentiometer*.

**rheumatic fever** A disease that involves inflammation of the joints. This disease is less common than it was a few decades ago. There are long-term complications, especially involving damage to the heart.

**rheumatism** Any of various diseases involving the joints and muscles. Characteristic symptoms are pain and inflammation. Examples are myalgia (muscular pain), myositis (muscular inflammation), and fibrositis (inflammation of connective tissue). Arthritis is sometimes called rheumatism, but this is technically a misnomer.

**Rh factor** A substance sometimes found in red blood cells. Some people have it, and their blood is called Rh positive; others lack it, and their blood is called Rh negative. Thus we might speak of a person having type A positive blood, or type O negative.

**rhizoid** In algae, liverworts, and certain other primitive plants a growth that assists in the absorbtion of nutrients, and also serves to hold the plant down. Similar to a root in larger plants.

**rhodium** Chemical symbol, Rh. An element with atomic number 45. The most common isotope has atomic weight 103. It occurs in nature along with platinum. It is used for various purposes in scientific work. In particular, it makes a good silvering for first-surface mirrors.

**rhyolite** A volcanic rock that is viscous when molten. It is ejected as pyroclastic. See *pyroclastic*.

**rib** Any of the bones connected to the spine and serving to house the lungs, heart, and other organs of the chest cavity and upper abdomen in vertebrates.

**rib cage** The structure formed by the ribs enclosing and protecting the organs of the chest cavity and upper abdomen in vertebrates.

**riboflavin** One of the B-complex of vitamins, also called vitamin B-2. See the VITAMINS AND MINERALS appendix.

**ribonucleic acid** A compound in organisms that is essential in the manufacture of proteins. It is made in cell nuclei. It resembles deoxyribonucleic acid (DNA) in structure, being a long molecule. The substance is responsible for genetic information contained in cells. RNA is sometimes called the "memory compound" because learning capacity can be transferred from one animal to another by removal and reinjection. See *deoxyribonucleic acid*.

**ribose** A simple sugar that is a part of ribonucleic acid (RNA). See *ribonucleic acid*.

**ribosome** A part of a cell that builds protein. A typical cell has many of them, found mostly in the cytoplasm. See *cell*.

**Richter scale** A means of measuring the intensity of earthquakes. See the RICHTER SCALE appendix.

**Rickenbacker, Edward** A famous aviator in World War I, and a major contributor to the development of aviation as a mode of transportation. He was also known for his bravery and physical endurance. He served the military as an advisor in World War II.

**rickets** A disease that occurs when children do not get enough vitamin D. Soft bones result from this deficiency, and this causes a characteristic bowing of the legs. The jaws might not develop enough to allow normal tooth formation. This disease is very rare today in civilized countries.

**rickettsia** A microorganism that is like a bacterium in some ways, and like a virus in other ways. They are carried by ticks, which in turn are carried by various animals such as rats. Rocky Mountain spotted fever, a potentially fatal disease, is caused by one variety of this microorganism.

**ridge** 1. A long cliff, such as might be formed when a large section of the earth's crust slides upward or downward along a fault. 2. A mountain range that forms on the ocean floor as it spreads outward; an example is the mid-Atlantic ridge. See *midocean ridge*.

**Riemann, Bernhard** A nineteenth-century mathematician. He is noted for his work in geometry of more than three dimensions. Einstein used his work in his general theory of relativity. He also did work in the theory of functions of complex variables.

**Riemannian geometry** A refined form of non-Euclidean geometry. Can involve any number of dimensions, not being limited to the three-spaces we can visualize. We might have four-sphere spaces; some scientists think our universe is of this shape. Or we might have saddle-shaped objects in seven dimensions. The possibilities are beyond imagining. See *non-Euclidean geometry*.

**Riemann surface** A continuum in Riemannian geometry. An example is the surface of a four-sphere; it has three dimensions and is finite yet unbounded. See *non-Euclidean geometry, Riemannian geometry*.

**riffle** An accumulation of small rocks in a stream. When the water level is low, the water speeds up when it flows over these rocks.

**rift valley** A system of faults that occurs when a continent begins to break apart. A large formation of this kind is found in eastern Africa. Volcanoes occur frequently in such zones. Eventually, in the case of eastern Africa, the valley will drop below sea level, and a new sea, similar to the Red Sea, will form.

**rift volcano** A type of volcano that gets its source material from spreading ridges. These volcanoes create underwater basalt flows, amounting to about 2.5 billion (2,500,000,000) cubic yards every year.

**right ascension** The angle along the celestial equator between the vernal equinox and the longitude of a given object. It is a form of measuring celestial longitude. See *celestial coordinates, celestial sphere, declination*.

**right-handed screw** A screw that goes in (tightens) when it is turned clockwise. This is the most common type of screw. See *left-handed screw*.

**ring** 1. Two concentric circles and the region in between them. 2. A short section of a cylinder. 3. One of the bands of fine particles orbiting Saturn, and also Uranus and Jupiter, in almost perfect circles. 4. A set whose elements have certain properties. The set is closed for two distinct operations; associativity and commutativity hold; one operation distributes over the other. See a text on theoretical algebra.

**ringback** In a telephone system, the impulses sent back along the line when a call is made, so that the caller can tell that the distant phone set is ringing.

**ringdown** In a telephone system, the impulses sent along the line that cause the distant set to ring.

**Ring Nebula** A celestial object in the constellation Lyra. It is a spherical cloud of gas surrounding a very active star. The gas glows because of radiation from the star. When we look at it through a telescope, we see it as a ring-shaped object with the star at the center.

**ring of fire** A geologically active region of the earth, following the rim of the Pacific Ocean (see drawing). There are numerous earthquakes and volcanoes in this region. There are frequent tsunamis on the shores around the Pacific.

RING OF FIRE

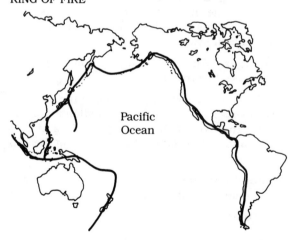

**ring theory** In theoretical algebra, the formal set of axioms, definitions and theorems concerning sets known as rings. See *ring*.

**rise** 1. An increase in level or amplitude. 2. To increase in level or amplitude. 4. The beginning part of a pulse or wave cycle, starting at zero and going to the maximum level. 4. A region of the ocean floor that is less deep than the surrounding region(s).

**rise time** 1. The time between the start of a pulse and its reaching maximum level. 2. The time between the start of a pulse and its reaching a certain percentage of maximum, such as 63 percent (electric time constant) or 99 percent (practically to the maximum). See *time constant*.

**r/min** Abbreviation for revolutions per minute.

**RMS or rms** Abbreviation for root mean square.

**Rn** Chemical symbol for radon.

**RNA** Abbreviation for ribonucleic acid.

**robot** 1. An electronically controlled mechanical device that can perform a specific function. 2. An electronically controlled set of machines that is built in the general form of a human being and that can do some of the mechanical jobs of a human being.

**robotics** The branch of engineering concerned with the design of robots. See *robot*.

**rock** Any of various materials in the crust or mantle of the earth. They contain many different possible combinations of elements and compounds. See *rock formation* and the ROCKS appendix.

**rock formation** 1. A region of the earth's crust, in which a certain type of rock predominates. 2. The process by which rocks are created. They may form from compressed deposits (sedimentary), from volcanic magma (igneous), or from earlier rocks that have been subjected to great pressures or stresses (metamorphic). See the ROCKS appendix.

**rocket** A vehicle that can travel through air or through outer space. It does not require an external source of oxygen to burn its fuel. These vehicles can carry weapons, launch satellites into orbit, and propel people to other planets. The vehicle might have fins for stabilization;

gyroscopes also can be used (see *gyroscope*). A typical rocket is built as shown in the simplified drawing.

ROCKET

A cross-section of a V-2 rocket: 1-war-head, 2-guidance system, 3-alcohol tank, 4-liquid oxygen tank, 5-turbine and pumps, 6-combustion chamber.

From Erickson, EXPLORING EARTH FROM SPACE (TAB Books, Inc., 1989).

**rocket engine** A thrust device that works without the need for external oxygen. Thus, it will operate in the near vacuum of outer space. The oxygen is carried along in the vehicle in liquid form, highly condensed. See the drawing. See *rocket*.

**rocket plane** An aircraft that employs rocket engines. This might someday make it possible for an aircraft to take off from a runway, fly into the upper atmosphere, and then accelerate enough to go into orbit.

**rockslide** A massive breaking-off of rock, especially on a mountainside or cliff. It gains momentum as falling rocks knock off more rocks. These can be dangerous to people on mountain roads. Such catastrophes may be precipitated by earthquakes, a heavy rain, or by human-made explosions.

**rod** 1. A solid, cylindrical object. 2. A receptor in the retina of the eye, highly sensitive to light but not responding to color.

ROCKET ENGINE

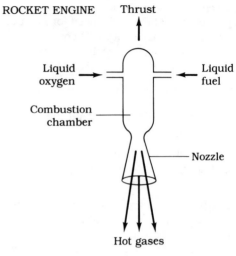

(Functional diagram)

**rodent** Any of various small mammals with powerful jaws and sharp teeth. Examples are mice, rats, chipmunks, and squirrels.

**Roe, Anne** A psychologist who studied the traits of scientists during the 1950s. She found that they have personalities similar to artists.

**Roebling, John** A German-American engineer of the nineteenth century. He is best known for his work in designing and building suspension bridges.

**roentgen** An obsolete unit of ionizing radiation dose. See *rad, rem*.

**roll** 1. A rotational motion along the axis of a cylinder. 2. In an aircraft, rotational motion along the axis of the fuselage. 3. In a ship, a back-and-forth rotational motion along the axis from bow to stern.

**Roman numerals** A method of writing number values used in ancient times, notably in the

ROMAN NUMERALS

| Symbols | Examples |
|---|---|
| I = 1 | III = 3 |
| V = 5 | IV = 4   VII = 7 |
| X = 10 | IX = 9   XXII = 22 |
| L = 50 | XL = 40   LXI = 61 |
| C = 100 | XCII = 92   CCI = 201 |
| D = 500 (not often used) | CD = 400 |
| M = 1000 | MCMLXXIII = 1973 |

Roman Empire. A history text should be consulted for details of the way in which the symbols were arranged, because the rules were rather complicated. The actual symbols, and a few examples, are shown in the table. There was no symbol for zero.

**root** 1. The part of a plant beneath the soil that gathers water and other nutrients. 2. A solution to an equation. 3. A number (x) that, when multiplied by itself an integral number (n) of times, yields a given result (y). We then say that (x) is the (nth) root of (y). The second root is called the square root; the third root is called the cube root. Roots of order 4 or more are called simply by their integer number, such as the tenth root.

**root mean square** Abbreviation, RMS or rms. A way of expressing the effective value of an alternating-current (ac) wave. If the peak value for an ac sine wave is E, then the RMS value is given by 0.707E. For complex waveforms the function is usually different.

**root system** The underground part of a plant, that gathers water and nutrients. This part of the plant may or may not resemble a "mirror image" of the part above the ground. Often the underground part of a plant is as large as, or larger than, the part above the ground.

**rotary dialer** An old-fashioned method of dialing a telephone unit. The device consisted of ten holes in a disk, one for each digit 1, 2, 3, ..., 9, 0. The numbers were dialed by placing a finger in the appropriate hole and turning the disk. The disk would rotate back, interrupting the circuit. These devices are still used in some areas of the United States, but have largely been replaced by Touchtone dialers. See *Touchtone dialer*.

**rotary engine** An engine first designed by Charles Manly for use in early aircraft. The cylinders were arranged in a circle, instead of in-line with each other. The pistons drove a crankshaft at the center of the circle. This type of engine provided more power for its weight than the conventional type. But it also had more air resistance because of its shape.

**rotation** Twisting motion of a solid or semi-solid object around a point or line axis. Examples are the turning of the earth (23 hours and 56 minutes per complete turn relative to the distant stars) and the turning of a phonograph disk. See *revolution*.

**rotational force** 1. See *Coriolis effect*. 2. See *centrifugal force*.

**rotifer** A tiny, primitive, water-dwelling creature with cilia. The cilia, or microscopic hairs, move in such a way that the creature seems to rotate. See the ANIMAL CLASSIFICATION appendix for the place of the rotifers on the evolutionary scale.

**roughage** See *dietary fiber*.

**roulette** See *cycloid*.

**RPM** Abbreviation for revolutions per minute.

**RPS** Abbreviation for revolutions per second.

**r/s** Abbreviation for revolutions per second.

**Ru** Chemical symbol for ruthenium.

**rubella** Also called German measles. A disease that often occurs in childhood. The main symptom is a skin rash. It is rarely serious in children, but in an adult, especially a pregnant woman, it can be. Babies born to women with this disease can have a variety of birth defects. Immunizations are available.

**rubeola** Also called measles or red measles. An acute, highly contagious viral disease. The main symptoms are fever, cough, and skin rash. Pneumonia sometimes occurs as a complication. Vaccines are available. After a person has had the measles, immunity is usually permanent.

**rubidium** Chemical symbol, Rb. An element with atomic number 37. The most common isotope has atomic weight 85. In its pure form it is a soft metal. It reacts very easily with elements such as oxygen and chlorine.

**ruby** A mineral corundum, with small amounts of chromium. The clear variety is useful in the manufacture of a certain type of laser (see *ruby laser*). It is also regarded as a gem.

**ruby laser** A special type of laser that produces coherent light in the red part of the visible spectrum (6943 Angstroms). Generally, aluminum oxide is used, with a small amount of chromium added, to make the crystal. The energy output is pulsed, and the efficiency is low. Typical ruby lasers produce just a few milliwatts of power. See the drawing. See also *laser*.

RUBY LASER

![Ruby laser diagram with helical flash tube, ruby crystal, 100-percent reflector, 95-percent reflector, and output]

From Gibilisco, UNDERSTANDING LASERS (TAB Books, Inc., 1989).

**rudder** The mechanism that enables a boat or ship to be steered. It works by deflecting the water. It is located at the stern of the boat or ship.

**running** 1. In sailing, moving in a direction exactly, or nearly, downwind (see drawing). 2. The execution of a program in a computer system.

RUNNING

**Russell, Bertrand** A well-known mathematician and philosopher who did his work during the latter part of the nineteenth century and the early part of the twentieth century. Along with Alfred North Whitehead, he wrote a massive book *Principia Mathematica* (Principles of Mathematics), in which he elaborated on his belief that all of the results of mathematics can be proven from just a handful of axioms.

**Russell, Henry** An American astronomer who, along with Ejnar Hertzsprung, found that the absolute brightness of stars usually follows a direct relationship with their surface temperatures, and therefore with their colors. See *Hertzsprung-Russell diagram*.

**Russell's Paradox** Best stated this way: "Suppose Sal, the barber, shaves all people, but only the people who do not shave themselves. Then who shaves Sal?" This question leads to a contradiction no matter how it is answered. A more sophisticated, general version of this paradox was demonstrated by Kurt Godel in 1930. See *Incompleteness Theorem*.

**rust** Corroded iron, in the form of iron oxide, sometimes along with iron chloride. Unprotected iron and iron alloys will corrode over a period of time when exposed to the atmosphere. Moisture, and especially salt spray from the ocean, accelerates the process.

**rustproofing** The coating of iron, or iron alloys, to protect the metal against rusting. There are various ways to do this. Special paints can be used; zinc or some other corrosion-resistant metal may be employed as a plating.

**ruthenium** Chemical symbol, Ru. An element with atomic number 44. The most common isotope has atomic weight 102. In its pure form it is a hard metal. It is used in certain alloys.

**Rutherford atom** An early, simplified model of atoms of matter developed by the physicist Ernest Rutherford in 1912. He was the first to suggest that atoms are like miniature solar systems, except that the attractive force is not caused by gravity, but by opposing electrical charges. See the drawing. Niels Bohr revised Rutherford's theory a year later. See *Bohr atom*. See also *Bohr, Niels* and *Rutherford, Ernest*.

**Rutherford, Ernest** A physicist of the early twentieth century. He is remembered for his model of the atom (see *Rutherford atom*). He also was the first to explain why radioactive materials decay. He was a chemist early in his career, winning a Nobel prize in this field in

RUTHERFORD ATOM

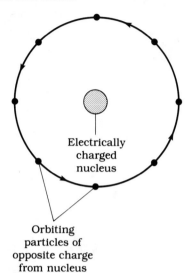

**Rydberg Constant** In atomic physics, a constant having to do with the energy that keeps electrons bound to nuclei. It is a sophisticated concept; a text on atomic physics is recommended for further study.

**Ryle, Martin** A radio astronomer who was among those to pioneer the interferometer, a type of radio telescope that made it possible to observe the radio sky in great detail. See *interferometer, radio astronomy.*

1903. He directed the Cavendish Laboratory at the University of Cambridge, England for a time.

**S** Abbreviation for siemens, chemical symbol for sulfur.

**s** Abbreviation for second.

**sac** Any small enclosure or gland, usually holding fluid, and with one or more ducts through which the fluid can go out or in.

**saccharide** A sugar or starch molecule. Sugars are called monosaccharides and disaccharides; more complex carbohydrate molecules are called polysaccharides. See *disaccharide, monosaccharide, polysaccharide.*

**saccharin** A substance used as an artificial sweetener. It has practically no calories. The most common forms are sodium saccharin and calcium saccharin. At one time, this was widely used in place of sugar. However, new artificial sweeteners have largely replaced it in recent years.

**sacral nerves** In the human body, a set of nerves in the lumbosacral plexus, branching off from the end of the spinal cord, located around the sacrum. These nerves occupy the region of the buttocks. See *sacrum*.

**sacrum** The bone at the lower end of the vertebral column to which the hipbones are attached. The extreme end of the sacrum is attached to the small coccyx, what remains in humans of a tail.

**safety engineering** 1. A branch of industrial engineering, concerned with the safety of personnel in an assembly plant. 2. The practice of ensuring that devices are as safe as possible to use.

**safety factor** A margin that is typically incorporated into the maximum ratings for a device or machine. This allows for the possibility that someone may use the device or machine at, or near, the specified limits. For example, a transistor might be rated at 100 watts maximum continuous power; actually, 90 percent of them would work at 120 watts.

**Saffir-Simpson scale** An intensity scale for expressing the severity of hurricanes. There exists a close relationship between central barometric pressure and wind speed, regardless of the diameter of the storm. See the illustration. "Minimal" hurricanes are called category 1; intensity ranges all the way up to the worst storms of category 5. See *hurricane*.

**Sagan, Carl** A well-known astronomer and author of the best-selling book *Cosmos*. He is

## SAFFIR-SIMPSON SCALE

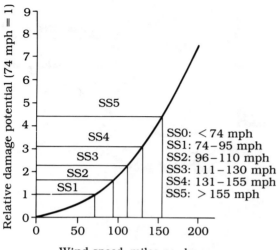

SS0: <74 mph
SS1: 74–95 mph
SS2: 96–110 mph
SS3: 111–130 mph
SS4: 131–155 mph
SS5: >155 mph

also an environmentalist, warning of various hazards that might be caused by the irresponsible behavior of humans.

**sailplane** See *glider*.

**Saint Elmo's fire** An electrostatic discharge that often is seen during thundershowers or during unsettled weather. Large metal objects with sharp points are most likely to have this glow at their ends. The name comes from centuries ago when it was seen at the ends of sailing ships' masts at night.

**Sakharov, Andrei** A Russian physicist of the twentieth century, and a well-known dissident. He made worldwide news by his defiance of the Soviet system of government. Russia has produced many scientific and literary geniuses.

**sal ammoniac** Ammonium chloride. In pure form it appears as white crystals. It is used in dry cells.

**salicylate** Any substance that is made using salicylic acid, a compound found in certain plants. Anti-inflammatory medications are commonly made with this compound or derivatives of it. Aspirin is one example; it is technically called acetylsalicylic acid. Certain medicines for diarrhea and nausea are another, using bismuth subsalicylate.

**saline** Containing, or comprised mainly of, a salt compound. Of course the most often thought-of is sodium chloride, but there are many other salts, such as sulfides and iodides. See *salt*.

**salinity** The extent to which water contains salt. Generally, this term is used in reference to natural salt water, and is compared with the average saltiness of seawater.

**saliva** A liquid produced by glands near the mouth, and secreted into the mouth. Amounts are increased when food is eaten. Enzymes in the liquid start the process of breaking down food for digestion.

**salivary gland** Any of the glands near the mouth, that secrete saliva. See *saliva*.

**salmonella** Any of about 1400 similar strains of bacteria, some of which are harmful to humans. These bacteria can cause various types of infections, including typhoid and gastroenteritis. See *salmonellosis*.

**salmonellosis** An infection with salmonella (see *salmonella*), especially of the digestive system. This can occur when contaminated food is eaten. The resulting gastroenteritis produces nausea and diarrhea, often severe, with accompanying dehydration and fever. This problem tends to occur in outbreaks, and can be a public health danger.

**salt** 1. Any compound that results when, in an acid, the hydrogen is replaced by a metallic element or compound. For example, hydrogen chloride (hydrochloric acid) might react with calcium to form calcium chloride, a salt. 2. Table salt, or sodium chloride. 3. Table salt substitute, usually potassium chloride or a mixture of sodium chloride and potassium chloride.

**saltation** A process in which grains of sand move across a desert floor. A similar effect occurs on the ocean bottom. See the drawing.

**salt bed** The remains of a body of saltwater, after it has dried up. In the western United States, there are several such regions, also known as salt flats.

SALTATION

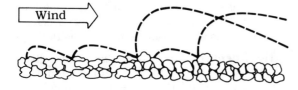

From Erickson, VIOLENT STORMS (TAB Books, Inc., 1988).

SAN ANDREAS FAULT

From Erickson, EXPLORING EARTH FROM SPACE (TAB Books, Inc., 1989).

**salt crystal**  The characteristic structure of a mineral salt. Sodium chloride, or table salt, is the most common example, with crystals roughly cubical in shape.

**salt lake**  A landlocked body of salt water. Examples are the Great Salt Lake in the western United States and the Dead Sea in Israel. These lakes can be much saltier than the oceans.

**samarium**  Chemical symbol, Sm. An element with atomic number 62. The most common isotope has atomic weight 152. In pure form it is a soft metal. It is used in alloys to make permanent magnets. It also absorbs neutrons, and is useful in nuclear reactors.

**Samios, Nicholas**  The scientist who first proposed the search for the omega minus subatomic particle. He instructed his team to look for certain particle tracks in a bubble chamber. He finally spotted the track himself in January, 1964. See *Eightfold Way*.

**sampling**  A means of selecting a representative cross section of a large number of individuals, for the purpose of statistical analysis. If we want to find out what proportion of the United States population has colon cancer, for example, we would test a certain number of people, not all the people in the entire country. But we would want to be sure that we chose a cross section that gave an accurate assessment.

**San Andreas fault**  A well-known system of faults in California (see drawing), responsible for many earthquakes in the region, some severe. Scientists predict that there will be a "big one" within a few years as the coastline of California slides northward along the fault line.

**Sanctorius**  An Italian physician of the seventeenth century who conducted one of the earliest experiments to find out what happens to food in the body. He weighed himself and his food together before eating the food, and then he weighed himself after eating all the food. He noticed that the total weight following the meal was less than the total weight before the meal. Exhaled moisture and carbon dioxide, and also evaporated sweat, accounted for the difference.

**sand**  Fine silicate particles found on the ocean floor and in deserts and many other places on the continents. It is used to make glass and optical fibers. It is also employed in construction, being part of concrete.

**Sandage, Alan**  An astronomer known for his work with unusual celestial objects. He was one of the first to observe a quasar; he also has theorized that X-ray stars are binary stars in close orbit around each other.

**sandglass**  An old-fashioned clock, used by seamen prior to the fifteenth century. Also called an "hourglass," although the time required for the sand to flow from top to bottom was $1/2$ hour. A bell was rung once at 12:30 P.M., twice at 1:00 P.M., and so on; this tradition is still used on ships today.

**sandstone**  Sand that has been compressed and cemented into rock by the weight of overlying materials. This is often found in regions that were once seafloor, such as certain parts of the Midwestern United States.

**Santos-Dumont, Alberto**  A Brazilian who flew an early aircraft in 1906 in France. He thought he was the first person to fly in a heavier-than-air craft. He did not know about the Wright Brothers' feat three years before. See *Wright Brothers*.

**sapphire**  A hard form of corundum, or aluminum oxide. Usually blue in color, but also can be various other shades. This is a precious stone and is used as a gem. It is also employed as the stylus in a phonograph transducer. See *stylus*.

**sarcoma**  A cancerous growth in bone, cartilage, muscle, or deep in the skin.

**sartorius**  A muscle in the thigh; the longest muscle in the human body. These muscles (one set in each leg) help in locomotion and balance.

**satellite**  An object that orbits around a heavier, central object. Usually, the central object is many times as massive as the satellite. Thus, the earth is a satellite of the sun; the moon is a satellite of the earth. There are thousands of artificial, human-made satellites also in orbit around the earth.

**satellite communications**  The use of repeaters, on board satellites in orbit around the earth, for radio and television communications. See the drawing. See *repeater*.

SATELLITE COMMUNICATIONS

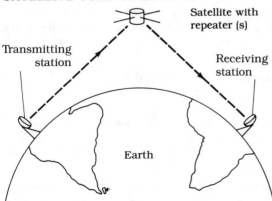

**satellite television**  Television broadcast in which the signals are sent through a satellite repeater. This is the way in which most overseas network links are established. See *repeater*.

**satiety center**  A part of the brain that gives you the sensation that you have had enough to eat. In some people this center does not operate well, and they might overeat or undereat, causing obesity or emaciation. See *appestat*.

**saturated**  1. For a given compound, a condition of having all of its possible bonds filled. A saturated fat, for example, has a hydrogen atom in every possible place. See *saturated fat*. 2. Holding as much water as possible. We might speak of a saturated sponge, or saturated air. See *relative humidity*.

**saturated adiabatic lapse rate**  The rate at which a volume of saturated air cools as it rises. Can be expressed in degrees Fahrenheit per 1000 feet, or in degrees centigrade per 1000 meters.

**saturated air**  Air with relative humidity of 100 percent. The higher the temperature of the air, the more water vapor it can hold per unit volume; that is, the greater the amount of water vapor needed for 100 percent relative humidity. See *relative humidity*.

**saturated fat**  A fat in which the carbon atoms have the largest possible number of hydrogen atoms attached. These fats tend to be semisolid at room temperature. The highest proportion of saturated fat is found in foods such as butter and lard.

**saturation**  1. A condition of being saturated. See *saturated*. 2. In an electronic transistor or tube, a condition in which the current flowing through the device is as large as it can possibly be for a given voltage.

**Saturn**  The sixth planet from the sun in our Solar System. It is known especially for its spectacular system of rings, easily seen through a small telescope. It is the second largest planet, and it is the only planet with a density less than that of water. See the SOLAR SYSTEM DATA appendix.

**Saturn rocket** The gigantic rocket that boosted the Apollo spacecraft into initial orbit around the earth. It is the largest and most powerful rocket that has carried any American astronauts into space.

**sawtooth wave** An alternating-current or pulsating direct-current waveform with a characteristic shape. There are three types, as shown in the drawing: (a) rapid rise and slow decay, (b) slow rise and rapid decay, (c) slow rise and slow decay. In this context, "slow" is a relative term, because the whole cycle can be completed in a tiny fraction of one second.

SAWTOOTH WAVE

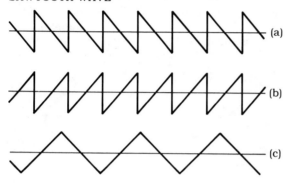

**Sb** Chemical symbol for antimony.

**Sc** Chemical symbol for scandium.

**scabies** A contagious skin infection caused by a small mite. The main symptom is itching. The infected person may scratch the lesions, aggravating them. Creams and lotions are available for treatment; it is usually necessary to see a physician.

**scalar** A quantity that can be fully expressed by means of a number line. That is, it needs one dimension. Such quantities show magnitude (amount or extent) only. Real numbers are the most common example. See *vector*.

**scalar product** See *dot product*.

**scale** 1. Relative size for a given system. We might speak of small-scale weather patterns, such as dust devils and downdrafts, or large-scale patterns such as high-pressure ridges. 2. Magnitude. 3. A device for determining mass or weight.

**scaling** A principle of physics, that volume increases with the cube of linear size, but surface area increases with the square of linear size (see drawing). If you suddenly became 10 times taller, you would have 100 times the cross-sectional area in your legs. But you would have 1000 times your present weight. Your legs would be under 1000/100, or 10, times as much strain, and you couldn't stand up.

SCALING

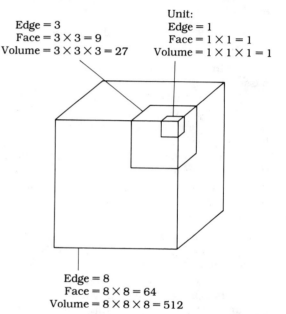

**scan** 1. To search in an orderly fashion for some characteristic among a large number of samples. 2. To search for an occupied or empty channel in a radio band. See *scanner*. 3. To analyze a tissue sample by means of X rays, ultrasound, nuclear magnetic resonance imaging, or some other method. 4. A means of displaying a radar image.

**scandium** Chemical symbol, Sc. An element with atomic number 21. The most common isotope has atomic weight 45. In pure form it is a soft metal. It is rare and expensive. There are no common industrial uses.

**scanner** A radio receiver that has many channels and searches through them for a busy channel. Alternatively, it can search for an empty channel.

**scapula** Either of two large bones in the upper back, to which the upper arm bone (humerus) is attached. There is one on the left side and one on the right side.

**scarlet fever** A rather uncommon infectious disease caused by a certain strain of streptococcus. The ailment gets its name from the reddening, or flushing, of the skin that accompanies the fever. Other symptoms include fatigue, skin rash, and inflamed tongue. Treatment is with antibiotics.

**scarp** A steep slope formed by movements in the earth's crust.

**scattering** 1. When an electromagnetic wave passes through a certain medium, the tendency for the direction of wave travel to become spread out. 2. In atomic physics, the tendency of certain heavy nuclei to cause high-speed particles to "bounce off" in many directions. 3. In radio communications, the spreading out of the direction of wave propagation, caused by refraction in the atmosphere or by reflection from the earth. 4. The spreading-out of light reflected from a rough surface.

**schematic diagram** A diagram that shows the interconnection of components in a device or electronic circuit, but not how the unit is mechanically constructed. Used by electronics engineers to show, in easy-to-read form, the nature and operation of a circuit. See the SCHEMATIC SYMBOLS IN ELECTRONICS appendix.

**schist** A finely layered type of metamorphic rock. It tends to break apart easily into thin flakes.

**schizophrenia** A severe form of mental illness, characterized by withdrawal and loss of contact with reality. There are various different forms. Hallucinations and delusions are common. People with this illness often cannot care for themselves, and might require hospitalization.

**Schleiden, Matthias** One of the biological scientists of the nineteenth century who first proposed that all living things are made up of units called cells. See *cell*.

**Schulze, Johann H.** A European scientist who discovered, early in the eighteenth century, that silver nitrate becomes darker when exposed to intense light. This ultimately led to the use of this compound in photographic films.

**Schwarzchild, Karl** A German astronomer and astrophysicist who first noticed that an extremely dense object might vanish because of the intensity of its gravitation. See *black hole, Schwarzchild radius*.

**Schwarzchild radius** Also sometimes called the gravitational radius. The radius that a given mass must be compressed within, in order to become a black hole (see drawing). This radius is directly proportional to the mass of the object. For the sun, it is about 3 kilometers. For the earth, it is about a centimeter. See *black hole*.

SCHWARZCHILD RADIUS

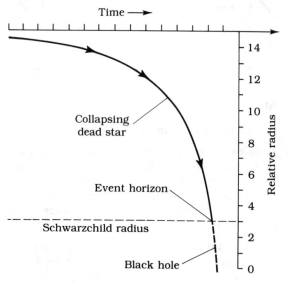

**sciatic nerve** Either of two large nerves leading from the end of the spinal cord into the legs.

**scientific method** The basic reasoning that a scientist employs to draw conclusions. Data must be gathered by careful observation, and analyzed according to logical principles. Theories are verified or rejected on the basis of whether or not they accurately predict events. It is often the case that a theory is close to being correct, and must be modified several times before it becomes accepted.

**scientific notation** Also called power-of-10 notation. A method of writing numbers that are very large or very small. A decimal is first written, having a value of at least 1 but less than 10. Then a times sign (×) is written. Finally, a power of 10 is written. We might say, for example, that $3853 = 3.853 \times 10^3$, or that $0.0033 = 3.3 \times 10^{-3}$, or that $1,100,000 = 1.1 \times 10^6$. This notation makes it easy to multiply and divide extreme number values. See *significant figures*.

**scintillation** 1. A change or fluctuation in the level of received radio signals; this is also called fading. 2. In radar, a fluctuation in the nature and intensity of the echo from a target. 3. The "twinkling" of a star, caused by refraction of light by the turbulent atmosphere of the earth. 4. A flash of light that occurs when a high-speed subatomic particle or photon hits a phosphor. See *scintillation counter*.

**scintillation counter** A device that is sometimes used to measure the level of ionizing radiation. It works by means of scintillation effect (see *scintillation*). The flashes produced are converted by a photocell into electrical signals that can be counted or observed on an oscilloscope.

**sclerosis** Hardening of certain tissues in the body, especially in the nervous system. The result is impaired nerve function.

**Scott, Robert** A British explorer who tried, in 1902, to reach the South Pole, but failed. He tried again in 1911 and succeeded, but a Norwegian, Roald Amundsen, had arrived a month earlier. On his way back to his supply base, Scott and two of his companions were caught in a blizzard and died.

**screw** One of the six simple machines, consisting of an inclined plane wound around a cone. It has a variety of uses; the most common is in construction, because it works well in binding objects together.

**scrotum** In male mammals, the cavity in which the testicles are contained. See *testicles*.

**SCUBA** Acronym for self-contained underwater breathing apparatus. The air tank or oxygen/helium tank, regulator, hose, and mouthpiece used by divers so that it is not necessary to keep surfacing for air.

**scud** 1. Also called fractocumulus clouds. Broken-off pieces of cloud that blow along at low altitude in a storm. 2. A short-range, Soviet-built ballistic missile.

**scurvy** A disease caused by a chronic deficiency of vitamin C (ascorbic acid). This ailment is very rare today, since even when fruits and vegetables are not available, supplements of the vitamin can be taken in tablet form. The main symptoms include fatigue, depression and bleeding gums.

**Se** Chemical symbol for selenium.

**seafloor** The bottom of the ocean, particularly well offshore. Much of this part of the world remains unexplored. About three-quarters of the solid surface of the earth is under the oceans.

**sea-floor spreading** The outward movement of the ocean bottom, away from central ridges. As the mantle rises to the sea bottom through the crust, it pushes the seafloor outward. This contributes to the movement of the continents. This is vividly apparent in the Atlantic Ocean, where the expanding seafloor is pushing North America and South America away from Europe and Africa. See *continental drift, mid-Atlantic ridge*.

**sea mount** An undersea volcano. Many new islands form from active underwater volcanoes; they grow until they poke through the surface. The Hawaiian Islands formed (and are still forming) in this way. These volcanoes are especially numerous in the Pacific Ocean.

**season** 1. Any of four times of year, according to the calendar. In the Northern Hemisphere, spring is from about March 21 to June 22; summer from about June 22 to September 23; fall from about September 23 till December 22; winter from about December 22 until March 21. In the Southern Hemisphere these are reversed, displaced by six months. 2. Any time of year during which a phenomenon, such as an eclipse or a tropical hurricane, is most likely to occur.

**seawater** Water from the ocean, having certain properties including a characteristic concentration of sodium chloride and traces of other minerals. The concentration of dissolved solids varies somewhat in different parts of the world. There are also microorganisms such as plankton, whose numbers vary greatly among different ocean locations.

**sebaceous gland** A gland that secretes sebum (oil) onto a hair shaft. There are glands of this type in all hair follicles. See *follicle*.

**seborrhea** A chronic skin disease with oiliness and lesions resembling pimples. Treatment is difficult, although modern methods are improving. A doctor should be consulted.

**sec** Abbreviation for secant, second.

**secant** Abbreviation, sec. The reciprocal of the cosine (see *cosine*). On the unit circle, the secant is given by $1/x$. See the drawing.

SECANT

Secant = sec $\theta$ = $1/x$

Unit circle $x^2 + y^2 = 1$

$(x,y)$

$\theta$

**sech** Abbreviation for hyperbolic secant.

**second** 1. The Standard International unit of time. It is 0.00001574 solar day, or 9,192,631,770 cycles of a certain emission line of the cesium atom. Light travels about 300,000 kilometers in this time. 2. A unit of angular measure, equal to $1/60$ minute of arc, or $1/3600$ of a degree.

**secondary colors** Any colors of visible light that are derived by combining the primary colors (red, blue, and green) in various intensities. This covers all imaginable colors of visible light. See *primary colors*.

**secondary pigments** Any pigments for reflected light that are derived by combining the primary pigments (red, yellow, and blue) in various concentrations. This covers all imaginable colors for reflected visible light. See *primary pigments*.

**Second Law of Motion** See *Newton's Laws of Motion*.

**Second Law of Thermodynamics** See *thermodynamics*.

**secretion** The material discharged by the duct in a gland. Examples are saliva, sweat (perspiration), and bile.

**sedative** A drug that causes drowsiness, intended for the purpose of inducing sleep. This might vary from something as mild as warm milk (quite effective, incidentally) to barbiturates. For some people, aspirin has a sedative effect. These drugs have the potential for abuse; some are addicting.

**sediment** Material that has precipitated to the bottom of a body of water over a long period of time. Also, material that has been washed by water action from one place to another. This might include dirt, fragments of shells, microorganisms and the dead bodies of various water-dwelling animals.

**sedimentary dome** A mound-shaped accumulation of sediment. The presence of this type of formation often indicates a deposit of minerals or fossil fuels.

**sedimentary rock** Rock formed from compressed minerals and debris that have been washed from one place and deposited someplace else. The compression occurs over millions of years.

**Seebeck Effect** The appearance of a voltage between two different metals when they are brought into contact. This is the basis for operation of the thermocouple. See *thermocouple*.

**seed** 1. The embryo and surrounding nutrients, in a protective structure, and capable of growing into a new plant given the right conditions. 2. A crystal of some mineral that, when placed in a solution of the same mineral, will grow larger. This is how certain semiconductor materials are made.

**seiche** Oscillation of the water in a bay, caused by the tidal action of the moon and sun, or by severe storms. This effect can result in tides that change far more than they do on a straight coastline. In certain cases, catastrophic flooding is the result, especially when a hurricane strikes from the right direction at the right time.

**seismic** Pertaining to vibrations in the earth, resulting from earthquakes, volcanoes, or meteorite impacts.

**seismic tomography** A method that geologists use to get a picture of the interior of the earth. It works much like certain techniques doctors use to look inside the body (CAT scan). By analyzing the way in which seismic waves change speed inside the earth, it is possible to tell where the various layers meet. See *seismic wave*.

**seismic wave** A shock wave on land or at sea, resulting from earthquakes, volcanoes, or meteorite impacts. Inside the earth, these waves make it possible to determine where the various layer boundaries are. The drawing shows such effects inside our planet, and also in the ocean. See *tsunami*.

**seismogram** A recording of an earthquake, made using a seismograph. See *seismograph*, *seismometer*.

**seismograph** A device that records earthquake vibrations, and also those earth vibrations from other causes, by means of a display of intensity versus time. The result is similar to the output of an electroencephalograph or barograph, that make graphs of brain waves and atmospheric pressure changes, respectively. It consists of an oscilloscope or pen recorder attached to a seismometer. See *seismometer*.

**seismology** The study of earthquake waves. Also, the study of the earth from recordings of earthquake waves. See *seismic wave*.

**seismometer** A device that detects earthquake waves, and also seismic waves from other causes, such as volcanoes and meteorite impacts. The drawing shows the basic principles on which modern earthquake detectors work.

**seizure** A sort of "electrical storm in the nervous system." The sufferer loses most or all control of the body. Often, unconsciousness occurs. Some people have these events regularly (see *epilepsy*). They vary in severity from fainting-like episodes to violent attacks. Sometimes they can be caused by overuse of drugs, or by withdrawal from certain drugs or alcohol.

**SELCAL** Acronym for selective calling.

SEISMIC WAVE

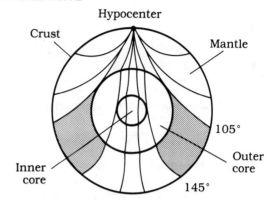

Seismic wave shadow zones in Earth's interior

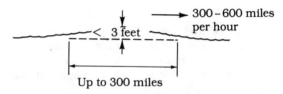

From Erickson, VOLCANOES AND EARTHQUAKES (TAB Books, Inc., 1988).

## SEISMOMETER

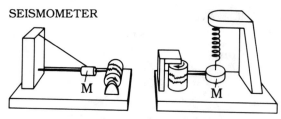

Principles of modern seismometers

From Erickson, VOLCANOES AND EARTHQUAKES (TAB Books, Inc., 1988).

**selective breeding** A method used in agriculture to obtain hardy grains that also yield large amounts of food. This is done by plant genetics. The best grains are obtained by crossing one strain with another; these are called hybrids.

**selective calling** In radio communications, a method in which a code is sent at the transmitting station, so that only the desired receiver gets the message. This is employed especially in aircraft communications.

**selective fading** In radio communications, a form of fading that takes place in the high-frequency (HF) spectrum between 3 megahertz (MHz) and 30 MHz. The fade moves in frequency across the signal's bandwidth, causing momentary severe distortion at the receiving end. See *fading*.

**selectivity** The extent to which a radio or communications receiver can distinguish between signals that are close to each other in frequency. This can be expressed in various ways, such as by a curve of frequency response, or by figures for the adjacent channel attenuation.

**selenium** Chemical symbol, Se. An element with atomic number 34. The most common isotope has atomic weight 80. It is important as a semiconductor in electronics applications. Its resistivity varies depending on how much light falls on it; therefore, it works well as a photocell. It is also used in a certain type of diode. See *selenium rectifier*.

**selenium rectifier** A diode made using selenium and some conducting metal, such as aluminum. This type of diode is used in high-current power supplies. It is not useful at high frequencies. See *rectifier, selenium*.

**self-purification** The removal of impurities in water by natural process. In reservoirs and lakes, particles settle to the bottom. Water might also flow through sand or gravel. Runoff dilutes rivers and streams. Microorganisms can metabolize wastes, rendering them harmless.

**semen** The fluid that serves as a medium for sperm cells in male animals. In mammals this fluid is produced by the prostate gland and is ejaculated along with the sperm during intercourse.

**semi-** A prefix meaning half or partly or somewhat.

**semiaquatic** 1. Pertaining to a plant that dwells near water, but not actually in the water; a marsh plant is a good example. 2. Pertaining to an animal that can function in or out of the water. Amphibians, such as frogs, are good examples.

**semicircular canal** A part of the inner ear that is important for keeping balance. It is filled with fluid, having the shape of three loops. When the head moves, this fluid flows, giving you the sensation of up and down.

**semiconductor** 1. A material that conducts electricity fairly well, but not very well. Selenium, silicon, and germanium are examples. These and other such materials are important in the electronics industry because they are used to manufacture diodes, integrated circuits, photocells, transistors, and other devices. 2. An electronic device manufactured using the aforementioned substances. Examples are diodes, transistors, and integrated circuits.

**semilog graph** A graph in which one axis is graduated linearly, and the other logarithmically (see drawing). This type of graph is useful in showing certain relationships that would seem grossly distorted on a conventional Cartesian plane.

**Senarens, Lu** A nineteenth-century fiction writer who came up with ideas for devices,

SEMILOG GRAPH

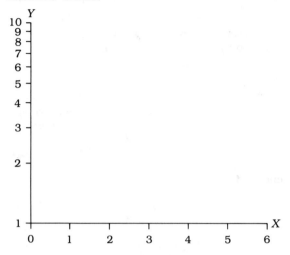

some of which were later realized. One such work was a book published in 1896, in which he depicted an "electric horse." He apparently didn't think of a "horseless carriage," but instead, envisioned a robot horse.

**senility**  See *dementia*.

**sense**  1. Detect. A device might sense a signal, for example. 2. Pertaining to one of the five senses. See *senses*. 3. Orientation. A circularly polarized wave might rotate in the clockwise sense or in the counterclockwise sense.

**senses**  The five modes of perception in humans: sight, hearing, touch, taste, and smell.

**sensible heat**  The amount of heat needed to raise one gram of a liquid from the melting point to the boiling point. In the case of water, this is about 100 calories per gram, raised from zero degrees centigrade to 100 degrees centigrade (32 degrees Fahrenheit to 212 degrees Fahrenheit) at normal atmospheric pressure.

**sensitivity**  1. The ease with which a certain stimulus is detected by the body. This might pertain to light, sound, touch, taste, or smell. 2. A measure of the amount of signal detectable by a transducer. 3. A measure of the ability of a radio or television receiver to pick up signals.

**sensor**  A device that samples some phenomenon, such as temperature or the brightness of visible light, and delivers a current or voltage proportional to the intensity of the phenomenon. This current or voltage can be measured to determine the relative level of the phenomenon. Or, it can be used to control some machine.

**sensory neuron**  A nerve cell that tells the brain about various events in the body, according to what it receives from sensory receptors (nerve endings). Thus one might feel pain when stung by a bee, or warmth when the sun shines on the skin.

**sensory perception**  The sensations resulting from stimulus of one or more of the five senses.

**sept-**  A prefix meaning seven or in groups of seven.

**septillion**  The number one followed by 24 zeros, or 1000 followed by seven sets of three zeros.

**septum**  A wall of muscle that separates the left and right halves of the heart. See *heart*.

**sequence**  1. A set of numbers, denumerable as a list. For example, we might have the finite sequence (2, 4, 6, 8) or the infinite sequence ($1/2$, $1/4$, $1/8$, $1/16$, ...). 2. A set of events that occur in an orderly progression.

**serial data transfer**  A method of transferring digital signals one after the other along a single line. This has the advantage of requiring only one line, but the disadvantage of being rather slow. See *parallel data transfer*.

**series**  1. The sum of a sequence. In the examples shown in the definition of sequence, the sum of the finite sequence $2+4+6+8=20$, and the sum of the infinite sequence $1/2+1/4+1/8+1/16+...=1$. 2. A method of connecting components in a circuit. See *series circuit*. 3. A set of phenomena with a specific order and relationship.

**series circuit**  A method of connecting electrical components, such that the same current flows through each component. The voltages across different components may be different. If one component fails (opens), current stops flowing in the entire circuit (see drawing). See *parallel circuit*.

SERIES CIRCUIT

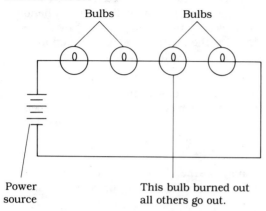

Power source

This bulb burned out all others go out.

**series-parallel circuit** A means of connecting electrical components together to get a higher power-dissipation rating. The number of components is usually four or nine, all identical (see drawing). The power rating is multiplied by the total number of components. The total value (usually resistance) is the same as that of the individual components.

SERIES-PARALLEL CIRCUIT

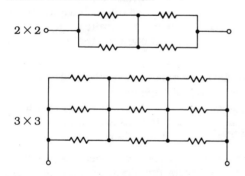

**serotonin** A compound whose levels appear to affect alertness. Higher levels are correlated with drowsiness. Warm milk causes an increase in the level of this substance in the body.

**servo system** An electromechanical control system that corrects itself. If an error in position occurs, a signal is sent to the controller to cause an adjustment that cancels the error out.

**set** A group of things. In theory these can be anything at all, but usually the elements, or members, of a set are numbers. The standard notation consists of curly brackets enclosing the list of set elements, such as {1,2,3,4}, the set containing the numbers 1, 2, 3, and 4.

**set intersection** For two sets A and B, the set containing all elements that belong to both A and B. See the illustration.

SET INTERSECTION
SET UNION

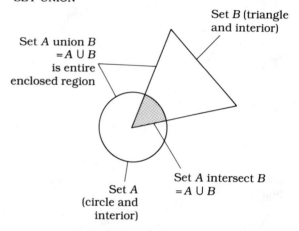

**set notation** The standard method of depicting a set: the use of curly brackets. See *set*. If a set has infinitely many elements, we might express the elements in a form such as A = {a: 2 < a < 5}. This would be the set of all real numbers greater than 2 and less than 5.

**set theory** A branch of mathematics, concerned with set operations. It is considered fundamental, and all other mathematics is based on it and on logic.

**set union** For two sets A and B, the set containing all elements that belong to A or to B, or to both A and B. See the illustration.
<**Production!** Insert Fig. set union>

**Severinus** One of the earliest scholars to have voiced the modern scientific spirit. In the seventh century he encouraged exploration for the furthering of knowledge. He was a Roman. It would be almost 1000 years before his spirit reawakened in Europe.

**sewage** Waste products resulting from human activity. This might include unused foodstuffs; it especially pertains to urine and feces.

**sewage treatment** Processing of sewage to render it less contaminating to the water supply. After it has been treated, it is generally dumped into rivers or into the ocean.

**sewer** A pipeline that carries sewage. Also, the system of waterways that provide drainage for rainwater.

**sex** 1. Gender, either male or female. 2. The complex behavior that results in fertilization of the female egg by the male sperm in animals.

**sex-** A prefix meaning six or in groups of six.

**sex chromosome** The characteristic pattern of chromosomes that determines the gender of an individual. For males the chromosomes look like the capital letters X and Y; for females they look like a pair of capital X's.

**sex hormone** Any of the hormones that regulate sexual maturation and activity. In the male, the main hormone is testosterone. In the female it is estrogen.

**sextant** A device that allows accurate determination of the angle of the sun above the horizon. Once used extensively in ship navigation, it has now been replaced by satellite navigation and radionavigation. See the illustration.

SEXTANT

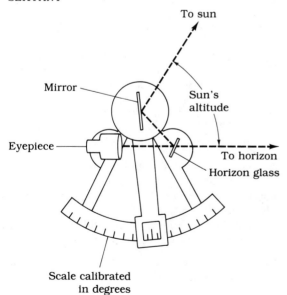

**sextillion** The number one followed by 21 zeros, or 1000 followed by six sets of three zeros.

**sexual reproduction** A method of producing offspring, in which two distinct genders (female and male) interact. Generally there are two individuals of the species involved—one male and one female. Self-pollinating plants are one exception; they contain both male and female components.

**shadow** The region in space where light from an object is partially or totally eclipsed. The region of partial eclipse is called the penumbra; the region of total eclipse is called the umbra.

**shale** Clay that has been compressed over a period of time. It tends to occur in thin layers.

**shale oil** A fossil fuel that can be extracted from shale. This is a largely untapped source of crude oil. The main challenge in utilizing this oil is lowering the cost of extracting it.

**Shannon, Claude** A mathematician who pioneered the field of information theory. This field has seen major advancements since World War II, and has applications in practical engineering, especially in radar, communications, and computers.

**Shapiro, I.** A radio astronomer who measured an apparent distortion in the orbit of Mercury around the sun. This distortion, caused by the sun's gravity, was found to be in close agreement with the extent of the effect as predicted by Einstein's general theory of relativity.

**Shapley, Harlow** An astronomer who, in 1917, first constructed a model showing the spherical "halo" of globular star clusters around the disk of our galaxy. He was also among those to first recognize that our Milky Way is an "island universe," similar to billions of other galaxies in the Cosmos.

**shear** 1. To cut off in a slicing fashion, along a plane surface. 2. A condition in which two regions of a substance move in different directions, with a sharp, plane-surface boundary. 3. See *wind shear*.

**shelf life** 1. The length of time that processed foods will remain safe to eat while in storage. 2.

The length of time that certain electronic devices, especially batteries, will last when stored.

**shell** 1. A hard outer coating, that serves to protect certain living things; an exoskeleton. 2. A supporting framework for a building or other structure. 3. The supporting functions and processes in a complex computer system.

**shellfish** Any of various marine creatures having an exoskeleton. Examples are lobsters, shrimp, and crabs.

**SHF** Abbreviation for superhigh frequency.

**shield** 1. A barrier that is intentionally installed to block some effect or radiation. 2. To prevent some effect or radiation from traveling from one place to another. 3. A broad, low-lying, geologically stable area of the interior of a continent. It gets this name from its rounded, shield-like appearance. Thought to be remnants of the original, earliest crust of the earth.

**shielding** 1. The blocking of an electric, magnetic or electromagnetic field, to prevent the field(s) from interfering with the operation of some device. 2. The use of dense materials, such as lead and various other heavy metals, to block ionizing radiation.

**shield volcano** A broad, low-lying volcanic cone, built up by lava flows having low viscosity.

**Shklovsky, Iosif** A Russian astronomer who has contributed numerous theories. One is that radio galaxies contain a large number of supernovae (see *radio galaxy, supernova*), and that this accounts for their high radio emission level. He has also theorized that photons can collide to produce X rays that we can detect coming from certain regions of the sky.

**shock** 1. A physiological condition that occurs in serious injuries or cases of extreme emotional tension. All of the body functions slow down to such an extent that the person may faint. Sometimes this condition can be fatal. 2. See *electric shock*. 3. A violent, jarring motion or vibration. See *shock wave*.

**shock absorber** 1. Any device that serves to impede the transmission of vibrations or shock waves. 2. A flexible, metal attachment in the suspension of a motor vehicle to reduce vibration as the vehicle moves over a rough road surface.

**shock hazard** A dangerous situation in which there is a high likelihood of electric shock to unprotected persons. See *electric shock*.

**shock wave** 1. An air blast that travels outward from the site of an explosion. It can actually move faster than the speed of sound in violent explosions, such as nuclear blasts. 2. See *seismic wave*.

**short circuit** In an electrical circuit, a condition in which current does not flow through the components as intended, but instead, takes a route through a circuit fault. This might affect part of the circuit (partial short) or all of the circuit (total or catastrophic short). Fuses and circuit breakers protect against the fire hazard that can accompany a short circuit.

**shunt** 1. To divert the flow of a fluid or current around a component. 2. To place one component across another, to alter the nature of the current flow or circuit behavior. 3. A component placed across another, to alter the nature of current flow or circuit behavior. 4. The connection or operation of circuit components in parallel. See *parallel circuit*.

**Shutt, Ralph P.** The physicist in charge of the group who searched for the omega minus particle at Brookhaven National Laboratory in New York. The particle was found in 1964. See *Eightfold Way* and *Samios, Nicholas*.

**SI system** Abbreviation for Standard International System of Units.

**Si** Chemical symbol for silicon.

**sibilant** Any of the highest-pitched components of speech. In the English language these sounds are produced in some or all cases by the consonants c (soft), g (soft), j, s, x, and z. In high fidelity, these sounds are often the most difficult to reproduce without distortion.

**sickle-cell anemia** A type of anemia in which the red blood cells are badly formed. Instead of the usual saucer shape, they appear as crescents or sickles. These cells do not carry oxygen as well as normal cells; the patient feels tired all the time.

**sideband** The frequency components that carry the modulating information in a radio signal. These components exist, no matter what the type of modulation, although their nature varies depending on the form of modulation. The drawing shows a spectral view of an amplitude-modulated signal.

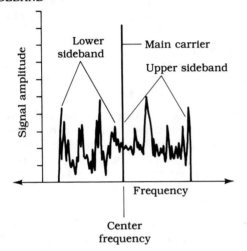

SIDEBAND

**sideral** Pertaining to the background of distant stars, considered as a whole on the celestial sphere.

**sideral day** The time needed for the earth to rotate once on its axis with respect to the distant stars. This is 23 hours and 56 minutes—slightly shorter than the 24-hour synodic day. See *synodic day*.

**side-wheeler** An early type of steam-powered ship. The propulsion was obtained from water wheels mounted on the sides of the vessel. The first such ship crossed the Atlantic Ocean in 1819.

**siemens** Abbreviation, S. The electrical unit of conductance. It is the reciprocal of the ohm. Thus, for example, if a component has 100 ohms of resistance, the conductance is 0.01 siemens. Often used when the resistance is very low, such as in specifying the conductivity of copper wire. Many people use the older term mho as the unit of conductance; the mho and the siemens are identical units.

**Siemens, Frederick** A German engineer who lived during the nineteenth century. He is known for having developed a special furnace for processing steel. It was more efficient than earlier furnaces. See *open-hearth process*.

**sievert** Abbreviation, Sv. A unit of dose for ionizing radiation.

**sigmoid** The last $1/3$ of the colon. In humans this is usually in the lower left part of the abdomen. It is also sometimes called the descending colon and leads into the rectum.

**sigmoidoscope** An instrument that doctors use to observe inside the sigmoid and rectum. It is a form of endoscope (see *endoscope*).

**sign** 1. The polarity, or quality of being positive or negative. Can apply to electric poles or to numbers. 2. The call designator for a radio, television or communications station. 3. In communications, to transmit the call designator. 4. An indicator of a phenomenon, either past, present or future. 5. A symptom of a device malfunction. 6. A symptom of a disease.

**signal** 1. In communications, the carrier wave and associated modulation that is used to transmit data from place to place. 2. An alternating-current waveform. 3. To send a message for some purpose.

**signal generator** An instrument that delivers signals of a precise nature, usually for laboratory testing of electronic equipment. Calibrated in frequency, amplitude and also sometimes according to test modulation characteristics.

**signal-to-noise ratio** In communications, a means of expressing the strength of a signal with respect to the background of random disturbances. Usually given in decibels, that is, logarithmically. See *decibel*. The weakest detectable signals are represented by ratios of about one decibel (1 dB).

**significant figures** A means of expressing precision or accuracy in scientific notation (see scientific notation). If we have a quantity of 1,230,000, for example, we can write $1.23 \times 10^6$, yielding three significant figures. Or we might write $1.230000 \times 10^6$, specifying seven significant figures.

**silica** A type of compound containing silicon and oxygen. Quartz is a common example. So are flint and certain types of sand, used to make glass. See *silicate*.

**silicate** A type of compound containing silicon and oxygen. Much of the earth's crust is made up of these compounds. There are numerous forms, all solids at standard atmospheric pressure and temperature. See *silica*.

**silicon** Chemical symbol, Si. An element with atomic number 14. The most common isotope has atomic weight 28. Next to oxygen, it is the most abundant element in the earth's crust. It is used in many types of electronic devices, including diodes, transistors, integrated circuits, and photocells.

**silicon chip** Also called a silicon wafer. A thin slice of processed silicon, used in the manufacture of certain electronic devices, especially integrated circuits. See *integrated circuit*.

**silicon crystal** A piece of silicon that has been set up with a point-contact wire so that it operates as a diode. This type of diode was used in early radio receivers.

**silicon dioxide** A compound consisting of silicon and oxygen; a special form of silicate (see *silicate*). It is used in making integrated circuits with a large number of components in a small area. The devices consume very little current, so they are ideal for portable, battery-powered devices.

**silicone** A form of silicon polymer. It consists of fine particles with a slippery texture. This material is useful in various types of lubricants. It is an electrical insulator but it conducts heat very well. It comes in a paste-like form that is easy to apply between an electronic component and heatsink.

**silicone rubber** A synthetic, rubber-like material made with silicones. This rubber has been used in artificial heart valves. It is extremely durable and resists attack from body fluids.

**silicon rectifier** A common type of diode rectifier, made using P-type and N-type silicon joined together. These devices can be made to work at a variety of frequencies. They are most often employed in low-cost power supplies. See *rectifier*.

**silicon steel** An alloy of about 96 percent steel and 4 percent silicon. It is used in transformer cores for alternating-current and low audio frequencies. See *transformer*.

**silicon transistor** A common type of transistor, made using P-type and N-type silicon. There are many different types for different electronic applications. See *transistor*.

**Silliman, Benjamin** A chemist and the founder of the *American Journal of Science*. He lived in the United States during the nineteenth century. He was known for his excellent lecturing. He toured the country speaking and teaching. He was a professor at Yale University in New Haven, Connecticut.

**silt** A fine material that settles out of streams and rivers when the water moves slowly, such as in the reservoirs behind dams. This material is often excellent for agriculture. It can be deposited during regular floods along some rivers, notably the Nile in ancient times.

**silt bed** A deposit of silt, remaining from a dried-up river. This type of deposit is often excellent for agriculture. See *silt*.

**silting** A gradual process in which silt builds up. This is how river deltas form. The buildup can change the course of a river, and often the river branches into many individual channels. The slow-flowing, meandering waterway encourages further deposition of silt. This process can eventually fill up a reservoir behind a dam.

**Silurean period** A part of the Paleozoic era, from about 435 million years ago to 405 million years ago. The first land plants appeared during this time. See the GEOLOGIC TIME appendix.

**silver** Chemical symbol, Ag. An element with atomic number 47. The most common isotope has atomic weight 107. In pure form it is an almost white metal. It is expensive and is considered a precious metal. It is an excellent conductor of electricity and of heat. Switch and relay contacts are sometimes plated with it, since it resists corrosion.

**silver nitrate** A compound of silver and nitrogen that darkens when exposed to visible light. It also darkens under infrared and ultraviolet radiation. This makes it useful in making camera film.

**silver plating** Coating of metal with silver to improve its electrical properties, and to reduce corrosion. This is done in high-grade switches and relay contacts, such as those found in computers. See *silver*.

**silver solder** Also called hard solder. An alloy of silver, copper and zinc, used as a solder when a high melting point is desirable. See *solder*.

**simple harmonic motion** A form of oscillating, back-and-forth motion. Imagine an object revolving in a circle around a center point, at constant angular speed. If you look at this edge-on, from a great distance, the object will seem to swing back and forth. This is simple harmonic motion. A weight suspended from a spring allowed to bounce up and down provides another example.

**simplex** 1. Communications in which both transmitters and receivers operate on a single frequency. 2. The transmission of just one signal at a time over a communications line.

**simulate** To imitate or mimic some set of conditions or parameters. For example, you might have the environment of a jungle in a greenhouse. See *simulation*.

**simulation** 1. An imitation of a real-life situation, but that does not involve any danger to human life or to machinery. An example is a video game used to teach pilots how to fly an aircraft. 2. An artificial environment, such as a greenhouse that allows plants to be grown at high latitudes during the winter.

**simulcasting** The transmission of the same program on more than one station or channel. An example is a symphony concert, broadcast on television and also on some frequency-modulation (FM) stereo station. One can then watch the concert on television and listen in full stereo on the radio.

**simultaneity** The occurrence of two events at the same time, as seen from a certain point of view. From most other reference frames, the events will not appear to take place at the same time, because of the finite speed of light.

**simultaneous** 1. Taking place at the same time. 2. Considered all at once. See, for example, *simultaneous equations*.

**simultaneous equations** A set of two or more equations in two or more variables. There may or may not be a unique solution. If there are n equations in m variables, then there can be a unique solution if n = m, but not necessarily.

**sin** Abbreviation for sine.

**sine** Abbreviation, sin. On the unit circle, the sine is the value given by y. See the drawing.

SINE

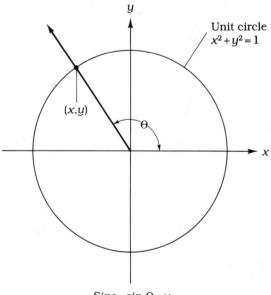

Sine = sin θ = y

**sine wave** See *sinusoid*.

**single inline package** A type of semiconductor component package, in which there are several contacts, or pins, all in line with each other on one side or edge of the package.

**sinh** Abbreviation for hyperbolic sine.

**sinkhole** A place where the ground surface collapses as a result of depletion of the groundwater. This process is accelerated during droughts and when an area is overdeveloped. It also can occur when acidic water percolates into the ground, dissolving limestone or the lime cement of sandstone layers.

**sinus** 1. Any cavity in a bone, especially in the skull, containing air. 2. A passageway or holding chamber for blood. 3. A gap between two adjacent structures.

**sinus node** A part of the right auricle of the heart. It sets the rate of the heartbeat; thus, it is the body's own natural pacemaker.

**sinusoid** A curve that depicts simple harmonic motion plotted against time (see *simple harmonic motion*); also the curve of the sine or cosine function plotted against angular measure. The waveshape is characteristic (see drawing). A sound wave or radio wave, with a single frequency component, has this shape.

SINUSOID

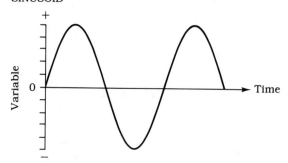

**sinusoidal** Having the shape of a sinusoid. See *sinusoid*.

**SI System** See *Standard International System of Units*.

**skeletal muscle** Any muscle that is attached to the bones, and that assists in moving the body. All such muscles are voluntary—that is, they operate according to conscious commands from the brain. They often are found in pairs.

**skeleton** 1. The supporting bone structure in vertebrates. 2. See *exoskeleton*. 3. A framework for a structure.

**skin effect** A tendency for high-frequency alternating current to flow mostly on the outside of a conductor. This causes an increase in the effective resistance of the conductor, compared with the resistance for direct current or low-frequency alternating current.

**skip** 1. Propagation of radio waves, especially at medium and high frequencies, by ionospheric refraction. 2. A phenomenon in which radio waves do not return to earth from the ionosphere, until they have traveled a certain distance from the transmitter. See *skip zone, sky wave*.

**skip zone** The region near a radio transmitter at high frequencies, within which signals are not returned from the ionosphere (see drawing). The size and shape of this region depend on the frequency, the time of day, the time of year, the latitude, and the level of sunspot activity. See *skip, sky wave*.

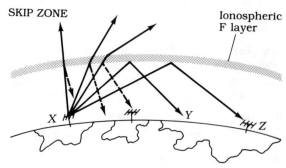

Z can hear X but Y can't.

**Skylab** An orbiting laboratory made from leftover hardware from the Apollo project. It was launched in May, 1974. It was largely devoted to looking at the sun from above the earth's atmosphere. It was also used for various space experiments and for observation of the earth. It came down in a fireball, burning up in the atmosphere over Australia in 1979.

**sky wave**  In radio communications, the wave that is returned by the ionosphere. A sky wave almost always exists at frequencies up to about 10 megahertz (MHz). Above this frequency, there is sometimes a sky wave and sometimes not, depending on several factors. There is almost never a sky wave above about 100 MHz. See *skip, skip zone*.

**slate**  A metamorphic rock, formed mostly from shale. It has a characteristic way of breaking off in slabs. It is usually dark gray to black in color. See *shale*.

**sleep**  A condition in which consciousness and reflexes are greatly slowed down. It takes place in many animals, and in humans, on a regular basis, usually in a daily cycle. It serves to rest the body and reduces the need for food. Many aspects of this condition are still not well understood.

**Slipher, Vesto M.**  The astronomer who first observed the red shift of spectral lines from distant celestial objects. He noticed this in 1913 when examining the calcium lines in the light from galaxies. See *red shift*.

**slug**  1. A lump of metal, such as refined crude ore, ready for further processing. 2. A small creature, similar to a snail, often with little or no shell. 3. An antiquated and little-used unit of mass, such that an applied force of one pound will cause an acceleration of one foot per second per second.

**slurry**  A mixture of ground ore and water, an intermediate stage in the refining process of various metals and minerals.

**Sm**  Chemical symbol for samarium.

**small intestine**  The long, narrow, coiled tube in which nutrients are absorbed by the body. In an adult human, this complex is about 20 feet long. It is located in the abdomen (see drawing). Food passes from the stomach into this tube, and after its course it moves into the colon, or large intestine.

SMALL INTESTINE

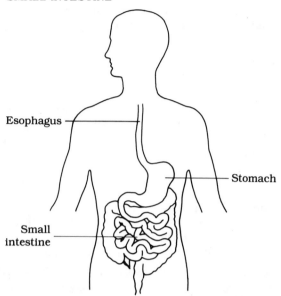

**smallpox**  A contagious disease caused by a virus. The main symptoms are fever and exhaustion at first, followed by skin eruptions. The lesions often result in permanent scars. In extreme cases, death occurs within four days. This disease has been largely eliminated by vaccinations.

**smart bomb**  Any bomb that is guided to its target by precise electronic means. One method uses a laser to heat the target for a heat-seeking missile. Another method uses a television camera and transmitter in the nose cone of a remotely controlled missile. Still another uses a computer programmed to find the target in complex terrain.

**Smeaton, John**  A British engineer of the eighteenth century. He is best remembered for his work in building dams and canals, and also for designing and building bridges, drainage systems and lighthouses.

**Smith, Adam**  An eighteenth-century scientist and mathematician who is known for his work in economics. His work is regarded by some as fundamental in modern economics. He was a strong proponent of the free-market system, believing that this is the best-working economic scheme for a society.

**Smith chart**  A curvilinear coordinate chart used to plot complex impedances. Used especially in antenna engineering.

**smog**  A contraction of the words smoke and fog. A form of air pollution in which various contaminants become concentrated. Most likely in large cities, during temperature inversions, and in valleys between mountain ranges. Especially severe if all three of these factors exist at the same time.

**Sn**  Chemical symbol for tin.

**Snell's Law**  A rule concerning the refraction of light, and also other electromagnetic radiation, when it passes from one medium to another. It relates the ratio of refractive indexes to the extent of refraction in a quantitative way. See a text on physics or optics for details.

**snowflake**  A complex ice crystal, that condenses forming a six-pointed or six-sided geometric shape, given the right weather conditions. There are billions of different possible patterns. In fact, it is very difficult to find two flakes that are alike.

**social science**  The broad field of science that is concerned with the behavior of individuals, groups, and societies. Also can be concerned with the fate of the whole human species; then it overlaps with environmental and medical sciences.

**sociology**  A branch of science concerned with the behavior of very large groups (societies) of people. Also concerned with the interaction among different societies.

**soda ash**  The compound sodium carbonate. A mildly alkaline substance, used in industry for producing materials such as glass, paper and various textiles. Also used to control the acid-base balance (pH) of water in swimming pools.

**sodium**  Chemical symbol, Na. An element with atomic number 11. The most common isotope has atomic weight 23. In pure form it is a soft metal that reacts so rapidly that it must be preserved in oil or kerosene. In the body, it is an important macromineral needed in substantial amounts and must be in a balance with potassium. Numerous compounds exist that contain this element.

**sodium bicarbonate**  A simple and abundant compound, the chief component of ordinary baking soda. It is mildly alkaline. In the old days it was used as toothpaste.

**sodium chloride**  Common table salt, and the compound that makes seawater "salty." Called halite in mineral form.

**sodium fluoride**  The "fluoride" that is most often added to toothpaste to protect against tooth decay. It is also added to drinking water in many communities. In recent years, the "fluoridation" of drinking water has been criticized by some scientists.

**sodium nitrate**  A compound used in curing and preserving processed foods, especially meats such as bacon. In recent years there has been some concern that excessive consumption of this compound may be a cancer risk. In the body, it can be converted to nitrosamine, a carcinogen.

**sodium nitrite**  A compound, similar to sodium nitrate and used for similar purposes. See *sodium nitrate*.

**sodium propionate**  A food additive, used to retard spoilage. It is commonly used in bread. It increases the shelf life of such foods.

**sodium-vapor lamp**  A highly efficient, brilliant light that works by energizing sodium vapor and causing it to fluoresce. These lamps have a characteristic candle-flame color. They are commonly used as street lamps.

**soft iron**  Iron in nearly pure form. It is malleable and is easy to magnetize. It is not very good for making permanent magnets, since it tends to be easily demagnetized.

**software**  The programs in a computer that are provided by the operators for specific tasks. These programs are often available commercially, for use with home computers. The programs are stored on sets of floppy disks that can be changed in seconds. This book was written, for example, with word-processing software and a personal computer. See *firmware*.

**software engineering** The branch of computer engineering that is concerned with the development and debugging of software programs for specific purposes. See *software*.

**soil profile** The various layers beneath the surface of the earth on dry land. The exact composition varies with location. An example is shown in the illustration.

SOIL PROFILE

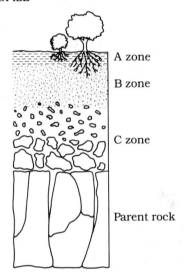

From Erickson, EXPLORING EARTH FROM SPACE (TAB Books, Inc., 1989).

**solar battery** A set of solar cells, connected in series to obtain a higher voltage and power rating than is possible with a single cell. Usually such a battery provides 6 or 12 volts. See *solar cell*.

**solar cell** Also called a photovoltaic cell. A semiconductor device that generates an electric current directly from sunlight. Large arrays of these cells can provide considerable power. They are commonly used to provide power for satellites. See *solar battery*.

**solar constant** The overall power output from the sun. It was once thought not to change, but recent studies have shown that it varies slightly, even over short periods. For example, the brilliance of the sun decreased almost 0.1 percent from 1981 to 1984. There is some evidence to suggest that solar brightness is greatest during sunspot maxima.

**solar cooling** 1. Air conditioning that uses electricity generated from solar cells. 2. Heat pumping that makes use of solar energy. Heat is removed from the space to be cooled (such as the interior of a building) and pumped outdoors. See *heat pump*. 3. The use of solar panels to radiate away excess heat at night; they absorb heat from the sun during the day.

**solar day** See *synodic day*.

**solar corona** The halo of hot gases surrounding the sun and extending out several million miles. It is visible during a total eclipse of the sun. It is largest during times of peak sunspot activity. It reaches temperatures of around 1 million degrees centigrade.

**solar distillation** A method of purifying water using solar energy. Water also can be extracted from the soil by this means. A greenhouse-like enclosure traps solar energy, heating the interior. Water condenses on the clear roof, which is made like an inverted cone, so that water will run down the inside of the roof and drip into a receptacle.

**solar eclipse** See *eclipse*.

**solar energy** Energy derived from sunlight, or from infrared radiation that comes from the sun.

**solar flare** A brilliant burst or flash that takes place on the sun in the vicinity of a violent disturbance, called a solar storm. Large numbers of charged particles are ejected during such an event. These particles arrive at the earth after several hours and cause the aurora when they are focused near the geomagnetic poles. See *aurora*.

**solar furnace** A furnace that operates using solar energy. This may be done by means of dark panels that absorb infrared and visible light, converting it to heat. Or the furnace may operate from electricity that has been obtained from photovoltaic cells. See *solar heating*.

**solar heating** 1. The heating of the earth's surface, caused by insolation. See *insolation*. 2. A method of warming water and/or the interior of a building by means of solar energy. The drawing shows an example, suitable for heating small swimming pools or water for home use. See *solar furnace*.

SOLAR HEATING

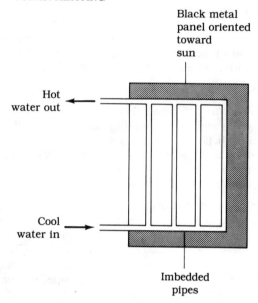

**solar panel** 1. See *solar battery*. 2. A large, dark, heat-collecting and/or radiating device for solar heating or cooling. See *solar cooling, solar heating*.

**solar prominence** A large streamer, often several thousand miles high, that billows upward from the surface of the sun during periods of intense solar activity. Some of these hot-gas streamers reach altitudes of about 1 million miles.

**Solar System** The sun and the region in which its gravitation holds influence. This includes the orbits of the planets, the asteroids and the comets. See *asteroid, asteroid belt, comet, Oort cloud, planet* and the SOLAR SYSTEM DATA appendix. Also see *Earth, Jupiter, Mars, Mercury, Neptune, Pluto, Saturn, Sun, Uranus, Venus*.

**solar wind** The constant stream of particles that is ejected from the sun. These particles exert a small but measurable outward force on objects in the Solar System. Charged particles are focused toward the magnetic poles of the earth, Jupiter, and other planets having magnetic fields.

**solder** An alloy of tin and lead, and sometimes of other metals such as zinc, copper and silver. Ordinary tin/lead solder melts at a rather low temperature. It is used to bond metals, especially in electronic equipment. See *silver solder*.

**solenoid** 1. A coil wound into a helix, often on a cylindrical form. 2. A coil of helical shape with a movable magnetic bar inside. When a current is applied to the coil, the bar is pulled inward or pushed outward, depending on the way it is magnetized and on the polarity of the current. This device has numerous applications in electromechanical equipment.

**solid** 1. One of the three states of matter. In this state, matter holds its shape, and does not conform to a container in which it is placed unless it is finely ground up. 2. Continuous and unbroken.

**solid angle** An angle in three dimensions, represented by a cone. The measure of the angle is given by a cross section of the cone, cut by a plane passing through the axis of the cone.

**solid fuel** A form of rocket fuel that is a solid at standard atmospheric pressure and temperature. It is mixed in a ready-to-burn form. This was the earliest type of rocket fuel, being used by the Chinese in their fireworks centuries ago. It is regaining application nowadays in booster rockets.

**solid-state** 1. Consisting of solid material. See *solid*. 2. Pertaining to the behavior of solid materials, particularly with respect to their electrical properties. 3. Pertaining to an electronic device consisting entirely of semiconductor components, without vacuum tubes.

**solid-state physics** The branch of physics that deals with the properties of solid materials, particularly the electrical behavior of semiconductors.

**solstice** The times of year when the earth's axis subtends its smallest angle relative to the line connecting the earth and the sun. These times are approximately on June 22 (summer solstice for the Northern Hemisphere) and December 22 (winter solstice for the Northern Hemisphere). See the illustration.

SOLSTICE

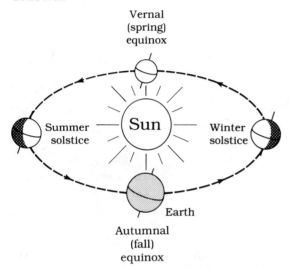

**solute** A substance that has been dissolved in some other substance. In salt water, for example, the salt is the solute. See *solution, solvent*.

**solution** 1. A combination of compounds or elements in which one is dissolved by the other. Salt water is a common example. 2. Any of the roots of an equation. 3. The answer to a complex question or problem.

**Solvay, Ernest** A Belgian engineer of the late nineteenth and early twentieth centuries. He is known for finding a new method of making soda ash. See *Solvay Process*.

**Solvay Process** A method of manufacturing soda ash, or sodium carbonate, from sea water, ammonia and rock (limestone). It was first invented by Ernest Solvay in 1861. Most of the needed ingredients are abundant and cheap; the expensive materials are recovered so they can be used again. The process is still used today.

**solvent** 1. A substance that dissolves many different things, and can therefore be used for cleaning, for preparing a surface to be painted, and for various other purposes. 2. The part of a solution in which the solute is dissolved. In salt water, the solvent is the water. See *solute, solution*.

**soma** 1. A vine found in certain parts of Asia. 2. The body of a living thing. 3. The various body cells and organs in a human being.

**somatic** Pertaining to the physical processes in the body, especially those that are not subject to conscious commands, but take place independently of the action of the mind.

**sonar** A method of underwater sounding used by ships and submarines for the purpose of depth finding. Also can be used to locate other vessels, schools of fish, and changes in water density caused by different concentrations of salt. See the illustration.

SONAR

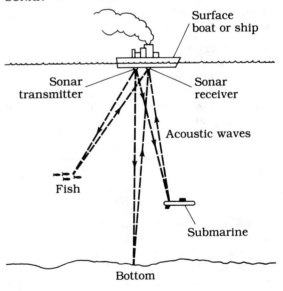

From Turner & Gibilisco, THE ILLUSTRATED DICTIONARY OF ELECTRONICS—4TH EDITION (TAB Professional and Reference Books, 1988).

**sonic** 1. Pertaining to sound, sound effects, and sound waves. See *sound*. 2. Having the effect of producing sound waves.

**sonic boom** A compression wave generated when an aircraft moves at, or faster than, the speed of sound. The sound wavefronts "pile up" and cause a loud noise along a cone-shaped region trailing behind the aircraft. See *Mach, sound.*

**sonography** The analysis of sound patterns, and the process of making visual displays (graphics) of sounds.

**sorbitol** A complex carbohydrate similar to sugar but metabolized more slowly. In some people it is not completely absorbed in the intestine, and draws water into the bowel, thereby acting as a laxative.

**sound** 1. A wave disturbance produced by mechanical vibration. Can travel through gases, liquids or solids. In air it occurs as a longitudinal compression wave (see *sound wave*). It has a frequency that falls within the range of human hearing, or about 20 to 20,000 hertz. 2. In mathematics, pertaining to a set of axioms that is consistent; that is, there are no contradictions. 3. Making logical sense, without any inherent contradictions.

**sound barrier** 1. The speed of sound. This varies depending on the nature of the substance through which the sound is traveling. In air at sea level it is about 700 miles per hour or 1100 feet per second. 2. The speed beyond which a sonic boom and other effects are produced. See *sonic boom.*

**sounding** 1. The use of sonar to determine the distance to, and possibly also the location of, an object underwater. 2. The data produced by the foregoing process. See *sonar.*

**sound pressure** The mechanical pressure, per unit area, induced on an object as a result of sound waves striking it. This can be measured and is directly proportional, at a given frequency, to the loudness of the sound. See *sound.*

**sound wave** 1. A longitudinal compression wave that travels through air (see drawing). At sea level the speed is about 700 miles per hour or 1100 feet per second. The individual molecules move back and forth locally, at a frequency ranging from about 20 to 20,000 hertz. 2. A mechanical disturbance in a liquid or solid medium, that can be heard as noise because of its frequency (between 20 and 20,000 hertz).

SOUND WAVE

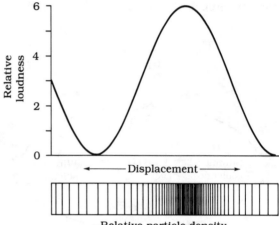

**source** 1. A point or small region from which energy radiates. 2. The place where a river begins. 3. The electrode in a field-effect transistor from which the charge carriers flow. 4. The origin of some phenomenon, effect or disease.

**Southern Cross** A constellation, visible generally in the Southern Hemisphere, and used to mark the direction of geographic south.

**Southern Lights** See *aurora.*

**space** 1. A continuum of three dimensions, all measurable in terms of distance. 2. A set of mathematical objects, within which certain processes can be carried out. 3. The regions beyond the atmosphere of the earth, not including the interiors or atmospheres of other celestial objects.

**Space Age** The time since the first space vehicles were launched. This is usually considered to have been after the first orbiting satellite, Sputnik, was successfully operated by the U.S.S.R. on October 4, 1957.

**spacecraft** Any vehicle that is designed for travel above the earth's atmosphere. In particular, this pertains to vehicles that can carry human passengers.

**space probe** 1. Any of the unmanned space vehicles that have explored the earth from above its atmosphere. 2. Unmanned space vehicles that observe distant celestial objects without the hindrance of the earth's atmosphere. 3. Unmanned space vehicles that have been sent to the moon and other planets. See the SPACE PROBES appendix.

**Space Shuttle** Any of the various vehicles recently constructed by the United States, launched with rockets and capable of returning to earth like an aircraft landing on a runway. Used to place satellites into orbit, to repair satellites, and to conduct various space experiments in earth orbit.

**space-time** The four-dimensional continuum consisting of the three "space" dimensions (such as height, width, and depth) and a single "time" dimension that can be thought of as perpendicular to all the space dimensions. A concept introduced by Albert Einstein in his theory of relativity early in the twentieth century. See *relativity*.

**spark** An electrical discharge that occurs across a gap for a very brief moment. In air, it is the result of the voltage being sufficient to cause ionization, allowing conduction for a short time.

**spark chamber** An enclosure filled with inert gases, and containing two sets of plates. A high voltage is applied to the plates. When a charged particle enters the region between the plates, a spark is produced. This device can be used to detect, track and count charged particles.

**spark gap** 1. An adjustable pair of point contacts, used for the purpose of allowing excessive voltages to be discharged as sparks before they can build up to dangerous levels. This is the principle by which a lightning arrestor works. 2. The gap between the contacts of a spark plug. See *spark plug*.

**spark plug** A device used in internal combustion engines to ignite the fuel vapors inside the cylinder. The resulting pressure pushes on the piston, providing the force that the engine uses to obtain its propulsion power.

**spasm** An uncontrollable, violent, and usually painful contraction of a muscle or muscle group. Common places are in the foot, the calf of the leg, and the abdomen.

**spatial distortion** 1. A shortening of distances in the direction of high-speed motion. It is a relativistic effect. 2. Changes in distances in intense gravitational fields, or during large accelerations. Also a relativistic effect. See *relativity*.

**spatial perception** The sensing of distances in three dimensions. Especially, the ability to sense the orientation of objects, and of oneself relative to those objects. This sense may be upset in outer space where there is no definable "up" or "down," and where great distances might seem small.

**speaker** A transducer that converts audio-frequency alternating current into sound waves. There are various different types of speakers that are used for different purposes, such as communications, high-fidelity, or underwater sound transmission.

**Special Creationism** A school of thought that teaches that each species was created in the form that we see it, and that no kind of evolution has ever occurred. Scientific evidence appears to contradict this theory. See *Creationism, Evolutionism*.

**specialist** A scientist who has much knowledge in certain fields. Most modern scientists must specialize because science has so many different aspects. To get an idea of how many specialties there are, see the table of science disciplines at the front of this book.

**specialization** Training in a particular field of science, in great detail. See *specialist*.

**special relativity** See *relativity*.

**species** 1. A category of living organisms with traits that make them all alike in certain ways. 2. A category of animals, particularly with reference to the way they might be affected by human activities. 3. The human race.

**specific** 1. Pertaining to a certain type of object or phenomenon. 2. Precise. 3. Pertaining to some characteristic of a substance. See, for example, *specific gravity*, *specific heat*.

**specifications** A set or list of characteristics, usually quantitative, for a machine, device, or system.

**specific gravity** Abbreviation, sp gr. The density of a substance compared with that of water. Pure water in air at sea level has sp gr = 1. Substances with sp gr < 1 will float in water; if sp gr > 1, a substance will sink.

**specific heat** The amount of heat energy needed to warm a substance, divided by the heat energy needed to warm the same mass of water to the same extent (one degree centigrade).

**specimen** 1. A sample of living tissue, isolated so that it can be examined under laboratory conditions. 2. A sample of blood, urine, or other body matter, used by doctors for detecting or diagnosing disease conditions.

**spectral class** A type of star based on the wavelength at which it gives off the most radiation. These classes are, from hottest (bluest) to coolest (reddest), called O, B, A, F, G, K, and M. These can be remembered by the sentence, "Oh, be a fine girl, kiss me." Or, "Only bad and foolish girls kiss men."

**spectroheliograph** A device that is used to record the spectral output of the sun in great detail.

**spectrophotometer** A sensitive type of spectroscope. Instead of a film or white surface on which the spectrum may fall, a movable sensing device is used. This allows the intensity to be precisely measured for selected wavelengths. See *spectroscope*.

**spectroscope** A device that splits visible light into its component wavelengths. A diffraction grating or prism is used (see the functional drawing). This enables astronomers to evaluate the composition of distant celestial objects. It also made possible the detection of the red shift. See *red shift*.

**spectroscopy** The use of a spectroscope to analyze the light from distant celestial objects.

SPECTROSCOPE

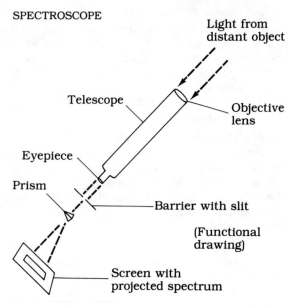

From Gibilisco, COMETS, METEORS AND ASTERIODS: HOW THEY AFFECT EARTH (TAB Books, Inc., 1985).

Also, the use of a spectroscope to examine light that has passed through various substances. These processes enable us to tell, even from a distance, what is in these objects or substances. See *spectroheliograph*, *spectrophotometer*, *spectroscope*.

**spectrum** 1. The "rainbow" of visible light, spread out by a prism or a diffraction grating. 2. A band of electromagnetic wavelengths. See the ELECTROMAGNETIC SPECTRUM, RADIO SPECTRUM appendices.

**spectrum analysis** The examination of a part of the electromagnetic spectrum using special instruments. In particular, this pertains to bands of the radio spectrum. See the ELECTROMAGNETIC SPECTRUM, RADIO SPECTRUM appendices.

**spectrum analyzer** An electronic device that uses an oscilloscope display to render a picture of a part of the radio spectrum. Frequency is shown along the base line, or horizontal scale. Amplitude is shown on the vertical scale. Signals appear as "pips" or vertical lines whose heights depend on how strong they are.

**speech compression** A technique that raises the average power of a voice communications signal, relative to the peak power. In some cases this improves the signal-to-noise ratio. See *signal-to-noise ratio*.

**speech formants** The bands of frequencies in which most voice energy is contained. These bands are separated by gaps, or ranges in which there is very little energy.

**speech recognition** 1. An electronic method of reading words. With this type of device, you might be able to give your computer commands by speaking to it. This technology is just getting started. See *speech synthesis*. 2. An electronic method of telling a person's identity by analyzing the "voice print," or audio patterns in the speech.

**speech synthesis** An electronic method of making word sounds. This makes it possible for a computer to actually talk to its operator. This technology is being improved, although it is farther along than that of speech recognition. See *speech recognition*.

**speed** The rate at which something moves with respect to a certain point of view. Direction is not specified. We measure speed as length per unit time, such as meters per second. See *velocity*.

**speed of light** The distance per unit time traveled by light, in a vacuum or in some transparent medium. In a vacuum, this speed is 299,792.5 kilometers per second. The more familiar, rounded-off figure is 186,000 miles per second. In media other than a vacuum, the speed is slower. This is also the speed of electromagnetic field propagation in a vacuum.

**speed of sound** The distance per unit time traveled by sound in air at standard sea-level temperature and pressure. In air this is 344 meters per second, or 770 miles per hour. It is much faster in water, and faster still in certain hard metals such as steel.

**Spencer, Herbert** A philosopher and scientist who wrote about the forces of nature, and how they can be so great and yet be found in such ordinary things as rocks and water droplets. The drop of water could be converted into a huge blast of energy if all its mass were liberated at once. The marks on a rock might have been left there 100,000 human generations ago. Science holds its own beauty, wrote Spencer—but only for those whose minds want to learn its secrets.

**spent fuel** See *irradiated fuel*.

**sperm** A cell produced by the male in many animal species. It fertilizes the female egg, forming a zygote. A human sperm has a flagellum (tail) and looks like a microscopic tadpole as it swims along. See *zygote*.

**sphere** 1. In three dimensions, the set of points that are all the same distance from some center point. 2. A solid ball, whose surface points are all the same distance from the center.

**spherical coordinates** A method of uniquely determining the location of a point in three-dimensional space. Two angles and a radial distance are specified (see drawing).

SPHERICAL COORDINATES

From Gibilisco, ENCYCLOPEDIA OF ELECTRONICS (TAB Professional and Reference Books, 1985).

**spherical geometry** Geometry as done on the surface of a sphere. The ordinary rules of Euclidean geometry are close to being accurate for very small distances. But when the distances are large, the Euclidean rules break down. See *non-Euclidean geometry*.

**spherical mirror** A mirror whose surface conforms exactly to a part of a sphere. For many applications, this is just as good as a parabolic mirror. But for some purposes a parabolic mirror is better, because it has a more precise focus. See *parabolic mirror*.

**spherical reflector** A reflector whose surface conforms exactly to a part of a sphere. Used in microwave radio communications, it is often just as good as a parabolic reflector. See *parabolic reflector*.

**spherical universe theory** The idea that our universe is a four-dimensional sphere (four-sphere), and that our three-dimensional space is the surface of that four-sphere. This is impossible to envision directly, but it has been mathematically defined. The four-sphere would probably be expanding, like a four-dimensional balloon. Such a universe would be finite but unbounded, like the surface of the earth. See *Big Bang*.

**spherule** Also called a spherulite. A small, roughly spherical type of volcanic rock with a crystalline structure.

**sphincter** A muscle that opens and closes a valve in the body. An example is found between the stomach and the start of the small intestine. See *pyloric sphincter*.

**spin** 1. Rotation, especially of a planet or star on its axis. 2. A property of subatomic particles. A particle with spin 0 looks the same from every direction, like a cue ball. A particle with spin 1 must be rotated one full turn (360 degrees) to look the same again, like an 8-ball. A particle with spin 2 must be rotated one-half turn (180 degrees) to look the same again, like a dumbbell. A particle with spin $1/2$ is strange, because in order for it to look the same again, it must be spun around two full turns (720 degrees). It cannot be represented by any common object. Yet this type of particle makes up all of the matter in the universe. The ones with spin 0, 1, and 2 are responsible for forces between material particles.

**spinal cord** The nerves that run from the brain down through the vertebrae. Numerous nerves branch off from this main cord, serving various body parts and organs.

**spinal tap** A method of anesthesia, in which numbing medicine is injected into the spinal cord. This blocks nerve impulses from the lower body, allowing surgery in the lower body without having to "put the patient to sleep" (general anesthesia).

**spine** 1. The structure containing the spinal nerves and the vertebrae in vertebrates. See *spinal cord, vertebra*. 2. A rigid, linear structure that holds an object together.

**spiral** 1. A curve whose radius decreases as one moves inward, and increases as one moves outward. 2. A pattern of motion that follows a curve as defined in (1). 3. To move along a curve as defined in (1).

**spiral galaxy** A congregation of stars with a spiral structure. Might contain tens or hundreds of billions of stars. Our Milky Way is thought to be a galaxy of this type; it would look like a pinwheel if seen face-on from about a million light years away. See *galaxy*.

**spiral nebula** A spiral galaxy as it was called before astronomers knew that they are outside our galaxy. Before about 1920, these objects were thought to be clouds of interstellar gas and dust, perhaps planetary systems in formation.

**spiral of Archimedes** A spiral in a geometric plane, whose radius is directly proportional to the extent of its curvature. See the drawing. In

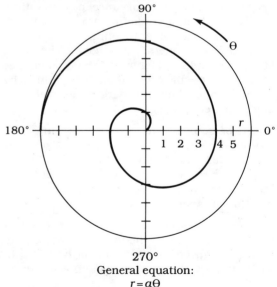

SPIRAL OF ARCHIMEDES

General equation:
$r = a\theta$

polar coordinates, the equation is simple; the constant, a, can be any positive real number.

**spirillum**  A form of bacterium with a spiral shape. It moves around by means of flagella.

**spirochete**  A microorganism that causes certain disease states in humans. Lyme disease, transmitted by a common tick, is caused by one strain of this organism. Syphilis, a form of sexually transmitted disease, is another malady caused by this spiral-shaped organism (see drawing).

SPIROCHETE

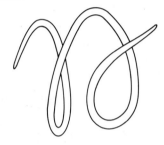

**spleen**  The organ in the body largely responsible for the manufacture of blood. It is located in the left-hand middle section of the abdomen, next to the stomach and just underneath the left lung.

**splitter**  1. A device that allows a signal on one line to be separated, so that it can be sent along two lines. 2. An audio device that allows two speakers to be used in place of one, with equal power distribution between speakers.

**sponge**  1. An ocean-dwelling animal with an internal skeleton that is complex and flexible. It exists as a colony and stays in one place. It is one of the most primitive animals. For its place on the evolutionary tree, see the ANIMAL CLASSIFICATION appendix. 2. A porous piece of metal, such as lead, for industrial use. In the case of sponge lead, the metal is useful for storage of electronic devices that are susceptible to damage from electrostatic charges.

**spontaneous combustion**  The ignition and burning of materials without the application of a spark, flame or other apparent source of heat. Chemical reactions can supply the needed heat. Oily rags have been known to catch fire with no apparent cause. It has even occurred, on isolated and very rare occasions, to people.

**spontaneous generation**  The notion that living things can arise from inanimate things. At one time this was thought to be a common process; for example, a glass of water without any sign of life might develop millions of microorganisms overnight. Nowadays this theory is not believed.

**spore**  A primitive type of seed, usually consisting of just one cell. Mushrooms and ferns are two types of plants that reproduce by means of these seeds.

**Sprague, Frank**  An American engineer who worked during the late nineteenth and early twentieth centuries. He is best known for development of the electric trolley and the electric train. He also worked on developing electrically operated elevators.

**sprain**  A form of injury that sometimes occurs to joints placed under stress, such as the knee or ankle. The ligaments are strained and sometimes actually torn. A severe sprain can seem at first like a bone fracture because of the pain and immobility.

**spread-spectrum communications**  A means of greatly increasing the bandwidth of a radio signal. This makes it possible to put many signals within the same frequency band. In such a system, a link cannot be "knocked out" by interference from only one signal. Instead, interference appears as noise that gets gradually louder, the more signals that occupy the band.

**spring**  1. The season of the year from March 20 to June 21 in the Northern Hemisphere, or from September 23 to December 21 in the Southern Hemisphere. 2. A coiled, elastic piece of metal, used as a shock absorber and for various other industrial purposes.

**spring balance**  A device that determines the weight of an object by the stretching or compression of a spring. It is not a very accurate device. It cannot measure true mass, because it works on the basis of the force of gravity.

**spring tide** The most extreme tides, that occur during new moon and full moon. The gravitational effects of the sun and the moon add together at these times, producing the greatest bulge in the oceans (see drawing). See *neap tide*.

SPRING TIDE

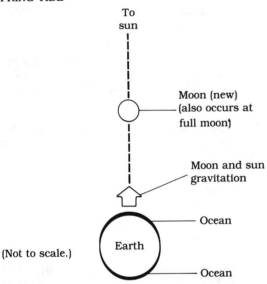

**spurious** 1. Spontaneous; arising without any apparent cause. 2. In communications and electronics practice, pertaining to transmitted signals on frequencies not intended.

**Sputnik** The first satellite (Sputnik I) placed into orbit. It was launched by the U.S.S.R. on October 4, 1957 by a ballistic missile, and had an orbit that carried it between 140 and 580 miles up. A month later, Sputnik II was launched, also by the U.S.S.R., carrying a live dog. The dog died in space because there was no means to recover the satellite at that time.

**squall** 1. A rain shower, accompanied by gusty winds, that occurs in the outer rain bands of a hurricane. See *hurricane*. 2. A storm at sea. 3. A heavy thunderstorm associated with a frontal system. See *squall line, weather front*.

**squall line** A group of heavy showers and thunderstorms, that merges together to form a continuous band. These often precede strong cold fronts in the summer months at temperate latitudes.

**square** 1. A plane polygon with four sides, all having the same length, and four angles, all measuring 90 degrees. 2. The second power of a number; that is, the number multiplied by itself. 3. A number whose square root is an integer. See *square root*.

**square-cube law** A rule of scaling in physics. When the linear dimensions of an object increase by some factor, the surface area increases by the square of that factor, while the volume increases by the cube of that factor. Therefore, large objects have more volume, in relation to their surface area, than small objects.

**square root** A value that when squared, or multiplied by itself, gives a specified number. The square roots of all negative numbers are called imaginary. See *imaginary number*. Often, the square root of a whole number is an irrational number. See *irrational number*.

**square wave** A wave with flat peaks and nearly instantaneous rise and fall times. On an oscilloscope, this type of wave appears as two sets of parallel, horizontal dotted lines, representing the flat peaks.

**Sr** Chemical symbol for strontium.

**sr** Abbreviation for steradian.

**SST** Abbreviation for supersonic transport.

**stability** 1. Proper operation of a ship or aircraft so that there is minimal turbulence. See *stabilization, stabilizer*. 2. A condition in which there is minimal chance of violent or erratic behavior. 3. Proper operation of an electronic amplifier, so that there is no oscillation. 4. A measure of the extent to which an electronic oscillator can maintain a constant frequency.

**stabilization** 1. In aviation, the prevention of turbulence in the air currents around a flying craft, ensuring level, smooth flight. 2. In boats and ships, the use of a rudder to keep the craft on course and on line. 3. In electronics, the prevention of unwanted oscillation in an amplifier. 4. In an electronic oscillator, the prevention of unwanted frequency changes.

**stabilizer** 1. The rear wings in an aircraft, along with the tail, that keep the plane on line and keep its flight smooth and level. 2. The rudder of a boat.

**stack** 1. A long vertical pipe that discharges smoke well above the ground, so that it will not cause high concentrations of pollutants anywhere on the surface. 2. A formation along some rocky coastlines where the ocean has caused erosion. At first a bridge-like structure is carved out; then the arch collapses, leaving a vertical pillar of rock standing offshore.

**stalactite** A formation that develops on the ceiling of a cavern. Water dripping from the rocks above leaves mineral deposits over the years, and they slowly build up. This creates a good place for the dripping to continue, so the formation grows. See drawing.

**stalagmite** A formation that develops on the floor of a cavern beneath a stalactite (see *stalactite*). The dripping, mineral-rich water leaves deposits that build up. See the drawing.

STALACTITE
STALAGMITE

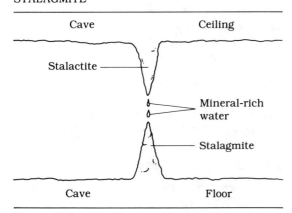

**stall** 1. In aviation, a condition in which lift fails because the wing angle becomes too steep. The smooth flow of air over the airfoil gives way to turbulence instead. 2. The failure of an internal combustion engine. There are various possible causes.

**stamen** In a flower, an organ inside the petals. At the tip is the anther. It produces pollen, which cause fertilization and the formation of the seed. The stalk is called the filament.

**Standard International System of Units** Once called the metric system. Also sometimes called the MKS system. The most common unit system used by scientists. The fundamental units are the meter, kilogram, and second. See the STANDARD INTERNATIONAL SYSTEM OF UNITS appendix.

**standard temperature and pressure** Abbreviation, STP or s.t.p. A temperature of zero degrees centigrade (32 degrees Fahrenheit), the freezing point of pure water; and 760 millimeters of mercury (29.9 inches), one atmosphere at sea level.

**standing wave** A stationary wave that occurs when forward-moving and reflected waves are in just the right relationship. This phenomenon occurs in mismatched transmission lines. It can be demonstrated with a rope attached to a wall (see drawing).

STANDING WAVE

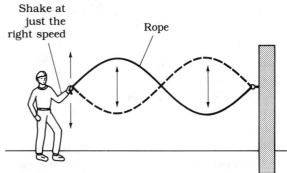

**Stanley Steamer** An early automobile that used a steam engine, rather than an internal combustion engine. Such cars were difficult to maintain, and became obsolete when internal combustion engines took over the market. The last of these cars was built in 1925. See *internal combustion engine, steam engine*.

**Stanley, William** One of the engineers who developed an alternating-current transformer during the early days of electricity (1885). Thomas Edison had been in favor of using direct current, but the work of Stanley and others convinced Edison that alternating current was better.

**stapes** One of the three bones in the middle ear that serve to transfer sound. Also called the stirrup.

**staphylococcus** A common form of bacterium, that causes a variety of infections in the human body. Many of these bacteria exist on and in the body without any harmful effects.

**star** A celestial object, consisting of condensed matter, mostly hydrogen. The compression is so great that nuclear fusion occurs at the core, changing hydrogen to helium and liberating radiant energy. Heavier elements are also created. All the elements except hydrogen are thought to have been generated in the cores of stars by nuclear fusion. See *fusion*.

**starboard** Pertaining to the right-hand side of a boat, or to directions that lie off towards the right as a boat passenger looks forward.

**starch** A polysaccharide consisting of several or many glucose molecules linked together. This is the most abundant, least expensive form of carbohydrate in the human diet. It is also a very efficient source of energy for the body.

**star cluster** See *globular star cluster, open star cluster*.

**starfish** A marine animal with five nearly identical limbs. It has a hard, tough skin, almost like a shell. It is easily recognizable by its shape.

**star formation** The congealing of matter in space, mostly hydrogen but also entrails from previous star explosions. Often this takes place along with rotation, so that the matter is flattened into a disk with a central spherical blob that becomes the star (see drawing). It is thought that planetary systems often arise within the rotating disk, by accretion of matter.

**starvation** 1. Prolonged lack of food with severe malnutrition as a result. The victim is extremely thin and loses resistance to disease. 2. Death from the aforementioned causes, either in one person or in a large group of people.

STAR FORMATION

*Interstellar gas and dust; Angular momentum; Disk may condense to form planets; Central condensation becomes star*

**state** 1. Condition of existing. We might speak of a state of constant change, or of stable temperature, for example. 2. A form in which matter can exist. See *gas, liquid, plasma, solid*.

**static** 1. Unchanging; remaining constant with time. 2. See *static electricity*.

**static behavior** 1. The behavior of a system when there is no motion. 2. The behavior of a system under conditions in which variable factors are held constant.

**static electricity** 1. A condition in which an object is electrically charged, either having a large surplus of electrons (negative charge) or a large deficiency of electrons (positive charge). 2. The phenomena, often annoying, that accompany electrostatic charge of common objects. Examples are the "static cling" of clothes after drying, and the little shocks you get when you touch metal objects after walking on a carpet.

**stationary front** See *weather front*.

**statistical analysis** A powerful mathematical technique used by scientists to evaluate complex systems of all kinds. This requires the use of computers to make models of systems, and to simulate huge numbers of different possible events.

**statistical mechanics** A branch of mechanics and also of thermodynamics. It is concerned with the behavior of large numbers of moving particles. While we cannot predict the movement of a single particle, the activity of the entire collection is often very predictable and regular.

**statistics** A branch of mathematics concerned with probability and with distributions of events. There are numerous applications in theoretical and practical science.

**statute mile** A unit of distance, equal to 5280 feet or about 1609 meters. Generally employed by lay people to indicate travel distances on land.

**steady state** 1. A condition in which no major changes in form occur; the condition tends to be self-perpetuating unless an outside force intervenes. 2. See *Steady State Model of Earth*. 3. See *Steady State Theory*.

**Steady State Model of Earth** A general theory concerning the nature of matter movement in the earth. Fluids and solids come from within the mantle to the oceans and crust. Gases in the atmosphere originated from within the earth. An equilibrium has been reached among these different flow patterns (see drawing).

STEADY STATE MODEL OF EARTH

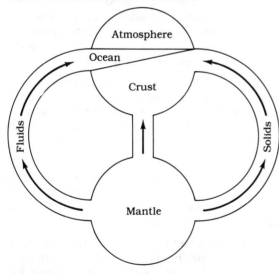

From Erickson, THE MYSTERIOUS OCEANS (TAB Books, Inc., 1988).

**Steady State Theory** A theory of the structure and evolution of the universe, invented by Thomas Gold, Herman Bondi, and Fred Hoyle to explain the apparent outward motion of galaxies. According to this theory, new matter is constantly being created, everywhere in the universe, in the form of hydrogen atoms. The Big Bang theory is more commonly accepted today. See *Big Bang*.

**steam** 1. Water vapor. 2. The condensation cloud that forms over a container of boiling water. Actually the water vapor is not visible, but the cloud consists of tiny droplets.

**steam energy** Energy obtained from the pressure that is exerted by hot water vapor. The heat applied to boil the water is converted into mechanical motion; the water and steam serve as the medium for this conversion. The water can be recycled again and again.

**steam engine** A device that converts heat into mechanical motion, with steam serving as the medium for the conversion. Water is boiled, perhaps by burning coal. The steam is under great pressure, and this can be used to drive a turbine or to move pistons.

**steam pump** An early version of the steam engine used to pump water from mines in England. Developed in the late seventeenth and early eighteenth centuries. See *steam engine*.

**steam turbine** A turbine driven by hot water vapor under pressure. The original idea was conceived in the third century B.C. In this system, steam forced a ball to go around and around. The first operational device was built in the nineteenth century to generate electric power. See *turbine*.

**steatite** Magnesium silicate, sometimes with small amounts of other ceramics. It is used to make electrical insulators at high frequency and high voltage. It looks like porcelain.

**Stefan-Boltzmann Law** For a blackbody, the total radiated energy, per unit surface area, is proportional to the fourth power of its absolute temperature. See *blackbody, blackbody radiation*.

**stem** 1. In plants, the vertical structure that branches off and terminates in leaves and flowers. 2. A supporting central structure or column; for example, the brain stem. 3. To cut off the flow in a stream, pipeline, or blood vessel.

**stenosis** In a body organ such as a vein, artery, or the intestine, a place where the diameter is decreased, resulting in impairment of the flow through the organ. This might be caused by a tumor, plaque or spasm, or by a disease state.

**Stephenson, George** A British engineer of the late eighteenth and early nineteenth centuries. Best remembered for his work in steam-powered rail transportation. He built steam locomotives and designed tracks and routes in England.

**steradian** A solid angle on a sphere, represented by a cone with its apex at the center of the sphere, and intersecting the surface of the sphere in a circle. Within this circle, the enclosed area on the sphere is equal to the square of the radius of the sphere. See *solid angle*.

**stereo** Slang for stereophonics.

**stereophonics** Recording and reproduction of sound in two channels, usually called left and right. This makes it possible to make sound effects far more realistic than with older monaural systems. More advanced systems use four channels. See *quadrophonics*.

**sterile** 1. Unable to produce offspring. 2. For a certain region or object, being without any living microorganisms. 3. Uncontaminated.

**sterility** 1. Lack of ability to produce offspring. 2. Having no living microorganisms.

**sterilization** 1. The rendering of a male incapable of getting a female pregnant or of fertilizing an egg. 2. The rendering of a female incapable of getting pregnant or of producing eggs. 3. Any surgical procedure that destroys the capacity for an animal to reproduce. 4. The killing of all microorganisms in a region or on an object. 5. Decontamination.

**sternum** Also called the breastbone. A flat bone-and-cartilage structure in the center of the chest, about midway between the nipples (see drawing). It serves to provide strength to the rib cage, protecting the lungs and heart.

STERNUM
STOMACH

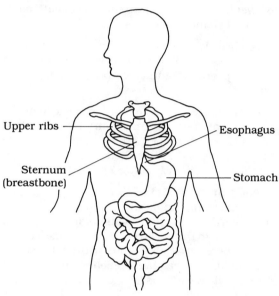

**steroid** A class of hormones that build body tissues and cause various other effects. There are health risks associated with this practice. The hormones have effects on metabolism, sex drive and immunity to disease, as well as the ability to increase muscle mass.

**stimulant** Any drug that increases alertness and quickens the reflexes. Caffeine and nicotine are common examples. Amphetamines are a stronger class of these drugs. Some such drugs have addictive properties; cocaine is an example.

**stimulus** Any action that causes some response, or reaction, in a living organism. Examples are light (visual), sound (auditory or aural), and heat (sensory).

**stimulus-response** 1. Pertaining to the reaction(s) of living things to various environmental and artificial factors. 2. The study of these factors and reactions.

**stochastic process** A process that has been made intentionally random. For example, we might select one out of 10 units to be tested at the factory for quality assurance. But we would not choose every tenth unit; rather, 1/10 of the units would be selected at random from the lot.

**stock ticker** One of the inventions of Thomas Edison. First developed in 1870, it kept track of the ups and downs in the stock market, providing almost up-to-the-minute information. It is still used today.

**Stokes-Adams disease** A condition in which the body's natural heart pacemaker fails. This problem can be practically cured by means of an artificial pacemaker. See *heart pacemaker*.

**Stokes' Law** A principle for determining the friction of a spherical object moving in a viscous material. The friction is directly proportional to the speed of the object, to its diameter, and to the viscosity of the material through which it moves.

**stoma** 1. An opening in a plant leaf. There are typically hundreds or thousands of these on an individual leaf. 2. A permanent, surgically cut hole in the stomach or intestines, leading to the skin.

**stomach** The body organ where food is prepared for nutrient absorption in the small intestine. Food passes into this organ from the esophagus after having been swallowed. Acids and enzymes help break down the complex molecules. In humans, this organ is located in the left middle abdomen (see drawing). Some animals, such as cows, have more than one stomach.

**stomach ulcer** A sore or lesion in the lining of the stomach. Often there are no symptoms. If severe, or if there is more than one such lesion, there could be pain (acid indigestion) and nausea. People who think they have stomach ulcers should see a doctor.

**stomata** The pores in plant leaves. See *stoma*.

**Stone Age** 1. The time before humans developed the ability to make metal tools. People were nomads and made their spears and other utensils from sticks and stones. 2. Pertaining to the time before the development of metal tools.

**Stonehenge** A rock structure in the British Isles, constructed in ancient times and apparently designed to predict astronomical events such as eclipses. The sun rises over a certain stone, as seen from an observation point, at the summer solstice (June 22). One of the remarkable things about this structure is that it will function properly only at the specific latitude where it is located.

**storage** 1. The placing of devices, equipment, or chemicals "on the shelf" for future use. 2. In computer practice, the placing of data in memory.

**storm** 1. A low-pressure system with unsettled weather. This usually includes precipitation and strong winds. 2. A thunderstorm. 3. A tornado. 4. A blizzard. 5. A sandstorm or duststorm. 6. A tropical disturbance with winds of at least 38 but less than 74 miles per hour. 7. A hurricane. 8. A disturbance on the sun. 9. A disturbance in the earth's ionosphere and/or geomagnetic field.

**storm surge** The rise in sea level that accompanies a tropical storm or hurricane. The winds drive the water out in front of the system, and the reduced pressure causes the water in and near the center to be pulled up several feet. With the right tidal conditions and certain coastal features, the sea level can rise 25 feet or more in extreme hurricanes. See *hurricane*.

**STP, s.t.p.** Abbreviation for standard temperature and pressure.

**strain** 1. A form of stress, caused by a pulling-apart force. Measured in pounds or kilograms per unit cross-sectional area, or in total pounds or kilograms. 2. A particular variety of microorganism, especially a bacterium or virus. 3. To filter. 4. A form of muscle injury caused by overexertion. Not usually serious, although it can be painful.

**strain gauge** A form of transducer that converts strain into a variable resistance or current. This usually is done by measuring small changes in the linear size of an object (stretching or compression). See *strain, transducer*.

**strange attractor** A pattern that develops in a certain form of chaotic movement called turbulence. It takes the form of a pair of loops. As a moving point follows the pattern, it approximately, but not exactly, retraces its path (see drawing).

STRANGE ATTRACTOR

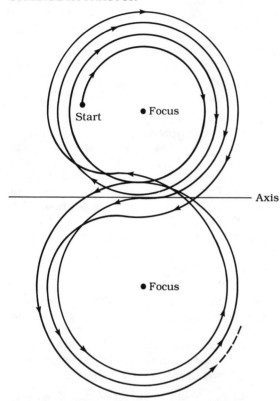

**strata** 1. Layers of rock beneath the surface of the earth, and also in exposed rock areas called outcrops. These layers provide information about the history of our planet. 2. Levels in any layered material or medium.

**stratigraphy** A branch of earth science concerned with the nature and origins of rock layers. See *strata*.

**strato-** 1. A prefix, denoting something that is flattened, or stratified. 2. Pertaining to stratified clouds. See *stratocumulus clouds, stratus clouds*.

**stratocumulus clouds** Flattened clouds at middle and low altitudes, usually having level bases and somewhat irregular tops. They are

STRATOCUMULUS CLOUDS
STRATUS CLOUDS

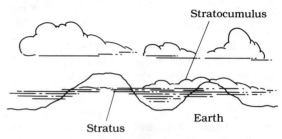

not very thick (see drawing). They generally do not produce precipitation.

**stratosphere** The level of the earth's atmosphere starting at about 8 to 12 miles up, and extending to about 40 or 50 miles. There is no weather in this part of the atmosphere, although turbulence sometimes occurs there. Some aircraft fly in this region.

**strato-volcano** A volcano whose cone has a layered, or stratified, structure, resulting from alternate ejection of lava and fragmented solid matter.

**stratus clouds** Clouds at low altitude, with flat bases and tops (see drawing). They ordinarily do not produce precipitation, although rain or snow may fall from clouds above. These clouds are associated with the "iron-gray" sky of dreary overcast. When they form at ground level, fog is the result.

**stream** 1. A small river or tributary. 2. A flow of fluid through a pipeline, blood vessel, the intestines or other defined medium. 3. A sequence of data bits in electronic communications.

**streamlining** Construction of a ship, motor vehicle, or aircraft to minimize drag against the water or air. This maximizes the fuel efficiency and also reduces turbulence.

**stream order** A definition of the size and volume of a stream. A stream of the first order is a small brook with no tributaries. A stream of the second order is any stream fed by a first-order stream. Third-order streams are fed by second-order ones, and so on, until the main river empties into the ocean.

**strep throat** A common streptococcal infection characterized by sore throat and fever. Treated with antibiotics such as penicillin. If left untreated this disease can become serious. Pneumonia is one possible complication. See *streptococcus*.

**streptococcus** A spherical bacterium that is responsible for a variety of infectious conditions. The most common is known as strep throat. Another, less common nowadays, is scarlet fever. The eyes, ears, and intestinal tract can also be affected. Treatment is with antibiotics.

**stress** 1. A force that tends to stretch or compress an object. Usually measured in pounds or kilograms per unit cross-sectional area. 2. Repeated mechanical force applied to a substance or object. 3. Any of various physical or emotional factors that place the human body under strain.

**striation** 1. Having minute grooves or channels. 2. A narrow line or band, especially when there are numerous such lines or bands in a parallel group. 3. A condition of being filament-like; an example is striated muscle.

**strip mining** A method of mining in which the land is stripped away mechanically, by blasting or by excavation. This method is effective when deposits are near the surface. It causes changes to the drainage pattern and can be environmentally destructive unless it is well managed.

**stroboscope** A device with a flashing light, called a strobe light, whose rate of flashing is adjustable and precisely measured. This allows accurate determination of the speed of fast-rotating objects (see drawing). Used in a variety of industrial situations.

**stroke** A blockage or rupture of one of the small arteries in the brain, causing death of brain cells in the region supplied by the artery. This often causes paralysis of a certain part of the body. Death can occur in severe cases. This is one of the leading causes of disability in the United States today.

**stromatolite** A fossil of ancient algae. It is found as a layered, sedimentary rock.

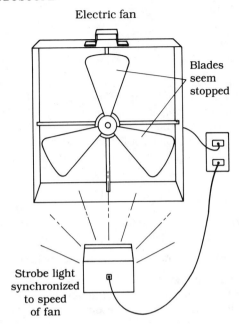

STROBOSCOPE

Electric fan

Blades seem stopped

Strobe light synchronized to speed of fan

**strong force** In an atom, the attractive force that keeps the particles in the nucleus from flying apart. The protons, all having positive electric charge, repel each other; the attractive force, however, is stronger.

**strontium** Chemical symbol, Sr. An element with atomic number 38. The most common isotope has atomic weight 88. In pure form it is a metal. It reacts easily with many substances. See *strontium-90*.

**strontium-90** A radioactive isotope of strontium present in the fallout from nuclear explosions. It became the focus of a controversy during the late 1950s and 1960s, when nuclear tests were conducted above ground. It tends to accumulate in the food supply, and is stored in the bones. The risk was overcome by the Test Ban Treaty. See *strontium*.

**structural engineering** A branch of engineering concerned with the details of building construction. In some regions, high winds and/or earthquakes present a special hazard, and buildings must be able to withstand most hurricanes or earthquakes. In other regions, ground creep, subsidence, and other factors must be considered.

**structural geology** The study of the formation and evolution of various features in the earth's crust, by analyzing their structure.

**stupor** A condition of diminished consciousness. The person is not asleep, but cannot respond normally to stimuli. A good example is a very drunken person. This condition can be a complication of infectious diseases, poisoning, and mental illness.

**stylus** The "needle" in a phonograph (common record player). It is a transducer, converting the mechanical vibrations into electric currents. The needle itself is made from a hard mineral, usually sapphire.

**sub-** A prefix meaning under or inferior to.

**subcarrier** A low-frequency carrier wave that modulates a higher-frequency wave. Often used in frequency-modulation (FM) broadcasting. The low-frequency carrier is itself modulated; this information can only be recovered by using special receivers. See *carrier*.

**subclavian artery** The major artery that leads into the arm. It branches off from the aortic arch. There is one such artery for either arm.

**subclavian vein** The major vein that leads from the arm into the superior vena cava, and from there to the heart. There is one such vein for either arm.

**subduction** The downward movement of a plate of the earth's crust, in a region where two plates are being compressed against each other (see drawing). The downward-moving plate is eventually assimilated by the mantle.

**subduction zone** A region where an oceanic plate dives beneath a continental plate in the earth's crust. This results in the formation of a trench, or very deep zone, in the ocean floor.

**subharmonic** A signal or wave, whose frequency is some integral fraction ($1/2$, $1/3$, $1/4$, etc.) of the frequency of the main signal or wave. See *harmonic*.

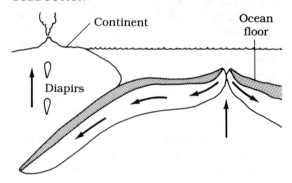

From Erickson, THE MYSTERIOUS OCEANS (TAB Books, Inc., 1988).

**sublimation** A change of state of matter from solid to gas, without any intervening liquid state. A good example is the evaporation of dry ice in air.

**submarine** 1. Pertaining to underwater activities or devices. 2. A vessel capable of being operated underwater.

**submarine cable** A communications cable that is laid underwater. Transatlantic telegraph cables were the earliest such lines. Nowadays, fiberoptic cables are replacing older wire cables. See *fiberoptics*.

**subroutine** A small program within a computer program, that performs a specific function. Often a given subprogram is run many times during the course of the main program.

**subsidence** Lowering of the ground level over a period of time. This often takes place in an irregular fashion, causing damage to buildings. It might take place suddenly during an earthquake, resulting in catastrophic structural damage.

**subsolar point** The point on the earth from which the sun appears directly overhead. This point is always in the tropics, that is, between 23.5 degrees north and south latitude. The exact latitude varies with the time of the year. It revolves around the planet once every 24 hours.

**subsonic** 1. Moving at a speed less than the speed of sound. 2. Pertaining to operation of aircraft at speeds less than that of sound. See *supersonic*.

**sucrose** A simple carbohydrate, consisting of a molecule of glucose attached to a molecule of fructose. This is common table sugar. It occurs naturally in many foods, especially fruits. See *carbohydrate*.

**sugar** 1. Any monosaccharide or disaccharide; that is, a simple carbohydrate. Common examples are glucose, fructose, sucrose and lactose. 2. See *sucrose*. See *carbohydrate*.

**sulfate** A salt of sulfuric acid. See *sulfuric acid*.

**sulfide** 1. A compound containing sulfur, usually attached to two carbon atoms. 2. A salt of hydrogen sulfide. 3. A compound containing sulfur and some other element that is more electropositive.

**sulfite** A salt of sulfurous acid. See *sulfurous acid*.

**sulfur** Chemical symbol, S. An element with atomic number 16. The most common isotope has atomic weight 32. In pure form it is a yellow solid, often occurring as a powder. It has a rotten-egg smell. It is an important component in certain proteins. It has numerous industrial uses. Sulfur-containing coal is known for causing air pollution.

**sulfur dioxide** A gas produced by burning (oxidation) of sulfur. It is an eye and lung irritant. An especially high amount of this gas is released when low-grade, high-sulfur coal is burned. It causes acid rain. See *acid rain*.

**sulfuric acid** Also called tetraoxosulfuric acid because its molecule has four oxygen atoms. An oily, caustic acid with numerous industrial uses.

**sulfurous acid** Also called trioxosulfuric acid since its molecule has three oxygen atoms. A fairly weak acid, it is thought to occur in acid rain from the dissolving of sulfur dioxide in atmospheric water droplets. See *acid rain*.

**Sullivan, Louis** Sometimes called the "pioneer of skyscrapers." An architect who developed methods of constructing large buildings.

**summer solstice** See *solstice*.

**sun** 1. The medium-sized, moderately "cool" star (as stars go) at the center of our Solar System. It is 864,000 miles in diameter at visible-light wavelengths and is about 93 million miles distant. 2. Radiation from this star. See *star*.

**sunblock** A cream or lotion that can be applied to the skin to shield against ultraviolet radiation, preventing sunburn and suntan. See *sunburn, sunscreen, suntan*.

**sunburn** A reddening of the skin that appears from two to four hours after exposure to ultraviolet radiation from the sun or from sunlamps. In severe cases, the skin blisters and peels. The condition can be extremely painful. Frequent sunburn, over a period of years, is correlated with skin cancer.

**sundial** An ancient device for timekeeping using the sun. A central post (gnomon) is set parallel to the earth's axis. A circular scale is calibrated in hours. The shadow cast by the gnomon tells the time to within about plus-or-minus 1/2 hour. See *gnomon*.

**sunscreen** A cream or lotion that can be applied to the skin to prevent sunburn, while still allowing some suntan. See *sunblock, sunburn, suntan*.

**sunspot** A region on the surface of the sun, that is darker than the surrounding surface and also is cooler. These occur in a cycle (see *sunspot cycle*). They reflect an increase in overall solar activity, and are thought to be gigantic storms.

**sunspot cycle** A variation in the number of observed sunspots. The count reaches a maximum about every 11 years (see drawing on the next page). The magnetic polarity of the spots alternates, so the actual complete cycle is 22 years. Peak levels vary from cycle to cycle. There are thought also to be supercycles, lasting about 100 years; the last major peak of this was in 1957–58. The variability in sunspot numbers is thought to be correlated with changes in overall solar activity, and therefore with climate fluctuations. The sunspot activity affects the behavior of the earth's ionosphere, causing

## SUNSPOT CYCLE

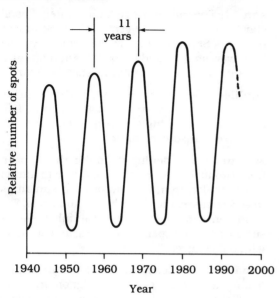

changes in the way radio waves are propagated. See *ionosphere, Maunder minimum, sunspot.*

**suntan** A darkening of the skin, caused by repeated exposure to ultraviolet from the sun or from sunlamps. It results from an increased amount of melanin (pigment) in the skin.

**super-** A prefix meaning over, extreme, high, or ranking above.

**superconductivity** 1. A condition that occurs in some substances at very low temperatures. The electrical resistance decreases to essentially zero, so that currents can flow in loops for a long time before dying down. 2. A field of research in which exciting new discoveries are being made, especially relating to electronic and electromagnetic devices, involving extremely low electrical resistances.

**superconductor** A substance with an electrical resistance that is practically zero. See *superconductivity.*

**supercontinent** 1. See *Pangaea.* 2. See *Gondwanaland.* 3. See *Laurasia.* 4. The huge land mass that might eventually result from the continents of the earth drifting back together again. See *continental drift.*

**supercooling** 1. Cooling of water droplets below the freezing point (zero degrees centigrade or 32 degrees Fahrenheit) without solidification. Occurs in some clouds. A sudden shock wave or other disturbance causes crystallization.

**superheterodyne radio receiver** Also called a superhet. A communications receiver, with a mixer that produces a constant output frequency, regardless of the frequency of the received signal. This constant output frequency is much easier to filter and amplify than the variable input frequencies.

**superhigh frequency** Abbreviation, SHF. The part of the radio spectrum from 3 to 30 gigahertz (GHz). The wavelength range is 10 centimeters (cm) to 1 cm. These are considered microwaves. They propagate essentially in straight lines, and are not affected by the ionosphere. See the RADIO SPECTRUM appendix.

**superior** 1. Of high quality. 2. On the far side of (see *superior conjunction*). 3. Higher than.

**superior conjunction** The alignment of Mercury or Venus with the sun, but on the far side (see drawing). See *inferior conjunction.*

## SUPERIOR CONJUNCTION

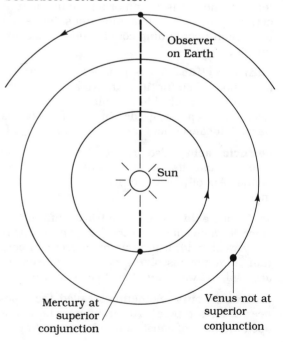

**superior vena cava** The major vein leading to the heart from the part of the body above the heart: the arms, shoulders, head, and neck. The jugular vein, subclavian veins, and axillary veins flow into it; in turn, blood flows into the right auricle of the heart. See *heart*.

**supernova** The explosion of a star. This occurs with many stars that are at least several times as massive as our sun. It probably will not happen to the sun. An exploding star can become as bright as all the stars in its galaxy combined. Our Solar System probably formed from the debris from a supernova that took place billions of years ago.

**superposition theorem** A rule for electronic circuits. It simplifies the determination of currents and voltages in complex direct-current (dc) circuits. See an electronics text for details.

**superregenerative radio receiver** An early type of radio receiver, used at medium and high frequencies. It employed positive feedback in the amplifying stages to increase the sensitivity. Superheterodyne receivers are used today. See *superheterodyne radio receiver*.

**supersaturation** A condition in which the atmosphere has a relative humidity greater than 100 percent. Condensation usually prevents this situation. See *relative humidity*.

**supersonic** 1. Moving at a speed greater than the speed of sound. 2. Pertaining to the operation of aircraft at speeds greater than that of sound. See *subsonic*.

**supersonic transport** An aircraft designed to carry passengers long distances, at high altitudes, and at speeds greater than the speed of sound. This aircraft requires a long runway. It is not commonly used because of the noise it causes, particularly sonic booms. There is also concern that its high-altitude flight could cause damage to the ozone layer. See *ozone layer, sonic boom*.

**superstructure** 1. The portion of a ship that is above the top, or highest, full deck. 2. A part of a building added above the roof, containing such hardware as communications antennas.

**surface** 1. The two-dimensional boundary, in a three-dimensional object, between the interior and the exterior. 2. In an object having n dimensions, the n−1-dimensional boundary between the interior and the exterior.

**surface area** The total number of square units (such as millimeters, meters, or miles) in the surface of a three-dimensional object. Increases in proportion to the square of the radius.

**surface-mount technology** A recently developed way to build electronic circuits in compact enclosures. Components are mounted right on the circuit board, without using wire leads. This minimizes interaction between the different parts of a circuit, as well as reducing the physical size of devices.

**surface tension** A tendency for water molecules to create a "cap" at the surface of a container. This produces a characteristic upturning of the surface (see drawing). It is responsible for capillary action. See *capillary action*.

SURFACE TENSION

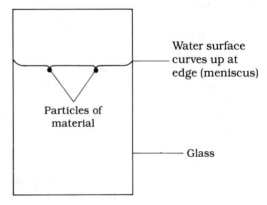

**surface water** Water in the oceans (including frozen portions of the Arctic and Antarctic), inland seas, lakes, rivers, and ponds. This accounts for about 70 percent of the surface of the earth.

**surface wave** In radio communications, a vertically polarized electromagnetic wave that follows the curvature of the earth. The earth serves as part of the circuit. This occurs only at frequencies up to a few megahertz.

**surfactant** A hydrocarbon that interferes with the exchange of atmospheric gases and water vapor at the surface of a body of water. Such substances are produced by human activity and can cause pollution. They are also used as soaps. Natural varieties help to keep nutrients and oxygen in the water.

**surgery** Any medical procedure in which a part of the body is removed or modified. Examples are liver transplants, face lifts, and the removal of tumors.

**surgical stapler** A precision device, resembling an ordinary paper stapler, used to hold nerves and blood vessels in place while they heal after surgery. Originally invented by Russian engineers.

**Surveyor** A series of space probes that landed on the moon. The first was launched on May 30, 1966 and sent 10,000 pictures back to Earth-based stations. Later probes analyzed chemicals in the lunar soil, and found that the texture of the lunar surface was suitable for human-piloted craft to land.

**susceptance** The reciprocal of reactance. It is a form of alternating-current conductance. See *conductance, reactance*.

**susceptibility** 1. The capacity for a substance to become magnetized. Generally expressed as the ratio of the magnetization to the amount of magnetizing force applied. 2. The ease with which a system is affected by some external factor, effect or force.

**suspension** 1. A mixture of a solid in a liquid, where the solid is not dissolved, but stays evenly distributed because it is ground up into tiny particles. 2. The condition of hanging under tension.

**suspension bridge** A form of bridge in which just two major supports are used. The bridge itself hangs from a set of cables (see drawing). This type of bridge is especially well suited for deep canyons and rivers where numerous supports cannot be built.

**Sv** Abbreviation for sievert.

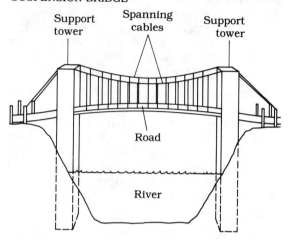

SUSPENSION BRIDGE

**swamp** A waterlogged area of land supporting a vast number of different species of plants and animals. These regions exist at practically any latitude where the ground can thaw during at least part of the year.

**swamp gas** Any volatile gas that can accumulate near the ground in swampy areas. Methane is a common example. A lightning discharge or a lighted match can ignite this gas and produce a flash or explosion.

**S-wave** A secondary, or shear type, earthquake wave. It penetrates well beneath the surface and is also called a body wave. It travels at about 10,000 miles per hour, and causes vibration from side to side.

**sweat** Technically called perspiration. The liquid, mostly water but also containing some salt and waste products, that is released from glands in the skin when the body temperature rises to a certain level. See *sweat gland*.

**sweat gland** Any of the numerous glands in the skin that produce perspiration to cool the body when its temperature rises too high. The cooling is accomplished by evaporation of the sweat from the skin.

**sweep** 1. A motion between two limits, such as from a minimum to a maximum frequency (see *sweep generator*). 2. A smooth, uninterrupted motion. 3. In television, the horizontal movement of the electron beam in the camera tube or picture tube. 4. The left-to-right movement of the electron beam in a cathode-ray tube.

**sweep generator** An oscillator that produces a variable-frequency output signal. The frequency starts at some preset lower limit, increases at a rapid and constant rate to some preset upper limit, and repeats this many times per second. Used for testing of electronic equipment, especially radios.

**symbiosis** A condition in which two or more different species of animals or plants exist together, mutually benefiting each other.

**symmetric property** See *trichotomy law*.

**symmetry** 1. Having identical shape on either side of a line. 2. Having identical shape or displacement in opposite directions from a point.

**sympathetic** 1. Occurring in phase with; following along. 2. Tending to work together. 3. Concerning a certain function and part of the nervous system. See *sympathetic nervous system*. 4. Producing or having resonance. See *resonance*.

**sympathetic nervous system** A part of the autonomic nervous system (see *autonomic*). The main functions are to relax muscles and to decrease the diameter of the blood vessels. These functions are involuntary; they are not affected by conscious effort.

**sympathetic vibration** Mechanical vibration that occurs as a result of resonance (see *resonance*). An object might be set in oscillatory motion by some nearby vibrating object, if the physical dimensions and the frequency of oscillation are just right.

**synapse** A junction, or gap, through which nerve impulses pass from the axon to the dendrite. See *axon, dendrite*. It bears some theoretical resemblance to an electrical spark gap. See the illustration.

**synchro** A pair of generator-motor devices used for mechanical control. When the rotor of one device is turned, the other follows along.

SYNAPSE

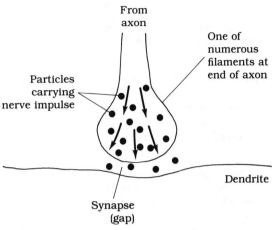

Employed in remote-control and robotic systems.

**synchrocyclotron** A type of cyclotron (see *cyclotron*) in which the mass of the accelerated particles increases relativistically. The accelerating charge polarities are switched back and forth, and are synchronized to follow the revolution of the particles, driving them to extreme speed.

**synchronization** The coinciding of one process, event or operation with another or others. For example, we might set several clocks according to the signal from an independent time standard. Then any one of the clocks is synchronized with any one of the others.

**synchronized communications** Any communications in which the receiver and transmitter operate together, based on signals from some independent time-and-frequency standard. This greatly enhances the sensitivity and accuracy of the system. The receiver "knows" when to "expect" data bits from the transmitter.

**synchronous** 1. A condition of following along, electrically or mechanically. 2. Having identical frequency. 3. Having an orbital period identical to the rotation rate of the earth. See *geostationary orbit, geostationary satellite*.

**synchronous orbit** See *geostationary orbit*.

**synchronous satellite** See *geostationary satellite*.

**synchro system** A device or machine that uses synchros (see *synchro*) for the purpose of remote control.

**synchrotron** An atomic particle accelerator, similar to a cyclotron (see *cyclotron*). Electric and magnetic fields operate in cyclic fashion, and they are synchronized with the circular motion of the subatomic particles to be accelerated. The result is extreme particle speed.

**synchrotron radiation** High-energy electromagnetic radiation that takes place when charged particles are accelerated in tight circular orbits. It occurs in narrow, intense beams. See *synchrotron*.

**syncline** 1. A huge bulge in the earth's crust beneath the ocean, caused by the weight of sedimentary deposits. 2. A U-shaped fold in rock strata on the surface.

**syncope** Sudden loss of consciousness with no warning symptoms. Can take place as a reaction to certain drugs, and especially if depressant drugs are taken with alcoholic drinks.

**synergy** A condition in which two or more effects act together, and the result is greater than the sum of the individual effects. A dangerous, and possibly lethal, example is the combination of alcohol with depressant drugs.

**synod** For a given celestial object, the time required for it to return to the same position, with respect to the sun as a reference standard.

**synodic** 1. Pertaining to the sun as an astronomical standard of reference. 2. Pertaining to the synod of a celestial object. See *synod*.

**synodic day** The time required for the earth to rotate exactly 360 degrees (one full circle) relative to the line connecting the earth with the sun (see drawing). This is the basis for timekeeping, and is 24 hours long.

**synoptic weather mapping** The analysis of weather observations, gathered from many different points over a large area, and the drawing of detailed maps showing weather systems.

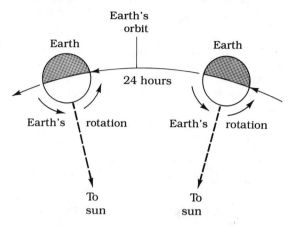

SYNODIC DAY

These maps are used by meteorologists in predicting weather phenomena.

**synovial fluid** The liquid in the joints of the body that serves as lubrication, minimizing friction and maximizing efficiency and durability. It has a consistency similar to raw egg white.

**synthesis** 1. Creation by artificial means. 2. The generation of a radio signal of precise frequency, based on a standard crystal oscillator and employing multipliers, dividers, and phase locking.

**synthesizer** 1. A variable-frequency oscillator that uses a fixed crystal standard, and that obtains its output by multiplying, dividing, and phase locking. 2. See *Moog synthesizer*.

**synthetic** Made artificially. This applies especially to mass-produced industrial substances such as plastics and textiles.

**synthetic compound** Any chemical compound produced by human manufacture, and not obtained from natural sources. We might synthesize sodium chloride (ordinary salt), for example, by allowing sodium and chlorine to react. More often, complex compounds are obtained by artificial means. See, for example, *synthetic fabric*.

**synthetic fabric** A human-made fabric, such as nylon. Others include dacron and rayon. These are trade names that have become so commonly used that they seem like regular words. But the fibers are synthesized; they do not occur in nature.

**syphilis** A sexually transmitted disease, caused by a spirochete (see *spirochete*). It is treated with antibiotics. It was once called *the great masquerader* because of its horrible and unpredictable effects such as paralysis and insanity.

**systematics** A branch of life science concerned with the natural way in which different species interact.

**systems analysis** A way of determining the factors in a problem, and feeding them into computers to find solutions. This is especially applicable when problems have several or many variables. It is used for such things as the design of spacecraft, plotting courses to the planets, and in economics.

**systems engineering** The design and perfection of complex machines, using systems analysis. See *systems analysis*.

**systole** The portion of the heartbeat during which the ventricles are contracted. This is when the blood pressure reaches a sharp maximum. See *diastole, systolic*.

**systolic** Maximum pressure; pertaining to the blood pressure when the left ventricle is contracted. Normal systolic pressure is about 100–130 millimeters of mercury (mmHg) above atmospheric pressure. See *diastolic*.

**syzygy** 1. The times of full moon or new moon when the earth, moon, and sun all lie along a straight line. This is when the tidal effects are greatest. See *spring tide*. 2. The near alignment of most or all of the planets on the same side of the sun. 3. Any time, for a particular planet other than the earth, when that planet, the earth, and the sun all lie along a straight line. See *inferior conjunction, opposition, superior conjunction*.

**Szilard, Leo** A Hungarian physicist and refugee in the mid-twentieth century. He was one of the people who convinced Albert Einstein, just before the start of World War II, to urge President Franklin Roosevelt to hire scientists to develop the atomic bomb.

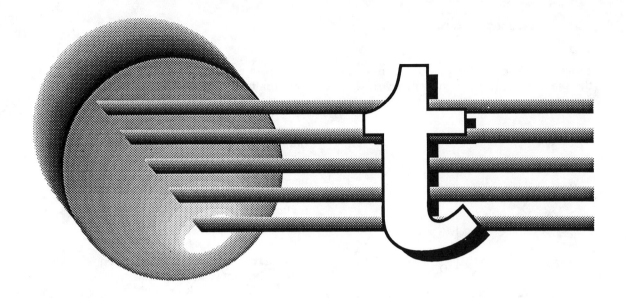

**T** Abbreviation for period, temperature, tera- (see the PREFIX MULTIPLIERS appendix), tesla.

**t** Abbreviation for time, ton, tonne.

**Ta** Chemical symbol for tantalum.

**tachometer** A device that measures angular speed. Usually calibrated in revolutions per minute (RPM) or per second (RPS). Commonly used to monitor the operation of high-performance internal combustion engines.

**tachycardia** Abnormally rapid heartbeat at rest. A rapid beat is normal during strenuous exercise, especially in young people.

**tachyon** A hypothetical subatomic particle that moves faster than the speed of light. It therefore has imaginary time; in a sense it goes "sideways" in time relative to the universe as we know it.

**tacking** In sailing, a way of navigating into the wind. The boat angles back and forth, never exactly against the wind, and makes progress by repeatedly coming about and changing the sail position (see drawing).

**taconite** Crude iron ore. It is more difficult to refine than higher grades, but there is abundant iron in it. Recent methods of ore processing have improved our ability to get iron from this ore.

TACKING

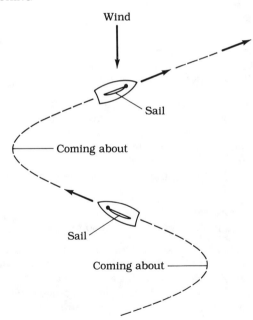

**tan** Abbreviation for tangent.

**tangent** Abbreviation, tan. On the unit circle, the tangent is given by the value y/x. See the drawing.

TANGENT

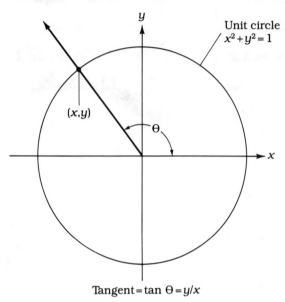

Tangent = tan θ = y/x

**tanh** Abbreviation for hyperbolic tangent.

**tank circuit** An electronic circuit containing an inductance and a capacitance in parallel. At a certain frequency the reactances cancel, and the circuit resonates. At this frequency, the impedance is very high. See *capacitive reactance, impedance, inductive reactance, resonance*.

**tannic acid** A mild acid found naturally in certain plants and used in some dyes.

**tannin** A chemical found naturally in certain plants. It is used to tan leather. It has a characteristic bitter taste. It can be used medicinally as an astringent. When the substance is refined, tannic acid is obtained. See *tannic acid*.

**tantalum** Chemical symbol, Ta. An element with atomic number 73. The most common isotope has atomic weight 181. In pure form it is a blue metal. A common use of this metal is in the manufacture of small, high-capacitance electronic components. See *tantalum capacitor*.

**tantalum capacitor** A type of capacitor that has small size, high capacitance and moderate operating voltage. It has excellent reliability. This type of capacitor has replaced bulkier electrolytic capacitors in many modern applications.

**tape** 1. A long strip of paper, now seldom used, once employed to store data by means of punched holes. 2. A long strip of plastic or mylar with magnetic particles fused on one side, used for audio and video. See *magnetic tape*.

**tape drive** A motor and associated regulating heads that move a magnetic tape through a recorder or playback device. The motor must run at a very precise, constant speed for the sound and/or video reproduction to be accurate and of high quality. See *tape recorder*.

**tape recorder** 1. An audio and/or video device that converts sounds and/or images into electrical impulses, and encodes these impulses on a magnetic tape. See *magnetic tape*. 2. An audio device that records sound on, and plays it back from, magnetic tape.

**tapeworm** A long, segmented parasite that sometimes grows in the intestinal tract. It may become several feet long. It causes malnutrition by robbing the body of calories, vitamins and minerals. The parasite can be killed by appropriate medication.

**tap root** A deep root system characteristic of certain types of plants. It enables the plant to obtain water even in arid regions. It has a main, carrot-shaped central shaft that sometimes reaches depths of several tens of feet (see drawing).

**taro** A vegetable that resembles a potato. It is a staple in the diets of the natives of some Pacific islands.

**tar pit** A pocket of bituminous material ("tarry ooze") that preserves fossils. When this pocket is exposed at the surface, fossils are easily recovered.

**tarsus** The bone that joins the shin bones (tibia and fibula) to the foot. Also can be called the "ankle bone."

TAP ROOT

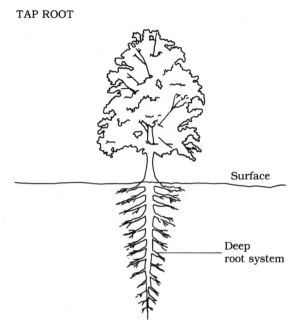

**Tartaglia, Niccolo** A sixteenth-century mathematician who first used his craft in military applications. He wrote textbooks as well, and specialized in arithmetic and algebra.

**taste buds** Tiny organs on the tongue that detect tastes: sweet, salty, bitter, and sour. In various combinations, these sensations give rise to all the tastes of food, like the way primary colors can be combined to produce any color.

**tautology** 1. The use of a term to define itself. 2. A logically true statement under all conditions.

**taxis** Motion of a cell, caused by some outside factor. Such factors might be heat, cold, light, darkness, oxygen, or some poisonous substance. Usually the motion is either towards or away from the source of the factor.

**taxonomy** 1. The classification of living things according to their characteristics. 2. The science of categorizing things or data.

**Taylor, Brook** A British mathematician of the eighteenth century. He graduated from Cambridge University. He contributed much to the theory of perspectives. He is remembered especially for his infinite-series expansions of functions. See *Taylor series*.

**Taylor, Frederick** An American engineer of the late nineteenth and early twentieth centuries. Best remembered for his new approach to industrial management. He improved working conditions and efficiency according to scientific principles, including statistics and psychology.

**Taylor series** An infinite series that expands the value of a function $f(x+k)$, where $x$ is a variable and $k$ is a constant. This is a rather complicated series, similar to other series in which derivatives appear. See a text on advanced calculus for details. See *Taylor, Brook*.

**Taylor, Theodore** An astronomer and engineer who helped to conceive the idea of using nuclear bombs to propel an interstellar spacecraft. The ship was named Orion and might attain speeds up to about $1/10$ the speed of light. Such craft have not yet been built.

**Tb** Chemical symbol for terbium.

**Tc** Chemical symbol for technetium.

**Te** Chemical symbol for tellurium.

**tear duct** The tube leading from the tear gland to the corner of the eye. Tears pass from the tear gland through this duct to lubricate the eyeball. From the eyeball they pass through another duct into the nasal cavity. This is why you get the sniffles when you cry or have irritated eyes. See *tear gland*.

**tear gland** A gland just above and to the outside of either eyeball (in humans), also called the lacrimal gland. Secretes a watery fluid that keeps the eye clean and lubricated.

**technetium** Chemical symbol, Tc. An element with atomic number 43. The most common isotope has atomic weight 99. In pure form it is a metal, and occurs when the uranium atom is split by fission.

**technician** A person who tests and/or repairs machines and electronic equipment.

**Technological Revolution** The past 100 years, and especially the time since 1900, when the pace of technological advances increased dramatically. Today, the changes happen so fast that individuals must specialize, and it is impossible for any one person to keep up with all changes in all fields.

**technology** 1. The development of new devices, machines and industrial techniques. 2. The hardware in industrial, consumer, scientific and military devices. 3. The general level of sophistication in hardware.

**tectonic activity** The formation of the earth's crust by large-scale geologic movements over periods of millions or billions of years. See *plate tectonics*.

**tectonic plate** A large section of the earth's crust, that moves with respect to other sections. See *continental drift, plate tectonics*.

**teflon** A trade name, now become generic, for a synthetic, plastic-like substance with high durability and excellent friction-reducing properties. Used in various industrial applications, such as bearings and wire coatings. The technical name is polytetrafluoroethane.

**tektite** A certain type of rock found at widely separated points on the earth. Their origin is somewhat a mystery. One theory is that a large meteorite struck the moon, and some of the debris fell all the way to the earth. Another theory is that a large meteorite hit the earth and threw debris over a wide region (see drawing).

**telecommunication** Communication between widely separated points, using electronic or electrical equipment. This includes wire telephones, two-way radios, satellite links and laser devices.

**telegraph** 1. An early wire communication system, in which direct current was switched on and off by means of a key, sending dots and dashes. Still used to a limited extent. 2. Communications by means of the aforementioned circuit.

**telemetry** 1. The signals sent back to earth stations from satellites and spacecraft. These signals tell ground personnel how the satellite or spacecraft is operating, and also send data concerning observations made. 2. The signals sent by an unmanned probe back to a ground-based station.

TEKTITE

**telephone** 1. A means of sending voices over wires, and in recent years, also by means of microwave links and optical fibers. 2. A public communications network with which any subscriber may talk to any other at any time. 3. A unit or set used by a subscriber to the aforementioned system.

**telephone dialer** See *rotary dialer, Touchtone dialer*.

**telephone network** The system of wires, optical fibers, microwave links, computer-controlled switches, and sets that are used for public communications. See *telephone*.

**telephonoscope** The name coined by the nineteenth-century writer Albert Robida, for a device that would allow people to talk to, and watch, each other. Nowadays the devices are just being put into common use, and are called picture telephones.

**telephony** The branch of electrical engineering concerned with the design and operation of telephone systems. See *telephone, telephone network*.

**telescope** A device consisting of lenses and /or mirrors, that enlarges distant images. These instruments are used mostly to observe celestial objects. The three most common types are shown in the drawing.

TELESCOPE

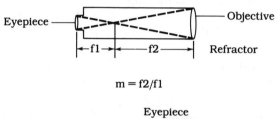

$m = f2/f1$

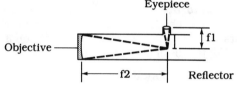

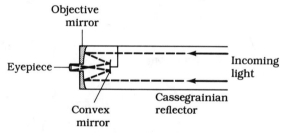

From Gibilisco, COMETS, METEORS AND ASTEROIDS (TAB Books, Inc., 1985).

**teletype** Trade name, now become generic, for typewritten messages sent by wire or radio.

**television** 1. The transmission of moving pictures by electronic means. 2. The modulation of a radio-frequency carrier with a picture signal. 3. A receiver for pictures sent electronically.

**Telex** Trade name for a method of sending teletype messages over a common telephone. This has been used extensively by corporations throughout the world. Nowadays, facsimile, or "fax," is more common. See *facsimile*.

**Telford, Thomas** A civil engineer of the late eighteenth and early nineteenth centuries. He is known for his work on the road system in the British Isles. In Wales, he constructed one of the first functional suspension bridges.

**tellurium** Chemical symbol, Te. An element with atomic number 52. The most common isotope has atomic weight 130. In pure form it is a metal, and is found along with other metals such as copper. It is used in certain electronic semiconductors.

**Telstar** The first communications satellite, launched on July 10, 1962. It was used to send television pictures overseas. It was in a low orbit, and it could be used only while it was in the right position. Modern satellites can be used all the time, because they are in geostationary orbits. See *geostationary satellite*.

**temperament** 1. Nature of nervous response. 2. Nature of response to stimuli, either for an organism or (increasingly) for a complex machine.

**temperate** Pertaining to a moderate climate, neither extremely hot nor extremely cold. See *temperate zone*.

**temperate zone** 1. The regions between 23.5 and 66.5 degrees north and south latitude. The drawing shows the zone in the Northern Hemisphere. 2. Either of the regions where the prevailing winds are mainly from the west, and the climate is neither tropical nor polar. These correspond approximately to the latitudes between 23.5 and 66.5 degrees north and south.

TEMPERATE ZONE

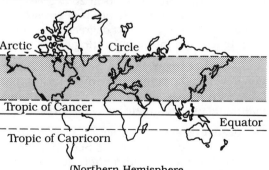

(Northern Hemisphere zone shown)

**temperature** 1. An expression of the heat energy content, and heat behavior, of something. There are four scales generally used for this. See *centigrade temperature scale, Fahrenheit temperature scale, Kelvin temperature scale, Rankine temperature scale*. 2. Thermodynamic temperature: a measure of the kinetic energy in molecules or atoms of a substance. The greater this energy, the faster the particles are moving, and the higher the temperature. This is the method that we most often use. 3. Spectral temperature: A measure of the peak wavelength of electromagnetic energy that an object emits. The shorter the wavelength, the higher the frequency, and the higher the temperature. This method is used by astronomers to gauge how hot the stars and nebulae are.

**temperature coefficient** A number that indicates how a device behaves with changes in temperature. This number might be positive, zero, or negative indicating an increase, no change, or a decrease in component value with a rise in temperature.

**temperature compensation** Canceling out the effects of temperature coefficient, so a device is stable over a range of temperatures. We might put components together, some with positive temperature coefficients and some with negative, to get a net coefficient of zero. See *temperature coefficient*.

**temperature derating** A method of adjusting the ratings of a component or device to compensate for temperature. A radio transmitter might be usable at 100 watts at 20 degrees centigrade, but only 80 watts at 60 degrees centigrade and 60 watts at 100 degrees centigrade. We might graph this as a derating curve.

**temperature inversion** A weather phenomenon in which the temperature rises, rather than falls, with increasing altitude. This can cause pollutants to accumulate near the surface.

**temperature scale** A method of measuring temperature quantitatively. See *centigrade temperature scale, Fahrenheit temperature scale, Kelvin temperature scale, Rankine temperature scale*.

**tempering** 1. A way of enhancing the durability of a metal that is to be put to heavy use. This is done by controlled heating when the alloy is processed. 2. A controlled heating- and cooling-process used in the manufacture of some glass panes, so that they will break into little squares without sharp edges. Used especially in glass doors and in motor-vehicle windshields.

**temporal** 1. Pertaining to time as a dimension. 2. Pertaining to the sides of the head or skull. See *temporal bone, temporomandibular joint*.

**temporal bone** Either of the bones on the sides of the skull, just behind the eyes, and in the vicinity of the ears.

**temporomandibular joint** The joint, on either side of the head, just below and in front of the ears, where the lower jaw (mandible) is attached to the skull (see drawing).

TEMPOROMANDIBULAR JOINT

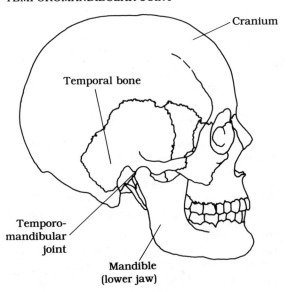

**tendon** A tough, fibrous tissue that serves to attach some muscles to the bones. Usually thicker, but less flexible, than ligaments. See *ligament*.

**tensile strength** The amount of pulling-apart force, or tension, that a wire or cord can withstand before it breaks. Usually measured in pounds or kilograms.

**tension** 1. The tautness of a wire or cord, usually measured in pounds or kilograms. 2. Nervous irritability. 3. The relative level of voltage on a power line.

**tephra** Solid matter that comes from an erupting volcano. This includes rock, ash and dust. The finer particles often reach the upper atmosphere, where they can encircle the earth.

**tera-** A prefix meaning 1 trillion ($10^{12}$ or 1,000,000,000,000). See the PREFIX MULTIPLIERS appendix.

**terbium** Chemical symbol, Tb. An element with atomic number 65. The most common isotope has atomic weight 159. In its pure form it is a silver-colored metal.

**Terman, Frederick E.** The first director of the communications laboratory at Stanford University in California. This facility was set up in 1924 after Lee De Forest invented the triode vacuum tube.

**terminal** 1. A device for sending and receiving messages, especially in written form, such as facsimile or teletype. 2. A teletype device that can be used via telephone to access a computer. 3. Resulting in death; fatal.

**terminal unit** 1. See *terminal*. 2. A device that demodulates radioteletype signals, so that a terminal can display them or print them out with a minimum of errors. Also converts the teleprinter impulses to a form suitable for transmission by radio. See *modem*.

**terminating decimal** A number in decimal form that has a finite number of digits. After this, all the rest of the digits are zero. These always represent rational numbers. See *nonterminating decimal, rational number*.

**termination** 1. The end point of a line segment. 2. The load of an electrical circuit; the device in which power is ultimately dissipated. 3. Conclusion.

**terminator** Also called the gray line. The great circle on the earth's surface, representing all the points from which the sun is on the horizon. The dividing line between the light and dark sides of the earth.

**terrace** A part of the floodplain of a river, where the river no longer rises. It is a flat, elevated platform on one or both sides of the river valley. See *floodplain*.

**terrestrial** 1. Pertaining to the earth's surface. 2. Pertaining to the land surface, rather than the water surface, of the earth.

**Tertiary period** The earlier part of the Cenozoic era, beginning about 65 million years ago and ending 2 million years ago. It encompasses the Paleocene, Eocene, Oligocene, Miocene, and Pliocene epochs. See the GEOLOGIC TIME appendix.

**tertiary rock** Rocks that contain oil and natural gas. These are found in the southeastern and western United States. The rocks also are sources of nonmetallic minerals such as clay, salt, limestone, gypsum, and phosphates.

**tesla** Abbreviation, T. A unit of magnetic flux density, equal to one weber per square meter, or 10,000 gauss. See *gauss, weber*.

**Tesla coil** An air-core, step-up transformer for developing high voltages at radio frequencies. It can produce a spark up to several inches long. But the current is quite low. See *transformer*.

**Tesla, Nikola** A Russian scientist and inventor, who made the first demonstration of radio-wave phenomena in 1893. He got a patent for his radio device in 1900—before Marconi. It was not until 1943 that Tesla was officially recognized for his primary role in the invention of radio.

**test** 1. A means of evaluating the performance of a device, to see if it meets specifications. 2. An experiment or set of experiments, conducted to find the problem when a device or system is not working correctly. 3. To do either of the foregoing. 4. See *exoskeleton*. 5. A slang expression for the performance-delivering ability of a fuel; generally refers to the octane level.

**testes** See *testicles*.

**testicles** The male sex glands, located just underneath the penis. These produce sperm and various hormones. For illustration see *endocrine glands*.

**test instrument** A laboratory device, used for the purpose of evaluating the performance of, or finding problems with, devices. Especially pertains to electronic apparatus.

**testosterone** The male sex hormone. Its production is increased during adolescence. In humans this begins at around age 12–14. It is responsible for hair growth, lowering of the voice, and sometimes acne, as well as representing the beginning of male fertility.

**test pattern** A standard set of color bands used for calibrating television transmitting and receiving equipment.

**tetanus** Also called lockjaw. An infectious disease that has been controlled in recent decades by vaccination. Can be acquired through breaks in the skin, as when cut by dirty or rusty metal. Symptoms include fever, fatigue and muscle spasms throughout the body. The exhaustion can be fatal.

**Tethys Sea** A sea in the middle latitudes of the earth, that existed between the Permian and early Tertiary periods between Gondwanaland and Laurasia. See *Gondwanaland, Laurasia*.

TETHYS SEA

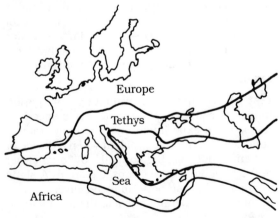

From Erickson, THE LIVING EARTH (TAB Books, Inc., 1989).

**tetra-** A prefix meaning four or in groups of four.

**tetraethyl lead** Also called ethyl. An additive used in gasoline until recently, to counteract engine knocking.

**tetrahedron** A polyhedron with four flat faces. This is the smallest number of flat faces a three-dimensional figure can have.

**tetrode** See *vacuum tube*.

**Th** Chemical symbol for thorium.

**thalamus** The center of the brain in vertebrates. This region acts as a center for relaying nerve messages. See *brain*.

**thallium** Chemical symbol, Tl. An element with atomic number 81. The most common isotope has atomic weight 205. It is found naturally along with zinc. It is used in some electronic semiconductors.

**theorem** 1. A logical result. 2. A logically valid fact, proven from definitions and axioms in a mathematical system.

**theoretical** 1. Pertaining to concepts that are arrived at by means of reasoning and logic, but have not necessarily been verified by experimentation. 2. Pertaining to the structure of a mathematical system; the definitions, axioms and theorems.

**theorist** A scientist who does theoretical work, but does not ordinarily do experiments to determine whether or not the theories fit the facts. Sometimes this is the only way to arrive at a scientific theory.

**theory** 1. A set of definitions, axioms and theorems in mathematics. 2. Pertaining to the theoretical, as opposed to the practical or experimental, aspects of a branch of science. 3. A scientific explanation or model for a phenomenon. 4. An idea.

**thermal** 1. Pertaining to heat and heat energy. 2. An atmospheric updraft, resulting from daytime heating of the surface of the earth.

**thermal breakdown** The failure of a component, device, or system because of excessive heat. This can occur from actual mechanical destruction of a component, or because of malfunction that is corrected when the temperature falls.

**thermal conductivity** The extent to which a material will transport heat energy. Often, but not always, good electrical conductors are also good thermal conductors, and vice-versa. Metals are usually good conductors of heat as well as of electric current. Some substances, such as silicone paste, are good conductors of heat but poor conductors of electric current.

**thermal energy** The kinetic energy in the atoms of matter. This is directly proportional to the temperature of a substance. See *kinetic energy, temperature*.

**thermal noise** Electromagnetic noise, produced over a band of wavelengths, resulting from the motion of atoms in matter. In general, the wavelength at which this noise is loudest gets shorter as the temperature rises.

**thermal runaway** A phenomenon that can occur in transistor amplifiers that are not properly designed. The current heats the transistor(s), and this lowers the resistance, in turn raising the current still more. The vicious circle ends with component failure.

**thermal shock** 1. See *heat exhaustion*. 2. See *hypothermia*. 3. The subjecting of a device to hot and cold temperatures, alternately. Sometimes used in laboratory testing to simulate conditions in real time.

**thermal spring** Also called a hot spring. Forms when an artesian water pocket dips deep enough beneath the surface to get warmed by the earth's interior heat. See *artesian spring*.

**thermal stability** A measure of the extent that a device parameter changes with temperature. For example, an oscillator might change frequency by +21 hertz per degree centigrade. The smaller the change per degree, the better the thermal stability.

**thermionic emission** The ejection of electrons by a negatively charged, heated electrode. This is the principle by which vacuum tubes operate. See *vacuum tube*.

**thermistor** An electrical resistor with a value that changes with temperature by a known, precise amount. Made from metal oxides, such a device is useful for temperature measurement and for temperature-controlled devices.

**thermocline** The boundary between warm surface water and cold deeper water in a lake or in the ocean. The transition is often abrupt. In the ocean it ranges from about 150 to 600 feet below the surface. In a spring-fed glacial lake it may be just a few feet down (see drawing).

THERMOCLINE

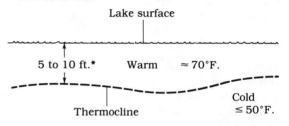

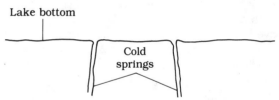

*in oceans it's much deeper

**thermocouple** A device for sensing temperature changes. It consists of a junction between two different metals. Temperature changes produce a voltage between the metals, called the Seebeck Effect.

**thermodynamics** A branch of physics involving the relationships between thermal and mechanical energy. There are three laws. First Law: Thermal energy can be converted into mechanical energy, and mechanical energy into thermal energy. Second Law: In a reversible system, not all of the available thermal energy is converted into mechanical energy. Third Law: As the temperature approaches absolute zero, the entropy approaches zero. See *entropy*.

**thermoelectric conversion** The changing of thermal energy into electric current. In power plants, this is usually done by boiling water, producing steam to drive turbines. On a small scale it can be done with semiconductor devices.

**thermoelectric effect** The generation of a voltage difference between dissimilar metals, joined together in a thermocouple. See *thermocouple*.

**thermogenesis** The generation of body heat in mammals and birds. Accomplished mainly by the metabolism of fats and carbohydrates.

**thermography** The mapping of the infrared that is radiated by the human skin. This technique is useful to medical doctors in the diagnosis of certain disease conditions, such as varices (enlarged veins) and cancer.

**thermoluminescence** A glow that occurs in solids when they are heated. The phenomenon is useful to scientists in determining the ages of some materials. See *luminescence*.

**thermometer** A device for measuring temperature. One common type uses a glass tube with a bulb at the base, filled with mercury or dyed alcohol. This expands and moves up the tube as the temperature rises. Another type uses a bimetallic strip that bends with temperature changes, moving a needle.

**thermonuclear** Pertaining to heat that is generated as atoms undergo radioactive decay, or fission or fusion. This heat might be of low intensity but long duration, such as the heat within the earth. Or it might be a sudden release of energy, such as occurs when an atomic bomb explodes.

**thermonuclear energy** Energy in the form of heat resulting from nuclear reactions. See *thermonuclear*.

**thermonuclear reactor** See *nuclear reactor*.

**thermophile** Literally, "heat lover." An organism that lives at high temperature. Such organisms are found near volcanoes. The upper limit of temperature is about 140 degrees Fahrenheit or 60 degrees centigrade for multicellular organisms. Some bacteria apparently can live even in water heated to boiling.

**thermosphere** A level of the earth's atmosphere in which temperature increases with increasing altitude. This begins at about 50 miles, near the top of the stratosphere and extends to outer space.

**thermostat** A device that opens and closes a circuit with changes in temperature. The circuit is connected to a heating and/or air-conditioning unit. The sensor is usually a bimetallic strip that bends in one direction when heated and in the other direction when cooled, closing the cooling and heating circuits, respectively (see drawing).

THERMOSTAT

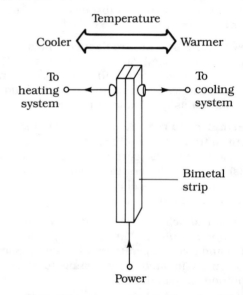

**Thevenin's Theorem** A rule for simplifying the analysis of a circuit having two or more loads. Any linear circuit can be mathematically reduced to an equivalent circuit, having just one power source and one load. For details, a text on intermediate or advanced electronics should be consulted.

**thiamine** Also called vitamin B-1. The first B-complex vitamin to be identified. Deficiencies are rare in developed countries, because this vitamin is added to most flours and cereals today. See the VITAMINS AND MINERALS appendix.

**thiol** Any of various organic compounds containing sulfur and hydrogen. They are similar to alcohols in their characteristics.

**Third Law of Motion** See *Newton's Laws of Motion*.

**Third Law of Thermodynamics** See *thermodynamics*.

**Thomson Effect** When a current flows in an electrical conductor, and when that conductor is not at the same temperature all along its length, heat is generated or absorbed. If heat is generated by a current flowing in one direction, then heat will be absorbed when the direction of the current is reversed.

**Thomson, J. J.** One of the early directors of the Cavendish Laboratory at the University of Cambridge (England). A physicist, he discovered the electron in 1897. For this he won the Nobel Prize in physics.

**Thomson, Sir George** A physicist and Nobel Prize winner. He speculated about how electrical activity in the brain is responsible for human emotions, giving rise to the arts as well as scientific knowledge.

**Thomson, Sir William** The man who is better known as Lord Kelvin. See *Kelvin, Lord*.

**thoracic nerves** The 12 nerves that branch off of the spinal cord in the midsection of the body, below the cervical nerves and above the lumbosacral plexus.

**thorax** 1. A section of the body in insects, between the head and the abdomen. 2. In humans, the region of the rib cage, in which many vital organs are located, including the heart and lungs.

**thorium** Chemical symbol, Th. A radioactive element with atomic number 90. The most common isotope has atomic weight 232. This natural isotope has an extremely long half-life, more than 10 billion (10,000,000,000) years. This is comparable with the age of the universe. It is used in nuclear reactors.

**threadworm** A long, extremely thin worm classified as a nematode. It is fairly primitive on the evolutionary scale of animals. See the ANIMAL CLASSIFICATION appendix.

**three-phase current** Alternating current in which there are three waves, each 120 degrees out of phase with the other two (see drawing). There are three conductors, each carrying one phase of the current. This is the most commonly used current for long-distance electric power transmission.

THREE-PHASE CURRENT

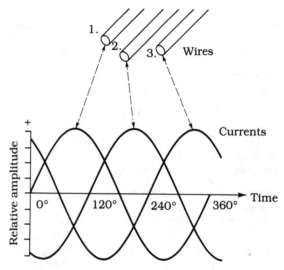

**three-way switch** An electric switch that can be used in conjunction with one or more other switches to control electric lights and appliances. This makes it possible to switch the current on and off from more than one location.

**three-wire circuit** An electrical circuit in which there is a ground wire along with two "live" wires. Outlets in such circuits have two vertical slots and one ground jack insert. Most modern appliances have three-prong plugs; the extra prong is for electrical safety grounding.

**threonine** An amino acid. This substance must be obtained in the diet, because the human body cannot synthesize it from other amino acids. See *amino acid*.

**threshold** 1. A critical level that, when exceeded, results in a change in the way a system behaves. For example, in a diode conduction begins when the threshold voltage is exceeded. This causes a dramatic decrease in resistance through the device. 2. The minimum detectable level for some phenomenon, such as sound or visible light.

**thrombin**  An enzyme needed for the formation of blood clots. It is formed from prothrombin. It causes fibrinogen to be converted to fibrin. See *fibrin, fibrinogen*.

**thrombosis**  A blood clot, especially in an artery. Such clots can be dangerous because they might break loose and lodge in some other part of the body, interrupting the flow of blood to a region. This is called a thromboembolism.

**throughput**  A measure of the amount of data processed in a certain time by a computer.

**thrush**  A yeast infection that can occur in the mouth. It looks like cauliflower on the tongue and throat. Sometimes this occurs when antibiotics are used in high doses. There are medications that can be used to treat it.

**thrust**  1. The force produced by a propeller as a result of the backward movement of air or water. 2. In a jet or rocket engine, the forward force produced as a reaction from the backwards motion of the exhaust. 3. See *thrust fault*.

**thrust fault**  A reverse fault whose plane is nearly flat, and in which the movement is mainly horizontal for great distances. See *fault, reverse fault*.

**thulium**  Chemical symbol, Tm. An element with atomic number 69. The most common isotope has atomic weight 169. It is a grayish metal in pure form.

**thunder**  The sound made by sudden heating of air in a lightning stroke. The heating causes the air to expand violently, and this produces the acoustic wave. Its noise is drawn out by echoing and by sound propagation delays (see drawing).

**thundershower**  A rain shower in which occasional lightning discharges occur. See *thunderstorm*.

THUNDER

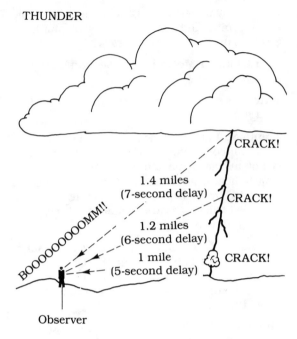

**thunderstorm**  A heavy rain shower, often with high winds, hail and/or frequent lightning. When the winds reach 55 miles per hour, it is classified a severe thunderstorm and warnings are issued for the regions it is expected to strike.

**thymine**  A protein important in DNA and RNA. It links together with adenine in the long chain molecules of these nucleic acids. See *deoxyribonucleic acid, ribonucleic acid*.

**thymus**  A gland that makes immune system cells early in the life of most vertebrates. There are two lobes, located just underneath the breastbone. After this job is done, the gland shrinks greatly.

**thyroid**  The gland in the neck that manufactures thyroxine. It needs a certain amount of iodine to do this, obtained from food or from iodized salt. See *endocrine glands, thyroxine*.

**thyroxine**  A hormone produced by the thyroid gland and responsible for regulating the metabolism of food. A component of this hormone is the element iodine. If the diet is deficient in iodine, the gland becomes enlarged. See *goiter, goiter belt*.

**Ti** Chemical symbol for titanium.

**tibia** The larger of the two bones in the shin. Runs parallel with the fibula. In animals, the thicker of the two bones between the ankle and the knee. See *fibula*.

**tidal bore** A wave that accompanies a rising tide in some rivers. It is a single wave, carrying the tide upstream from the sea. These waves occur especially when the moon is full or new.

**tidal current** The movement of ocean water as the tides rise and fall. This current often occurs parallel to the shore and can be quite strong. It can make swimming dangerous; it can also be used as an energy source. See *tidal power*.

**tidal flood** An unusually high tide with the result that low-lying areas along a coastline are submerged. This can occur during tropical storms and hurricanes, and also with onshore winds taking place over a vast surface area of the ocean.

**tidal power** 1. Electricity generated from tidal effects. The tidal currents can drive turbines directly, or a reservoir can be filled and drained as the tides rise and fall, producing a water flow to drive turbines. See *tidal current*. 2. The use of tidal currents to generate electric power. See *tidal current*.

**tidal wave** See *tsunami*.

**tide** The rise and fall of sea level, usually occurring twice a day, in a cycle caused by the gravitational pull of the sun and moon. See *neap tide, spring tide*.

**time** 1. A dimension, through which we can be considered to be traveling at a rate that depends on point of view. According to Einstein, time could be thought of as represented by an axis in a coordinate system. 2. The aspect of the universe that makes two events distinguishable from each other when they occur at the same point in space. See *space-time, time dilation*.

**time constant** 1. The time between a cause and effect for a given system. 2. The time required for an electronic circuit to charge or discharge. This applies especially to circuits containing resistance along with inductance or capacitance.

**time dilation** The change in the rate at which time seems to move, resulting from relative motion and/or acceleration. This effect arises from the fact that the speed of light is constant and always appears the same, regardless of the reference frame. It was derived mathematically by Einstein when he developed the special theory of relativity in the early twentieth century. It has been demonstrated by experiment. See *relativity*.

**time lapse** A method of making motion pictures that show things greatly speeded up. Frames are taken at regular intervals, such as each minute or each hour. When the film is run at normal speed, we can observe things like the evolution of a thunderstorm or the growth of a plant.

**time sharing** A method of using many terminals with a computer. If there are n terminals, each is connected to the computer for 1/n of the time. The computer scans the terminals rapidly, so each user gets the impression of being on line continuously.

**time-space** See *space-time*.

**time standard** A time source established by the government (primary) or synchronized with a source established by the government (secondary). A primary time standard is the transmission of radio station WWV in Colorado, WWVH in Hawaii or CHU in Canada. A secondary standard might be a quartz clock that you set by WWV every day.

**time travel** Any means by which a change is made in the rate at which time passes. This can be done by traveling at relativistic speed, causing forward displacement in time. The means for traveling backward in time is not known. See the illustration.

**time zone** A designated region on the earth where all clocks are synchronized. Astronomically, these regions are 15 degrees of longitude in width. But because of political boundaries, they are irregular.

TIME TRAVEL

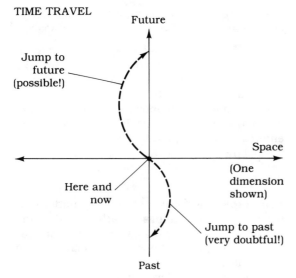

**tin**  Chemical symbol, Sn. An element with atomic number 50. The most common isotope has atomic weight 120. In pure form it is a metal, grayish or silver in color. It is used in various alloys, especially in solder. It is corrosion-resistant and can be used for protecting metals against oxidation.

**TIROS**  Acronym for television infrared observation satellite. An early series of American weather satellites. The first of these was launched in 1960. It orbited over the poles.

**TIROS-N**  The latest version of the TIROS weather satellites. The "N" stands for NOAA, the National Oceanic and Atmospheric Administration. See *TIROS*.

**tissue**  A medical term for body cells and structures, particularly muscle, ligaments, tendons, fat, arteries, veins, cartilage, nerves, and other soft parts.

**titanium**  Chemical symbol, Ti. An element with atomic number 22. The most common isotope has atomic weight 48. In pure form it is a metal that resembles aluminum. It is ideal for use in aircraft, because it is strong and light.

**Titius-Bode Law**  A rule discovered in 1766, concerning the distances of the planets from the sun. We start with 3, 6, 12, 24, and so on, with each number twice its predecessor. Zero is placed at the beginning, and 4 is then added, giving 4, 7, 10, 16, 28, 52, 95, 196. The earth corresponds to 10. There is no planet at 28, but the asteroid belt is there. Thus the sequence holds through the orbit of Saturn. Nowadays this rule is thought to be coincidence.

**Tl**  Chemical symbol for thallium.

**Tm**  Chemical symbol for thulium.

**TNT**  Abbreviation for trinitrotoluene.

**tocopherol**  Also known as vitamin E. It is found in various forms. It behaves as an antioxidant. It is sometimes added to vegetable oils to keep them from becoming rancid. See the VITAMINS AND MINERALS appendix.

**tolerance**  1. For a device, especially a machine part or electronic component, the maximum amount by which its actual value differs from the stated value. Usually given as a percentage. 2. The extent to which a machine or device will withstand operating conditions different from those under which it is meant to be used. 3. A physiological effect that can occur with prolonged use of certain drugs. It takes larger and larger doses to get the same effect.

**tomography**  A method of using computers to get a three-dimensional image, especially in X rays and nuclear magnetic resonance viewing. Can also be applied to radar and seismic systems. Many "bread-slice" images are obtained, each representing a cross section of the object to be analyzed. See *computer-aided medical imaging*.

**tonne**  Also called a metric ton. A unit of mass equal to 1000 kilograms. On the earth's surface this is about 2200 pounds, a little more than an English ton.

**tonsil**  Either of two glands in the throat. They frequently become enlarged and inflamed during childhood. At one time, most people had these glands removed. They produce blood cells.

**topaz**  A hard, transparent, yellowish silicate mineral, used mainly as a precious stone.

**topographical map** A map of a region, with lines that connect points having equal elevation. The drawing shows an example of a map of this kind.

TOPOGRAPHICAL MAP

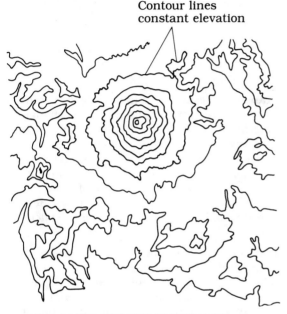

MOUNT SAINT HELENS BEFORE ITS ERUPTION IN 1980.

From Erickson, EXPLORING THE EARTH FROM SPACE (TAB Books, Inc., 1989).

**topography** The nature of the terrain in a given area, and in particular, the degree of irregularity in elevation.

**topology** A branch of mathematics, concerning point sets and the nature of shapes.

**topsoil** The uppermost, fertile layer of earth on land in which plants grow and agriculture is possible. In some regions there is very little or none of this layer; in other regions the layer is several feet deep.

**tornado** A violent storm associated with severe thunderstorms and hurricanes, producing the strongest winds known on the earth, up to 300 miles per hour. The diameter ranges from about 100 feet to about a mile. Most spin counterclockwise in the Northern Hemisphere and clockwise in the Southern Hemisphere.

**tornado warning** Notice given to a certain region that a tornado has actually been sighted, and that there is immediate danger. See *tornado watch*.

**tornado watch** Notice given to a certain region that conditions are favorable for the development of tornadoes. See *tornado warning*.

**toroid** 1. A coil wound on a form that is shaped like a doughnut, or torus. See *torus*. 2. Any object with the general shape of a torus.

**torque** 1. A rotational force. 2. Force times radial distance from the center of rotation. See *force*.

**torr** A very small unit of pressure, equivalent to 1 millimeter of mercury (1 mmHg), or $1/760$ of an atmosphere.

**torrid zone** The region of the earth also called the tropics, between 23.5 degrees north latitude and 23.5 degrees south latitude (see drawing).

TORRID ZONE

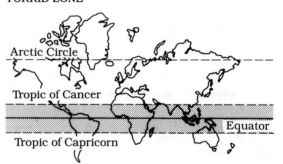

**torsion balance** A device for measuring force or weight in very small amounts. It works by twisting of a fine wire. Even microscopic displacements can be detected by the use of a laser and a set of mirrors. The laser beam serves as a very long, massless meter needle.

**torus** A doughnut-shaped geometric figure in three dimensions. This can be thought of as a cylinder bent so that its axis becomes a circle, and its ends are joined together.

**total eclipse** See *eclipse*.

**total internal reflection** For a ray of energy, a condition that occurs when the angle of incidence is near grazing at a boundary between two substances having different indexes of refraction. The medium in which the ray travels must have a higher refractive index than that on the other side of the boundary. This effect makes possible the transmission of light within optical fibers. See *index of refraction*.

**Touchtone dialer** Trade name, now become generic, for a telephone keypad and the associated tones that make it possible to dial telephone numbers quickly. Some pads have 12 keys (digits 0 through 9 and the symbols * and #). There are also some pads with 16 keys, including the usual 12 plus the letters A, B, C, and D.

**toxin** A poisonous chemical. Can occur naturally or be made artificially. Can also be secreted by bacteria in certain disease conditions.

**trace** 1. A very small, almost undetectable amount. 2. A fall of rain or snow that is too light to be measured. 3. Needed in minute amounts. See *trace element*. 4. The path followed by the electron beam in a cathode-ray oscilloscope. 5. The line drawn by a pen recorder. 6. To locate the source of a telephone connection. 7. To locate the source of a radio transmission.

**trace element** Also called a trace mineral. Any elemental substance, such as iodine or chromium, that is needed in the human diet, but only in very small amounts, measured in micrograms.

**trachea** Also called the windpipe. The passageway leading from the throat to the bronchi, through which air passes when breathing.

**trade winds** Prevailing winds in the tropics, that blow mainly from the east (see drawing). See *torrid zone*.

**tranquilizer** Any substance that depresses the function of the central nervous system. The minor tranquilizers include such drugs as Valium. The major tranquilizers are such drugs as Thorazine. These substances require a doctor's supervision and prescription.

**transceiver** A radio or television transmitter and receiver, all in a single cabinet. This is the most common type of "two-way radio."

**transcendental number** An irrational number that cannot be derived from algebraic equations with rational-number coefficients. An example is the natural logarithm base, usually denoted e and approximately equal to 2.718. Another is the ratio of the circumference of a circle to its radius. See *irrational number, rational number*.

**transconductance** 1. In a field-effect transistor (FET), the ratio of drain-current change to gate-voltage change. 2. In a bipolar junction transistor, the ratio of collector-current change to base-voltage change. 3. In a vacuum tube, the ratio of plate-current change to grid-voltage change. This is an indication of how much amplification is possible with a given device. See *field-effect transistor, transistor, vacuum tube*.

**transducer** Any device that converts one form of signal, or variable quantity, into another. Common examples are microphones (sound into electric current), speakers (electric current into sound), and antennas (high-frequency current into electromagnetic waves or vice-versa).

**transfinite cardinal** The number of elements in an infinite set. *Aleph-null* is the number of counting numbers in the set {0, 1, 2, 3, ...}. This is the same as the number of integers, and also the same as the number of rational numbers. But the number of irrationals is a

TRADE WINDS

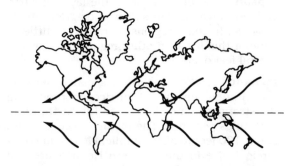

From Erickson, VIOLENT STORMS (TAB Books, Inc., 1988).

"greater" infinity, called *aleph-one*. There are infinitely many larger infinities, called *aleph-two* and up. First discovered by Georg Cantor. See *aleph-null, aleph-one, and Cantor, Georg*.

**transfinite ordinal** See *infinite ordinal*.

**transform** 1. To change from one state into another, or from one substance into another. Coal, for example, changes into diamond under extreme pressure. 2. The result of a mathematical transformation. See *transformation*.

**transformation** 1. A change of state or chemical composition. 2. A mathematical function that maps the points of one space into the points of another space. See *function*. 3. A form of genetic change that can occur in certain cells by adding deoxyribonucleic acid (DNA).

**transformer** 1. A device intended to increase or decrease an alternating-current voltage by a constant ratio. A step-up device increases the voltage; a step-down device decreases the voltage. 2. A device intended to match the impedance of a load to that of a transmission line or amplifier output. See *impedance*.

**transform fault** A lateral fault in the earth's crust where one plate slides along sideways relative to the other.

**transgression** The depositing of sediment on old rocks as a result of the ocean moving over the land. This can happen because of subsidence of the land, or because of a rise in the sea level, or both.

**transient** 1. Lasting for only a very short time. 2. A very brief, high-voltage "spike" that often occurs on the alternating-current (ac) utility lines. Also called a surge. Some devices, especially computers, can malfunction because of such a surge.

**transistor** 1. A semiconductor device with three layers, either an N-type semiconductor between two P-type layers (PNP) or vice-versa (NPN). Also called a bipolar transistor. 2. See *field-effect transistor*. 3. Any of various three-terminal or four-terminal semiconductor devices, intended as amplifiers, high-speed switches, oscillators and for use in various other electronics applications.

**transition** The process of changing state, chemical composition, or general characteristics. Might be gradual or abrupt, small, or large in extent.

**transitive property** In mathematics, a property that a relation might or might not have. Let us say we have the relation * ("star"). Suppose that if a*b and b*c, then a*c for all a, b, and c. Then * is transitive. Also, if * is transitive, then a*b and b*c logically implies a*c. The relation of equality (=) has this property. So do the greater-than (>) and less-than (<) relations.

**translation** 1. Conversion of data from one language (code) into another. 2. Motion in which all parts of the object travel together; that is, the paths of all the points are parallel. 3. A process by which amino acids are put together in cells.

**translucent** Passing light, but without preserving images. The light is scattered on its way through such a medium (see drawing). Frosted glass and wax paper are good examples.

TRANSLUCENT
TRANSPARENT

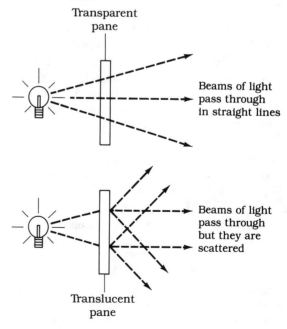

**transmission line** 1. A set of two or three wires for carrying electrical energy for long distances. 2. A set of two wires, or a single wire surrounded by a tubular shield, for carrying radio-frequency energy from a transmitter to an antenna, or from an antenna to a receiver. 3. See *waveguide*.

**transmittance** The ratio of the power passing out of a substance, to the power entering that substance. Usually expressed as a percentage. A perfectly transparent material has a transmittance of 100 percent. See *absorptance*.

**transmitter** A device that generates radio or television signals. There is always an oscillator and a set of one or more amplifiers. There usually is also a modulator. The exact circuit depends on the type of signal generated. See *amplifier, modulation, modulator, oscillator*.

**trans-oceanic cable** See *submarine cable*.

**transparent** Passing light, and also preserving images. Clear pane glass (see drawing) and pure water are good examples. Although the images are not obliterated, they are sometimes distorted, such as when coming from under water.

**transpiration** In plants, the evaporation of water from the leaves. This takes place as a sort of exhaling of vapor.

**transonic** Speeds from about Mach 0.8 to Mach 1.2, at which some of the airflow around an aircraft is subsonic (slower than sound) and some is supersonic (faster than sound). See *Mach*.

**transplant** 1. To move a plant from one place to another. 2. To remove an organ from the body of one person and surgically implant it into the body of another person.

**transponder** Acronym for transmitter/responder. A transceiver (see *transceiver*) that automatically sends a response signal whenever it receives an interrogating signal.

**transuranic element** Any element with an atomic number higher than that of uranium; that is, any element with 93 or more protons. These elements do not occur in nature, but only as a result of human-made nuclear reactions.

**transverse** 1. Sideways, or from side to side. 2. At right angles to the direction of motion, force or propagation. See, for example, *transverse wave*.

**transverse wave** A wave whose displacement is sideways relative to its direction of propagation. A good example is the train of ripples set up when an object is dropped into a pond (see drawing).

TRANSVERSE WAVE

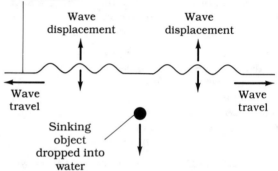

**trapezius** The muscle group on either side of the neck, connecting the deltoids with the base of the skull.

**trapezoid** A four-sided plane polygon in which one pair of opposite sides is parallel, but shorter than the other. Generally takes the shape of a triangle that has been truncated.

**treble** 1. The upper part of the musical scale, consisting of all notes with pitch higher than middle C. 2. Sound of high frequency, above the midrange and up to the limit of hearing, or 20,000 hertz. 3. Pertaining to high-frequency sound.

**trench** 1. A deep region in the ocean caused by one crustal plate sliding underneath another. See *subduction*.

**tri-** A prefix meaning three or in groups of three.

**trial** A test, conducted during an experiment. Usually refers to one of several or many identical tests, done to evaluate the behavior of a system or material in real time.

**trial-and-error** 1. A method of doing scientific research. Also called experimentalism. Facts are discovered based on what happens in real time. 2. A means of perfecting a device. Different designs and parameters are tried, until one is found that works best.

**triangle** 1. A plane polygon with three straight sides. 2. A drafting or drawing instrument with three straight sides, and usually having angles of 45, 45, and 90 degrees or 30, 60, and 90 degrees. 3. A region of the earth contained within three geodesics, connecting three points on the surface.

**triangulation** 1. A means of locating the position of an object based on signals or waves arriving at three different observing stations. 2. A method of determining great distances, based on parallax from two observing points separated by a certain distance. See *parallax*.

**Triassic period** The first period of the Mesozoic era, starting about 230 million years ago and lasting 50 million years. This was when the dinosaurs began to rule the earth. See the GEOLOGIC TIME appendix.

**triceps brachii** The group of muscles along the back of the upper arm. These muscles are used in pushing, and whenever the arm is to be straightened out.

**trichinosis** An infection caused by a parasite roundworm. It occurs when pork is eaten after having been inadequately cooked. Symptoms vary greatly from case to case; nausea and fever almost always occur.

**trichotomy law** 1. For any two real numbers x and y, exactly one of the following is true: $x = y$, $x < y$ or $x > y$. 2. A rule that is said to apply in a mathematical system in which the reflexive, symmetric, and transitive properties all hold true.

**trigger** 1. A small event that causes some much larger event to occur. 2. To cause a large event to happen. 3. An input on an oscilloscope that synchronizes the sweep with the frequency of the displayed wave, so the waveform remains fixed on the screen.

**triglyceride** A substance in the body, especially of interest because its level in the blood is thought to be related to the risk of heart disease. Chemically it is a form of fat. All fats consist mostly of this substance.

**trigonometric function** 1. Any of the functions represented by points on a unit circle. See *cosecant, cosine, cotangent, secant, sine, tangent*. 2. Any of the functions represented by points on a unit hyperbola. See *hyperbolic cosecant, hyperbolic cosine, hyperbolic cotangent, hyperbolic secant, hyperbolic sine, hyperbolic tangent*.

**trigonometric identity** A fact in trigonometry that holds true for all possible values of a variable angle, for which the functions are defined. For example, for any angle, the sum of its sine and cosine is equal to 1. See the TRIGONOMETRIC IDENTITIES appendix.

**trigonometry** A branch of mathematics dealing with angles and lengths, and the way in which they are interrelated in two-dimensional and three-dimensional Euclidean space.

**trillion** The number 1,000,000,000,000—a 1 followed by 12 zeros, or 1000 followed by three sets of three zeros.

**trilobite** An ocean-dwelling creature, now extinct. Some believe it was among the earliest complex living organisms on the earth. Fossils of this arthropod are abundant (see drawing). They lived during the early Paleozoic. The horseshoe crab is the only descendant alive today.

**trimester** 1. For a mammal, any $1/3$ of its pregnancy period, after conception and before birth. 2. In humans, either the first three months, the second three months, or the third three months of pregnancy. 3. For a given period, either the first, second or third $1/3$.

**trinitrotoluene** Abbreviation, TNT. A compound that is used as an explosive, especially in construction of buildings and highways when blasting is necessary.

TRILOBITE

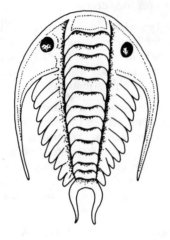

From Erickson, THE LIVING EARTH (TAB Books, Inc., 1989).

TRITIUM

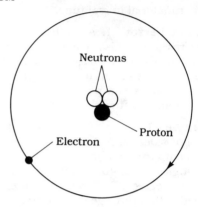

**triphosphate**  A food preservative used in processed meats and cheeses.

**trisection**  The dividing of a unit into three parts of equal size. This applies in geometric constructions. It is possible to construct the trisection of a line segment, but not of an angle, using only a compass and a straightedge.

**triode**  See *vacuum tube*.

**triple point**  For a given substance, the temperature and pressure at which the three phases (gas, liquid, and solid) exist in equilibrium.

**tritium**  A form of "heavy hydrogen," in which the nucleus contains two neutrons and one proton (see drawing).

**tropical cyclone**  A low-pressure system in the tropics, in which there is a circular flow of wind around the center. The most severe type is a hurricane. See *hurricane*.

**tropical depression**  A tropical cyclone of low intensity, with winds of less than gale force (about 38 miles per hour). See *tropical cyclone*.

**tropical storm**  A tropical cyclone of moderate intensity, with winds of gale force or storm force (39 to 73 miles per hour). When winds reach gale force, the cyclone is given a name from a list made ahead of time by the National Hurricane Center. See *tropical cyclone*.

**tropical wave**  A disturbance in the normal flow of the trade winds, corresponding to a low-pressure zone. The winds do not flow in a complete circle around the center, but are deflected towards the pole. Showers and thunderstorms usually occur.

**tropopause**  The level of the atmosphere at which the troposphere ends, and above which there is no weather as we know it. This is at an altitude of about 8 to 12 miles. See *atmosphere*.

**troposphere**  The lowest layer of the atmosphere, from the surface to an altitude of 8 to 12 miles. This is where the weather takes place. See *atmosphere*.

**tropospheric propagation**  The long-distance transmission of a radio or television wave, by means of bending or ducting in the lower atmosphere. This occurs especially along weather fronts. It is sometimes noticeable on the frequency-modulation (FM) broadcast or television broadcast bands, when unusually distant stations are heard.

**troubleshooting**  A methodical way of finding the nature and location of a malfunction in a complex machine, circuit, or system. The situation is "narrowed down" by tracing and by a process of elimination.

**trough**  1. A low-pressure region in the atmosphere, not intense enough to make winds flow in a circle around the center. In temperate latitudes, the prevailing westerlies are deflected towards the equator. 2. The minimum point in a wave cycle.

**true north** Also called geographic north. The direction along a great circle between the observer and the North Geographic Pole. This is almost exactly in the direction towards a point on the horizon underneath Polaris, the North Star (in the Northern Hemisphere).

**true power** In an alternating-current (ac) electrical circuit, the amount of power absorbed by the load; that is, the power actually dissipated by resistance. This can consist of heat, mechanical energy, sound, or various forms of electromagnetism.

**trunk line** In a telephone network, a line that carries many calls (signals) between central-office switching systems. This might be a wire cable, an optical fiber bundle, or a microwave link, perhaps via satellite.

**truth table** A logical diagram showing the breakdown of a complex sentence, so that all possible truth values can be easily seen.

**tryptophan** An amino acid. It is not synthesized in the body from other amino acids, and therefore must be obtained in the diet. Supplements are not normally needed if the diet is adequate. It is plentiful in milk and various meats. See *amino acid*.

**tsunami** Japanese for tidal wave. An ocean wave caused by undersea earthquakes, island volcanoes, coastal earthquakes, and meteorite impacts. It moves at 300 to 600 miles per hour and is about 3 feet high. When it reaches a coastline, the drag magnifies its height tremendously (see drawing). The resulting huge breaker(s) can cause great devastation.

**tube** See *vacuum tube*.

**tuber** A plant root that swells, storing water during times of little or no rainfall. Also serves for plant reproduction. Some are useful as sources of food. Sweet potatoes are a common example.

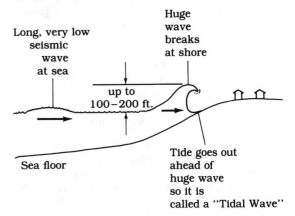

TSUNAMI

**tubercle** 1. A nodule in the central nervous system. 2. A capsule in the lung of a patient with tuberculosis, containing the disease-causing bacteria. See *tuberculosis*.

**tubercle bacillus** See *tuberculin, tuberculosis*.

**tuberculin** A substance used to diagnose tuberculosis. In the tuberculin test, a small amount of this liquid is injected in the skin. If the patient has tuberculosis, a lump will appear after about two days. The converse does not always hold; the lump does not necessarily mean that the patient has tuberculosis. See *tuberculosis*.

**tuberculosis** A contagious disease, caused by tubercle bacillus, a bacterium. In humans, it affects mainly the lungs. Symptoms include congestion, chronic cough, fatigue, and weight loss. The loss of weight can be spectacular, resulting in emaciation. This is why the disease is sometimes called consumption. It is not common in developed countries. See *tubercle, tuberculin*.

**tube worm** A type of worm that lives on the ocean floor. Sulfur-eating bacteria live inside of these worms, and the two organisms exist in symbiosis (see *symbiosis*).

**tubule** Any elongated, cylindrical path in the body, such as a vein, artery, or urinary passage.

**tumor** A growth of cells in the body, not normal and not serving any function. A benign tumor is noncancerous, and is easy to remove if found early enough. A malignant tumor is cancerous, and treatment is often more complicated than simple surgical removal.

**tundra** Treeless terrain, found in the Arctic, near the coasts in Antarctica, and at high elevations in some mountainous regions. Many types of life abound despite the harsh environment. The ecology of these regions is fragile.

**tungsten** Chemical symbol, W. Also called wolfram. An element with atomic number 74. The most common isotope has atomic weight 184. It is used in various alloys because of its hardness and durability. It works well as filaments in incandescent bulbs.

**tuning** 1. The adjustment of a radio or television receiver or transmitter for the desired frequency. 2. The adjustment of a radio or television transmitter for optimum power output at the intended frequency. 3. The adjustment of the pitch of a musical instrument or synthesizer.

**tuning fork** A two-pronged, metal object that vibrates at a known audio frequency when struck. Used for adjusting audio oscillators. Also used to tune musical instruments or synthesizers.

**turbidity** 1. Cloudiness or murkiness of water, caused by particles in suspension. 2. The extent to which the water in a river or stream is clouded by suspended silt and dirt.

**turbine** A wheel with blades, shaped and slanted so that water, air, or steam passing through causes the wheel to turn (see drawing). Water turbines are used in hydroelectric plants. Air turbines are used in jet engines. Steam turbines are employed in electric generators at oil-burning, gas-burning, and coal-burning power plants. These are only a few examples of applications.

**turbo-** A prefix describing machines, and especially engines, using turbines. See *turbine, turbofan engine, turbojet engine, turboprop engine, turboramjet engine*.

**turbofan engine** A jet engine that uses a fan, run by a turbine, to enhance the intake of air for

TURBINE

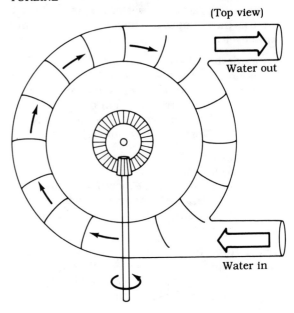

combustion. Also called a bypass engine. A more efficient version of the turbojet. See *turbojet engine*.

**turbojet engine** A jet engine that uses compressed, heated air to provide propulsion. Air enters at the front, is compressed, and then is mixed with fuel and ignited. The extreme pressure causes hot air to be ejected rearward at high speed. The reaction provides thrust. The compressed air also drives the turbine to keep the engine running.

**turboprop engine** Acronym for turbojet/propeller engine. A combination turbojet and conventional airscrew (propeller) aircraft engine. Nowadays it has been largely replaced by jet engines. See *turbofan engine, turbojet engine*.

**turboramjet engine** A ramjet engine that incorporates a turbine. See *ramjet*.

**turbulence** Unrest in a gas or liquid, so that the flow is not smooth, but contains many eddies, updrafts, and downdrafts.

**tweeter** A speaker designed to reproduce high-frequency audio, especially that above the voice frequencies. Many tweeters can respond to frequencies well above the maximum for human hearing. Used in high-fidelity equipment.

**twin-lead** A two-wire transmission line with a plastic webbing or ribbon that keeps the conductors at a uniform distance from each other. Used mainly with television receiving antennas.

**two-channel stereo** See *stereophonics*.

**Tycho** 1. See *Brahe, Tycho*. 2. A large crater on the moon prominent enough to be visible with binoculars.

**tympanic membrane** See *eardrum*.

**typhoid fever** A serious infectious disease caused by a bacterium and spread mainly by water contaminated with excrement. It can also be transmitted by careless handling of food. Symptoms include abdominal pain, fever, slow heartbeat, and sometimes a rash. Treatment is with antibiotics.

**typhoon** See *hurricane*.

**typhus** An infectious disease, caused by a rickettsia. It is transmitted mainly by lice and fleas, in turn often carried by rats. Symptoms include headache, fever, and rash. Treatment is with antibiotics.

**tyrosine** An amino acid that can be synthesized by the human body from other amino acids. It is therefore not likely that deficiencies will occur. See *amino acid*.

**U**  Chemical symbol for uranium.

**u**  Abbreviation for atomic mass unit, micro- (formally the lowercase Greek mu; see the PREFIX MULTIPLIERS appendix).

**UFO**  Abbreviation for unidentified flying object.

**UHF**  Abbreviation for ultrahigh frequency.

**ulcer**  A sore inside the body. These can occur especially in the digestive tract, from the mouth to the large intestine and rectum. They might cause no symptoms; sometimes they bleed and cause pain.

**ulna**  One of the two bones connecting the elbow to the wrist.

**ulnar nerve**  One of the nerves in the arm. The others are called the radial nerve and the median nerve.

**ULSI**  Abbreviation for ultra-large-scale integration.

**ultra-large-scale integration**  Abbreviation, ULSI. An electronic technology that is rapidly developing, in which integrated circuits are made with more than 1000 logic gates per chip. See *integrated circuit*.

**ultrasonic**  Pertaining to acoustic waves at frequencies higher than the maximum that can be heard by a person. Normally 20,000 hertz and above.

**ultrasonic depth sounding**  A means of determining the distance to the bottom of a body of water in a given location, using ultrasound waves. It works like sonar except at higher frequencies. See *sonar*.

**ultrasonic intrusion detector**  A device that detects motion and sets off an alarm, using ultrasound. If something moves, a phase disturbance occurs, triggering the alarm (see drawing). The intruder cannot hear the acoustic waves because their frequency is beyond the range of hearing.

**ultrasonic transducer**  A speaker or microphone that works at acoustic frequencies above 20,000 hertz, up to several hundred kilohertz. Converts alternating currents at these frequencies to acoustic vibrations, and vice-versa.

ULTRASONIC INTRUSION DETECTOR

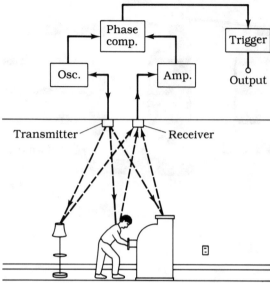

From Gibilisco, ENCYCLOPEDIA OF ELECTRONICS (TAB Professional and Reference Books, Inc., 1985).

**ultrasound** Acoustic energy at 20,000 hertz or more. There are various applications. See *ultrasonic depth sounding, ultrasonic intrusion detector, ultrasonic transducer.*

**ultraviolet** Electromagnetic radiation with wavelengths shorter than those of visible light, but longer than those of X rays. Longwave ultraviolet, abbreviated UV, has wavelengths from 390 nanometers (nm) to 50 nm. Shortwave ultraviolet, sometimes abbravated XUV, has wavelengths from 50 nm to 4 nm. Much of this radiation is kept from the earth's surface by the ozone layer. See *ozone, ozone depletion, ozone hole, ozone layer.*

**ultraviolet astronomy** A branch of astronomy concerned with observing the universe at wavelengths shorter than those of visible light, but longer than those of X rays. Much of this observation must be done from above the earth's atmosphere.

**ultraviolet lamp** A device that generates ultraviolet (and also usually a large amount of visible light). There are various types. A common method of obtaining ultraviolet is by excitation of mercury vapor. A carbon arc will also generate ultraviolet radiation.

**umbilical cord** The vascular system that connects a mammal fetus to its mother during pregnancy. After birth this cord serves no purpose and shrivels. Nowadays, in humans, it is surgically removed at birth, leaving the navel.

**umbra** 1. The darkest part of a shadow, where the light source is completely eclipsed by the object casting the shadow. 2. The dark, central part of a sunspot. See *penumbra.*

**uncertainty principle** Also called the Heisenberg Principle, after its discoverer, who defined it in 1927. Trying to determine the speed of an electron confuses the issue of its position, and vice-versa. We can predict where electrons are likely to be, but we can never pinpoint them exactly. A corollary of this is that causes and effects are not always paired off one-to-one. This principle has far-reaching philosophical consequences as well as scientific significance.

**underflow** A flow of water in the rock underneath a river. This flow is much slower than the movement of the water in the river itself, but it takes place in the same direction. The speed depends on how porous the rocks are.

**underwater housing** The means for long-term dwelling beneath the surface of a lake or ocean. This requires special structures, usually semispherical in shape, because a sphere is the strongest of any three-dimensional shape (see drawing). The pressure is usually greater than 1 atmosphere, but may be considerably less than the outside underwater pressure.

**Unge, Wilhelm** The assistant to Alfred Nobel, the inventor of dynamite. See *Nobel, Alfred.*

**Unh** Chemical symbol for unnilhexium.

**uni-** A prefix meaning one or occurring singly.

**unidentified flying object** Any object that appears to be in flight, and that cannot be recognized as anything familiar. Usually such objects

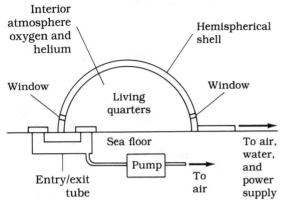

UNDERWATER HOUSING

(Hypothetical arrangement)

turn out to be ordinary aircraft or phenomena, whose appearance is affected by atmospheric conditions.

**unidirectional** Working mostly in one direction. A dish antenna, many types of microphones, and most speakers are of this type.

**unified field theory** An as-yet unproven scientific theory that provides a single explanation for all the forces in the universe, including gravitation, electromagnetism, the strong nuclear force and the weak nuclear force.

**Uniformitarianism** A theory concerning the forces that shape the surface of the earth. According to this idea, these slow processes have acted in the same way for the entire history of our planet.

**unijunction transistor** Abbreviation, UJT. A semiconductor device consisting of a bar of N-type material, with a small bit of P-type material near one end. The electrodes are called the emitter, base 1 and base 2. Also called a double-base diode. It is especially useful as an oscillator and timer.

**unimolecular** Pertaining to any chemical reaction that involves just one molecule.

**unisexual** Having reproductive organs for either the female sex or the male sex. Most animals are of this type.

**unit** 1. A device or component within a system. 2. A self-contained device or machine. 3. The "ones" place in a decimal number. 4. The number one. 5. A standard criterion for measuring some quantity.

**unit circle** A circle on the Cartesian plane, with radius equal to 1, and with its center at the origin. If the coordinates are x and y, then the equation for this circle is $x^2 + y^2 = 1$.

**unit hyperbola** A hyperbola on the Cartesian plane, centered at the origin. If the coordinates are x and y, then the equation for this hyperbola is $x^2 - y^2 = 1$.

**universal receiver/transmitter** An integrated circuit used in data communications. There are many different types with various applications, such as printer-to-computer interfacing, test instruments, modems, and multiplexers.

**universal set** 1. The set of all things. 2. The set of all objects or elements that are used in a mathematical system.

**universe** 1. All things that exist, seen and unseen, from the tiniest particles to the entire realm of outer space. See the UNIVERSE DATA appendix to get an idea of the range of distances involved. 2. Everything that lies within the range of any type of astronomical telescope. 3. A set within which a given mathematical system operates.

**unnilhexium** Chemical symbol, Unh. An element with atomic number 106. It does not occur in nature.

**unnilpentium** Chemical symbol, Unp. An element with atomic number 105. It does not occur in nature.

**unnilquadium** Chemical symbol, Unq. An element with atomic number 104. It does not occur in nature.

**Unp** Chemical symbol for unnilpentium.

**Unq** Chemical symbol for unnilquadium.

**uplink** The signal sent up to a communications satellite. It is received by the satellite and retransmitted, usually on a different frequency at the same time, by a repeater on the satellite. See *downlink, repeater*.

**upwelling** 1. The rising-up of water from deep in the ocean, to a level near the surface. This is an important part of the circulation in the Earth's oceans. It often brings nutrients on which fish can feed. 2. The rising-up of magma from the Earth's interior.

**uranium** Chemical symbol, U. An element with atomic number 92. The most common isotope has atomic weight 238. This is the heaviest naturally occurring element. In pure form it is a white mineral. It is radioactive, and its isotope U-235 is important because of its applications in nuclear power plants and in atomic bombs.

**Uranus** The eighth planet from the sun in the solar system. It is unique because of the extreme tilt of its axis, almost 90 degrees with respect to the plane of its orbit (see drawing). See the SOLAR SYSTEM DATA appendix.

URANUS

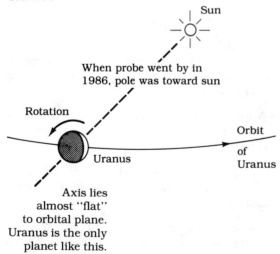

**urban engineering** A branch of civil engineering concerned with the construction of cities. Various factors are involved, such as how to arrange the roadways, where to put the largest buildings, and how best to minimize air and water pollution effects.

**urban planning** The branch of urban engineering that involves new construction in cities. Basically, it consists of finding the most efficient way to utilize available funds. See *urban engineering*.

**urbanization** The concentration of human populations in cities. This has given rise to a variety of transportation, social, and environmental problems that are dealt with by civil engineers.

**urea** A by-product of the metabolism of protein, contained in the urine of humans and other mammals. It has some industrial uses.

**uremia** A disease in which the kidneys do not excrete toxins normally, and the toxins therefore accumulate in the blood instead of passing out of the body in the urine.

**ureter** The tubule that runs from the kidneys to the bladder, carrying urine on its way out of the body. There are two of these passageways in humans, one for each kidney.

**urethra** The tubule that runs from the bladder out of the body, and through which urine passes. In males it also carries sperm during sexual intercourse.

**Urey, Harold** One of the scientists who conducted an experiment to synthesize amino acids under conditions similar to those of the primordial earth. See *Miller-Urey Experiment*.

**uric acid** A compound present in small amounts in the urine. It can accumulate in the kidneys when there is an abnormally high level of it in the blood. This creates kidney stones because the compound does not readily dissolve in water.

**urinary system** The part of the body's waste-eliminating system that involves urine. See *bladder, kidney, ureter, urethra*.

**urine** The waste liquid that passes out of the kidneys, through the ureters, bladder and urethra, and out of the body. It is mostly water, with various dissolved minerals, toxins, and by-products of metabolism.

**urology** A branch of medicine concerned with the diagnosis and treatment of urinary-system disorders. See *urinary system*.

**UTC** Abbreviation for Coordinated Universal Time.

**uterus** In the female mammal, the cavity in which a fetus grows during pregnancy. See the drawing.

**UV** Abbreviation for ultraviolet.

**uvula** The dangling tag of flesh in the throat. It is visible through the open mouth.

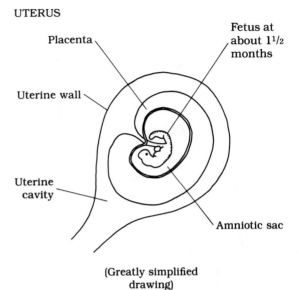

UTERUS

(Greatly simplified drawing)

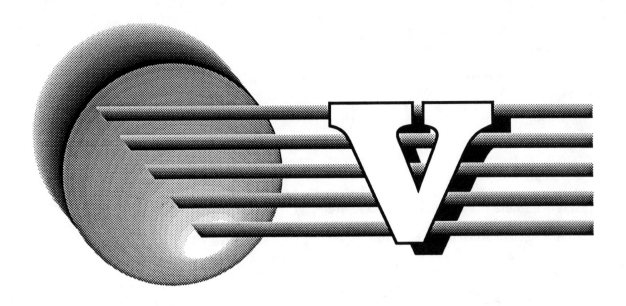

**V** Chemical symbol for vanadium; abbreviation for volt, voltage, volume.

**v** Abbreviation for velocity.

**vaccination** 1. The administration of a vaccine to a person or population, with the intent of protecting that person or population from the risk of getting a certain disease. 2. The same procedure, but with animals rather than people. See *vaccine*.

**vaccine** A substance administered to prevent a person or animal from getting a disease. It works by enhancing the immune system against certain bacteria or viruses. Some such treatments can protect against several diseases, caused by bacteria or viruses that resemble each other.

**vacuum** 1. The complete absence of any matter in a given enclosure or region. 2. A condition of extremely low gas pressure, such that, for all practical purposes, the pressure can be considered as zero.

**vacuum packing** The packaging of food in containers from which most or all of the air has been removed. This greatly slows the process of spoilage by oxidation.

**vacuum pump** A device that works as an air pump, but in reverse, removing the air from a container and pumping it into the atmosphere. These machines are commonly found in physics and chemistry laboratories.

**vacuum tube** Also called simply a tube. In England it is often called a valve. An electronic device consisting of an evacuated glass chamber with two or more electrodes. These devices have been mostly replaced by semiconductor components. A diode has two elements, called the cathode and the anode or plate. A triode has a third element, the grid. A tetrode has a second grid; a pentode has a third grid. There are even tubes with four or more grids. Sometimes two tubes are enclosed in a single glass envelope. Nowadays, tubes are still used in certain specialized applications, such as microwave oscillators, television picture tubes and high-power radio-frequency amplifiers.

**vagina** In female mammals, a tubule running into the uterus. It is where the sperm travels on its way to fertilizing the egg in the uterus.

**vagus** A nerve running from the base of the brain to various body organs such as the stomach. It is part of the autonomic nervous system. See *autonomic*.

**valence** An expression of the extent to which a chemical element will react with other elements. See *valence band, valence electron, valence number*.

**valence band** A range of energy levels that an electron can have when it is in the outermost shell of an atom.

**valence electron** Any electron in an atom that is in the outermost shell, when that shell is not completely filled. Under these conditions the outer shell of an atom is called the valence shell.

**valence number** An expression of the "excess" or "shortage" of electrons in the outermost shell of an atom. When the outer shell is full, the valence number is zero. If there exist n extra electrons, the valence number is +n. If there is a shortage of n electrons, the valence number is −n. These numbers are also called oxidation numbers.

**valine** An amino acid, not synthesized in the body from others. Thus it must be obtained from dietary protein. See *amino acid*.

**value** 1. Quantity or measure; amount. 2. Usefulness. 3. Magnitude or size.

**valve** 1. A device that controls fluid flow. Some are either fully on or fully off; others can vary the amount of flow continuously from zero to maximum. 2. English term for vacuum tube. 3. A segment of a legume. See *legume*. 4. The covering on a diatom. See *diatom*.

**vanadium** Chemical symbol, V. An element with atomic number 23. The most common isotope has atomic weight 51. In pure form it is a silver-colored metal. It has various industrial uses, especially in manufacture of alloys.

**Van Allen belts** Zones of radiation, caused by movement of charged particles in the earth's magnetic field. These zones are in outer space near the earth (see drawing). At one time, it was thought that the radiation would harm or kill astronauts. But spacecraft pass rapidly through the zones on their way into outer space.

**Van Allen, James** A mid-twentieth-century physicist. He worked with projects to study the upper atmosphere by launching rockets from high-altitude balloons. He is best remembered for his discovery of radiation belts that surround the earth. See *Van Allen belts*.

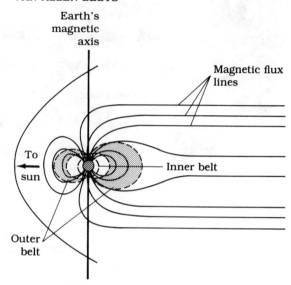

VAN ALLEN BELTS

**van de Graaff generator** A device that generates large electrostatic voltages. It works by friction. The charge builds up on a metal globe at the top of a column. The charge can exceed 1,000,000 volts, causing a spark that jumps several feet.

**van de Hulst, Henrik** The Dutch astronomer who predicted that a strong radio emission from space would occur at a wavelength of 21 centimeters. It was also predicted by the Russian, Iosif Shklovsky. The radiation has been observed with radio telescopes. See *radio telescope* and *Shklovsky, Iosif*.

**van der Walls force** A force among atoms, causing them to attract each other. It is a weak force, and diminishes very rapidly with distance. It arises from electrical charge. This force is the main reason that gases do not behave exactly as they should in theory.

**vapor pressure** For a substance in the gaseous state, the pressure it has when it is in equilibrium with its liquid and/or solid states.

**vapor trail** Condensation left in the wake of an aircraft at high altitude, when atmospheric conditions are favorable for cloud formation. These trails are, in fact, clouds.

**var** The unit of potential power in a reactance. See *volt-ampere*.

**variable** 1. In mathematics, a quantity whose value is not constant, but can change over a given range of values. 2. Not fixed; moving, fluctuating, or changing. 3. See *variable star*.

**variable star** A star that changes brightness. Eclipsing binaries fluctuate as the dimmer star passes in front of the brighter one. Cepheid variables pulsate with a period that is a function of their absolute brightness. Mira type stars are variable red giants. Supernovae literally explode, sometimes becoming as bright as a whole galaxy. See *Cepheid variable star, eclipsing binary star, Mira type star, supernova*.

**variac** A variable transformer used in household alternating-current (ac) circuits. It has a toroid-shaped winding and a movable contact, so that the output can be continuously varied from zero to the input voltage, usually 117 volts ac.

**variation** 1. Change, usually of a continuous nature. 2. Type or variety. 3. The difference between true north and magnetic north.

**varices** Swollen, fragile veins that develop in certain disease conditions. A common example occurs in cirrhosis of the liver, when these veins develop alongside the digestive organs. If such a vein breaks, gastrointestinal bleeding results.

**varicose** Pertaining to swollen, tender, fragile veins. They sometimes form in the legs, especially in older people, and in various other locations in the body. See *varices*.

**variometer** A form of variable inductor consisting of two coils. One coil is inside the other and can be rotated; the outside coil is fixed. The two coils are connected in series or parallel. When the fields from the coils act together, the inductance is maximum; when they act against each other, it is minimum.

**varve** Lake-bed sediments from glacial meltwater. Found at the outlets of glaciers throughout the world. The sediment occurs in layers, whose thickness is thought to be a function of cycles in the earth's average temperature. These temperature changes could have been caused by variations in the brilliance of the sun over periods of thousands of years.

**vascular plant** A plant with a vascular system (see drawing). One set of tubules carries fluid and nutrients upwards from the root system. Another carries fluid back down. See *phloem, xylem*.

VASCULAR PLANT

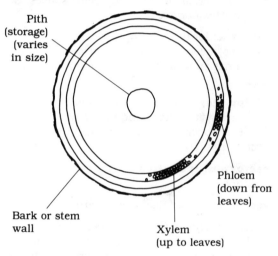

**vascular system** 1. The system of veins and arteries in an animal's body; the circulatory system. 2. The lymphatic system in an animal's body. 3. The system of vessels in a plant. See *vascular plant*.

**vasectomy** Sterilization of a male by surgical closing of the duct through which the sperm are ejaculated.

**vasoconstrictor** Any substance or drug that causes the blood vessels to become narrower. This tends to raise the blood pressure.

**vasodilator** Any substance or drug that causes the blood vessels to become larger in diameter. This tends to lower the blood pressure. It may also produce a flushing reaction.

**VCR** Abbreviation for videocassette recorder.

**VDU** Abbreviation for visual display unit.

**vector** A quantity that has both magnitude and direction. If a quantity has only magnitude (size), it is a scalar. Speed is a scalar, such as 55 miles per hour. Velocity is a vector, such as 55 miles per hour going northwest. Vectors are often represented by arrows, pointing in a specific direction, with length proportional to magnitude. See the following several definitions.

**vector addition** See *vector sum.*

**vector analysis** A branch of mathematics concerned with vectors and vector spaces. See *vector, vector space.*

**vector calculus** A sophisticated branch of calculus involving functions of vector quantities. It has applications in physics and engineering.

**vector diagram** A drawing showing vectors as arrows with length and direction. This technique is especially useful in vector addition. See *vector sum.*

**vector multiplication** See *cross product, dot product.*

**vector product** See *cross product.*

**vector space** A set of vectors and operations, such that certain properties hold. The simplest example is the set of real numbers (one dimension) and the familiar operations of addition and multiplication. Vector mathematics lets us generalize this in two or more dimensions. See a text on vector analysis. See *cross product, dot product, vector, vector sum.*

**vector sum** The sum of the components of a vector in a Cartesian coordinate system. The drawing shows a two-dimensional case. The endpoints of the vectors are $A = (x_a, y_a)$ and $B = (x_b, y_b)$. The sum is obtained by adding the x and y components. The method is similar in spaces of three or more dimensions.

**vegan** A vegetarian whose eating habits are so strict that even eggs and milk products are excluded. A person on such a diet runs the risk of vitamin B-12 deficiency unless supplements are taken. See *vegetable protein, vegetarian.*

**vegetable protein** Proteins obtained from such foods as grains, beans, and nuts. When properly combined (rice and beans, for example) vegetable proteins are complete, supplying all of the essential amino acids. Thus, even the strictest vegetarians do not go short of protein if they are careful to combine their foods the right way, and provided they get enough calories. See *vegan, vegetarian.*

**vegetarian** A person who does not eat flesh of any kind. The less strict diets allow eggs and milk products. The strictest diets exclude even these (See *vegan*). Protein deficiency is not likely with a vegetarian diet, as long as enough calories are consumed. See *vegetable protein.*

**vein** Any of the tubules into which capillaries merge, carrying blood from body tissues to the heart. Blood in most veins is dusky bluish in color because its oxygen supply has been used up. The exception is in the pulmonary vessels, where blood has just been enriched by oxygen in the lungs.

**velocity** Speed and direction. This is a vector quantity. We might say that a car is moving 55 miles per hour in a northwesterly direction, for example. In equations, the quantity is often abbreviated v.

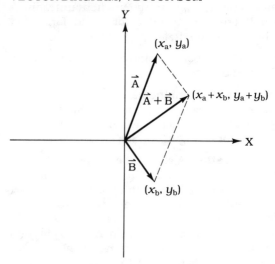

VECTOR DIAGRAM, VECTOR SUM

**vena cava** The major veins leading to the heart. The inferior vena cava comes from the part of the body below the heart; the superior vena cava comes from the part above the heart. They both lead into the right auricle. See *heart*.

**Venera** A series of Russian space probes sent to Venus. In 1966, Venera 3 landed on Venus, and told us it is hot enough to melt lead, and the pressure is equivalent to that at 3000 feet underwater on earth. In 1982, Venera 13 and Venera 14 photographed the surface, showing a barren landscape strewn with rocks. Later probes were put into orbit.

**venereal disease** Any disease that is sexually transmitted. Examples are acquired immune deficiency syndrome, gonorrhea, herpes, and syphilis.

**Venn diagram** A method of illustrating set unions and intersections. In the drawing, the upside-down U symbol denotes set intersection. The universal set is denoted U. Circles denote three sets X, Y, and Z.

VENN DIAGRAM

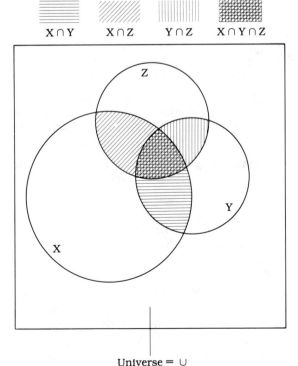

**venom** The toxic substance secreted by glands in some animals, such as stingrays and snakes. It serves two purposes: to disable prey, and for self defense.

**ventral** Pertaining to the underside or front side of an animal; the side of the chest and belly. See *dorsal*.

**ventricle** Either of the large chambers in the heart. The right ventricle pumps blood to the lungs to be enriched with oxygen, after it has come back from body tissues. The left ventricle pumps blood to the body tissues after it has come back from the lungs. See *heart*.

**Venus** The second planet from the sun in the Solar System. It is about the size of the earth, but with a much different climate (See *Venera*). See also the SOLAR SYSTEM DATA appendix.

**vernal equinox** See *equinox*.

**Verne, Jules** A science-fiction writer who romanticized space and undersea travel. One of his most famous works, *From the Earth to the Moon*, published in 1856, chronicled an attempt to shoot a man to the moon in a gigantic cannon shell.

**vernier** A method of obtaining high accuracy from a rotatable dial scale. A system of gears is used to slow down the rate of adjustment. A pair of scales, graduated from 0 to 9, allow precise interpolation. Used in micrometer calipers and in some electronics test equipment.

**vertebra** Any of the bones that form the spine in vertebrates. Together they are called vertebrae (Latin plural). The nerves and blood vessels run through holes in the centers of these bones.

**vertebrate** Any animal with a defined backbone. See *invertebrate*.

**vertigo** A prolonged state of disorientation or dizziness. There are numerous possible causes, such as middle-ear infection or high fever. Some drugs produce this effect. It can also occur in outer space or underwater when up/down sense is lost.

**very high frequency** Abbreviation, VHF. The range of radio frequencies from 30 megahertz

(MHz) to 300 MHz. The wavelengths range from 1 to 10 meters (about 3.3 to 33 feet). Signals at VHF may interact with the ionosphere, but not always.

**very-large-scale integration** Abbreviation, VLSI. A technology of integrated circuits, in which there are 100 to 1000 logic gates on a single chip. See *integrated circuit*.

**very low frequency** Abbreviation, VLF. The range of radio frequencies from 3 kilohertz (kHz) to 30 kHz. Signals above about 9 kHz propagate worldwide on a reliable basis, but extremely high power and huge antennas are needed.

**vestigial organ** An organ of an animal's or human's body that has evolved out of use. The appendix is a good example.

**VHF** Abbreviation for very high frequency.

**vibration** 1. Rapid back-and-forth movement, sometimes at hundreds, thousands or millions of times per second. 2. See *oscillation*.

**videocassette recorder** Abbreviation, VCR. A television camera and cassette recorder combined. The video information is directly recorded on the tape in the cassette. This tape can be played back with a system attached to a conventional television set. Sometimes the whole set, including the camera, recorder and playback system, is called a VCR or VCR system.

**video display unit** See *visual display unit*.

**videophone** A telephone that transmits pictures along with the voice. Slow-scan devices are already available. They send one still picture every few seconds.

**videotape recorder** Abbreviation, VTR. A television camera and tape system. The video is directly recorded on the tape. The most common such device uses cassettes. See *videocassette recorder*.

**vidicon** A compact, simple television camera (see drawing). It is quite sensitive, and is commonly used in surveillance systems. But it does

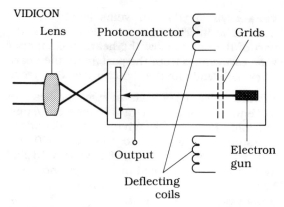

VIDICON

not have a very rapid image response. The image orthicon is better when fast-moving images must be recorded. See *image orthicon*.

**Viking** American space probes that landed on Mars. They touched down on opposite sides of the planet in 1976. They photographed a surface of red sand, strewn with boulders. No definite signs of life were found.

**villi** 1. Tiny protuberances in the small intestine, through which proteins, fats, sugars, vitamins, minerals, and water are absorbed. 2. A membrane on the egg of a mammal that becomes part of the placenta.

**virga** Rain that evaporates before it reaches the ground. It can often be seen on radar, and can be hard to distinguish from actual rainfall.

**Virgo cluster** A complex and vast cluster of galaxies in the constellation Virgo. This is a distant cluster, and it provides astronomers with clues as to the structure of the universe.

**virtual image** An unfocused image. An example is the reflection of light from a flat mirror. See *real image*.

**virus** A living thing that is somewhere between a complicated molecule and a simple microorganism. Such organisms cause diseases, such as the common cold, influenza, and some pneumonias. Acquired immune deficiency syndrome (AIDS) is a viral disease.

**viscosity** A property of a fluid that causes it to tend to hold together as it flows. The greater the viscosity, the more the liquid resists shearing, or local variations in the speed of its flow. Water has low viscosity; motor oil is more viscous. Liquid rubber cement has very high viscosity.

**visibility** The distance from which images can be clearly seen with the unaided eye. This is often specified by meteorologists and aviators; for example, 1/2 mile or 10 miles.

**visible spectrum** The range of electromagnetic wavelengths that can be detected by the human eye. This ranges from about 3900 to 7500 Angstroms, where one Angstrom is equal to a ten-billionth ($10^{-10}$) of a meter. See the illustration. See also the ELECTROMAGNETIC SPECTRUM appendix.

VISIBLE SPECTRUM

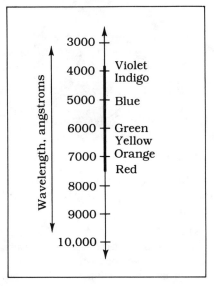

**vision** 1. The ability to detect electromagnetic energy at or near wavelengths of approximately 3900 to 7500 Angstroms (see *visible spectrum*). 2. The ability to resolve images at wavelengths at or near the visible spectrum. 3. The general extent of visibility. See *visibility*.

**visual display unit** Also called video monitor, video display unit. A device resembling a television set, but with higher resolution of detail. Used with computers and communications terminals to display data.

**vitamins** Chemicals that are not minerals, protein, carbohydrate, or fat but that occur in foods and are needed by the body. They are designated by letters. See the VITAMINS AND MINERALS appendix.

**vitamin deficiency** An unhealthy condition resulting from lack of one or more vitamins in the diet. See the RECOMMENDED DAILY ALLOWANCES and VITAMINS AND MINERALS appendices.

**vitamin toxicity** An unhealthy condition resulting from getting too much of a certain vitamin. This occurs mainly with the fat-soluble vitamins A and D. See the RECOMMENDED DAILY ALLOWANCES and VITAMINS AND MINERALS appendices.

**vitreous humor** The clear fluid inside the eyeball. This fluid keeps the tissues bathed and nourished, and keeps the eyeball from collapsing.

**Vitruvian wheel** A paddlewheel with a horizontal axle, suspended so that water flowed through the bottom of the wheel, causing it to rotate. A right-angle gearing system connected the axle to a millstone shaft. The gears were in a ratio so that the millstone turned several times for each rotation of the water wheel.

**Vitruvius** An ancient Roman author and scientist who is known for his water wheel and water clock. See *Vitruvian wheel*.

**VLF** Abbreviation for very low frequency.

**VLSI** Abbreviation for very-large-scale integration.

**vocal cords** The fibers in the larynx, or voice box, that vibrate in various combinations, making the complex audio waveforms of the lower-frequency and midrange parts of the human voice. These fibers exist in most mammals.

**vocoder** An electronic circuit that reduces the bandwidth needed by a voice signal. It changes the voice into impulses. The output is easy to understand but lacks inflection. See *voder*.

**voder** An electronic speech synthesizer. A keyboard is used to select voice sounds. Almost all inflection is lost, so that the output sounds monotonous and unemotional. See *speech synthesis*.

**voice box** See *larynx*.

**volcanic action** The eruption, ejection of lava and other materials, and general effects of volcanoes. They build islands, contribute gases to the atmosphere, provide fertile soil, and can also cause global cooling if great amounts of dust get high enough into the atmosphere.

**volcanic ash** Fragments of the hardened plug in the throat of a volcano, broken up by the blast of the eruption. These fall back to earth and can accumulate near the volcano to a considerable depth.

**volcanic cone** The buildup of ash, dust and lava that occurs around some volcanoes. Over long periods this can grow to the size of a small mountain. See *volcano*.

**volcanic crater** The hollowed-out, concave depression at the top of a volcano. Fluid magma moves up into the crater until it overflows. During inactive times, the crater drains. If the crater is more than a mile wide, it is called a caldera. See *volcano*.

**volcanic dust** Fine solid particles ejected during some volcanic eruptions. If these rise high enough in the atmosphere, in great enough quantities, global cooling can occur because the dust blocks out some of the sun's energy. This took place after the eruption of Tambora in 1815.

**volcanic energy** 1. The energy released during a volcanic eruption. 2. The energy from volcanism that is available for harnessing by humans. Hot lava, for example, could be used to boil water to drive steam turbines.

**volcanic rock** Any rock that has come from a volcano, and therefore from the earth's mantle, below the crust. Also called igneous rock. Basalt and granite are common examples.

**volcanism** Volcanic activity, especially in reference to a particular time or a certain region of the earth. See *volcano*.

**volcano** A place where fluid magma and gases from the earth's mantle break through the crust and are ejected onto the surface and into the atmosphere. See the drawing. See also the preceding several definitions.

VOLCANO

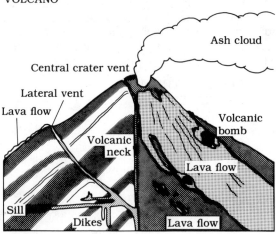

After NOAA Publication 1985-579-010/25212.

**volcanology** A branch of earth science concerned with the study of volcanoes, their behavior, and their effects.

**volt** Abbreviation, V. A unit of electrical potential. When a current of 1 ampere flows through a resistance of 1 ohm, the potential difference across the resistance is 1 volt. Conversely, a potential of 1 volt will drive a current of 1 ampere through a resistance of 1 ohm. See *Ohm's Law*.

**Volta, Alessandro** An Italian physicist of the late eighteenth and early nineteenth centuries. His voltaic pile, an early version of the electrochemical battery, led to the development of modern batteries.

**voltage** 1. Electrical potential or potential difference. 2. The number of volts at a given pair of terminals, representing the potential difference between the terminals. Might be alternating-current (ac) or direct-current (dc) voltage. Can also be expressed for radio frequencies (RF). See *volt*.

**voltage feed**  In an antenna, the connection of the feed line at a point on the radiating element where the voltage is maximum. See *current feed*.

**voltage gradient**  1. The change in voltage along an electrical conductor, measured in volts per unit distance. 2. The voltage change per unit distance between two points having different electrical potential.

**Voltaire**  An eighteenth-century philosopher and mathematician who was also politically active in France. His real name was Francois-Marie Arouet. He was a prolific author.

**volt-ampere**  A unit of potential power, equal to a product of voltage and current (volts and amperes) yielding 1. When power is dissipated, this unit is more often called the watt. In a reactance, it is called the var, for volt-ampere reactive. See *watt*.

**voltmeter**  A device for measuring electrical potential. The simplest device consists of a microammeter or sensitive galvanometer, with a large resistance in series. More sophisticated meters use field-effect transistors to greatly reduce the flow of current through the device.

**volume**  1. Displacement in three dimensions. Generally given in cubic units, such as cubic centimeters or cubic feet. 2. Amount or bulk. See *mass*. 3. See *loudness*. 4. Large numbers or amounts, as in manufacturing.

**voluntary muscle**  A muscle that can be moved on conscious command. These include the large muscles of the arms and legs, the muscles in the torso, and numerous smaller muscles in the hands, feet, and face.

**von Braun, Wernher**  A German rocket scientist of the twentieth century. He did extensive work for the United States space program in its early years.

**von Neumann, John**  A prominent twentieth-century mathematician, physicist, and computer scientist. He worked in quantum mechanics and fluid flow early in his career. Later he worked with the scientists developing the atomic bomb. Still later, he developed early versions of computers. He is known for the breadth of his knowledge and the wide variety of his contributions.

**vortex**  1. A whirlwind, caused by Coriolis force or by wind shear. Dust devils and tornadoes are two examples. See *Coriolis effect, wind shear*. 2. An eddy in a body of water. 3. Any inward, rotating effect with a central attractive force. 4. Less commonly, an outward, rotating effect.

VORTEX

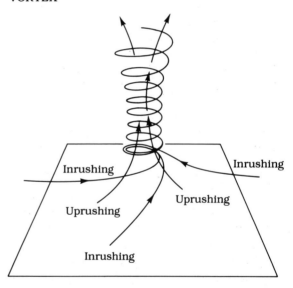

Hot surface

**Voyager**  American space probes that sent back pictures of the outer planets. These probes provided detailed information, some unexpected and spectacular, of Jupiter, Saturn, Uranus, and Neptune, and also of their moons.

**VTR**  Abbreviation for videotape recorder.

**Vulcan**  A hypothetical planet, once thought to orbit closer to the sun than Mercury. Today we know that no such planet exists, at least not of any significant size.

**W** Abbreviation for power, watt, work; chemical symbol for tungsten.

**wall cloud** The intense, ring-shaped band of stormy weather immediately around the eye of a hurricane. Also called the eyewall. See *hurricane*.

**Walton, Ernest** One of the builders of the first particle accelerator. The other was Sir John Cockcroft. See *particle accelerator*.

**warm-blooded** In animals, the ability to generate body heat, maintaining a constant internal temperature. Mammals and birds have this ability. See *cold-blooded*.

**warm front** See *weather front*.

**warm sector** The wedge-shaped part of a midlatitude low-pressure system between the warm front and the cold front. This is the portion where most of the violent weather occurs, especially in the spring and summer.

**wart** A skin lesion caused by a virus. The most common locations are the hands and feet. The lesion is usually skin-colored at first, but can turn gray or yellow as it grows. Treatment is by freezing or by surgical removal.

**wastage** 1. Unnecessary waste by-products from a process, resulting from inefficiency. 2. Inefficiency. 3. Impaired food absorption by the body, resulting in emaciation (extreme thinness).

**waste** 1. Useless by-products of industrial processes. 2. Urine and feces of animals and humans. 3. Radioactive by-products of nuclear reactor operation. See *high-level radioactive waste, low-level radioactive waste*.

**waste accumulation** An undesirable and unhealthy buildup of useless by-products, caused by some combination of inefficiency, inadequate disposal, or poor management.

**waste disposal** Any means of getting rid of waste. Some ways are environmentally sound, and some are not. See *waste management*.

**waste management** A branch of civil engineering concerned with the efficient and environmentally safe disposal of waste products. This includes by-products of industrial processes, sewage and radioactive materials.

**water** The compound resulting from two hydrogen atoms being bonded to one oxygen atom. As found in nature, it usually contains dissolved minerals. It can exist as a solid, liquid, or gas. All three states are found on the Earth. This compound was important in the development of life, and is necessary for practically all life on this planet.

**water clock** Any timekeeping device that uses running or dripping water. The earliest such device was the clepsydra, built by Ctesibius in ancient Greece. See *clepsydra*.

**water cycle** The journey of water from lakes, rivers, and oceans into the atmosphere and back. The average time for a single molecule to complete this cycle is about 10 days. It is only a few hours in the tropics, but at the poles it can be as long as 10,000 years. The illustration shows the basic nature of this cycle, also called hydrologic cycle.

WATER CYCLE

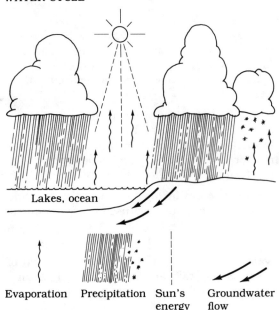

Evaporation   Precipitation   Sun's energy   Groundwater flow

**water droplet** A tiny, spherical accumulation of liquid water that stays suspended in the atmosphere, held up by the motion of air molecules around it. Clouds, steam, vapor trails, and fog are comprised of these particles. They condense around grains of dust or pollutants.

**waterfall** 1. A place along a river or stream where the elevation abruptly drops. 2. A human-made, abrupt drop in the elevation of a river. This is generally done by damming. The force of the falling water can be used to generate useful power. See *hydroelectric energy, hydroelectric power plant, water mill*.

**water mill** A facility that uses water power to turn millstones to refine grain into flour. Water wheels can be directly connected to the millstones. In England in the early nineteenth century, there were about 5600 such mills. This method of making flour is still used in some places today.

**water pollution** Contamination of lakes, rivers, and oceans by unhealthful by-products of human activity. This can have long-term environmental and climatic impact. It can also harm the human population, mainly by poisoning, and also by interfering with the food chain and by encouraging the growth of unwanted organisms.

**water purification** Any process by which harmful natural and human-made substances are removed from water, so that the water is safe for drinking and bathing.

**water screw** A device first invented by Archimedes in the third century B.C. It is a hollow cylinder with an inclined plane wrapped around inside, like an inside-out screw. When it is turned, it brings water up from a lake or river, so that the water can be used for irrigation. In some parts of the Nile River valley, these devices are still used today.

**waterspout** A tornado over water. Usually, these are less violent than midlatitude tornadoes over land. They are often seen over the oceans in the tropics and temperate regions.

**water storage** Any method of accumulating water for long-term, controlled use. Reservoirs are one common method. Water tanks and towers are another.

**water table** The top of the underground region where the earth is saturated with water. This is normally the level that the water attains in a well, or in lakes in a given region.

**water tower** An elevated water tank. The water is kept at a level higher than that of all the users in the area served by the tank. This provides water pressure for the residents, without the need for pumps. Of course, the water must be pumped up into the tower in the first place.

**water turbine** A specialized form of water wheel, used in hydroelectric power plants. The force generated by flowing water, from a reservoir through a sluice in a dam, is efficiently harnessed by the turbine, to power electric generators. See *hydroelectric power plant*.

**water vapor** Water in the gaseous state. This is not to be confused with steam, which consists of microscopic droplets of liquid water.

**waterwheel** A paddlewheel turned by flowing or falling water (see drawing). The torque from the axle of the wheel can be used to generate electricity, to mill grain, or for other purposes.

WATER WHEEL

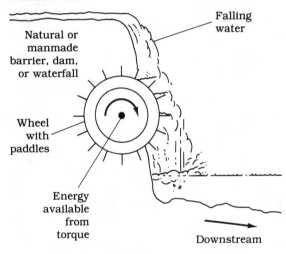

**waterworks** The system of reservoirs, storage tanks, pumps, pipelines, and purification apparatus that supplies the public with water for drinking and bathing.

**WATS** Abbreviation for wide-area telephone service.

**watt** The unit of power, most commonly given for electric or radiant energy. One watt is the equivalent of one volt at one ampere of current. It is also equal to one joule of energy expended per second. See *power*.

**watt hour** Abbreviation, Wh. A unit of energy, equal to 3600 joules. It is the equivalent of one watt of power expended for one hour.

**watt hour meter** A device that measures electrical energy, in units of watt hours. More often, in utility circuits, the meter measures kilowatt hours. See *kilowatt hour*.

**Watt, James** An instrument-maker in Scotland who improved the design of the steam engine in the eighteenth century. He developed a device called a condenser to make the engine more efficient. He also ran a factory that produced steam engines. See *steam engine, steam pump*.

**wattmeter** A meter that measures electrical power, usually in watts, but sometimes in milliwatts or kilowatts. In electrical circuits, the power in watts can be determined by multiplying the current in amperes by the voltage in volts. Some meters measure radio-frequency (RF) power.

**wave** 1. Undulations in a body of water caused by friction with the wind. 2. A pattern caused by periodic oscillation of some quantity, usually electricity and/or magnetism. 3. See *electromagnetic field, electromagnetic wave*. 4. A period of unusually hot or cold weather.

**wave crest** 1. The peak of a wave on a body of water. 2. The maximum positive point on an electrical wave disturbance. 3. The highest point on a displayed waveform.

**wave drag** A type of friction experienced by aircraft traveling near, at, or above the speed of sound. The shock wave at these speeds robs the aircraft of power that could otherwise go into propulsion. Swept-back wings and other features of high-speed aircraft reduce this drag.

**waveform** The general shape of a wave disturbance. The purest wave, representing a component at just one frequency, is the sinusoid. Voices and music have very complicated wave shapes. See *sawtooth wave, sinusoid, square wave*.

**wavefront** The fast-moving surface representing points on a transmitted wave that are all in the same phase and all on the same cycle. From a light source, for example, these can be thought of as concentric spheres, expanding outward at 186,000 miles per second through space.

**waveguide** A type of transmission line for ultra-high radio frequencies and microwaves. It connects the antenna to the transmitter and/or receiver. Physically it is a circular or rectangular pipe. The electromagnetic field is literally guided along inside this pipe.

**wavelength** For a wave disturbance, the distance between points in adjacent cycles, that exist at the exact same place on the waveform. See the drawing. The product of the wavelength and the frequency is a constant, representing the speed at which the wave disturbance moves through a substance or through space. Thus, the wavelength is inversely proportional to the frequency. See *frequency*.

WAVELENGTH

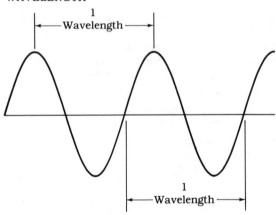

**wave model** The theory that explains electromagnetic phenomena as a wave disturbance. This applies particularly to visible light. See *particle model, particle/wave dichotomy*.

**wave/particle dichotomy** See *particle/wave dichotomy*.

**wave trough** 1. The lowest part of a wave on a body of water. 2. The maximum negative point on an electrical wave disturbance. 3. The lowest point on a displayed waveform.

**Wb** Abbreviation for **weber**.

**weather** Climatic conditions in a given location at a given time, or over a specific period of time. Includes variables such as the temperature, humidity, barometric pressure, extent of cloud cover, precipitation, and wind velocity. See *climate*.

**weather balloon** A large, helium-filled balloon, usually at least six feet in diameter and ranging up to more than 100 feet across, used to lift instruments into the upper atmosphere. Data is radioed back to surface stations for analysis. See *radiosonde*.

**weather forecasting** The science of weather prediction. Short-term forecasting is for periods less than a few days. Long-term forecasting is for periods of weeks or even months. A forecast can be made for a specific location or region, or for locations worldwide.

**weather front** A region in which two air masses, having different temperatures and humidity levels, meet. The drawing shows the common types of fronts and their symbols on weather maps. Almost all fronts are associated with circulation around low-pressure systems at temperate latitudes. Often, precipitation occurs in the vicinity of a front. Cold front: Advancing cool or cold, dry air pushes under warmer, moister air. Warm front: Advancing warm, moist air overrides cool, dry air. Occluded front: Cold front catches up with warm front in circulation around low-pressure system. Stationary front: Cold front or warm front slows down and stalls. Precipitation might last for several days in a given place under these conditions. Violent weather is most often associated with cold fronts, although it can occur with any type of front.

**weather map** A map of a region, showing weather conditions at various points, the locations of low-pressure and high-pressure systems, the locations and orientations of fronts, and usually lines of equal barometric pressure (isobars). Also might show features such as

## WEATHER FRONT

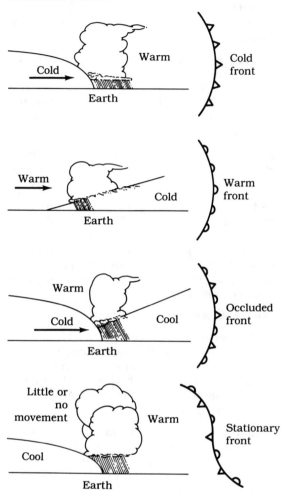

clouds, precipitation, tropical storms, and blowing snow, dust, or sand.

**weatherproofing** The insulation of a building to minimize the energy consumed to maintain the interior at a comfortable temperature, no matter what the outside weather.

**weather radar** A radar unit used to track the paths of precipitation areas. Rain and snow reflect the beams, and regions of precipitation appear as bright blobs. Certain features are distinctive, such as tornadoes (hook-shaped or comma-shaped echo) and the eyes of hurricanes (dark circle surrounded by bright spiral bands).

**weather satellite** An orbiting satellite, equipped with cameras that photograph the Earth at infrared, visible and/or ultraviolet wavelengths, and transmitters that send the pictures back to Earth at regular intervals. The drawing shows a typical satellite used by the United States National Oceanic and Atmospheric Administration (NOAA).

## WEATHER SATELLITE

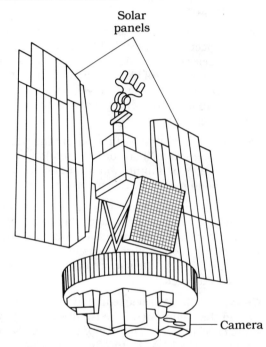

THE NIMBUS, A WEATHER POLAR ORBIT DURING THE 1960s. From Erickson, EXPLORING EARTH FROM SPACE (TAB Books, Inc., 1989).

**weber** Abbreviation, Wb. The Standard International unit of magnetic flux. It is equivalent to one volt-second, or 100 million ($10^8$) maxwells.

**wedge** A simple machine with a characteristic shape like a piece of cheese. It has a narrow triangular cross section in one plane, and a rectangular cross section in the other. It has various uses that make work easier.

**weed** Any plant that is not wanted or intended. Pertains especially to lawns, gardens and farm crops.

**weed control** Use of herbicides (plant killers) and other methods to kill weeds without harming the desired crops or plants.

**weight** The force exerted by a mass in a gravitational field or acceleration field. A mass of one kilogram weighs about 2.2 pounds in the earth's gravitational field. On Mars, the same mass would weigh just 0.81 pound. On Jupiter it would weigh 5.5 pounds. Weight varies with the intensity of the force field, but mass does not. See *mass*.

**weightlessness** The condition of zero gravity. This occurs in space flight when the vessel is not accelerating. Some ordinary tasks, such as drinking a glass of water, become difficult or impossible with zero gravity. Astronauts have endured months of zero gravity with few ill effects.

**welding** The bonding of metals by the use of heat to melt one or both of the metals. Sometimes an alloy of aluminum and magnesium is used to bond metals by this means.

**well** A hole drilled or dug into the ground, extending down past the water table (see *water table*). Used to supply water for drinking, bathing and irrigation. A pump is usually needed to bring the water to the surface, but not always. See *artesian well*.

**Wells, H. G.** A science-fiction writer who, around the beginning of the twentieth century, wrote novels such as *The War of the Worlds* and *The Time Machine*. In the first novel, he portrayed an invasion by Martians. The aliens died because they had no immunity to earthly bacterial disease. In the second novel, he depicted such events as nuclear war as an inventor traveled into the future.

**Werner, Abraham** An eighteenth-century scientist who was among the pioneers of geology as we know it today. Before his time, there were several loosely related fields such as mineralogy and paleontology.

**Westinghouse, George** A nineteenth-century engineer who employed pressurized air to make an improved system of brakes for trains. These were called pneumatic brakes. The method was adopted by all the railroads. He also played a role in pioneering the use of alternating current (ac) as the best means of transmitting electrical power.

**Weston Standard Cell** An electrochemical cell providing a standard of 1.018 volts at 20 degrees centigrade. It has a liquid electrolyte of cadmium sulfate. The electrodes are of mercurous sulfate (positive) and a mixture of mercury and cadmium (negative).

**wet brain** A condition sometimes seen in chronic alcoholics. It occurs only after years of serious disease. It is somewhat like senility.

**wet-bulb** A thermometer whose bulb is covered by a saturated wick. Evaporation from the wick cools the bulb. The lower the relative humidity, the greater this effect; wet-bulb and dry-bulb readings can be compared, and the relative humidity read from a standard table. See *relative humidity*.

**Wh** Abbreviation for watt hour.

**wheat germ** The part of the wheat kernel that contains the embryo for a new plant. It also contains the highest concentration of protein, vitamins and minerals, especially the B complex and vitamin E. See the VITAMINS AND MINERALS appendix.

**wheat germ oil** Oil extracted from wheat germ (see *wheat germ*). It is mainly polyunsaturated, and contains significant vitamin E (tocopherol). See also the VITAMINS AND MINERALS appendix.

**wheel and axle** One of the simple machines invented presumably during prehistoric times. See the drawing. It has many uses, the most common and significant being to reduce the effort needed to haul objects over land.

**Whewell, William** The man who is credited with having invented the word "scientist" in 1840.

**whirlwind** Any revolving wind system, such as a dust devil, tornado, low-pressure system, high-pressure system, or hurricane. Also included are tiny vortices that pick up dust and leaves.

WHEEL AND AXEL

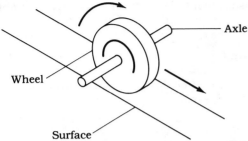

WHOLE GRAIN

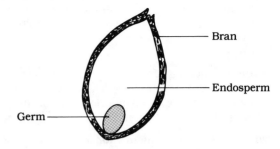

**white blood cell** Any blood cell that does not have hemoglobin. In particular, these are the cells that help the body fight infection.

**white dwarf** A small, white, dim, extremely dense star near the end of its life cycle. After the lighter elements have been exhausted, our sun will eventually collapse to a star of this type, only about the size of the earth, but still having nearly the mass of the sun today.

**white hole** A hypothetical place into which matter might be introduced into the universe from some other universe. The opposite of a black hole, it would have negative gravitation. No such objects have actually been proven to exist.

**white smoker** An undersea phenomenon associated with volcanic activity. It looks like it is spewing white smoke, hence the name. See *black smoker*.

**Whitney, Eli** An American engineer of the late eighteenth and early nineteenth centuries. He is remembered mainly for his development of the cotton gin. But just as important, he introduced the concept of interchangeability in machine parts.

**whole grain** The bran, endosperm, and germ of common grains such as rice, wheat, oats, and rye. See the drawing. In recent years, whole grains have been shown to have nutritional benefits over refined grains, from which the bran and germ have been removed. However, whole grains tend to spoil more rapidly than refined grains.

**whole number** Any nonnegative integer; that is, an element of the set {0, 1, 2, 3, ...}.

**whooping cough** Also called pertussis. A contagious disease caused by bacteria. The main feature is a cough that takes place in violent paroxysms, or episodes, so severe that they can cause vomiting. It is especially serious in infants. The disease lasts about six weeks. Pneumonia and middle-ear infections can occur as complications.

**wide-area telephone service** Abbreviation (acronym), WATS. A long-distance telephone service that is billed at a flat monthly rate, up to a certain amount of usage time each month. The familiar "800" numbers are incoming WATS lines. Outgoing WATS lines are also used by many corporations.

**Wien Displacement Law** The peak wavelength of emission from a blackbody is inversely proportional to the absolute temperature of the object. The frequency is therefore directly proportional to the absolute temperature. See *blackbody, blackbody radiation*.

**Wiener, Norbert** One of the prominent mathematicians of the twentieth century, pioneering the development of information theory, especially after World War II. See *information theory*.

**Wigner, Eugene** One of the physicists, a Hungarian refugee, who persuaded Albert Einstein, in 1939, to write to President Franklin D. Roosevelt, urging development of an atomic bomb. He and Leo Szilard feared that the Nazis might develop it first.

**willy-willy** The Australian term for a hurricane. See *hurricane*.

**Wilson cloud chamber** See *cloud chamber*.

**Wilson, Robert** One of the engineers for the Bell Laboratories who, in the 1960s, inadvertently discovered the background radiation from the Big Bang. See *Big Bang* and *Penzias, Arno*.

**wind** 1. Moving air, caused by differences in pressure over large areas. 2. The effects of moving air. 3. Moving atmospheric gases on other planets. 4. See *solar wind*.

**wind chill** The tendency for wind to increase the rate at which heat is removed from exposed flesh. This effect is most pronounced at very low temperatures and high wind speeds. No additional effect is observed at wind speeds of more than about 40 miles per hour.

**wind erosion** Wearing-down of the surface features of the earth, or of any planet with a surface and an atmosphere, as a result of atmospheric motion. This occurs as fine particles are carried by the wind, "sanding down" the rocks and other irregularities over many years.

**windmill** A device for obtaining power from the wind. Basically it consists of a propeller connected via gears to whatever device is to be operated. The term arose originally because farmers used them to mill grain (and often still do). It can apply nowadays to any wind-power generator (see drawing).

**window** 1. A period of time during which a space mission can ideally be launched. 2. A period of time during which a course correction must be made for a space vessel. 3. A range of wavelengths or frequencies for which a given medium, usually the atmosphere, appears transparent to electromagnetic energy.

**windpipe** See *trachea*.

**wind power** Power obtained directly from the force exerted by the wind. See *windmill*.

**wind pressure** The pressure exerted against an object by the wind. It increases with the square of the wind speed. Thus, a gale of 50 miles per hour has only $1/4$ the destructive power of a hurricane at 100 miles per hour, and just $1/16$ the force of a killer hurricane or tornado at 200 miles per hour. See *wind, wind speed*.

**wind propulsion** The use of the wind to drive ships over the sea. This method was used for many centuries before the advent of steam, internal combustion, and nuclear power. It is still used today, mainly for sport (sailing and windsurfing).

**wind resistance** 1. The drag that an object offers to the movement of the atmosphere. 2. The drag, or friction, of an object as it moves through the air.

**wind shear** An abrupt change in wind direction and/or speed. Usually takes place along a plane (flat) boundary. This is a major source of concern for aircraft pilots because of the turbulence it causes. It can also contribute to the development of tornadoes.

**wind speed** The average speed of air molecules with respect to some fixed object, usually the ground or ocean surface. This is never constant, and is difficult to measure except as an average for a period of time, such as 30 seconds to one minute. Specified in miles per hour (mph or mi/hr) or meters per second (m/s). Also may be given in knots, or nautical miles per hour. See *knot* and the BEAUFORT SCALE appendix.

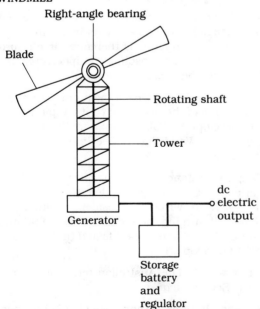

WINDMILL

**wind tunnel** A large, tube-shaped enclosure with gigantic fans or jets that generate extreme wind speeds. Used mainly for testing prototype aircraft designs. Can also be used to simulate conditions in tornadoes and hurricanes.

**wind velocity** Wind speed and direction. The direction is usually given as the compass point from which the wind is coming, with 0 or 360 degrees being geographic north, 90 degrees being due east, 180 degrees due south and 270 degrees due west. See *wind speed*.

**windward** 1. The side of an object, such as a ship or island, facing the wind. 2. In the direction from which the wind is coming. See *leeward*.

**wing** 1. Either of the limbs that enable birds and certain insects to fly. 2. The airfoil devices that enable an aircraft to fly. 3. A hydrofoil. See *airfoil, hydrofoil*.

**winter solstice** See *solstice*.

**wire** 1. An electrical conductor made of metal or alloy, usually copper or aluminum. It has a circular cross section and can consist of a single, thick strand or many fine strands woven together. 2. To send a message by means of a wire circuit. 3. To assemble or set up an electronic circuit. 4. A cable that provides mechanical support to an object.

**wiring** 1. The interconnection of an electrical or electronic circuit by means of wires. 2. The method or scheme for interconnecting the components of an electrical or electronic circuit.

**wiring diagram** A schematic diagram of an electronic circuit, showing all of the pin numbers and other data needed by technicians to build and repair the equipment. See *schematic diagram* and the SCHEMATIC SYMBOLS IN ELECTRONICS appendix.

**withdrawal** 1. The removal or discontinuance of the use of some substance, especially a drug. 2. The set of symptoms associated with sudden discontinuance of drug use. This syndrome is especially severe with depressant drugs such as alcohol and barbiturates, and with some tranquilizers.

**withdrawal use** Human use or consumption of subsurface water, later returned to the water cycle. Most water utilization is of this type.

**Wohler, Friedrich** A nineteenth-century chemist who created the organic compound, urea, in a laboratory. This was one of the first synthetic organic compounds. He demonstrated that some organic compounds can, in fact, be manufactured artificially.

**wolfram** See *tungsten*.

**woofer** A speaker that responds to the bass (low-frequency) component of high-fidelity audio. Such speakers must be fairly large because of the long wavelengths involved. See *bass*.

**word** 1. A unit of digital data, consisting of six characters. A space counts as one character. Therefore, the number of words in a given sample of data is equal to $1/6$ the number of characters. 2. A unit of data, agreed on by convention for a certain system or application.

**word processing** A technique for writing and editing using a computer or a specialized memory typewriter. Software features allow easy changing, adding or deleting of letters, words, sentences, paragraphs, and whole pages or sections.

**words per minute** A measure of the speed of a digital transmission. It is simply the number of words sent in one minute. See *word*.

**work** Mechanical energy, expressed as a weight lifted a certain distance, or a force operating over a specified distance. The common English unit is the foot-pound; the Standard International unit is the newton-meter or joule. See the ENERGY UNITS appendix.

**wormhole** A hypothetical space-time passageway between universes, or between widely separated parts of our universe. The drawing shows a dimensionally reduced, greatly simplified concept. No such passages have been created or found; they remain in the dreams of science-fiction writers.

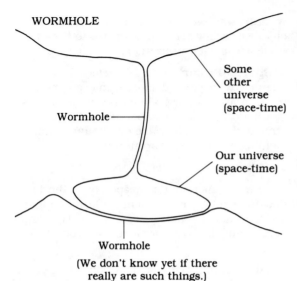
(We don't know yet if there really are such things.)

**Wright brothers** Wilbur and Orville Wright, engineers of the late nineteenth and early twentieth centuries. Best remembered for their work in aerodynamics. They are credited with the successful flight of the first powered aircraft.

**Wright, Irving** A physician who first pioneered the use of anticoagulant drugs to combat circulatory disease, especially strokes. The drugs help slow or prevent the formation of blood clots in the arteries of some people. This work, begun in the mid-twentieth century, continues today.

**wrought iron** An almost pure form of iron. What little impurity exists occurs as thin filaments. It is well suited to industrial use because of its durability.

**X** Abbreviation for reactance.

**x axis** In a Cartesian coordinate system, the axis on which the x variable is depicted. This is usually the independent-variable axis, or abscissa. See *abscissa, Cartesian coordinates, Cartesian plane.*

**X band** 1. The range of frequencies from 8 to 12 gigahertz (GHz), or wavelengths 2.5 to 3.8 centimeters (cm). It is used for radar, microwave links, and satellite links. 2. The range of wavelengths corresponding to X rays. See *X rays*.

**x chromosome** A chromosome that determines the sex of an organism. It is shaped like an x. Two x chromosomes together result in a female animal. See *y chromosome*.

**Xe** Chemical symbol for xenon.

**xenon** Chemical symbol, Xe. An element with atomic number 54. The most common isotope has atomic weight 132. In pure form it is an inert gas. It has various industrial uses, mainly for producing brilliant light, and also in voltage-regulator devices. See *xenon flash tube*.

**xenon flash tube** A device that uses xenon gas to obtain a bright, white flash of light. This flash can be used to excite lasers. Most of the light comes from ionization of the gas at visible red and blue wavelengths. The combination of these colors produces near-white light.

**xerophyte** Literal meaning, dry plant. It can store large amounts of water, so it can survive in places where rainfall is sporadic and light. The cactus is a good example of this type of plant.

**xerox** Trade name (of Xerox Corporation), now become generic, for photocopies and photocopying. This process is extensively used in offices today.

**X rays** Electromagnetic waves having wavelength shorter than ultraviolet. This ranges from about 40 Angstroms to 0.1 Angstroms, where 1 Angstrom = $10^{-10}$ (one ten-billionth) meter. These rays have considerable penetrating power, and are of use in medicine. Special instruments are used by astronomers to observe the Cosmos at these wavelengths. See the following several definitions.

**X-ray astronomy** The observation of space at X-ray wavelengths. This must be carried out from satellites above the earth's atmosphere, because the atmosphere blocks energy at these wavelengths.

**X-ray photography** The techniques of exposing film at X-ray wavelengths. Ordinary film works for this purpose, but the focusing devices (if any) must be of a special type. Glass lenses will not work. X-ray images are of use in medicine, and are also of interest to astronomers.

**X-ray source** 1. Any object that emits radiation in the form of X rays. Usually this occurs along with other wavelengths. See *X rays*. 2. A celestial object that is an unusually strong source of X rays. See *X-ray star*.

**X-ray star** A star that emits an unusual level of radiation in the X-ray band (see *X rays*). There are various theories that have been formulated to explain this. One theory holds that such stars are binary systems, and one star is sucking matter out of the other. This exposes the deeper layers of the latter star, and because these layers are extremely hot, the wavelengths emitted are shorter than those that come from the surfaces of stars. See *binary star*.

**X-ray telescope** An astronomical device for observing the Cosmos at X-ray wavelengths. Must be placed on a satellite to get it above the earth's atmosphere. It uses a special kind of mirror to focus the X rays and obtain an image (see drawing). The resolving power is not as good as that obtained with large optical telescopes. But significant discoveries have been made using these devices.

X-RAY TELESCOPE

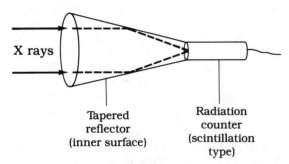

From Gibilisco, BLACK HOLES, QUASARS AND OTHER MYSTERIES OF THE UNIVERSE (TAB Books, Inc., 1984).

**X-ray tube** A device for generating X rays, especially in medical applications. A high voltage causes electrons to be accelerated to high speed. The electrons strike a tungsten plate, angled so that the resulting radiation comes out of the tube perpendicular to the path of the electrons (see drawing).

X-RAY TUBE

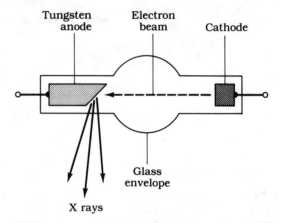

**xylem** In a vascular plant, the tissue that makes up the vessels that transport nutrients, and also serves to support the stem or trunk. The woody part of a plant is composed of this tissue. See *phloem*.

**Y** Chemical symbol for yttrium.

**Yagi antenna** A unidirectional antenna recognizable by its shape (see drawing). It radiates and receives well in one direction. The transmitter or receiver is connected to the driven element. Other elements, called directors and a reflector, produce the directional response.

**yang** In Oriental philosophy, the masculine element. In nutrition, foods such as red meat and grains are considered masculine. See *yin*, *yin/yang*.

**Yang, Chen Ning** A physicist who won a Nobel prize for his work concerning the behavior of subatomic particles. His complicated theory suggested that the universe might be twisted instead of symmetrical.

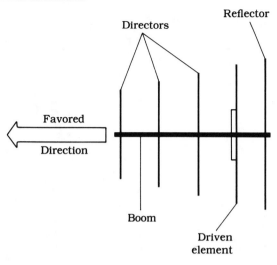

YAGI ANTENNA

**yard** Abbreviation, yd. A unit of distance equal to 36 inches or 3 feet. It is about 0.9144 meters or 91.44 centimeters.

**yardang** A formation that sometimes occurs in a desert as a result of wind erosion. It is like a long, narrow, small hill, sculpted of rock. These formations develop parallel to each other over a large region.

**yaw** In a ship or aircraft, a swaying of the bow or nose of the craft from side to side. This might occur in a jet aircraft if one of the engines fails, especially if the engine is located on a wing some distance from the fuselage. Also sometimes called fishtailing.

**y axis** In a Cartesian coordinate system, the axis on which the y variable is depicted. This is usually the dependent-variable axis, or ordinate, in a two-dimensional system. In three dimensions it is usually one of the independent variables. See *Cartesian coordinates, Cartesian plane, ordinate*.

**Yb** Chemical symbol for ytterbium.

**y chromosome** A chromosome that determines the sex of an organism. It is shaped like a small letter y. A male animal results from a combination of one x and one y chromosome. See *x chromosome*.

**yd** Abbreviation for yard.

**year** 1. Sidereal year: The time required for the earth to complete one orbit around the sun, with respect to the distant stars. Equal to 365.2564 mean solar days. 2. Synodic year: The time between successive appearances of the sun at the point of Aries. Equal to 365.2422 mean solar days. 3. Lunar year: A period of exactly 12 lunar months, equal to 365.3671 mean solar days. 4. A period of about 31,560,000 seconds, based on some standard independent of the motion of the earth around the sun.

**yeast** A primitive form of fungus that some scientists believe existed early in the evolution of life on earth. It reproduces by budding. It is important as a nutritional supplement (brewer's yeast) because it has a high concentration of B-complex vitamins. It is also used for making bread because it produces gas as a by-product of its metabolism.

**yeast infection** An illness in which certain yeast organisms proliferate in the digestive tract. These infections seem to be troublesome in certain people, while not affecting others. Symptoms include abdominal bloating, gas and pain. Various treatments are available. A doctor should be consulted. This infection has nothing to do with the use of yeast in foods or as a nutritional supplement.

**yin** In Oriental philosophy, the feminine element. In nutrition, foods such as sugar and fruit are considered feminine. See *yang, yin/yang*.

**yin/yang** In Oriental philosophy, the balance between feminine and masculine elements. An example is in nutrition, where it is thought that the ideal diet should be balanced between yin foods like vegetables and fruit, and yang foods like meat, grains and beans. Modern nutrition tells us the same thing in a different way. This is an example of how ancient traditions are often scientifically sound. See *yang, yin*.

**ylem** The stuff of the primordial fireball, from which some scientists believe our whole universe came as a result of the Big Bang. This term was coined by the astronomer George Gamow. See *Big Bang*.

**yogurt** Milk that has been fermented with a bacteria that sours it, helps to preserve it, and makes it easier for some people to digest. The bacteria also cause the milk to congeal so that it has the consistency of custard.

**yoke** 1. A structure in a motor or generator that holds the assembly together and provides magnetic coupling. 2. The set of electromagnetic deflection coils in a cathode-ray tube. See *cathode-ray tube*.

**yr** Abbreviation for year.

**ytterbium** Chemical symbol, Yb. An element with atomic number 70. The most common isotope has atomic weight 174. In pure form it is a silver metal.

**yttrium** Chemical symbol, Y. An element with atomic number 39. The most common isotope has atomic weight 89. In pure form it is a grayish metal. It is used to make permanent magnets and also to make superconducting materials. It is also used in certain types of lasers. See *superconductivity, superconductor*.

**Yukawa, Hideki** The physicist who, in 1935, predicted the existence of subatomic particles called mesons. See *meson*.

**Z** Abbreviation for electrochemical equivalent, impedance.

**z axis** In a three-dimensional Cartesian coordinate system, the axis that usually depicts the dependent variable. See *Cartesian coordinates*.

**Zener diode** A silicon diode that has a known, constant reverse breakdown voltage. When this voltage is exceeded, the diode conducts. This makes the device useful as a voltage regulator, because it can be connected across the output terminals of a power supply to prevent the voltage from exceeding the breakdown or Zener voltage.

**zenith** The point directly overhead, exactly opposite the direction toward the center of the earth (see drawing). This point has elevation 90 degrees with respect to the horizon. See *nadir*.

**zeolite** An aluminum-silicon substance with the ability to hold water and other molecules. It

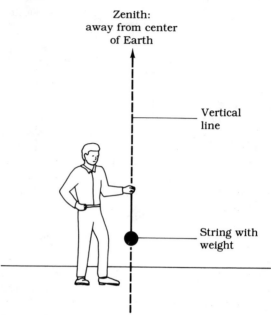

ZENITH

is used to remove minerals from water for residential use, so that "water hardness" problems are minimized.

**zeppelin** A lighter-than-air craft extensively used before modern airplanes proved more effective and faster. Similar to a blimp or dirigible. The immense, cigar-shaped balloon was filled with hydrogen at first, and later with helium, because helium is not flammable.

**zeppelin antenna** An end-fed wire antenna that needs no direct-current (dc) ground connection. Invented originally for use in zeppelins, where there is no earth ground. A parallel-wire transmission line is connected to a wire antenna 1/2 wavelength long.

**Zeno** Also known as Zeno the Eleatic, who lived around 450 B.C. He was a mathematician who liked to bring up paradoxes and inconsistencies. He enjoyed debate by means of showing that his opponent's premises led logically to contradictions. See *Zeno's Paradoxes*.

**Zeno's Paradoxes** A set of arguments contradicting observed events, particularly with respect to motion. One argument was that a runner can never get started, because he must go half the way to his goal before he can go all the way; $1/4$ of the way before $1/2$ the way; $1/8$ of the way before $1/4$ of the way, and so on, ad infinitum. He must therefore pass through an infinite number of steps before he can move at all, and this is impossible. Another paradox uses a similar argument to prove that a hare can never catch a tortoise if the tortoise has a head start. By the time the hare gets to where the tortoise was at the start, the tortoise has moved a little; by the time the hare catches up again, the tortoise has moved some more, and so on, ad infinitum.

**zero** 1. The number corresponding to the empty set. 2. A digit used to indicate the absence of value in a certain decimal place. 3. The numeral used to denote the number whose value represents a set containing no elements.

**zinc** Chemical symbol, Zn. An element with atomic number 30. The most common isotope has atomic weight 64. In pure form it is a grayish or grayish-blue metal. It is used to coat steel (galvanizing) to retard rusting. It reacts with various acids to yield hydrogen and compounds such as zinc chloride, zinc sulfate, etc.

**zipper** The familiar device used as a fastener. First thought of by W. Judson at the end of the nineteenth century. It took more than 20 years to perfect. The earliest fasteners were made of metal. Nowadays they are still sometimes made of metal, but also are fabricated from synthetic plastics.

**zirconium** Chemical symbol, Zr. An element with atomic number 40. The most common isotope has atomic weight 90. In pure form it is a grayish metal. It is used in nuclear reactors because it absorbs neutrons.

**Zn** Chemical symbol for zinc.

**zodiac** The twelve constellations that lie roughly along the ecliptic, and through which the sun passes at a rate of about one constellation per month. These constellations are also called *signs of the zodiac*. As of the end of the twentieth century, the sun enters the signs on the following dates: Aquarius, January 20; Pisces, February 19; Aries, March 21; Taurus, April 20; Gemini, May 21; Cancer, June 22; Leo, July 23; Virgo, August 23; Libra, September 23; Scorpio, October 24; Sagittarius, November 22; Capricorn, December 22. These dates are changing because of precession of the equinoxes in the earth's orbit. See *ecliptic*.

**zodiacal light** A faint glow observed along the zodiac, caused by residual interplanetary particles that are concentrated in the plane of the ecliptic. See *ecliptic*.

**zone of aeration** A layer in the soil above the level of the water table, through which water percolates down. See *water table, zone of saturation*.

**zone of saturation** A layer in the soil that lies below the zone of aeration, beneath the water table. The soil is completely saturated with water in this zone. See *water table, zone of aeration*.

**zone of silence** See *skip zone*.

**zoogeography** A branch of zoology and also of geography concerning the places where various animals live. This involves the long-term migrations of animals, having taken place over millions of years along with continental drift. It also concerns short-term migrations, such as we see in the behavior of birds annually.

**zoology** The branch of life science concerned with the study of animals, and their evolution, characteristics, and behavior.

**zoom lens** In a video camera or photographic camera, a lens that has variable magnification, allowing the image to be enlarged and reduced without having to adjust the focusing.

**zooplankton** Animal plankton, abundant in the oceans of the world. They are essential to the food chain. They are usually too small to be seen without a magnifying glass. Various marine animals, such as whales, feed on them in great quantities.

**Z particle** A subatomic particle that is responsible, together with W particles, for the weak nuclear force.

**Zr** Chemical symbol for zirconium.

**Zwicky** The first astronomer to study supernovae, or exploding stars. He began his work in 1933. He found that a supernova occurs in a given galaxy on an average of once every 360 years. See *supernova*.

**Zworykin, Vladimir** A Russian engineer of the twentieth century. He is best known for his work leading to television cameras and transmission. He invented the iconoscope, a camera tube still used today. See *iconoscope*.

**zygote** A fertilized egg having combined with the male sperm or pollen. The drawing shows an example. In animals, including humans, this eventually becomes an embryo, then a fetus, and finally a fully independent living organism.

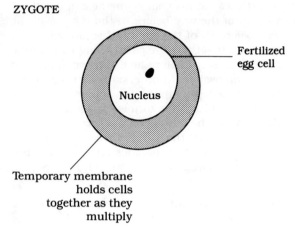

ZYGOTE

# Animal classification

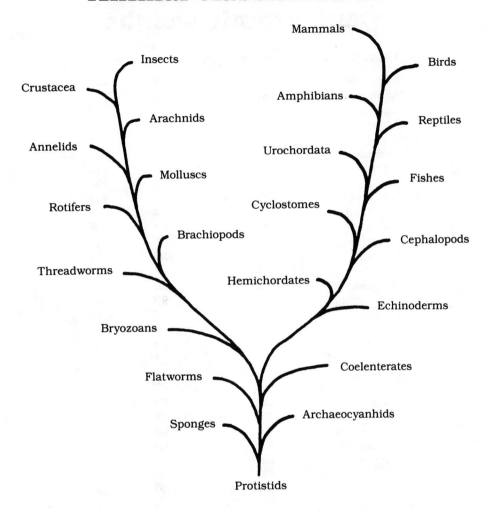

From Erickson. THE LIVING EARTH (TAB Books, Inc., 1989)

# Atomic number and atomic weight

| Element name | Abbreviation | Atomic number | Atomic weight* |
|---|---|---|---|
| Actinium | Ac | 89 | 227 |
| Aluminum | Al | 13 | 27 |
| Americium** | Am | 95 | 243 |
| Antimony | Sb | 51 | 121 |
| Argon | Ar | 18 | 40 |
| Arsenic | As | 33 | 75 |
| Astatine | At | 85 | 210 |
| Barium | Ba | 56 | 138 |
| Berkelium** | Bk | 97 | 247 |
| Beryllium | Be | 4 | 9 |
| Bismuth | Bi | 83 | 209 |
| Boron | B | 5 | 11 |
| Bromine | Br | 35 | 79 |
| Cadmium | Cd | 48 | 114 |
| Calcium | Ca | 20 | 40 |
| Californium** | Cf | 98 | 251 |
| Carbon | C | 6 | 12 |
| Cerium | Ce | 58 | 140 |
| Cesium | Cs | 55 | 133 |
| Chlorine | Cl | 17 | 35 |
| Chromium | Cr | 24 | 52 |
| Cobalt | Co | 27 | 59 |
| Copper | Cu | 29 | 63 |
| Curium** | Cm | 96 | 247 |
| Dysprosium | Dy | 66 | 164 |
| Einsteinium** | Es | 99 | 254 |
| Erbium | Er | 68 | 166 |
| Europium | Eu | 63 | 153 |
| Fermium | Fm | 100 | 257 |
| Fluorine | F | 9 | 19 |
| Francium | Fr | 87 | 223 |
| Gadolinium | Gd | 64 | 158 |
| Gallium | Ga | 31 | 69 |
| Germanium | Ge | 32 | 74 |
| Gold | Au | 79 | 197 |
| Hafnium | Hf | 72 | 180 |
| Helium | He | 2 | 4 |
| Holmium | Ho | 67 | 165 |
| Hydrogen | H | 1 | 1 |
| Indium | In | 49 | 115 |
| Iodine | I | 53 | 127 |
| Iridium | Ir | 77 | 193 |
| Iron | Fe | 26 | 56 |
| Krypton | Kr | 36 | 84 |

| Element name | Abbreviation | Atomic number | Atomic weight* |
|---|---|---|---|
| Lanthanum | La | 57 | 139 |
| Lawrencium** | Lr or Lw | 103 | 257 |
| Lead | Pb | 82 | 208 |
| Lithium | Li | 3 | 7 |
| Lutetium | Lu | 71 | 175 |
| Magnesium | Mg | 12 | 24 |
| Manganese | Mn | 25 | 55 |
| Mendelevium** | Md | 101 | 256 |
| Mercury | Hg | 80 | 202 |
| Molybdenum | Mo | 42 | 98 |
| Neodymium | Nd | 60 | 142 |
| Neon | Ne | 10 | 20 |
| Neptunium** | Np | 93 | 237 |
| Nickel | Ni | 28 | 58 |
| Niobium | Nb | 41 | 93 |
| Nitrogen | N | 7 | 14 |
| Nobelium** | No | 102 | 254 |
| Osmium | Os | 76 | 192 |
| Oxygen | O | 8 | 16 |
| Palladium | Pd | 46 | 108 |
| Phosphorus | P | 15 | 31 |
| Platinum | Pt | 78 | 195 |
| Plutonium** | Pu | 94 | 242 |
| Polonium | Po | 84 | 209 |
| Potassium | K | 19 | 39 |
| Praseodymium | Pr | 59 | 141 |
| Promethium | Pm | 61 | 145 |
| Proactinium | Pa | 91 | 231 |
| Radium | Ra | 88 | 226 |
| Radon | Rn | 86 | 222 |
| Rhenium | Re | 75 | 187 |
| Rhodium | Rh | 45 | 103 |
| Rubidium | Rb | 37 | 85 |
| Ruthenium | Ru | 44 | 102 |
| Samarium | Sm | 62 | 152 |
| Scandium | Sc | 21 | 45 |
| Selenium | Se | 34 | 80 |
| Silicon | Si | 14 | 28 |
| Silver | Ag | 47 | 107 |
| Sodium | Na | 11 | 23 |
| Strontium | Sr | 38 | 88 |
| Sulfur | S | 16 | 32 |
| Tantalum | Ta | 73 | 181 |
| Technetium | Tc | 43 | 99 |
| Tellurium | Te | 52 | 130 |
| Terbium | Tb | 65 | 159 |
| Thallium | Tl | 81 | 205 |
| Thorium | Th | 90 | 232 |

| Element name | Abbreviation | Atomic number | Atomic weight* |
|---|---|---|---|
| Thulium | Tm | 69 | 169 |
| Tin | Sn | 50 | 120 |
| Titanium | Ti | 22 | 48 |
| Tungsten | W | 74 | 184 |
| Unnilhexium** | Unh | 106 | ---- |
| Unnilpentium** | Unp | 105 | ---- |
| Unnilquadium** | Unq | 104 | ---- |
| Uranium | U | 92 | 238 |
| Vanadium | V | 23 | 51 |
| Xenon | Xe | 54 | 132 |
| Ytterbium | Yb | 70 | 174 |
| Yttrium | Y | 39 | 89 |
| Zinc | Zn | 30 | 64 |
| Zirconium | Zr | 40 | 90 |

*Most common isotope. The sum of the number of protons and the number of neutrons in the nucleus. Most elements have other isotopes with differnt atomic weights.

**These elements (atomic numbers 93 or larger) are not found in nature, but are human-made.

# Beaufort Scale

| Beaufort number | Wind speed, mph | Definition | Description of visible effects |
|---|---|---|---|
| 0 | 0 | Calm | Smoke rises vertically. |
| 1 | 1–3 | Light air | Wind felt on face. |
| 2 | 4–6 | Light breeze | Leaves rustle. |
| 3 | 7–12 | Gentle breeze | Twigs in motion, loose papers scatter. |
| 4 | 13–18 | Moderate breeze | Small branches move, pine trees "whisper." |
| 5 | 19–24 | Fresh breeze | Small branches move, dust blows. |
| 6 | 25–31 | Strong breeze | Large branches move, snow reduces visibility. |
| 7 | 32–38 | Near gale | Whole trees in motion, ground blizzard occurs. |
| 8 | 39–46 | Gale | Small branches break, driving becomes difficult. |
| 9 | 47–54 | Whole gale | Large branches break, severe blizzard occurs, walking is difficult. |
| 10 | 55–64 | Storm | Trees uprooted in wet soil, mild structural damage occurs. |
| 11 | 65–73 | Severe storm | Large trees blown down, moderate structural damage takes place. |
| 12 | 74 or more | Hurricane | Extensive damage occurs. |

# Boolean algebra

**Theorems in Boolean algebra.**
**Multiplication = AND; Addition = OR;**
**True = 1; False = 0.**

| | |
|---|---|
| $X + 0 = X$ | Additive Identity |
| $X \bullet 1 = X$ | Multiplicative Identity |
| $X + 1 = 1$ | |
| $X \bullet 0 = 0$ | |
| $X + X = X$ | |
| $X \bullet X = X$ | |
| $-(-X) = X$ | Double Negation |
| $X + (-X) = 1$ | |
| $X \bullet (-X) = 0$ | |
| $X + Y = Y + X$ | Commutativity of Addition |
| $X \bullet Y = Y \bullet X$ | Commutativity of Multiplication |
| $X + (X \bullet Y) = X$ | |
| $X + Y + Z = (X + Y) + Z$ | |
| $\quad = X + (Y + Z)$ | Associativity of Addition |
| $X \bullet Y \bullet Z = (X \bullet Y) \bullet Z$ | |
| $\quad = X \bullet (Y \bullet Z)$ | Associativity of Multiplication |
| $X \bullet (Y + Z) = (X \bullet Y) + (X \bullet Z)$ | Distributivity |

# Constants

| Constant | Abbreviation | Value |
|---|---|---|
| Atomic Mass Unit | amu | $1.661 \times 10^{-27}$ kilogram |
| Avogadro constant | N | $6.022 \times 10^{23}$ /mole |
| Electron charge | e | $1.602 \times 10^{-19}$ Coulomb |
| Electron radius | $r_o$ | $2.818 \times 10^{-15}$ meter |
| Electron rest mass | $m_e$ | $9.110 \times 10^{-31}$ kilogram |
| Faraday constant | F | $9.649 \times 10^{7}$ Coulomb/kilomole |
| Gas constant | $R_o$ | $8.314 \times 10^{3}$ joule/(kilomole Kelvin) |
| Gravitational constant | G | $6.673 \times 10^{-11}$ Newton meter squared per kilogram squared |
| Proton rest mass | $M_p$ | $1.673 \times 10^{-27}$ kilogram |
| Neutron rest mass | $M_n$ | $1.675 \times 10^{-27}$ kilogram |
| Radiation constant, first | $c_1$ | $4.993 \times 10^{-24}$ joule meter |
| Radiation constant, second | $c_2$ | $1.439 \times 10^{-2}$ meter Kelvin |
| Speed of electromagnetic field in vacuum | c | $2.998 \times 10^{8}$ meter/second |

# Conversion factors

| To convert | To | Multiply by | Conversely, multiply by |
|---|---|---|---|
| Cubic Feet | Cubic meters | $2.83 \times 10^{-2}$ | 35.3 |
| Cubic Inches | Cubic meters | $1.64 \times 10^{-5}$ | $6.10 \times 10^4$ |
| Cubic Feet | Liters | 28.3 | $3.53 \times 10^{-2}$ |
| Cubic Inches | Liters | $1.64 \times 10^{-2}$ | 61.0 |
| Dynes | Newtons | $10^{-5}$ | $10^5$ |
| Feet | Meters | 0.305 | 3.28 |
| Foot pounds | Newtons | 4.45 | 0.225 |
| Inches | Meters | $2.54 \times 10^{-2}$ | 39.4 |
| Gallons | Liters | 4.55 | 0.220 |
| Gallons | Cubic meters | $4.55 \times 10^{-3}$ | 220 |
| Miles per hour | Meters per second | 0.477 | 2.10 |
| Ounces | Milliliters | 35.5 | $2.82 \times 10^{-2}$ |
| Poundals | Newtons | 0.138 | 7.25 |
| Pounds | Kilograms | 0.454 | 2.20 |
| Pounds per cubic inch | Kilograms per cubic meter | $2.77 \times 10^4$ | $3.61 \times 10^{-5}$ |
| Square feet | Square meters | $9.29 \times 10^{-2}$ | 10.8 |
| Square inches | Square meters | $6.45 \times 10^{-4}$ | $1.55 \times 10^3$ |
| Watthours | Joules | $3.6 \times 10^3$ | $2.78 \times 10^{-4}$ |

# Electromagnetic spectrum

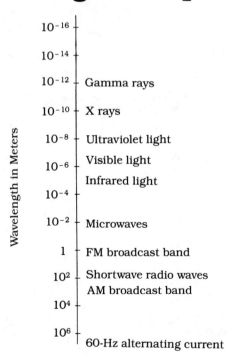

# Energy Units

| Unit | To convert to joules, multiply by | Conversely, multiply by |
|---|---|---|
| British Thermal Unit (Btu) | $1.055 \times 10^3$ | $9.480 \times 10^{-4}$ |
| electronvolt | $1.602 \times 10^{-19}$ | $6.242 \times 10^{18}$ |
| erg | $10^{-7}$ | $10^7$ |
| foot pound | $1.356$ | $7.376 \times 10^{-1}$ |
| kilowatt hour | $3.6 \times 10^6$ | $2.8 \times 10^{-7}$ |
| watt hour | $3.6 \times 10^3$ | $2.8 \times 10^{-4}$ |

# Geologic time

| Eon | Era | Period | Epoch | No. years ago began (in millions of years) | First life-forms | Geology |
|---|---|---|---|---|---|---|
| Phanerozoic | Cenozoic | Quaternary | Holocene | .01 | | |
| | | | Pleistocene | 2 | Man | Ice age |
| | | | Pliocene | 5 | Mastodons | Cascades |
| | | Tertiary | Miocene | 24 | Apes | Alps |
| | | | Oligocene | 37 | Saber-tooth tigers | |
| | | | Eocene | 54 | Whales | |
| | | | | | Horses | Rockies |
| | | | Paleocene | 65 | Alligators | |
| | Mesozoic | Cretaceous | | 137 | Birds | Sierra Nevadas |
| | | Jurassic | | 180 | Mammals | |
| | | Triassic | | 230 | Dinosaurs | Atlantic |
| | Paleozoic | Permian[1] | | 290 | Reptiles | Pangaea |
| | | Pennsylvanian [1,2] | | 320 | Trees | |
| | | Mississippian [1,2] | | 350 | Amphibians | |
| | | | | | Insects | |
| | | Devonian | | 405 | Sharks | |
| | | Silurian | | 435 | Land plants | |
| | | Ordovician | | 500 | Fish | |
| | | Cambrian[3] | | 575 | Sea plants | |
| Protozoic | | Proterozoic | | 2500 | Invertebrates | Ice age |
| | | Archean[4] | | 4600 | Prokaryotes | First rocks |

[1] Time between these two periods is called Permocarboniferous time.
[2] These two periods are also together called the Carboniferous period.
[3] All time before this period is known as Precambrian time.
[4] This eon is also called the Archeozoic eon.

From Erickson, THE LIVING EARTH (TAB Books, Inc., 1989).

# Melting points

| Substance | Melting point, degrees Fahrenheit | Melting point, degrees Celsius |
|---|---|---|
| Aluminum | 1220 | 660 |
| Calcium | 1550 | 842 |
| Copper | 1980 | 1080 |
| Gold | 1950 | 1060 |
| Iron | 2800 | 1540 |
| Lead | 621 | 327 |
| Mercury | −38 | −39 |
| Oxygen | −361 | −218 |
| Silver | 1760 | 961 |
| Tin | 449 | 232 |
| Water | 32 | 0 |
| Zinc | 787 | 419 |

# Metals

| Metal (elemental) | Mass of one cubic meter, kilograms |
|---|---|
| Aluminum | 2700 |
| Beryllium | 1800 |
| Copper | 8900 |
| Gold | 19,000 |
| Iron | 7900 |
| Lead | 11,000 |
| Magnesium | 1700 |
| Mercury | 14,000 |
| Molybdenum | 10,000 |
| Nickel | 8900 |
| Platinum | 21,000 |
| Silver | 11,000 |
| Tin | 7300 |
| Titanium | 4500 |
| Tungsten | 19,000 |
| Uranium | 19,000 |
| Zinc | 7100 |
| Zirconium | 6400 |

# Morse Code

## American Morse

| Character | Symbol | Character | Symbol |
|---|---|---|---|
| A | ·— | U | ··— |
| B | —··· | V | ···— |
| C | ·· · | W | ·—— |
| D | —·· | X | ·—·· |
| E | · | Y | ·· ·· |
| F | ·—· | Z | ··· · |
| G | ——· | 1 | ·——· |
| H | ···· | 2 | ··—·· |
| I | ·· | 3 | ···—· |
| J | —·—· | 4 | ····— |
| K | —·— | 5 | — — — |
| L | ⎯ | 6 | ······ |
| M | —— | 7 | ——·· |
| N | —· | 8 | —···· |
| O | · · | 9 | —··— |
| P | ····· | 0 | ⎯⎯ |
| Q | ··—· | period | ··——·· |
| R | · ·· | comma | ·—·— |
| S | ··· | question mark | —··—· |
| T | — | | |

## International Morse

| Character | Symbol | Character | Symbol |
|---|---|---|---|
| A | ·— | U | ··— |
| B | —··· | V | ···— |
| C | —·—· | W | ·—— |
| D | —·· | X | —··— |
| E | · | Y | —·—— |
| F | ··—· | Z | ——·· |
| G | ——· | 1 | ·———— |
| H | ···· | 2 | ··——— |
| I | ·· | 3 | ···—— |
| J | ·——— | 4 | ····— |
| K | —·— | 5 | ····· |
| L | ·—·· | 6 | —···· |
| M | —— | 7 | ——··· |
| N | —· | 8 | ———·· |
| O | ——— | 9 | ————· |
| P | ·——· | 0 | ————— |
| Q | ——·— | period | ·—·—·— |
| R | ·—· | comma | ——··—— |
| S | ··· | question mark | ··——·· |
| T | — | | |

# Multiplication table

| ×  | 0 | 1  | 2  | 3  | 4  | 5  | 6  | 7  | 8  | 9   | 10  | 11  | 12  |
|----|---|----|----|----|----|----|----|----|----|-----|-----|-----|-----|
| 1  |   | 1  | 2  | 3  | 4  | 5  | 6  | 7  | 8  | 9   | 10  | 11  | 12  |
| 2  |   | 2  | 4  | 6  | 8  | 10 | 12 | 14 | 16 | 18  | 20  | 22  | 24  |
| 3  |   | 3  | 6  | 9  | 12 | 15 | 18 | 21 | 24 | 27  | 30  | 33  | 36  |
| 4  |   | 4  | 8  | 12 | 16 | 20 | 24 | 28 | 32 | 36  | 40  | 44  | 48  |
| 5  |   | 5  | 10 | 15 | 20 | 25 | 30 | 35 | 40 | 45  | 50  | 55  | 60  |
| 6  |   | 6  | 12 | 18 | 24 | 30 | 36 | 42 | 48 | 54  | 60  | 66  | 72  |
| 7  |   | 7  | 14 | 21 | 28 | 35 | 42 | 49 | 56 | 63  | 70  | 77  | 84  |
| 8  |   | 8  | 16 | 24 | 32 | 40 | 48 | 56 | 64 | 72  | 80  | 88  | 96  |
| 9  |   | 9  | 18 | 27 | 36 | 45 | 54 | 63 | 72 | 81  | 90  | 99  | 108 |
| 10 |   | 10 | 20 | 30 | 40 | 50 | 60 | 70 | 80 | 90  | 100 | 110 | 120 |
| 11 |   | 11 | 22 | 33 | 44 | 55 | 66 | 77 | 88 | 99  | 110 | 121 | 132 |
| 12 |   | 12 | 24 | 36 | 48 | 60 | 72 | 84 | 96 | 108 | 120 | 132 | 144 |

# Periodic table of the elements

| | I A | II A | III B | IV B | V B | VI B | VII B | VIII | | | I B | II B | III A | IV A | V A | VI A | VII A | 0 | |
|---|---|---|---|---|---|---|---|---|---|---|---|---|---|---|---|---|---|---|---|
| 1 | H 1 | | | | | | | | | | | | | | | | | H 1 He 2 | 1 |
| 2 | Li 3 | Be 4 | | | | | | | | | | | B 5 | C 6 | N 7 | O 8 | F 9 | Ne 10 | 2 |
| 3 | Na 11 | Mg 12 | | | | | | | | | | | Al 13 | Si 14 | P 15 | S 16 | Cl 17 | Ar 18 | 3 |
| 4 | K 19 | Ca 20 | Sc 21 | Ti 22 | V 23 | Cr 24 | Mn 25 | Fe 26 | Co 27 | Ni 28 | Cu 29 | Zn 30 | Ga 31 | Ge 32 | As 33 | Se 34 | Br 35 | Kr 36 | 4 |
| 5 | Rb 37 | Sr 38 | Y 39 | Zr 40 | Nb 41 | Mo 42 | Tc 43 | Ru 44 | Rh 45 | Pd 46 | Ag 47 | Cd 48 | In 49 | Sn 50 | Sb 51 | Te 52 | I 53 | Xe 54 | 5 |
| 6 | Cs 55 | Ba 56 | La 57 | Hf 72 | Ta 73 | W 74 | Re 75 | Os 76 | Ir 77 | Pt 78 | Au 79 | Hg 80 | Tl 81 | Pb 82 | Bi 83 | Po 84 | At 85 | Rn 86 | 6 |
| 7 | Fr 87 | Ra 88 | Ac 89 | | | | | | | | | | | | | | | | 7 |

| Lanthanide | Ce 58 | Pr 59 | Nd 60 | Pm 61 | Sm 62 | Eu 63 | Gd 64 | Tb 65 | Dy 66 | Ho 67 | Er 68 | Tm 69 | Yb 70 | Lu 71 |
|---|---|---|---|---|---|---|---|---|---|---|---|---|---|---|
| Actinide | Th 90 | Pa 91 | U 92 | Np 93 | Pu 94 | Am 95 | Cm 96 | Bk 97 | Cf 98 | Es 99 | Fm 100 | Md 101 | No 102 | Lw 103 |

# Periodic table of the elements continued.

| | | | | | |
|---|---|---|---|---|---|
| Ac | Actinium | Gd | Gadolinium | Pm | Promethium |
| Ag | Silver | Ge | Germanium | Po | Polonium |
| Al | Aluminum | H | Hydrogen | Pr | Praseodymium |
| Am | Americium | He | Helium | Pt | Platinum |
| Ar | Argon | Hf | Hafnium | Pu | Plutonium |
| As | Arsenic | Hg | Mercury | Ra | Radium |
| At | Astatine | Ho | Holmium | Rb | Rubidium |
| Au | Gold | I | Iodine | Re | Rhenium |
| B | Boron | In | Indium | Rh | Rhodium |
| Ba | Barium | Ir | Iridium | Rn | Radon |
| Be | Beryllium | K | Potassium | Ru | Ruthenium |
| Bi | Bismuth | Kr | Krypton | S | Sulfur |
| Bk | Berkelium | La | Lanthanum | Sb | Antimony |
| Br | Bromine | Li | Lithium | Sc | Scandium |
| C | Carbon | Lu | Lutetium | Se | Selenium |
| Ca | Calcium | Lw | Lawrencium | Si | Silicon |
| Cb | Columbium | Md | Mendelevium | Sm | Samarium |
| Cd | Cadmium | Mg | Magnesium | Sn | Tin |
| Ce | Cerium | Mn | Manganese | Sr | Strontium |
| Cf | Californium | Mo | Molybdenum | Ta | Tantalum |
| Cm | Curium | N | Nitrogen | Tb | Terbium |
| Co | Cobalt | Na | Sodium | Tc | Technetium |
| Cr | Chromium | Nb | Niobium | Te | Tellurium |
| Cs | Cesium | Nd | Neodymium | Th | Thorium |
| Cu | Copper | Ne | Neon | Ti | Titanium |
| Dy | Dysprosium | Ni | Nickel | Tl | Thallium |
| Er | Erbium | No | Nobelium | Tm | Thulium |
| Es | Einsteinium | Np | Neptunium | U | Uranium |
| Eu | Europium | O | Oxygen | V | Vanadium |
| F | Fluorine | Os | Osmium | W | Tungsten |
| Fe | Iron | P | Phosphorus | Xe | Xenon |
| Fm | Fermium | Pa | Proactinium | Y | Yttrium |
| Fr | Francium | Pb | Lead | Yb | Ytterbium |
| Ga | Gallium | Pd | Palladium | Zn | Zinc |
| | | | | Zr | Zirconium |

# Plant classification

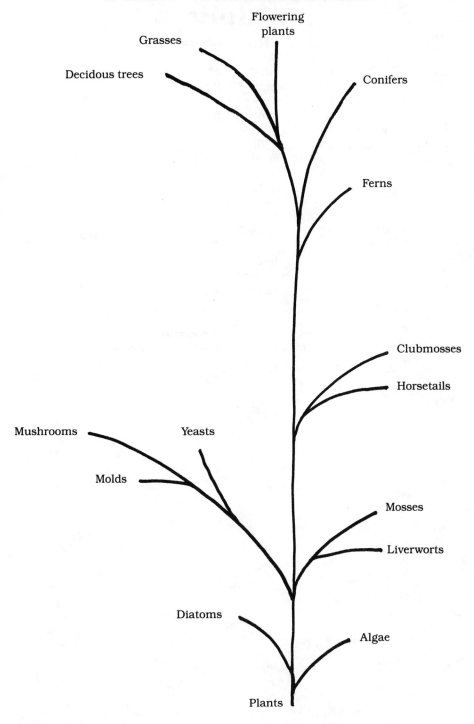

# Prefix multipliers

| Factor | Prefix | Abbreviation |
|---|---|---|
| $10^{-18}$ | atto-* | a* |
| $10^{-15}$ | femto-* | f* |
| $10^{-12}$ | pico- | p |
| $10^{-9}$ | nano- | n |
| $10^{-6}$ | micro- | u |
| $10^{-3}$ | milli- | m |
| $10^{-2}$ | centi- | c |
| $10^{-1}$ | deci-* | d* |
| $10$ | deca-* | da* |
| $10^{2}$ | hecto-* | h* |
| $10^{3}$ | kilo- | k |
| $10^{6}$ | mega- | M |
| $10^{9}$ | giga- | G |
| $10^{12}$ | tera- | T |
| $10^{15}$ | peta-* | P* |
| $10^{18}$ | exa-* | E* |

*Not often used. For exponents $-1$, $0$, $1$, and $2$, the numbers are generally written out. For extreme numbers, scientific notation is preferred.

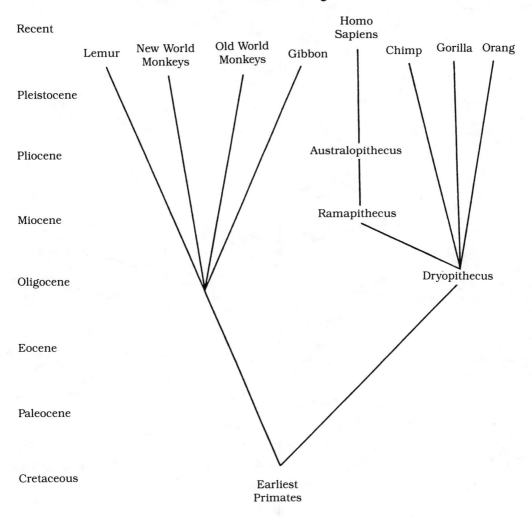

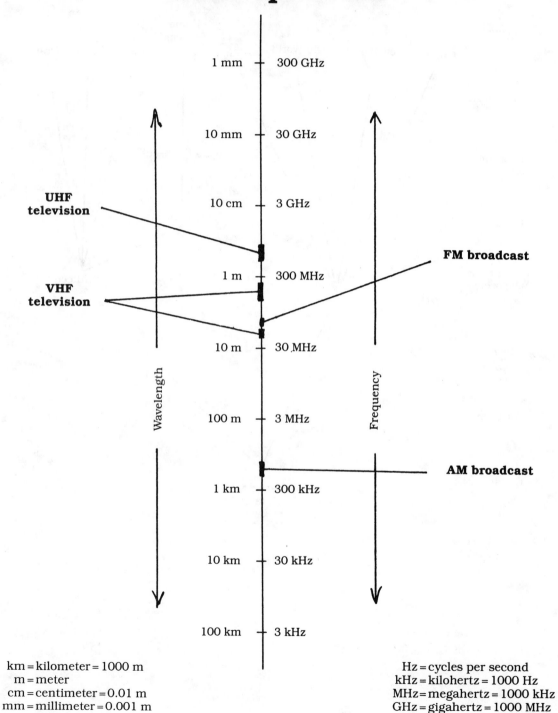

# Recommended Daily Allowances

| Nutrient | Recommended Daily Allowance (RDA) |
|---|---|
| Vitamin A | 5000 International Units (IU) |
| Vitamin B-1 (Thiamine) | 1.5 milligrams (mg) |
| Vitamin B-2 (Riboflavin) | 1.7 mg |
| Vitamin B-3 (Niacin) | 20 mg |
| Vitamin B-6 (Pyridoxine) | 2 mg |
| Vitamin B-12 (Cyanocobalamin) | 6 micrograms (mcg) |
| Biotin | 300 mcg |
| Folic Acid | 400 mcg |
| Pantothenic Acid | 10 mg |
| Vitamin C (Ascorbic Acid) | 60 mg |
| Vitamin D | 400 IU |
| Vitamin E (Tocopherol) | 30 IU |
| Iron | 12–18 mg |
| Calcium | 1000–1200 mg |
| Protein | 40–60 grams (g) |

These values may vary for specific individuals. A doctor should be consulted regarding special needs.

# Richter Scale

| Richter Scale | Earthquake effects | Average per year |
|---|---|---|
| <2.0 | Microearthquake. Imperceptible. | +600,000 |
| 2.0 to 2.9 | Generally not felt but recorded. | 300,000 |
| 3.0 to 3.9 | Felt by most people if nearby. | 49,000 |
| 4.0 to 4.9 | Minor shock. Damage slight and localized. | 6,000 |
| 5.0 to 5.9 | Moderate shock. Energy released equivalent to atomic bomb. | 1,000 |
| 6.0 to 6.9 | Large shock can be destructive in populous regions. | 120 |
| 7.0 to 7.9 | Major earthquake. Inflicts serious damage. Recorded over whole world. | 14 |
| 8.0 to 8.9 | Great earthquake. Produces total destruction to nearby communities. Energy released is millions of times first atomic bomb. | Once every 5–10 years. |
| 9.0 and up | Largest earthquakes. | One or two per century. |

Note: The Richter scale is logarithmic; an increase of 1 magnitude signifies 10 times the ground motion and the release of roughly 30 times the energy.

From Erickson, VOLCANOES AND EARTHQUAKES (TAB Books, Inc., 1988).

# Rocks
**Abundance of various rock types in the crust of the earth**

| Rock type | Percent volume | Minerals | Percent volume |
|---|---|---|---|
| Sandstone | 1.7 | Quartz | 12.0 |
| Clays and shales | 4.2 | Potassium feldspar | 12.0 |
| Carbonates | 2.0 | Plagioclase | 39.0 |
| Granites | 10.4 | Micas | 5.0 |
| Grandiorite | | Amphiboles | 5.0 |
| Quartz diorite | 11.2 | Pyroxenes | 11.0 |
| Syenites | 0.4 | Olivine | 3.0 |
| Basalts | | Sheet silicates | 4.6 |
| Gabbros | | Calcite | 1.5 |
| Amphibolites | | Dolomite | 0.5 |
| Granulites | 42.5 | Magnetite | 1.5 |
| Ultramafics | 0.2 | Other | 4.9 |
| Gneisses | 21.4 | | |
| Schists | 5.1 | | |
| Marbles | 0.9 | | |

# Schematic symbols in electronics

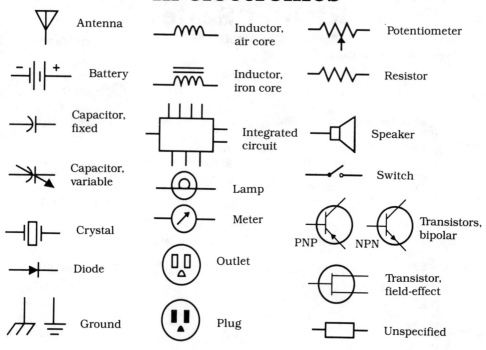

# Solar system data

| Object Name | Mean distance from sun, kilometers | Period of revolution, days, or years | Diameter, kilometers |
|---|---|---|---|
| Sun | --- | Rotates in 25 days at equator and 33 days at poles | 1,400,000 |
| Mercury | 58,000,000 | 88 days | 4,900 |
| Venus | 108,000,000 | 225 days | 12,100 |
| Earth | 150,000,000 | 365.26 days | 12,800 |
| Mars | 223,000,000 | 687 days | 6,800 |
| Jupiter | 778,000,000 | 11.9 years | 143,000 |
| Saturn | 1,430,000,000 | 29.5 years | 121,000 |
| Uranus | 2,870,000,000 | 84 years | 52,000 |
| Neptune | 4,500,000,000 | 165 years | 49,000 |
| Pluto | 5,900,000,000 | 248 years | 3,500 |

# Space probes

**Some major unmanned interplanetary space probes launched by United States unless otherwise stated.**

| Year | Probe name | Description of goals and mission |
|---|---|---|
| 1962 | Mariner II | First probe to any planet besides the Earth's moon. Flew by Venus and obtained data about temperature. |
| 1965 | Mariner IV | First fly-by mission to Mars. Showed a planet pocked with craters. |
| 1969 | Mariner VI, VII | Flew by Mars. Obtained further data. |
| 1971 | Mariner IX | Went into orbit around Mars. |
| 1974 | Mariner X | Flew by Venus and Mercury. Revealed that Mercury has terrain similar to our own moon. |
| 1973 | Pioneer X | Flew by Jupiter, showing cloud details and system of rings. |
| 1972 | Venera VIII, IX, X (USSR) | Landed on Venus, showing surface strewn with boulders. Failed after a short time because of extreme pressure and temperature. |
| 1976 | Viking I, II | Landed on Mars. Failed to show that there was life there. Sent pictures of desert terrain, with wind, sand and rocks. |
| 1979 | Voyager I | Flew by Jupiter. |
| 1980 | Voyager II | Flew by Jupiter, Uranus (1986) and Neptune (1989). Showed system of rings around Uranus, and revealed fascinating details about the moons of Uranus and Neptune. |

# Standard International System of Units

| Quantity | Unit name | Informal definition of unit |
|---|---|---|
| Amount | Mole | $6.022 \times 10^{23}$ |
| Current | Ampere | 1 Coulomb/second |
| Electric quantity | Coulomb | $6.3 \times 10^{18}$ electrons |
| Length | Meter | Originally, 1 meter was defined as 0.0000001 of the way from the equator to the North Pole. Now it is defined according to the wavelengths of certain atomic resonances. |
| Luminous intensity | Candela | Roughly the brightness of a candle flame. It is more formally defined in terms of the blackbody radiation at the temperature of the solidification of platinum. |
| Mass | Kilogram | A liter of pure water at room temperature masses very nearly 1 kilogram. |
| Temperature | Kelvin | Absolute zero = 0 Kelvin. Pure water at one atmosphere freezes at 273 Kelvin. |
| Time | Second | Defined according to the vibrations of certain atoms. |

# Trigonometric Identities

$\csc \theta = 1/\sin \theta$

$\sec \theta = 1/\cos \theta$

$\tan \theta = \sin \theta / \cos \theta$

$\cot \theta = \cos \theta / \sin \theta$

$\sin^2 \theta + \cos^2 \theta = 1$

$\sec^2 \theta - \tan^2 \theta = 1$

$\csc^2 \theta - \cot^2 \theta = 1$

$\sin(\theta + \phi) = \sin \theta \cos \phi + \cos \theta \sin \phi$

$\sin(\theta - \phi) = \sin \theta \cos \phi - \cos \theta \sin \phi$

$\cos(\theta + \phi) = \cos \theta \cos \phi - \sin \theta \sin \phi$

$\cos(\theta - \phi) = \cos \theta \cos \phi + \sin \theta \sin \phi$

$\tan(\theta + \phi) = (\tan \theta + \tan \phi)/(1 - \tan \theta \tan \phi)$

$\tan(\theta - \phi) = (\tan \theta - \tan \phi)/(1 + \tan \theta \tan \phi)$

$\sin -\theta = -\sin \theta$

$\cos -\theta = \cos \theta$

$\tan -\theta = -\tan \theta$

$\csc -\theta = -\csc \theta$

$\sec -\theta = \sec \theta$

$\cot -\theta = -\cot \theta$

$\sin 2\theta = 2 \sin \theta \cos \theta$

$\cos 2\theta = \cos^2 \theta - \sin^2 \theta$

$\tan 2\theta = 2 \tan \theta / (1 - \tan^2 \theta)$

$\sin \theta + \sin \phi = 2 \sin(\theta/2 + \phi/2) \cos(\theta/2 - \phi/2)$

$\sin \theta - \sin \phi = 2 \cos(\theta/2 + \phi/2) \sin(\theta/2 - \phi/2)$

$\cos \theta + \cos \phi = 2 \cos(\theta/2 + \phi/2) \cos(\theta/2 - \phi/2)$

$\cos \theta - \cos \phi = 2 \sin(\theta/2 + \phi/2) \sin(\theta/2 - \phi/2)$

# Universe data

| Object or span | Size or distance, meters |
|---|---|
| Radius of electron | 0.000000000000002818 |
| Diameter of cell (pneumococcus) | 0.0000001 |
| Diameter of dust grain | 0.00001 |
| Diameter of coarse grain of sand | 0.001 |
| Diameter of human heart | 0.09 |
| Height of a man | 1.8 |
| Thickness of lower atmosphere (troposphere) | 16,000 |
| Diameter of Earth | 12,800,000 |
| Distance to moon | 400,000,000 |
| Distance to sun | 150,000,000,000 |
| Radius of known solar system | 5,900,000,000,000 |
| Distance to nearest known star | 38,000,000,000,000,000 |
| Diameter of Milky Way | 950,000,000,000,000,000,000 |
| Distance to Great Nebula in Andromeda | 19,000,000,000,000,000,000,000 |
| Distance to Virgo cluster of galaxies | 480,000,000,000,000,000,000,000 |
| Distance to typical quasar | 48,000,000,000,000,000,000,000,000 |
| Radius of known universe | 200,000,000,000,000,000,000,000,000 |

# Vitamins and minerals

| Name of substance | Letter | Some functions of the substance |
|---|---|---|
| Ascorbic acid | C | Promotes health and mucous membranes, connective tissue and bones. |
| Bioton | * | Synthesis of enzymes. |
| Calciferol | D | Absorption of calcium and phosphorus. |
| Calcium | | Bone and tooth formation, muscle contraction. Balanced with magnesium. |
| Choline | * | Numerous biochemical reactions. |
| Copper | | Important in enzyme synthesis. Balanced with zinc. |
| Cyanocobalamin | B-12 | Manufacture of red blood cells, nerve function. |
| Essential fatty acids | * | Manufacture of certain hormones and enzymes. |
| Folic acid | * | Manufacture of red blood cells. |
| Inositol | * | Metabolism, nerve and brain function. |
| Iodine | | Important in thyroid function. |
| Iron | | Used to make hemoglobin and enzymes. |
| Magnesium | | Bone and tooth formation, muscle contraction. Balanced with calcium. |
| Niacin | B-3 | Metabolism, nerve and brain function. |
| Pantothenic acid | * | Metabolism, adrenal function. |
| Para-aminobenzoic acid | * | Health of skin, bones and joints. |
| Phosphorus | | Used to make bones, teeth and nucleic acids. Important in metabolism. |
| Phytonadione | K | Blood coagulation. |
| Potassium | | Electrolyte and water balance. Balanced with sodium. |
| Pyridoxine | B-6 | Cell function, metabolism, nervous system function. |
| Retinol | A | Night vision, health of mucous membranes. |
| Riboflavin | B-2 | Metabolism, health of mucous membranes. |
| Sodium | | Electrolyte and water balance. Balanced with potassium. |
| Thiamine | B-1 | Metabolism, nerve function. |
| Tocopherol | E | Antioxidant, health of mucous membranes. |
| Zinc | | Manufacture of enzymes, healing of wounds. Balanced with copper. |

*Indicates a vitamin without a letter designation. If the Letter column is blank, it indicates that the substance is a mineral nutrient.